PHYSICS RESEARCH AND TECHNOLOGY

EXCITON QUASIPARTICLES: THEORY, DYNAMICS AND APPLICATIONS

PHYSICS RESEARCH AND TECHNOLOGY

Additional books in this series can be found on Nova's website under the Series tab.

Additional E-books in this series can be found on Nova's website under the E-book tab.

PHYSICS RESEARCH AND TECHNOLOGY

EXCITON QUASIPARTICLES: THEORY, DYNAMICS AND APPLICATIONS

RANDY M. BERGIN
EDITOR

Nova Science Publishers, Inc.
New York

Copyright © 2011 by Nova Science Publishers, Inc.

All rights reserved. No part of this book may be reproduced, stored in a retrieval system or transmitted in any form or by any means: electronic, electrostatic, magnetic, tape, mechanical photocopying, recording or otherwise without the written permission of the Publisher.

For permission to use material from this book please contact us:
Telephone 631-231-7269; Fax 631-231-8175
Web Site: http://www.novapublishers.com

NOTICE TO THE READER

The Publisher has taken reasonable care in the preparation of this book, but makes no expressed or implied warranty of any kind and assumes no responsibility for any errors or omissions. No liability is assumed for incidental or consequential damages in connection with or arising out of information contained in this book. The Publisher shall not be liable for any special, consequential, or exemplary damages resulting, in whole or in part, from the readers' use of, or reliance upon, this material. Any parts of this book based on government reports are so indicated and copyright is claimed for those parts to the extent applicable to compilations of such works.

Independent verification should be sought for any data, advice or recommendations contained in this book. In addition, no responsibility is assumed by the publisher for any injury and/or damage to persons or property arising from any methods, products, instructions, ideas or otherwise contained in this publication.

This publication is designed to provide accurate and authoritative information with regard to the subject matter covered herein. It is sold with the clear understanding that the Publisher is not engaged in rendering legal or any other professional services. If legal or any other expert assistance is required, the services of a competent person should be sought. FROM A DECLARATION OF PARTICIPANTS JOINTLY ADOPTED BY A COMMITTEE OF THE AMERICAN BAR ASSOCIATION AND A COMMITTEE OF PUBLISHERS.

Additional color graphics may be available in the e-book version of this book.

LIBRARY OF CONGRESS CATALOGING-IN-PUBLICATION DATA

Exciton quasiparticles : theory, dynamics, and applications / editor, Randy M. Bergin.
   p. cm.
  Includes index.
  ISBN 978-1-61122-318-7 (hardcover)
 1. Exciton theory. 2. Quasiparticles (Physics) I. Bergin, Randy M.
  QC176.8.E9E93 2010
  530.4'16--dc22
                          2010041298

Published by Nova Science Publishers, Inc. † New York

CONTENTS

Preface vii

Chapter 1 Exciton Polariton Dispersion in Multinary Compounds 1
N. N. Syrbu and V. V. Ursaki

Chapter 2 Exciton Relaxation Dynamics in Colloidal Core and Core/Shell CdSe Nanorods 131
A. Cretí and M. Lomascolo

Chapter 3 Radiation-Assisted Preparation of Powder Materials and Their Exciton Luminescence 181
Václav Čuba and Martin Nikl

Chapter 4 The Semiclassical Molecular Exciton: Path/Time Ordered Propagators vs. Stochastic Dynamics 211
William R. Kirk

Chapter 5 Solitons and Exitons in Seismotectonic Problems 235
A. V. Vikulin

Chapter 6 Exciton Dynamics Study of InAs/GaAs Quantum Dot Heterostructures 255
Ya-Fen Wu, Jiunn-Chyi Lee, Jen-Cheng Wang and Tzer-En Nee

Chapter 7 Exciton Diffusion Length in Titanyl Phthalocyanine Thin Films as Determined by the Surface Photovoltage Method 275
Jiří Toušek, Jana Toušková, Martin Drábik, Zdeněk Remeš, Jan Hanuš, Věra Cimrová, Danka Slavinská, Hynek Biederman, Adam Zachary and Luke Hanley

Chapter 8 Accuracy of the Coherent Potential Approximation for Frenkel Excitons in One-Dimensional Arrays with Gaussian Diagonal Disorder and Nearest-Neighbor Transfer 291
I. Avgin and D. L. Huber

Chapter 9 Exciton Fano Resonance in Semiconductor Heterostructures and its Related Phenomena 303
Ken-ichi Hino, Muneaki Hase and Nobuya Maeshima

Chapter 10	Magnetoexciton Binding Energy in Polar Crystals, Quantum Wells and Graphene Bilayers *Z.G. Koinov*	**357**
Chapter 11	Wannier-Mott-Frenkel Hybrid Exciton in Semiconductor-Organic Systems Containing Quantum Dots *Nguyen Que Huong*	**375**

Short Communication

Excitons in CdO Parabolic Quantum Dots *M. A. Grado-Caffaro and M. Grado-Caffaro*	**397** |//
Index	**401**

PREFACE

Exciton can be regarded as an elementary excitation of condensed matter which can transport energy without transporting net electric charge. This book presents research in the study of exciton quasiparticles, including exciton polariton dispersion in multinary compounds; exciton relaxation dynamics in colloidal core and core/shell CdSe nanorods; solitons and excitons in seismotectonic problems; exciton diffusion length measured by surface photovoltage method; and excitons in CdO parabolic quantum dots.

Chapter 1 - Exciton resonance spectra are analyzed in the case of strong and weak polariton effects. Most strong polariton effects ($\hbar\omega_{LT} > \hbar\Gamma$) are realized in ZnP$_2$ crystals. The shape of optical spectra are analyzed in details on the basis of the theory of spectral-spatial polariton diffusion taking into account the data related to boundary conditions. The contribution of upper (UPB) and lower (LPB) polariton branches to the external emission is revealed. It is shown that the contribution to the UPB emission is significantly determined by the rate of interband relaxation of LPB in the UPB states, and the spatial-energetic distribution of LPB.

The exciton resonances with a weak polariton effect are investigated in the reflectivity, absorption, and luminescence experimental spectra of chalcopyrite crystals with D_{2d}^{12} symmetry such as CuAlSe$_2$, CuGaS$_2$ and CuGaSe$_2$. These crystals demonstrate strong anisotropy of optical properties in the region of exciton transitions. The most strong oscillator strength is inherent to the Γ_4 excitons (the A-exciton series). Optical effects related to the anisotropy of the exciton translation mass are observed in CuGaS$_2$ crystals, which demonstrate the spatial dispersion in this medium. The Fabry-Perot interference and the interference of additional polariton waves are investigated in CuGaS$_2$ crystals which reflect the polariton dispersion in the crystal. The interference is observed for the Γ_5 excitons with weak oscillator strength. Due to the mixing of states, the Γ_4 excitons with strong oscillator strength excite additional waves of Γ_5 excitons with weak oscillator strength, which interfere and determine the fine structure of reflectivity spectra in the E∥c polarization. The dispersion of the LPB and UPB for the Γ_4 and Γ_5 excitons is determined from the calculation of the reflectivity spectra in the E∥c polarization and the interference spectra in the E⊥c polarization. Strong emission lines due to excitons bound to neutral donor (acceptor) are observed in CuGaS$_2$ crystals. The transitions between the ground state of neutral donor (acceptor) and the excited states of the bound exciton complex are revealed in the luminescence spectra. Apart from that, transitions between the ground and excited states of the bound exciton complex as well as the excited states of the neutral donor (acceptor) are observed in the emission spectra.

The main parameters of the neutral donor (acceptor) bound excitons, such as the binding energy and the energy intervals between the 2S, $2P_x$, $2P_y$ and $2P_z$ excited states are determined. The excitation of luminescence has a resonance character. A scheme of electron transitions between the energy levels of neutral acceptor bound exciton is proposed.

Resonance Raman scattering spectra in the region of one-phonon and two-phonon scattering are measured in $CuGa_xAl_{1-x}S_2$ crystals with x=1.0, 0.95 and 0.9 at the temperature of 10K under the excitation by different lines of an argon laser with energy above the ground state of Γ_4 long-wavelength excitons. The observed change of the degree of linear and circular depolarization of LO- and 2LO-phonon emission lines are due to the resonance Raman scattering on the two polariton branches: the photon-like and the exciton-like ones. The polarization of the emission lines demonstrates the optical orientation of excitons.

The optical spectra in the region of exciton resonances allow one to determine the main parameters of excitons and energy bands such as the crystal field (Δ_{cf}) and the spin-orbital (Δ_{SO}) splitting of the valence band, the effective electron and hole masses, as well as many phenomena related to the behavior of light waves in the near-surface region of crystals.

Chapter 2 - In this chapter the authors focus on the exciton relaxation dynamics of colloidal CdSe core and core/shell nanorods. In particular they show how confinement effects and defect states affect linear and non linear optical properties. The authors present a brief introduction on the colloidal nanocrystals (NCs), namely on their electronic and optical properties, and on the recombination processes in these materials. Then the authors discuss the results of our systematic study of ultrafast exciton relaxation dynamics in CdSe core and CdSe/CdS/ZnS core/shell nanorods with a radius of few nm, different length (20-40 nm) and different shell thickness.

Femtosecond pump-probe spectroscopy, in the visible spectral range with non resonant / resonant pump energy has been performed to investigate the fast processes in NCs.

The effect of the shell thickness on Stimulated Emission (SE) and Photoinduced Absorption (PA) transitions in core and core/shell CdSe nanorods is exposed and the role of surface/interface defect states is pointed out. The authors show that the defect states distribution depends on the shell thickness and that the interface defects can be negligible for thin ones, resulting in a longer lifetime of SE. Furthermore the authors demonstrate that a resonant pumping increases the SE lifetime and enhances Auger scattering, clarifying that PA processes, involving defect states, are the main obstacle to sustain the SE. In the case of resonant pumping measurements the role of defect states on Auger processes and the presence of coherent confined acoustic phonons in CdSe core nanorods are also discussed. In particular the authors find that the modulation frequency observed in photobleaching and photoabsorption dynamics of core NCs, corresponds to the coherent radial breathing modes of the nanorods. Finally, the authors quantitatively investigate, by time-resolved photoluminescence (PL) spectroscopy, the shell thickness dependence of exciton trapping and its effects on the PL quantum yield.

Chapter 3 - This chapter provides overview of preparation of nanopowder materials using ionizing and UV irradiation. Basic principles and current status of research in this field are briefly discussed. The processes related to the radiation reduction of metal ions to lower valences or to metallic particles and radiation induced oxidation are described. Stabilization of nanoparticles in solutions and preparation of materials from aqueous or micellar solutions is overviewed. Exciton luminescence in nanopowders is discussed, including their advantages/disadvantages and comparison with the bulk materials. Preparation of selected nanomaterials with excitonic luminescence via irradiation route is described in detail, followed by discussion of their characterization and luminescence properties.

Chapter 4 - The semi-classical model of the molecular exciton as the short-time resonance ("evanescence") between electromagnetic and matter-field is reviewed. From a simple 'bubble diagram' quasi-physical model a richer and more evocative physical picture emerges of an electron/hole loop around a set of current sources (nuclear normal modes) which effectively trap the excitation into an electronic excited state. In reverse, the exciton is the source for the emergent electromagnetic field during the emission process. This picture leads to an account of 'mapping' vibrational modes vs. 'bath' modes, and an account of gaussian broadening at a fundamental level is obtained. The recent exposition by Adler of stochastic dynamics in the Schroedinger picture leading instead to effective Lorentzian profiles (arXiv:quant-ph/0208123v4) is contrasted. The 'loop'='bubble' picture makes other contacts with nonequilibrium quantum dynamics, gauge theory and time dependent density functional theory as well.

The word 'exciton' often means different things to different kinds of investigators. Herein the authors will not be concerned with the localized storage of electromagnetic field excitation in a medium, e.g. a crystal lattice, as is very commonly meant, still less shall they be concerned with the localized cross-sectional excitation of a (super)string!; instead, the authors will mean the excitation of an isolated molecule, a *fluorophore*, which is expected to be able to re-emit this energy back to the e-m field as light. That is, both 'up' and 'dowm' processes are of interest to us. The symmetry between these two processes is revealed in a deep and cogent manner in the molecular exciton model.

Chapter 5 - The problem of an elastic stress field in a rotating medium is formulated and solved analytically within the limits of the classical theory of elasticity with a symmetrical stress tensor. This is a rotation elastic field of action at a distance. There are two specific types of elastic waves with a moment in rotating media: solitons and excitons, or rotation waves. The soliton solutions to the wave equation represent waves of global earthquake migration (slow tectonic waves) which are no faster than (1-10) cm/s, i.e., approach the migration velocity of large and great earthquakes ($M \approx 8$ and more). The exciton solutions correspond to waves of local migration of foreshocks and aftershocks in earthquake sources (fast tectonic waves) and have their maximum velocity comparable to S-P-wave velocities.

Chapter 6 - The elementary excitation dynamics differ qualitatively from those in higher-dimensional systems, since the density of states in the zero-dimensional quantum dot (QD) systems is a series of δ-functions. Many unique phenomena, including electronic, optical, magnetic, and thermal characteristics, have been observed. As far as the optical properties of the semiconductor QDs are concerned, the excitonic process has attracted a lot of investigations because it is expected to realize very high-efficiency photonic devices due to the Bosonic character of excitons. A key issue is to attain a profound understanding of the corresponding dynamics to facilitate the research for innovative heterodevice architectures. In this work, a steady-state thermal model taking into account the dot size distribution, the random population of density of states, and all of the important mechanisms of exciton dynamics, including radiative and nonradiative recombination, thermal escaping and relaxing, and state filling effects is proposed. These mathematical analyses successfully explain the abnormality of the exciton-related emissions observed in the low dimensional nanostructures.

Not only the temperature- and excitation-dependent luminescence measurement systems, but also the metal-organic chemical vapor epitaxy is systematically discussed.

Chapter 7 - Exciton diffusion length was measured by the generalized surface photovoltage (SPV) method. The experiment needs no junction; it uses a spontaneously created space charge region (SCR) at the surface. The measurement is contactless and non-destructive. The SPV signal comes partly from excitons diffusing from neutral bulk towards the interface with the SCR, partly by excitons generated directly in the SCR. Both are separated in its electric field and subsequently generate the photovoltage. The SPV technique was modified to be applicable to arbitrary thickness of the layers and to samples with arbitrary thickness of the space charge region at the surface. Theoretical calculations of the photocurrents from the SCR and from the bulk of the layers were carried out and illustrated to show how the different parameters influence the form and relative size of the photogenerated signal. Experimental spectra were compared with the theoretical ones allowing determination of the exciton diffusion length and the thickness of the SCR. The authors studied titanyl phthalocyanine (TiOPc) thin films prepared using evaporation and surface polymerization by ion-assisted deposition (SPIAD). Bilayer (TiO_2/TiOPc) thin films were also prepared, where the TiO_2 layer was sputtered from TiO_2 target. All films were characterized by the surface photovoltage method using absorption coefficients evaluated from measurement of the optical transmission and reflection. The authors found that the drift lengths of the charge carriers were shorter than the SCR thickness, which means recombination in this depletion region. Typically, the thickness of the SCR was higher than that of the bulk and the diffusion length of excitons was ~15 nm.

Chapter 8 - This chapter is a report of the results of an assessment of the accuracy of the coherent potential approximation (CPA) when applied to a one-dimensional Frenkel exciton array with Gaussian diagonal disorder and nearest-neighbor

transfer. The integrated density of states, the density of states, the inverse localization length, and the optical absorption are compared with data obtained using mode-counting techniques (integrated density of states, density of states, and inverse localization length) and matrix diagonalization (optical absorption) applied to large arrays. The CPA is in excellent agreement with the numerical data thus providing a rare example of a theory that yields accurate results for a realistic model of a disordered system.

Chapter 9 - Excitonic Fano resonance (EX-FR) effects in semiconductor heterostructures, such as quantum wells, superlattices, and biased superlattices, are highlighted.

Here, these effects are classified into three types of linear EX-FR, non-linear EX-FR, and dynamic EX-FR.

In the first two effects, a built-in interaction such as Coulomb coupling and hole-subband mixing cause Fano interference between discrete and continuum states.

The linear EX-FR represents a Fano effect probed by means of conventional linear spectroscopy, while in the non-linear EX-FR, this effect is investigated from the viewpoint of non-linear spectroscopy.

On the other hand, in the third effect, a Fano coupling is created dynamically by laser irradiation, without which this effect turns off.

Therefore, both of the linear and non-linear EX-FRs are attributable to the static interactions inherent to the systems concerned, whereas the dynamic EX-FR literally has the dynamic origin.

Moreover, other non-excitonic dynamic FR phenomena, somewhat relevant to the dynamic EX-FR, are also presented.

One of these effects is FR of dynamic localization states realized in laser-driven biased superlattices, and the other is FR observed at the initial stage of coherent phonon generation; these two FRs are caused by an intersubband ac-Zener tunneling and a laser-induced electron-phonon interaction, respectively.

Because a continuum state plays a key role in FR, the above-mentioned various types of FRs are understood in a unified manner based on the framework of the multichannel scattering theory; a conventional variational method seems incorrect because of taking no account of significance of the continuum.

The R-matrix theory is considered as a powerful numerical method for solving such a FR problem in both accuracy and efficiency.

Applying it to the FR phenomena concerned here, a great number of intriguing characters are revealed.

Chapter 10 - Strong magnetic fields can dramatically change the dynamical properties of excitons because in this regime the electrons and holes are confined primarily to the lowest Landau Level (LLL), and the Coulomb energy is much smaller than the exciton cyclotron energy. In this chapter the authors apply the Bethe-Salpeter (BS) formalism to study magnetoexciton binding energy in bulk polar crystals and quantum wells assuming the existence of Froehlich interaction between the electrons and the longitudinal optical phonons. In the case of a bulk material the BS equation in the LLL approximation is reduced to a one-dimensional Schröedinger equation which has a nonlocal potential. It is shown that the magnetoexciton binding energy in polar quantum wells has the same form as in the non-polar structures but with an effective dielectric constant which depends on the strength of the magnetic field. Since the unique electronic behaviors of graphene is a result of the unusual quantum-relativistic characteristics of the so-called Dirac fermions, The authors study magnetoexciton binding energy in grapheme bilayers embedded in a dielectric by applying the relativistic BS equation in the LLL approximation. It is shown that in grapheme bilayer structures the magnetoexciton mass (binding energy) is four times lower (higher) than the corresponding magnetoexciton mass (binding energy) in coupled quantum wells with parabolic dispersion.

Chapter 11 - The chapter will describe the theory of the formation of a hybridization state of Wannier Mott exciton and Frenkel exciton in different hetero-structure configurations involving quantum dot. The hybrid excitons exist at the interfaces of the semiconductors quantum dots and the organic medium, having unique properties and a large optical non-linearity. The coupling at resonance is very strong and tunable by changing the parameters of the systems (dot radius, dot-dot distance, generation of the organic dendrites and the materials of the system etc). Different semiconductor quantum dot-organic material combination systems have been considered such as a semiconductor quantum dot lattice embedded in an organic host, a semiconductor quantum dot at the center of an organic dendrite, a semiconductor quantum dot coated by an organic shell.

The formation and the properties of the organic-semiconductor hybrid excitons have been modulated by electric and magnetic fields. The hybrid excitons are as sensitive to external perturbation as Wannier-Mott excitons. Upon the application of the magnetic and electric fields the coupling term between the two kinds of excitons increases.

The most important feature of this system is, by adjusting the system parameters as well as the external fields and their orientation, one can tune the resonance between the two kinds of excitons to get different regions of mixing to obtain the expected high non-linearity.

Short Communication - The main aspects of the physics of excitons confined into a parabolic quantum dot of cadmium oxide are discussed starting from considering in the dot an exciton gas of harmonic oscillators whose Fermi energy coincides with the band-gap shift experienced by cadmium oxide in the visible region. At this point, the authors must remark that the role of CdO as transparent material in the visible range is highly significant. In particular, the frequency of oscillation of the excitons is determined. In addition, some issues related to the plasma-optical effect are examined.

In: Exciton Quasiparticles
Editor: Randy M. Bergin

ISBN: 978-1-61122-318-7
© 2011 Nova Science Publishers, Inc.

Chapter 1

EXCITON POLARITON DISPERSION IN MULTINARY COMPOUNDS

N. N. Syrbu[1] and V. V. Ursaki[2]

[1]Technical University of Moldova, Stefan cel Mare str. 168, Chisinau, MD-2004, Moldova
[2] Institute of Applied Physics, Academy of Sciences of Moldova, Academy str. 5, Chisinau, MD-2028, Moldova

ABSTRACT

Exciton resonance spectra are analyzed in the case of strong and weak polariton effects. Most strong polariton effects ($\hbar\omega_{LT} > \hbar\Gamma$) are realized in ZnP_2 crystals. The shape of optical spectra are analyzed in details on the basis of the theory of spectral-spatial polariton diffusion taking into account the data related to boundary conditions. The contribution of upper (UPB) and lower (LPB) polariton branches to the external emission is revealed. It is shown that the contribution to the UPB emission is significantly determined by the rate of interband relaxation of LPB in the UPB states, and the spatial-energetic distribution of LPB.

The exciton resonances with a weak polariton effect are investigated in the reflectivity, absorption, and luminescence experimental spectra of chalcopyrite crystals with D_{2d}^{12} symmetry such as $CuAlSe_2$, $CuGaS_2$ and $CuGaSe_2$. These crystals demonstrate strong anisotropy of optical properties in the region of exciton transitions. The most strong oscillator strength is inherent to the Γ_4 excitons (the A-exciton series). Optical effects related to the anisotropy of the exciton translation mass are observed in $CuGaS_2$ crystals, which demonstrate the spatial dispersion in this medium. The Fabry-Perot interference and the interference of additional polariton waves are investigated in $CuGaS_2$ crystals which reflect the polariton dispersion in the crystal. The interference is observed for the Γ_5 excitons with weak oscillator strength. Due to the mixing of states, the Γ_4 excitons with strong oscillator strength excite additional waves of Γ_5 excitons with weak oscillator strength, which interfere and determine the fine structure of reflectivity spectra in the E||c polarization. The dispersion of the LPB and UPB for the Γ_4 and Γ_5 excitons is determined from the calculation of the reflectivity spectra in the E||c polarization and the interference spectra in the E⊥c polarization. Strong emission lines due to excitons bound to neutral donor (acceptor) are observed in CuGaS$_2$ crystals. The transitions between the ground state of neutral donor (acceptor) and the excited states of the bound exciton complex are revealed in the luminescence spectra. Apart from that, transitions between the ground and excited states of the bound exciton complex as well as the excited states of the neutral donor (acceptor) are observed in the emission spectra. The main parameters of the neutral donor (acceptor) bound excitons, such as the binding energy and the energy intervals between the 2S, 2P$_x$, 2P$_y$ and 2P$_z$ excited states are determined. The excitation of luminescence has a resonance character. A scheme of electron transitions between the energy levels of neutral acceptor bound exciton is proposed.

Resonance Raman scattering spectra in the region of one-phonon and two-phonon scattering are measured in CuGa$_x$Al$_{1-x}$S$_2$ crystals with x=1.0, 0.95 and 0.9 at the temperature of 10K under the excitation by different lines of an argon laser with energy above the ground state of Γ_4 long-wavelength excitons. The observed change of the degree of linear and circular depolarization of LO- and 2LO-phonon emission lines are due to the resonance Raman scattering on the two polariton branches: the photon-like and the exciton-like ones. The polarization of the emission lines demonstrates the optical orientation of excitons.

The optical spectra in the region of exciton resonances allow one to determine the main parameters of excitons and energy bands such as the crystal field (Δ_{cf}) and the spin-orbital (Δ_{SO}) splitting of the valence band, the effective electron and hole masses, as well as many phenomena related to the behavior of light waves in the near-surface region of crystals.

1. STRONG AND WEAK POLARITON EFFECTS IN THE REGION OF EXCITON RESONANCES IN MULTINARY COMPOUNDS

The concept of photon-exciton interaction formulated more than 50 years ago by Pekar [1,2] and Hopfield [3] leads to numerous experimental demonstration of one of the most interesting prediction of this model concerning the possibility of simultaneous propagation in a crystal of two or more wave with the same polarization in a definite interval of frequencies. This possibility occurs due to the dependence of the exciton energy upon the wave vector (the spatial dispersion). The detailed theoretical consideration of the photon-exciton interaction which leads to spatial dispersion is discussed in many books. An important point in the concept of propagation of additional waves is the problem of additional boundary conditions (ABC).

The theoretical considerations [1-11] stimulated a lot of experimental work [12-19]. The exciton resonances are investigated basically by standard optical methods. Detailed investigations of optical reflectivity, absorption, and luminescence spectra have been performed in order to prove the polariton nature of excitons. These investigations have been basically aimed to the understanding of differences between the experimental spectra and the theoretical calculations based on the model of exciton-photon interaction. Numerous confirmations of the spatial dispersion have been obtained. The form of the exciton-polariton reflectivity spectra is very sensitive the so-called "dead layer", in which the excitons are absent (see for instance [20] and refs there in). The thickness of this layer can be changed by different methods, including thermal treatment and annealing. Since the calculated reflectivity spectra are sensitive to the parameters of the exciton dispersion as well as to the additional boundary conditions, reliable results are obtained only in samples with good surface characteristics, for instance non-polished naturals surfaces.

In the first approximation the exciton in a dielectric or semiconductor crystal can be treated as an electron-hole pair bound by the Coulomb interaction. The exciton is characterized by a finite effective mass equal to the sum of electron and hole masses, and it can freely propagate through the crystal. The properties of the exciton, first of all the symmetry and the mass, are determined by the characteristics of the conduction and valence bands from which the quasi-particles forming the exciton originate.

The exciton polaritons are complex quasi-particles produced by mixing the photons with the dipole-active excitons. The dashed lines in Figure 1a show (in the form of the dependence of energy upon the wave vector) the dispersion curves of non-interacting photons (the photon line $\omega = ck/\sqrt{\varepsilon}$) and excitons (in the simplest case with a parabolic band $\omega_{ex} = \omega T + h^2 k^2 / 2M^*$). Here ω is the photon frequency, k is the wave vector, ε_b is the background dielectric constant containing the income from all the resonances except for the considered exciton transition, ω_T is the exciton frequency at $k = 0$, M^* is the effective exciton mass, $h/2\pi$ is the Planck constant, c is the velocity of light in the vacuum. If the excitons are dipole-active and are mixed with photons, the dispersion curve of the exciton polaritons is described by the following equation giving the dependence of the dielectric function upon the frequency:

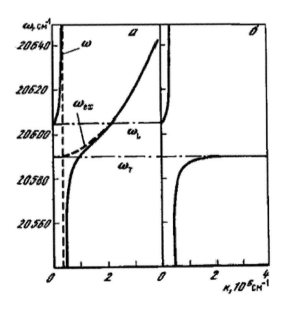

Figure 1. The dependence of the ftrequency upon the wave vector K for the exciton polaritons. (a) The dispersion curve of photons ($\omega = ck/\sqrt{\varepsilon_b}$) and excitons ($\omega_{ex} = \omega_T + hk^2/4\pi M^*$) in the absence of interaction (dashed curves), and in the case of interaction (full curves). The parameters correspond to the $n = 1$ exciton states in CdS crystals. (b) The dispersion curve of exciton polaritons in the absence of spatial dispersion ($M^* = \infty$).

$$\frac{c^2 k^2}{\omega^2} = \varepsilon_b + \Sigma \frac{4\pi\beta(\kappa)}{1 - \omega^2/\omega_{ex}^2(k)}$$

Here $4\pi\beta(k)$ is the oscillator strength which depends upon the wave vector. In this approach, the damping of exciton polaritons is not taken into consideration, since it does not significantly influence the form of the dispersion curves. The dispersion curve for an isolated parabolic exciton band is shown by a full curve in Figure 1a. Two waves with the same polarization can propagate above the energy ω_L. One can deduce from (1) that at k = 0 the value of $4\pi\beta(0)$ is connected to the longitudinal-transversal splitting of exciton states in the centre of the zone; $4\pi\beta(0) \cong 2\varepsilon_b\omega_{LT}/\omega_T$, where $\omega_{LT} = \omega_L - \omega_T$. The exciton polaritons have a photon character at frequencies below the resonance exciton frequency ω_T. The polariton states acquire a more pronounced exciton character with increasing the frequency, and they are exciton-like at frequencies above ω_T. An upper polariton branch emerges above the ω_L freaquency, which acquires a photon character with the frequency increase. The simultaneous presence of two waves at the same frequency is due to the limited value of the effective exciton mass M^*, which results in the dependence of the exciton frequency upon the wave vector. In the case of infinite mass (i. e. the absence of the spatial dispersion), one should have only one polariton state at each frequency, and an energy gap would arise between the ω_L and ω_T frequencies, in which the propagation of waves would be forbidden (see Figure 1b). In order to experimentally demonstrate the presence of spatial dispersion in the region of the polariton resonance, it is necessary at least to prove the bending of the lower branch of the polariton dispersion curve (or the change of the curvature of the upper branch). The existence of the upper branch is not a sufficient demonstration.

The described character of the polariton dispersion is realized in one of the most investigated systems corresponding to the n = 1A exciton state in CdS crystals. This state is formed by the electron from the conduction band and the hole from the valence band. However, for many crystals there are several sub-bands in the valence band at K=0. This determines the presence of several exciton series in a narrow energy interval. These series interact with each other and change the contour of the dielectric function, and respectively the contours of the reflectivity spectra. The contours of the reflectivity spectra can de analyzed on the basis of the following relations:

$$R(\omega) = \left|\frac{n-1}{n+1}\right|^2$$

$$\varepsilon(\omega, \vec{q}) = \varepsilon_f + \frac{\varepsilon_f \omega_{LT}}{\omega_0 + \frac{\hbar^2 q^2}{2M} - \omega - i\gamma}$$

$$\omega_{LT} = \omega_L - \omega_0$$

$$M = m_C^* + m_V^*$$

$$M \Rightarrow \infty$$

$$n^2 = \varepsilon(\omega, \vec{q}) = \left(\frac{cq}{\omega}\right)^2$$

In the case of a weak polariton effect, i. e. when

$$M \to \infty, \ \omega_{LT} \ll \gamma, \ \omega_{LT}/\gamma \ll 1$$

$$\varepsilon = \varepsilon_f + \frac{\varepsilon_f \omega_{LT}}{\omega_0 - \omega - i\gamma}$$

the coefficient of reflection (Figure 2) of exciton polaritons without taking into account the spatial dispersion of the transversal waves is determined by the following expression:

$$n^2 = \frac{cq}{\omega}; \ n_0 = \sqrt{\varepsilon_f}$$

$$n = n_0 \left[1 + \frac{\omega_{LT}}{\omega_0 - \omega - i\gamma}\right]^{1/2}$$

$$R_0 = \left|\frac{n_0 - 1}{n_0 + 1}\right|^2; \ R_0 = \left|\frac{\sqrt{\varepsilon_f} - 1}{\sqrt{\varepsilon_f} + 1}\right|^2$$

$$R(\omega) = R_0 \left[1 + \frac{2n_0}{n_0^2 - 1} \frac{\omega_{LT}(\omega_0 - \omega)}{(\omega_0 - \omega)^2 + \gamma^2}\right]$$

The change of the coefficient of reflection if one does not take into account the spatial dispersion is shown on Figure 3 for the case of strong polariton effect, i e. when the flowing conditions are fulfilled:

$$M \to \infty, \ \omega_{LT} \gg \gamma, \ \gamma = 0$$

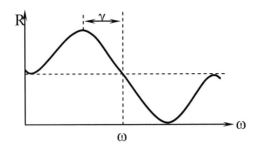

Figure 2. The reflectivity spectra in the region of the exciton polariton in the case of weak polariton effect.

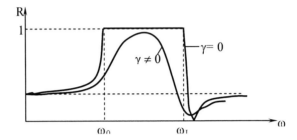

Figure 3. The reflectivity spectra in the region of exciton poalritons in the case of strong polariton effect without taking into account the spatial dispersion.

$$\varepsilon = \varepsilon_f \frac{\omega_{LT}}{\omega_0 - \omega} = \varepsilon_f \frac{\omega_L - \omega_0}{\omega_0 - \omega}$$

$$n = n_0 \sqrt{\frac{\omega_L - \omega_0}{\omega_0 - \omega}}$$

The coefficient of reflection in the case of strong polariton effect taking into account the spatial dispersion is shown in Figure 4, i e. when the flowing conditions are fulfilled:

$$M, \omega_{LT} \gg \gamma$$

1) The damping parameter is equal to zero ($\gamma = 0$)

$$\varepsilon(\omega, \vec{q}) = \varepsilon_f \left(1 + \frac{\omega_{LT}}{\omega_0 - \omega + \frac{\hbar^2 q^2}{2M}} \right)$$

By taking into account the additional boundary conditions of Maxwell one obtains

$$n_{1,2}^2 = \frac{1}{2}\varepsilon_f\left[\left[\varepsilon_f + \frac{2Mc^2(\omega-\omega_0)}{\hbar\omega_0^2}\right] \pm \left[\left(\frac{2Mc^2(\omega-\omega_0)}{\hbar\omega_0^2} - \varepsilon_f\right)^2 + \frac{8Mc^2\varepsilon_f\omega_{LT}}{\hbar\omega_0^2}\right]^{1/2}\right]^{1/2}$$

$$n_3^2 = \frac{n_1^2 \cdot n_2^2}{\varepsilon_f} = \frac{2M(\omega-\omega_L)}{\hbar}\left(\frac{c}{\omega_0}\right)^2$$

$$R = \left|\frac{1-n^*}{1+n^*}\right|^2$$

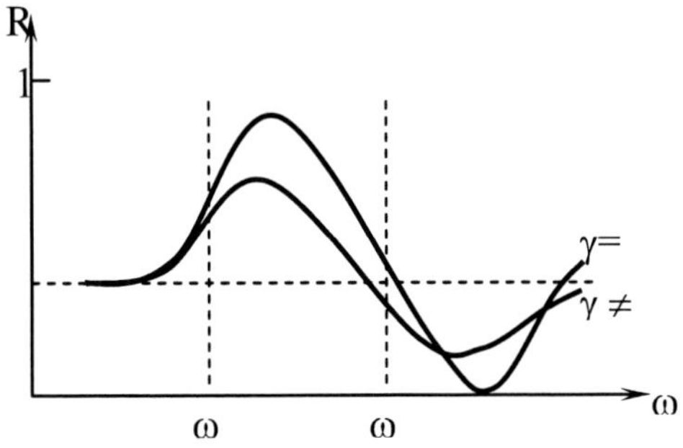

Figure 4. The reflectivity spectra of exciton poalritons in the case of strong polariton effect taking into account the spatial dispersion.

$$n^* = \frac{n_1 n_2 + \varepsilon_f}{n_1 + n_2}$$

2) The damping parameter is not equal to zero $\gamma \neq 0$

$$R = \left|\frac{1-n^*}{1+n^*}\right|^2 \quad n^* = \frac{n_1 n_2 + \varepsilon_f}{n_1 + n_2}$$

$$(n_1 \cdot n_2)^2 = -\varepsilon_f\left(1 - \frac{\omega^2}{\omega_0^2} - i\frac{\gamma\omega}{\omega_0^2}\right)\frac{Mc^2\omega_0}{\hbar\omega^2} - \frac{2\omega_{LT}\varepsilon_f Mc^2}{\hbar\omega^2}$$

$$(n_1 + n_2)^2 = \varepsilon_f - \left(1 - \frac{\omega^2}{\omega_0^2} - i\frac{\gamma\omega}{\omega_0^2}\right)\frac{Mc^2\omega_0}{\hbar\omega^2} + 2\left[-\varepsilon_f\left(1 - \frac{\omega_2}{\omega_0^2} - i\frac{\gamma\omega}{\omega_0^2}\right)\frac{Mc^2\omega_0}{\hbar\omega^2} - 2\frac{\omega_{LT}\varepsilon_f Mc^2}{\hbar\omega^2}\right]^{1/2}$$

$$R = \left| \frac{\left(\dfrac{1-n_0}{1+n_0}\right) + \left(\dfrac{n_0 - n^*}{n_0 + n^*}\right) e^{i2kn_0 l}}{1 + \left(\dfrac{1-n_0}{1+n_0}\right) \cdot \left(\dfrac{n_0 - n^*}{n_0 + n^*}\right) e^{i2kn_0 l}} \right|^2$$

$$k = \frac{\omega}{c}\left[\varepsilon_f\left(1 + \frac{2\omega_{LT}/\omega_0}{1 - \dfrac{\omega^2}{\omega_0^2} - \dfrac{i\gamma\omega}{\omega_0^2}}\right)\right]$$

In the case of two nearby situated exciton series, for instance A and C, the coefficient of reflection of the C-exciton taking into account the spatial dispersion and the presence of the dead layer is influenced by the dielectric function of the A-exciton (Figure 5).

The dielectric function taking into account the 1(A) and 2(C) states has the following form:

$$\varepsilon = \varepsilon_0 + \frac{2\pi\beta_{01}\omega_{01}^2}{\omega_{01}^2 - \omega^2 + \dfrac{\hbar\omega_{01}}{M_1}k^2 - i\omega\gamma_1} + \frac{2\pi\beta_{02}\omega_{02}^2}{\omega_{02}^2 - \omega^2 + \dfrac{\hbar\omega_{02}}{M_2}k^2 - i\omega\gamma_2}$$

$$\omega_{LT} = \frac{2\pi\beta_0\omega_0}{\varepsilon_0}$$

The spectral dependence of the dielectric function (Figure 6) is determined from the calculations of experimental reflectivity spectra in the region of the 1(A) state. This dependence is taken into account in calculations of the contour of reflectivity spectra of the 2(C) state.

$$|\omega - \omega_0| \ll \omega_0$$

$$\omega_0^2 - \omega^2 \approx 2\omega_0(\omega_0 - \omega)$$

$$i\omega\gamma \approx i\omega_0\gamma$$

$$\varepsilon = \varepsilon_0 + \frac{2\pi\beta_{01}\cdot\omega_{01}}{\omega_{01} - \omega + \dfrac{\hbar^2 k^2}{2M_1} - \dfrac{i\gamma_1}{2}} + \frac{4\pi\beta_{02}\omega_{02}^2}{\omega_{02}^2 - \omega^2}$$

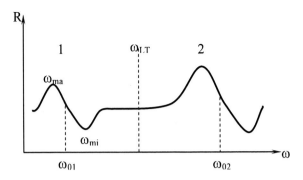

Figure 5. The reflectivity spectrum of exciton polaritons with two oscillators.

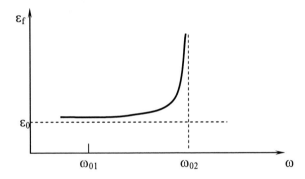

Figure 6. Spectral dependence of the dielectric function of the 1(A) state in the long-wavelength region of the 2(C) state.

$$\omega_{LT2} = 2\pi\beta_{02}\omega_{02}/\varepsilon_0$$

The dielectric function for the 2(C) state contains the gradient of the contour of dielectric function of the 1(A) oscillator (Figure 7).

$$\varepsilon = \varepsilon_0 + \frac{2\pi\beta_{02}\cdot\omega_{02}}{\omega_{02}-\omega+\frac{\hbar^2 k^2}{2M_2}-\frac{i\gamma_2}{2}} + \frac{4\pi\beta_{01}\omega_{01}^2}{\omega_{01}^2-\omega^2}$$

Taking into account the above mentioned one can calculate the spectral dependence of the ε_f without taking into consideration the spatial dispersion, but taking into account the gradient of the dielectric function in the region of $\varepsilon_1 \div \varepsilon_2$. The calculations of the dielectric function of the oscillator 2(C) with $l = 60$ Å give a contour shown in Figure 8.

$$\varepsilon = \varepsilon_0 + \frac{\varepsilon_0^{(1)}\omega_{LT}^{(1)}}{\omega_{01}-\omega-i\gamma_1/2} + \frac{\varepsilon_0^{(2)}\omega_{LT}^{(2)}}{\omega_{02}-\omega-i\gamma_2/2}$$

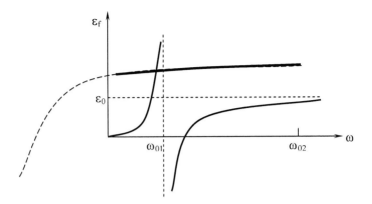

Figure 7. The influence of the dielectric function of the oscillator 1(A) upon the dielectric function of the oscillator 2(C).

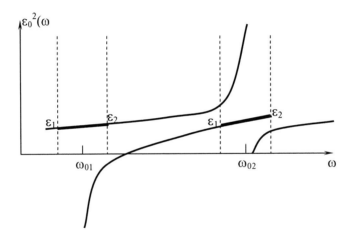

Figure 8. The dielectric function of the oscillator 2(C) taking into account the influence of the oscillator 1(A).

The numerical calculations in the region of 1(A) and 2(C) resonances in the case of two nearby situated exciton series should be conducted by taking ε_{10} and ε_{20}, approximating by a linear function, and introducing into the program $\varepsilon_0 = f(\omega)$ for each oscillator separately.

2. EXCITON SPECTRA IN CRYSTALS WITH STRONG POLARITON EFFECT ($\omega_{LT} \gg \Gamma$)

Many papers have been devoted to the theoretical and experimental investigation of optical phenomena (reflectivity, absorption, scattering and photoluminescence) in the region of exciton resonances, including the spatial dispersion of the dielectric function [1-7]. Especially detailed comparison of experimental and theoretical data has been conducted in A^2B^6 semiconductors, where the spatial dispersion is well pronounced [12-22]. The existence of additional waves caused by limited value of exciton effective mass was established in a series of experiments. Investigation of exciton resonances and the spatial dispersion is continued in less studied semiconductors, in which strong and weak polariton effects are observed.

The strong polariton effect is observed in crystals with the value of the longitudinal-transversal splitting much higher than the damping parameter. Such exciton spectra are observed in ZnP_2 crystals [23-43]. Monoclinic zinc diphosphide β-ZnP_2, with the space group $P2_1/c(C_{2h}^5)$, a direct-bandgap, optically biaxial semiconductor, in which dipole forbidden electronic transitions as well as electric dipole transitions (E1) are realized [23,42] is a beautiful model material for studying Wannier-Mott excitons and associated effects of spatial dispersion in low-symmetry crystals. In addition, the exciton states in monoclinic crystals should be influenced both by the anisotropy of the crystal field and by the dependence of the directions of the principal axes of the exciton polarizability and effective mass tensors on the direction of the wave vectors. But up to the present, exciton states in β-ZnP_2 have been examined insufficiently. This is due, first of all, to the incompleteness of the experimental data that has been obtained on high-quality single crystals for various directions of the wave vector of the exciton, the lack of quantitative calculations of the band structure, and also the absence of theoretical studies of exciton states in crystals whose symmetry is lower than rhombic.

For normal incidence of radiation on one of the principal crystal-forming faces bc or the (100) plane of β-ZnP_2 crystals (and the overwhelming majority of studies have been performed in this geometry) the two well-known hydrogen-like exciton series are observed. The electric-dipole *nS* series of the singlet exciton, called the *C* series, is manifested in the exciton reflection spectra for the radiation polarized parallel to the crystallographic c axis (E∥c) and has a pronounced polariton character of its dispersion contours [23-27,42-44]. Because of the large oscillator strength of the exciton transition [23-27] it has been possible to observe this series in the absorption spectra only in specially prepared thin samples. For E∥b the relative weak *B* series is observed in the absorption spectra. The authors of Ref. [23-27] identify the $n \sim 1$ line, the so-called *B* line, with the *1S* state of the ortho-exciton, and the authors of Refs. [23] and [24] identify the entire *B* series with high-energy components of the doublet lines for $n \geq 3$ with transitions to the *nS* ortho-states. The authors of Ref. [23] consider the lines with n≥2 to be due to transitions to the *nP* states of the exciton.

A structure characteristic for exciton transitions has been found in the reflectivity spectra of ZnP_2 crystals in the region of 1,550 – 1,600 eV (Figure 9, full curve). Apart from a very intensive long-wavelength reflectivity band I, three additional short-wavelength lines labeled as 2, 3, and 4 are revealed in the spectrum. The reflectivity spectra shown in Figure 9 were measured from the (110) face of the crystal in the polarization parallel to the second order 001 axis (the C-axis). The structure is absent in the reflectivity spectrum measured in the E⊥c polarization. As one can see from the figure, the 1 – 4 lines represent a convergent series in the short-wavelength side with decreasing intensity. It can be identified as a hydrogen-like exciton series corresponding to states with the n = 1 – 4 main quantum numbers [23-27].

Due to the strong absorption in the E∥c, one can not observe the structure of the absorption spectrum in this frequency range. Nevertheless, in the "forbidden" polarization E⊥c with the thickness of crystal plates of 0,5 – 1 mm one can observe a series of narrow lines converging to the short-wavelength side of the series. Up to 9 lines are observed in the spectrum, from which the lines 3 and 4 represent doublets. The long-wavelength line I' in the E⊥c polarization of the transmission spectrum is much narrower as compared to the respective line in the reflectivity spectrum measured in the E||c polarization, and it is shifted to lower energies. This line is probably associated with the dipole-forbidden state of the n = 1 exciton, which is decoupled from the dipole-active state under the action of the short-range exchange interaction, which lifts the degeneration between the singlet and triplet exciton states. The doublet structure of the 3 and 4 lines can be considered as a result of lifting the orbital degeneracy of the higher exciton states under the action of the crystal field. The series of n = 2 – 9 [23,31] lines in the E⊥c polarization is well fitted to the hydrogen-like dependence $E_i = E_g - R/n^2$ with the exciton binding energy of R = 46,2 meV and the bandgap of E_g = 1, 6028 eV.

The extrema of the bands in β-ZnP_2 can be located at the points Γ, A, R and D of the Brillouin zone (BZ). Since E1 transitions to S-type exciton states occur in β-ZnP_2 [23-27] but the group of the crystal class contains an inversion, the wave functions of the bottom of the conduction band (CB) and the top of the valence band (VB) have opposite parity if the corresponding group of the wave vector also contains an inversion. The valence band in β-ZnP_2, is formed by the 3p states of phosphorus, whereas the conduction band is most probably formed by the 4s states of zinc. In Ref. [45] the parity of the conduction-band wave functions was assumed *a priori* to be negative. It is significant that the triply orbital-degenerate valence band formed by the 3p states of phosphorus in a monoclinic β-ZnP_2 crystal is completely split by the anisotropic crystal field with symmetry 2/m (C_{2h}), while the magnetic moments associated with the orbital motion are to first order equal to zero (the so-called "freezing" of orbital angular momenta by the crystal field [45-51] and the relativistic interaction is expected to be weak. Data on the transmission spectra of thin crystals indicate, in all probability, that the largest value of the crystal splitting of the valence band in β-ZnP_2 is around 0.15eV. In Refs. [46] and [47,51] on the basis of magneto-optical measurements of the *B* series it was also concluded that the crystal splitting of the valence band in β-ZnP_2 significantly exceeds the spin-orbit splitting.

Since the monoclinic C_2 axis is aligned with the *Y* axis, for a correct group-theoretical analysis of the selection rules, it is necessary correspondingly to align the symmetry elements in the point group of the crystal class 2/m, specifically, the C_2 axis should be aligned with the *Y* axis, not with the *Z* axis as is customary in group theory. In this case, the z and y coordinates change places in the basis functions of the irreducible representations. In the basis functions of the irreducible spinor representations the projections of the total angular momentum *J* onto the quantization axis, which is now the *Y* axis, changes sign since in the new representation the matrix $\sigma_y' = -\sigma_z$, becomes the diagonal Pauli matrix. In the double group 2/m the one-dimensional spinor representations Γ_3^+ and Γ_4^+, and also Γ_3^- and Γ_4^-, belong to the case "b" according to their symmetry to time inversion; therefore they are combined and in what follows denoted as Γ_{34}^+ and Γ_{34}^-.

Assuming that the extrema of the bands are found in the center of the BZ, i.e., at k=0, four one-dimensional terms arise for the *nS* states of the exciton for the fully symmetric envelope wave function:

$$\Gamma_e \times \Gamma_v \times \Gamma_{env} = \Gamma_{34}^+ \times \Gamma_{34}^- \times \Gamma_1^+ = 2\Gamma_1^- + 2\Gamma_2^-.$$

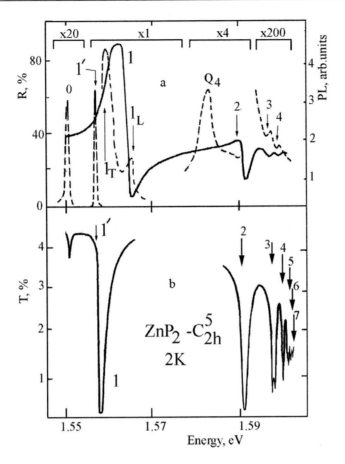

Figure 9. The reflectivity, luminescence, and transmission spectra of orthorhombic ZnP$_2$ crystals; (a) reflectivity spectrum in the $E\|c$ polarization (full line), luminescence spectrum in the $E\|c$ polarization (dashed line); (b) transmission spectrum in the $E\perp c$ polarization for a 1,2 mm crystal plate (full line). The multiplying coefficient in the upper part of the figure are related to the emission spectra.

The group symmetry of the wave function in the Γ and A points of the BZ does not allow degenerate states with integer spin except for the points R and D, where a double degeneracy exists, associated with the spatial symmetry [52]. The analytical part of the short-range exchange interaction leads to singlet and triplet excitons, or para- and ortho-excitons. The non-analytical part of the exchange interaction with its long-range character taken into account can lead to removal of degeneracy of the ortho-state and to a dependence of the energy on the wave vector.

E1 transitions to singlet *nS* states of symmetry Γ_1^- are allowed in the polarization $E\|b\|Y$, and to states with symmetry Γ_2^- in the polarization $E\|b,c\|X$ or $E\|c\|Z$ according to the orientation of the dipole moment P$_m$(s) for the given exciton band. The rezults [52] indicate that P$_m$(s) is aligned with the c (Z) axis for the lowest exciton band in β- ZnP$_2$. The singlet *nS* states have symmetry Γ_2^- with basis functions transforming in z. The three remaining symmetry states $2\Gamma_1^- + \Gamma_2^-$ must belong to the ortho-exciton, and in this case the basis functions of the Γ_2^- state most likely transform in *x*. Optical transitions to the ortho-states for systems with spherical symmetry are intercombinationally forbidden. This prohibition can be relaxed by the spin-orbit interaction and—this is especially important—in systems with a noncentral field, such as, probably, excitons in crystals with low symmetry. Transitions to ortho-states with symmetry $2\Gamma_1^-(y) + \Gamma_2^-(x)$ are forbidden in the electric quadrupole and magnetic dipole approximations by parity $(PP^l = -1)$.

For s||P$_m$(z) the long-range nature and the non-analytical part of the exchange interaction [45] lead to the appearance of a longitudinal exciton, also with symmetry $\Gamma_2^-(z)$.

Let us consider other possible excited states of an exciton. For nP states 12 terms arise: $\Gamma_{34}^+ \times \Gamma_{34}^- \times (\Gamma_1^- + 2\Gamma_2^-) = 6\Gamma_1^+ + 6\Gamma_2^+$. But for excitons with "p-like" envelope wave functions the exchange interaction at zero can be neglected; therefore only three states are of interest: $\Gamma_2^- \times (\Gamma_1^- + 2\Gamma_2^-) = 2\Gamma_1^+ + \Gamma_2^+$. These states do not have a simple analog, as is the case for crystals with axial symmetry, i.e., p_0 and $P_\pm$ but are completely split by the anisotropic crystal field into states of the type $P_a - \Gamma_1^+(xz)$, $P_b - \Gamma_2^+(yz)$ and $P_c - \Gamma_1^+(zz)$, where the coordinates indicated in parentheses are the coordinates in which the wave functions of the irreducible representations transform. Transitions to the nP states are split in the quadrupole approximation for scalar and magnetic-dipole (*M1*) and electric quadrupole (E2) transitions. However, the transition probabilities here may be expected to be several orders of magnitude smaller than for E1 transitions to nP states. Optical transitions to nP exciton states of β-ZnP$_2$ have been observed for two-photon absorption [44].

The analysis of the reflectivity band in the region of 1,550 – 1,570 eV has been performed in order to determine the parameters of the exciton state. A good agreement between the theory and experiment has been achieved in the frame of the Hopfieled-Thomas model of the exciton reflectivity [1-3]. The following main parameters of the n = 1 exciton resonance in ZnP$_2$ crystals have been obtained: the resonance energy $\hbar\omega_0 = 1,5606$ eV, the longitudinal-transversal splitting $\hbar\omega_{LT} = 4,5$ meV, the exciton translation mass M = 3m$_0$ (m$_0$ is the mass of the free electron), the "dead" layer thickness l = 80 Å, the damping parameter $\hbar\Gamma = 0,1$ meV), and the background dielectric constant $\varepsilon_{zz} = 12,0$.

The dashed line in Figure 9 represents the luminescence spectrum of a ZnP$_2$ crystal measured in the E||c polarization under the excitation by the 5145Å line of an Ar$^+$ laser. A polariton doublet is observed in the region of the n = 1 state. The short-wavelength component of this doublet I$_L$ is due to the maximum of the surface transmission coefficient of lower branch polaritons, while the I$_T$ component is due to the "bottle-neck" effect. In the region of higher exciton states one can observe the n=2,3,4 emission lines (marked with arrows). A narrow dipole-forbidden exciton luminescence line (the dashed line in Figure 9) is observed in the E⊥c polarization. Due to the exchange interaction, this line is shifted to the long wavelengths with respect to the resonance line of the dipole-active I$_T$ state. The spectral position of the luminescence line I' coincides with the corresponding absorption line I. The narrow emission line in the long-wavelength region of the spectrum is related to radiative annihilation of the weak dipole-active exciton discussed above. Somehow surprising is the emergence of the Q$_4$ line between the n = 1 and n = 2 states in the luminescence spectrum measured in the E||c polarization. The analysis of the emission spectrum in the region of energies above E$_g$ (see Figure 10, dashed line) shows that the Q$_4$ line is the most long-wavelength in the system of equidistant Q$_1$, Q$_2$, Q$_3$, and Q$_4$ lines which are separated by about 500 cm^{-1}. It was suggested [23] that the emission maximum Q$_4$ is due to hot luminescence caused by the consecutive relaxation of electron excitation energy of the crystal with the emission of LO-phonons [53,54].

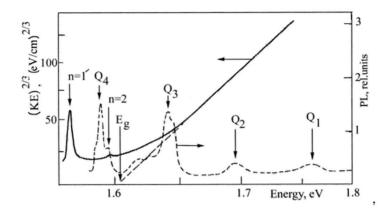

Figure 10. The absorption and luminescence spectra of ZnP$_2$ crystals on the fundamental band edge; the full line is the absorption spectrum of a crystal plate with thickness of 60 μm in the E⊥c polarization (the ordinate axis correspond to (KE)$^{2/3}$, where K is the absorption coefficient); the dashed line is the luminescence spectrum in the E⊥c polarization.

Then, one can consider that the most short-wavelength emission line Q$_1$ is a result of the electron transition from a deeper Q$_0$ band with emission of a photon and a LO-phonon. In order to clarify if the equidistant emission lines are not associated with the features of the band structure in the frequency interval of Q$_1$-Q$_3$ [23], the absorption coefficient spectra of crystals with the thickness d = 60 μm have been measured in the "forbidden" E⊥c polarization.

The short-wavelength segment of the absorption spectrum is linear in the (K·E)$^{2/3}$ coordinates, which indicates on the direct forbidden character of the transition in the E⊥c polarization. If one extrapolates the linear dependence (the dashed line) to the intersection with the abscissa axis, the value of 1,603 eV is obtained which corresponds to the E$_g$ obtained form the analysis of the exciton series in the absorption spectrum. The n =1 and n =2 states of the exciton series are well observed in the long-wavelength region of the absorption spectrum. The character of the spectral dependence of K in the short-wavelength region suggests that there are no transitions to higher energy bands up to the energy of 1,750 eV.

The reflectivity spectra of orthorhombic type 2 ZnP$_2$ crystals measured at 2K are discussed in Refs [24,27,28]. The n = 1, 2, and 3 lines of the C-exciton series are observed at 1,5644; 1,5909 and 1,5976 eV energies in the E∥c polarization. The Rydberg cionstant R=0,048 eV, and the bandgap energy E$_g$ =1,603 are determined. The absorption spectra of these crystals in the E⊥c polarization contain the B and D series with n = 1, 2, 3, and 4 at energies of 1,5744; 1,5939; 1,5997 and 1,6011 eV. The Rydberg constant for the D series R=0,042 eV, and the bandgap E$_g$=1,604 eV. A more detailed description of exciton spectra for this series can be found in refs [23-43,54-65].

2.1. Intraband and Interband Relaxation of Exciton Polaritons. Determination of Parameters of Polariton Disperssion

The polariton effects play an important role in the formation of the low-temperature emission spectra of crystals near the exciton resonances [1-7]. A series of common spectral-temporal peculiarities of the exciton luminescence are described in the frame of the polariton theory [7-23]. The main phenomena of polariton luminescence (PL) associated with the polariton dispersion, the conditions of radiation transmission through the boundary of the crystal in the region of exciton resonance, the character of the spatial and spectral distribution of polaritons have been intensively discussed (see for instance [15] and refs therein). Nevertheless, the qualitative analysis of specific experimental data in the frame of a consecutive approach presents still interest. It is not still clear which polariton states (upper and lower branches) give a main income to the emission on the short-wavelength wing of the PL spectrum, and how the spatial distribution of polaritons reveals itself. Especial attention should be paid to the experimental investigation of the role of the crystal boundary (concerning the additional boundary conditions [4] and the role of surface states [9]). The investigation of the energy relaxation of polaritons in the case of strong spatial diffusion presents also especial interest [66].

Two basic approaches are used in the practical analysis of PL spectra. One of them is based on the solution of a simplified kinetic equation, in which the integral of collisions, or a part is substituted by some empirical expressions selected from qualitative considerations [8,15,67-70]. The limits for the application of this approach should be argued. The second approach, in which the integral of collisions is accurately calculated, is considered to be more consistent [9,13,14,66]. The resonance exciton luminescence spectra of the black modification of ZnP_2 crystals have been analyzed with the second approach in the frame of the theory of spectral-spatial polariton diffusion. These crystals demonstrate a series of interesting and unusual properties in the region of the fundamental absorption edge [23,24], such as a very reach spectrum of hydrogen-like exciton states (up to the state with the main quantum number n = 9 [23,25-28]. The lowest n = 1 state in these crystals demonstrates a strong longitudinal-transversal splitting. The effective exciton mass does not significantly differ from the free exciton mass. ZnP_2 crystals are considered to be very convenient objects for the purposes of investigating polariton excitons, including the phenomena related to the additional waves.

Especial attention is paid to the problem of the relative income of the lower polariton branch (LPB) and upper polariton branch (UPB) to the external emission. A complex investigation of amplitude-phase reflectivity spectra of ZnP_2 crystals at different angles of incidence is discussed in the subchapter 2, since this problem is tightly connected to the analysis of additional boundary conditions. All the parameters necessary for the construction of the polariton dispersion curves are determined. The experimental data on the luminescence presented in the subchapter 2 are analyzed in the subchapter 4 on the basis of the theory of spectral-spatial polariton diffusion, which is described in the subchapter 3. The analysis is performed by taking into account the boundary conditions (BC), corresponding to the experimental data on reflectivity.

Single crystalline ZnP_2 platelets grown by chemical vapor transport have been used in investigations [25]. Their dimensions and quality of their natural surfaces are suitable for measuring the reflectivity spectra, as well as the luminescence spectra. The geometry of the experiment with the second order axis of the crystal C_2 coinciding with the dipole momentum vector of the exciton (non-degenerated state) parallel to the reflecting or emitting surface, and the light propagating in the $K \perp C_2$ direction was chosen. The thickness of the crystal was 0,1 – 1 mm, so that the influence of the back face of the crystal was excluded. The measurements have been conducted with different angles φ between the direction of propagation of the registered light and normal to the crystal surface. The spectra were measured at T = 2K (the crystals were immersed in the liquid He under the evaporation) with a DFS-24 spectrometer. The spectral slit of the spectrometer was 0.05 meV at φ = 0, and 0.1 meV at φ > 70°. The luminescence was excited by the 647,1 nm line of a Kr+ laser with the excitation power less than 500 mW in a spot with the diameter of 0,5 mm. The reflectivity was measured with an incandescent lamp.

The luminescence spectrum of polaritons is determined by the parameters of their dispersion. The amplitude-phase exciton reflectivity spectra of ZnP_2 crystals have been measured with different angles of incidence according to the method described in [25], in order to determine the dispersion polariton curves. The energy coefficients of reflectivity $R_δ(ψ_o, φ, ψ)$, depending on the azimuth $ψ_o$, and $ψ$ of the polarizer and analyzer, the angle of incidence φ and the additional phase shift δ between the p- and s-amplitudes of the reflected light, introduced by the phase-shifting wedge have been measured in the experiment. The $ψ_o$, and $ψ$ angles were measured in respect to the plane of incidence (perpendicular to the C_{2z} axis), in the direction of the light propagation.

The measured spectra have been analyzed in the frame of theory of additional wave [1,2] taking into account the existence of the "deal" layer (without excitons) at the crystal boundary [3]. The best correspondence between the theory (full lines in Figure 11) and experiment is achieved with the following values of parameters: the resonance frequency $ω_o$=1,5606 eV, the longitudinal-transversal splitting $ω_{LT}$=4,5 meV, the effective exciton mass M = $3m_0$, the background dielectric constant $ε_b$=12, the damping parameter Γ=0,1 meV, the thickness of the "dead" layer l=60 Å.

Note that in the reflectivity spectrum of ZnP_2 crystals measured at nearly normal incidence of the light (Figure 11, a_1, s-component of polarization) there are no sharp anomalies in the region of the reflectivity minimum (for instance spike-like in CdS [3]), which could allow one to identify the longitudinal frequency $ω_L=ω_o+ω_{LT}$. Nevertheless, as one can see from Figure 11, by increasing the angle of incidence, or by changing the azimuths $ψ_o$, $ψ$ and the phase shift δ one can cause arising such anomalies, and therefore allowing one to determine the energy spectrum of the polariton.

As concerns the Δ phase spectra, they bear independent information (with regards to usual reflectivity spectra), which demonstrate the relevance of the applied model.

The luminescence spectra of ZnP_2 crystals have been measured in the same points of the crystal surface as those measured in the reflectivity spectra, since a reliable comparison was necessary. Figure 12 presents the polariton emission spectra of ZnP_2 crystals measured with different directions of the emitted light φ=0° (a) and φ=76° (b) in the E∥c_{2z} polarization. The vertical arrows indicate the $ω_o$ and $ω_L$ frequencies.

One can see from Figure 12 that the PL spectrum of ZnP_2 demonstrates a doublet structure. The change of the output angle with unchanged E∥C_{2z} polarization leads to a qualitative restructuring of the spectrum: the short-wavelength component of the doublet is significantly narrowed and intensified as compared to the long-wavelength one. At the same time, the features of the PL spectrum are reproduced with enormous resolution for semiconductor crystals due to a big value of $ω_{LT}$ and low value of Γ. The analysis of ZnP_2 spectra in ref. [25] is performed more detailed as compared to other semiconductor crystals.

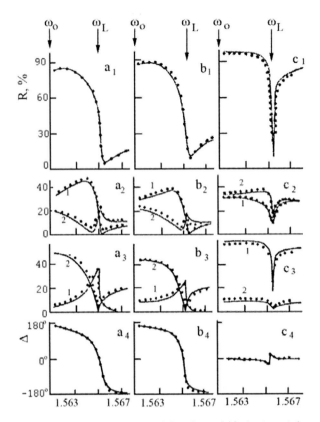

Figure 11. The spectra of reflection $R_\delta(\psi_o, \varphi, \psi)$, and the phase shift $\Delta = \Delta_P - \Delta_s$ between the p- and s-amplitudes of the reflected wave at the angles of incidence $\varphi = 8°$ (a), $65°$ (b), $85°$ (c): $\psi_o = \psi = 90°$ (i=1), $\psi_o = -\psi = 45°$ (i=2,3 - curves 2), $\psi_o = \psi = 45°$ (i=2, 3 - curves 1), $\delta = -\pi/2$ (i=2), $\delta = 0$, (i=3); the value of Δ is calculated on the basis of spectra i=2, 3 (see. [17]).

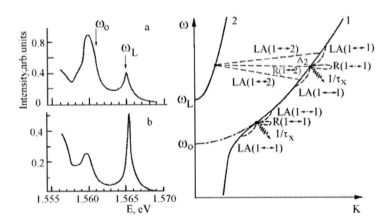

Figure 12. Polariton luminescence spectra of ZnP$_2$ crystals (T = 2K) measured with different angles between the direction of the emitted light and the normal to the surface $a \sim \varphi = 0°$, $b \sim \varphi = 76°$. The polariton dispersion curves of the lower branch (1) and the upper branch (2); the dashed curves show schematically the processes of inelastic polariton scattering by longitudinal acoustical phonons LA, and the elastic scattering by the defects and impurities; $1/\tau_x$ is the non-radiative polariton recombination.

2.2. Spectral-Spatial Diffusion of Polaritons

The emission of light from the crystal in the frame of the polariton mechanism is considered as a process of transformation of the polariton which is incident inside the crystal on its boundary to the out-coming photon [7]. The probability of this transformation is determined by transmission coefficient of the emission at the crystal boundary [18,19]. One considers that it is possible to use the Boltzmann equation for the polariton distribution function $f_j(\lambda)(K, r, t)$, which depends upon the polariton wave-vector K, its spatial coordinate r and the time t (j is the number of the polariton branch, λ is the index of polarization). The value $f_j^{(\lambda)}(K, r, t)d^3Kd^3r$ determines the number of polaritons in the elementary phase volume d^3Kd^3r with K, r coordinates at the moment of time t.

The system of kinetic equations for the $f_j^{(\lambda)}$ functions is considered in a simple case of an isolated dipole-active exciton resonance in a cubic crystal, without taking into account the contribution from the longitudinal excitons, i. e. in a two-band model (j=1, 2). In this case, the dependence of the polariton frequency ω upon the wave vector K has a usual form shown in Figure 12, where the lower- and upper- polariton branches are shown by full curves 1 and 2. The zone of mechanical excitons [4] is shown by the dashed curve, which would characterize the intrinsic excitations of the crystals in the case when the photon-exciton interaction is absent.

Of especial interest is the continuum wave photoluminescence in the spectral interval of $(3-4)\omega_{LT}$ in the region of longitudinal-transversal splitting. It is supposed that the contribution to the integral of collisions comes from processes of inelastic polariton scattering with the participation of longitudinal (LA) phonons [9] and from the mechanisms of elastic scattering (R) on defects and impurities (these processes are schematically illustrated by dashed lines in Figure 12). Apart from that, the following peculiarities are taken into account: (a) due to the relatively high spectral density of states in the branch 1, multiple intra-band polariton scattering (of the type 1↔1) occurs, so that the distribution function $f_j^{(\lambda)}$ becomes almost isotropic and independent on the index of polariton polarization λ, (b) due to the low density of states in the branch 2, the intra-band scattering of the type 2↔2 does not play an important role: the population of the polariton states in the branch 2 is mainly determined by inter-band scattering of the type *1↔2*. In these conditions, the kinetic equation for the LPB can be considered independently on the equation for the branch 2, and by integrating over the directions of propagation (for instance, as in the case of radiation transport theory [71-75]) it is reduced to the diffusion-type equation.

$$D\Delta f_1 - f_1/\tau_x + G + Qf_1 = 0 \qquad (1)$$

where f_1 is the isotropic component of the distribution function of the LPB, $D=v_1^2\tau_p$ is the coefficient of diffusion (not averaged over the energetic distribution), which depends on the frequency via the group velocity v_i, and the momentum relaxation time is equal to $\tau_p = (1/\tau_{ac}+1/\tau_{el})^{-1}$ (τ_{ac} and τ_{el} are the relaxation times caused by inelastic scattering on acoustic LA-phonons and elastic scattering on defects and impurities, respectively); τ_x is the life-time related to the non-radiative polariton annihilation; G is the source function averaged over the directions of radiation; Q is the operator of inelastic collisions, which is calculated with the accuracy of the second order corrections with respect to the u/v_i≪ 1 parameter (u is the velocity of sound, v_i is the group velocity of the LBP [25].

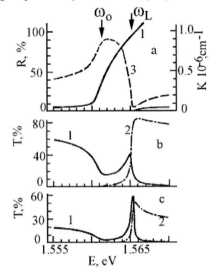

Figure 13. Theoretical spectral dependences of the reflection coefficients (a, dashed line), and transmission of the emission through the crystal boundary (b, c) at angles of incidence φ=0° (a, b) and φ=76° (c): 1 is for the LPB, 2 is for the UPB (the respective dispersion curves of the polaritons are shown by full lines 1 and 2 in Figure a). The calculations are performed using the values of parameters corresponding to experimental data on reflectivity from Figure 11.

$$Qf_1 = \frac{2(E_c - E_v)^2 v_1 d}{3\pi\rho\hbar K_1^2 d\omega} \left\{ \frac{K_1^6}{v_1^2} \left[1 + \left(\frac{86}{105} uK_1 + \frac{T}{\hbar} \right) \frac{d}{d\omega} \right] f_1(K_1(\omega), r) \right\} \quad (2)$$

E_c и E_v are the deformation potential constants for the conduction and valence band, respectively; ρ is the density of the crystal; T is the temperature in the energy units. When deducing the equation (2), the temperature dependence of the photon population numbers N_q (q is the wave vector of the phonon) is approximated by a linear function N≈T/ℏuq, which gives an adequate result for T=0 and T≫ℏuq. In the coordinate system with the y-axis directed inside the semi-infinite crystal perpendicularly to the plane of the crystal surface y = 0, the boundary condition for the equation (1) (which reflects the energy balance on the surface) takes the shape

$$\left(D\frac{df_1}{dy} - \gamma_s f_1 \right)_{y=0} = 0 \quad (3)$$

where γ_s is the rate of the surface annihilation of polaritons (including the effects of transmission, reflection, and non-radiative surface annihilation); apart from that $f_1(K_1(\omega),y)_{y=\infty} =0$.

The differential equation (1) describes the spectral-spatial diffusion of the LPB. The equation of such a type was solved in ref. [66] when calculating the PL spectrum of an anisotropic crystal platelet with a special-type source function G, and certain boundary conditions with respect to the frequency.

In the actual experimental conditions, the frequency of the excitation light is significantly higher than the frequency corresponding to the absorption band edge. One can consider that spatial distribution of the LPB is basically constituted outside the spectral region of the emission.

With a deep enough distribution, the energy relaxation of polaritons in a relatively narrow region of the longitudinal-transversal splitting should not result in a significant change of the character of the spatial distribution (the influence of the crystal boundary can have a significant influence only in the region $\omega<\omega_0$ due to the sharp increase of the rate of surface radiative annihilation of polaritons, which is mainly inversely proportional to the spectral density of states.

In such a case, the distribution function $f_1(K_1(\omega),y)$ can be presented in a factorized form with respect to the ω and y variables, with the spatial part f_1 presented in an approximation form $\sim\exp(-y/L)$, where L is the distribution depth. By setting G=0 in the region of emission and by integrating the equation (1) with respect to y in the limits from 0 to ∞, and taking into account the above mentioned boundary conditions one can obtain

$$Qf_1 - \Gamma_{xs}f_1 = 0 \qquad (4)$$

where $\Gamma_{xs} = 1/\tau_x + \gamma_s/L$ is the total inverse polariton lifetime in the region 1 (on the lower branch).

The last equation actually represents the balance equation for the polaritons written in the differential form (compare with [9, 72]). However, in contrast to [9,72] it takes into account the non-uniform spatial distribution of polaritons via the effective distribution depth L (in [9] the thickness of the crystal platelet plays the role of L). As one can see from (4), in the considered approximation it is impossible to distinguish between the probability of the bulk non-radiative decay, and the probability of surface annihilation.

The solution of the equation (4) with certain boundary conditions with respect to frequency allows one to find the frequency dependence of the LPB distribution function $f_1(K_1(\omega),y)$. As concerns the UPB, the corresponding distribution function $f_2(K_2,y)$ with the above mentioned assumptions satisfies the kinetic equation

$$\vartheta_{2y}\frac{\partial f_2}{\partial y} = -\Gamma_{12}f_2 + \sum_{K_1',h} W_{12}^{(h)}(K_2,K_1')f_1(K_1',y) \qquad (5)$$

$$\Gamma_{12} = \sum_{K_1',h} W_{12}^{(h)}(K_1',K_2)$$

Γ_{12} is the inverse UPB lifetime, caused by the inter-band scattering in the LPB states, and $W_{ji}^{(h)}(K_j,K_i')$ is the probability of polariton scattering from the $|K_i'\rangle$ state to the $|K_j\rangle$ state due to the h-type mechanism. The non-radiative lifetime τ_x does not enter in (5), since $\Gamma_{12} >> \tau_x^{-1}$ in the approximation for which the diffusion equation (1) is valid.

$$\Gamma_{12} \approx \Gamma_{11} = \sum_{K_1,h} W_{11}^{(h)}(K_1, K_1)$$

If one disregards the inelastic character of the inter-band polariton scattering on LA phonons, the $f_2^-(K_2,y)$ solution of the equation (5) for the polaritons propagating in the direction of the crystal boundary ($v_{2y}<0$) takes a simple form

$$f_2^-(K_2,y) = \frac{L\Gamma_{12}}{|v_{2n}|+L\Gamma_{12}} f_1(K_1,y) \qquad (6)$$

One can see from the relation (6) that the function f_2^- is determined not only by the probability of the inter-band scattering (via Γ_{12}), but also by the effective depth L, f_2^- being generally anisotropic in spite of the isotropic character of the inter-band scattering ($K_2 \ll K_1$). This anisotropy is connected to the presence in the denominator (6) of the projection of the velocity on the direction of the normal to the crystal surface, and finally it is caused by the non-uniformity of the spatial distribution of LPB.

The final solution of the emission problem is determined by the interdependence between the intensity of the external emission and the boundary values of the distribution function $f_1(K_j, +0)$. If $I^{(\lambda)}(\omega,\varphi)$ is the spectral density of the intensity of the outgoing emission from the crystal in λ-polarization ($\lambda = s, p$) under the output angle φ relative to the external normal, then

$$I^{(\lambda)}(\omega,\varphi) = \frac{\hbar\omega^3}{c^2} \sum_j T_j^{(\lambda)}(\omega,\varphi) f_j(K_j + 0) \qquad (7)$$

Where $T_j^{(\lambda)}$ is the energy transmission coefficient of the emission on the crystal boundary in the direction "crystal-vacuum" for the λ-polarization of the measured external light, c is the light velocity in the vacuum. The direction of the K vector relative to the direction of emission is determined by the diffraction rules of the beam at the crystal boundary.

2.3. Relative Contribution to the Emission of Upper and Lower Polariton Branches

In the analysis of the experimental emission spectra one should take into account that the main income to the emission in the region of $\omega < \omega_L$ comes form the LPB [25]. Then, the spectral form of the function f_1 in this region can be obtained by dividing the measured luminescence spectrum $I^{(\lambda)}(\omega,\varphi)$ to $T_i^{(\lambda)}(\omega,\varphi)$, and by using the equation of spectral diffusion (4) extrapolated to the region $\omega > \omega_L$. Further, the function f_2 is calculated on the basis of the kinetic equation (5), and the results of the theoretical analysis according to (7) are compaed with experimental data.

The results of the calculation of polariton dispersion curves are presented in Figure 13 (curves 1, 2). The reflection coefficients $R^{(s)}(\omega,\varphi)$ (a, curve 3, $\varphi = 0°$), and the transmission $T_{1,2}^{(s)}(\omega,\varphi)$ (b, curves 1, 2, $\varphi = 0°$; c, curves 1, 2 $\varphi = 76°$) are also shown in Figure 13. The coefficients are calculated for the case of propagation of the emission from the vaccum to the crystal in the frame of the dead layer model used for the analysis of reflectivity spectra. These coefficients should have the same values for the reverse process of the emission from the crystal, if the absorption is weak enough. This reversibility is due to the symmetry of the system with respect to the time inversion [19].

The weak absorption in this case determines also the criterion of the validity of the kinetic equation: the polariton wavelength should not exceed at least the mean free path. With the obtained values of the damping parameter Γ, this criterion is not valid only in the vicinity of the $K=0$ point, i. e. for the polaritons of the branch 2 near the ω_L frequency. Nevertheless, the coefficient $T_2^{(s)}$ maintains his macroscopic meaning of the ratio of energy fluxes in the region $\omega \leq \omega_L$, where $T_2^{(s)} \neq 0$ at $\Gamma \neq 0$ [18] (see. Figure 13,b, c, curves 2).

Figure 14. presents the results of dividing the emission spectra measured with two different output angles (see Figure 13) by transmission coefficients $T_1^{(s)}(\omega, \varphi)$ corresponding to these angels: $\varphi=0°$ with full circles, $\varphi=76°$ with open circles. One can consider that in the region $\omega<\omega_L$ (except for a small segment in the long-wavelength side of ω_L) the ratio $\sigma(\omega,\varphi)= I^{(s)}(\omega,\varphi)/T_1^{(s)}(\omega,\varphi)$ reflects the spectral dependence of the function $f_1(K_1(\omega),+0)$, since the states of the branch 2 in these spectral range do not participate in the emission $(T_2^{(s)}, f_2 \approx 0)$, and the factor before the sum symbol in (7) is not significantly changed at $|\omega - \omega_0| \ll \omega_0$. Moreover, the value of $\sigma(\omega, \varphi)$ should not depend significantly upon the angle φ, even for the anisotropic f_1 function. Due to a big refraction coefficient $n_1=cK_1/\omega$ in the region $\omega \geq \omega_0$, the polaritons from the LPB contributing to the emission propagate practically in the direction of the normal in a narrow solid angle $2\pi/n_1^2 \ll 1$.

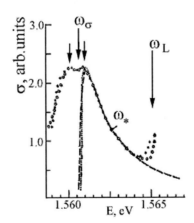

Figure 14. Spectral dependence of the ratio $\sigma(\omega, \varphi)=I^{(s)}(\omega,\varphi)/T_1^{(s)}(\omega,\varphi)$ of the intensity $I^{(1)}$ of the external emission to the transmission coefficient $T_1^{(s)+}$ for the output angles of the emission $\varphi=0°$ (full circles) and $\varphi=76°$ (open circles). The dashed and the dotted lines represent the numerical solution of the equation (5) in the approximation of parabolic polariton dispersion in the region $\omega>\omega_o$ for the three-dimensional and two-dimensional migration, respectively. ω_* is the frequency corresponding to the boundary conditions of the distribution function $f_1 = \sigma$ and its derivative $\partial f_1/\partial \omega = d\sigma/d\omega$.

One can consider the equality $\sigma(\omega, 0°)=\sigma(\omega, 76°)$ in the frequency interval $\omega_0 \leq \omega < \omega_L$ as a natural condition of the relative normalization of a and b spectra in Figure 12. The form of the $\sigma(\omega,\varphi)$ curves is similar in the spectral interval $1{,}5590<\omega<1{,}5640$ eV including the resonance frequency ω_0. Both curves demonstrate a deep in the vicinity of ω_0. This deep is not revealed in initial emission spectra (Figure 12). Both curves are bent up and are significantly different in the short-wavelength region. The difference between the $\sigma(\omega, 0°)$ and $\sigma(\omega, 76°)$ curves in the short-wavelength region of the spectrum indicates the income to the emission from the UPB which should significantly depend upon the output angle of the emission in the vicinity of ω_L due to the strong angular dependence of the $T_2^{(s)}(\omega,\varphi)$ coefficients. One should mention that the UPB emission emerges also at $\omega<\omega_L$.

The doublet structure in the maximum of the distribution function $f_1(K_1, +0) = \sigma(\omega,\varphi)$, is probably due to the polariton scattering (with the emission of LA phonons) from the region of the short-wavelength component of the doublet with a fast increasing of the radiative polariton emission in the long-wavelength side (it is compared with the situation realized in crystals of anthracene [73]). Therefore, one can consider that the long-wavelength component of the doublet of the distribution function is a "phonon replica" (on LA phonon) of the primary short-wavelength component. One should mention here that in the considered ZnP_2 crystals the minimum group velocity of the polariton $\upsilon_{1\ min} \approx 9,1 \cdot 10^5$ cm/s does not exceed significantly the velocity of the longitudinal sound $u \approx 6,1 \cdot 10^5$ cm/s [74].

Due to these values of velocities, the second approximation with respect to the u/υ_1 parameter in the operator of inelastic collisions (2) could be insufficient near the ω_0 frequency. Therefore, the equation (4) was used for quantitative analysis of the form of luminescence spectra only in region of parabolic polariton dispersion of the branch 1 ($\omega>\omega_0$). The experimental values of the f_1 function and its derivative $\partial f_1/\partial \omega$ in the point $\omega_0=1,5626$ eV situated in the region of longitudinal-transversal splitting were used as boundary conditions (see Figure 14). The ratio $\eta(\omega)=\Gamma_{xs}/a_0(\omega)$, where $a_0(\omega) \approx 5u/[\upsilon(\omega)\tau_{ac}(\omega)]$ is the coefficient at zero derivative in the differential operator Q (the case $T=0$ K) was used as fitting parameter. By changing the value of $\eta(\omega)$ at ω_L the authors [25] reached a good coincidence of the solution of equation (5) and the experimental $\sigma(\omega,\varphi)$ curve in a wide spectral interval. The best coincidence was obtained with $\eta(\omega_L)=0,14$ (dashed curve in Figure 14).

However, one should take into account that the investigated exciton state in ZnP_2 is non-degenerated, and the polariton migration in the region $\omega<\omega_L$ is basically two-dimensional [4]. In this case, the application of the three-dimensional relaxation model is incorrect. The solution of the equation (4) with the operator Q corresponding to the case of two-dimensional polariton relaxation (in the plane perpendicular to the C_{2z} crystal axis) is presented in Figure 14 by the dotted curve. The obtained optimal values of $n=0,01<<1$ allows one, from the one side, to make a conclusion about the absence of the surface non-radiative polariton annihilation, which is consistent with boundary condition system used for the calculation of reflection and transmission coefficients. On the other hand, such a small value of η indicates on a high enough value of the polariton lifetime in the band 1 as compared to τ_{ac}, which confirms applicability of the diffusion approach. As one can see from the comparison of dashed and dotted lines, the extrapolation in the region $\omega>\omega_L$ of the equations of two-dimensional and three-dimensional realaxation gives practically the same results (except for the value η, and consequently Γ_{xs}). As concerns the processes of radiative exit of polaritons due to the transparency of the boundary, in the region $\omega>\omega_0$ these processes do not give a significant income to Γ_{xs}. This is due to a relatively small value of υ_1 of the exciton-like polaritons from the one hand, and due to the big value of the refraction coefficient n_1 which determines a very narrow internal solid angle from which the polaritons from the LPB can be emitted, from the other hand. Due to the effect of the narrow solid angle in which the LPB propagate, the scattering resulting from the internal reflection in the UPB states is insignificant for the parameters of ZnP$_2$, as well the readsorption induced by the surface (in the region $\omega>\omega_L$) [4,11].

The contribution of the $I_1^{(s)}$ polaritons of the branch 1 in the intensity $I^{(s)}$ of the external emission is determined from the extrapolated f_1 function by means of equation (7). Then, the contribution $I_2^{(s)}=I^{(s)}-I_1^{(s)}$ from the branch 2 was determined on the basis of experimental data in Figure 12. The spectral dependencies of $I_1^{(s)}$ and $I_2^{(s)}$ intensities are presented in Figure 15. As one can see from this figure, the contribution of LPB and UPB states to the emission in the vicinity of the longitudinal frequency ω_L are comparable.

Figure 16 presents the short-wavelength wings of the function $f_2=f_2^-$ which contributes to the external emission in the directions $\varphi=0°$ (full circles) and $\varphi=76°$ (open circles). The spectral dependence of the function $f_2^-(K_2,+0)$ was plotted by dividing the corresponding partial intensities $I_2^{(s)}$ (see Figure 15) to the transmission coefficients $T_2^{(s)}(\omega,\varphi)$. This approach is suitable for the frequency range, where the application of kinetic equation for the UPB is reasonable, i. e. in the frequency range $\omega>\omega_L$, for the small output angles φ and in the region $\omega>\omega_1$, ($n_2(\omega_1)=1$) for very big φ. Revealing in the spectra (Figure 15) of emitting states from the branch 2 in the long-wavelength side of ω_L and ω_1 is due to the uncertainty of the wave vectors of these states near ω_L, ω_1 at finite values of the reverse lifetime Γ_{12}. The momentum conservation law for the scattering of the type 1→2 is not exactly accomplished for $\Gamma_{12}\neq0$. This leads to inter-band transitions in the region $\omega<\omega_L$, ω_1. A strong reconstruction of spectrum of normal waves of the type 2 also occurs near ω_L with increasing Γ_{12}. Therefore, namely the Γ_{12} value determines the character of the reflectivity spectrum in the region of the minimum value of the reflection coefficient, and the measured according to the reflectivity spectra value of the phenomenologic parameter Γ concides with the value of Γ_{12} in the region of the ω_L frequency.

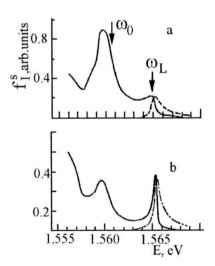

Figure 15. The partial contributions $I_1^{(s)}$ (full curves) and $I_2^{(s)}$ (dash-dot lines), respectively, to the intensity of the external emission of LPB and UPB determined by means of the reconstructed distribution function for $\varphi=0°$ (a), and $\varphi=76°$ (b).

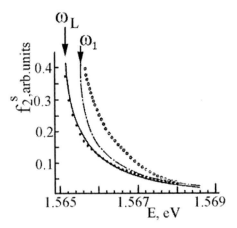

Figure 16. Distribution function of the UPB contributing to the external emission under the following output angle: $\varphi = 0°$ (full circles), $\varphi = 70°$ (open circles). The full and the dash-dot lines were calculated according to (7) for the same values of φ; ω_1 is the frequency corresponding to the value of the refractive index of the UPB equal to 1.

The effective distribution depth L of the polaritons from the branch1 was estimated by introducing the values of $\Gamma_{12} = \Gamma = 0,1$ meV in the equation (6) on the basis of experimental data for $f_1 = \sigma$ (Figure 14) and f_2^- (Figure 16). Thus, the function f_2^- for $\varphi = 0°$ calculated according to the equation (6) for $L = 5,4$ μm corresponds well to the experimental dependence. In the case of $\varphi = 76°$ at the same value of L there is a difference between the theoretical (dashed) and the experimental curves. However, the character of the anisotropy of the f_2^- predicted by the equation (6) is reproduced. This difference could be due to the insufficient precision of the approximation of the LPB spatial distribution, as well as due to the coordinative dependence of the parameter Γ. Some errors to the results of the analysis could be also introduced by the ignoration of the relaxation on the mixed polariton states (including the longitudinal excitons) in the region $\omega > \omega_L$. However, in the considered case, the frequency ω is not far enough from the ω_L $(\omega - \omega_L \leq 0,7\omega_{LT})$, and the spectral density of states $g(\omega, \theta)$ in an solid angle equal to 1 still strongly depends upon the angle θ between the optical crystal axis and the polariton wave vector. The maximum value of $g(\omega, \theta)$ for a fixed frequency is reached at $\theta = \pi/2$, i. e. for the states with the maximum values of $|K|$. Apart from that, the square of the absolute value of the matrix element entering the probability of the polariton scattering on LA phonons between the states with the wave vector K and K' linearly depends upon the value $q = |K' - K|$ [9], i. e. it takes a maximum value at the maximum possible (allowed by the energy-momentum conservation law) values of $|K|$ and $|K'|$. Therefore, one can consider that in the investigated frequency range the polariton relaxation occurs mainly via the states of the transversal (or nearly transversal) polaritons, for which the wave vectors are maximal. In connection with this, the two-band model of the polariton relaxation in a cubic crystal (which takes not into account the income from the longitudinal excitons) seems to be quite justified, provided that the ration $(\omega - \omega_L)/\omega_{LT}$ is small enough.

In conclusions, the performed investigations of the resonance luminescence spectra in ZnP_2 crystals revealed the most important peculiarities of polariton emission spectra. Due to exceptional (from the point of view of polariton effects) properties of these crystals, it was possible for the first time to analyze the question of the relative contribution of LPB and UPB to the external emission [25]. It was demonstrated that the emission in the vicinity of the longitudinal frequency ω_L, emerges due to the polaritons of both branches (upper, and lower ones). The spectral form and the intensity of this emission is determined by the spatial-energetic distribution of LPB and by the probability of the interband scattering of the LPB to the UPB states. The character of boundary conditions which determine the spectral dependence of transmission coefficients of the emission at the crystal boundary is very important for the formation of PL spectra. The role of the transmission coefficients is evidently demonstrated by a strong dependence of the shape of the luminescence spectrum upon the output angle of the emission from the crystal. The boundary conditions have been determined as a result of a complex investigation of amplitude-phase reflectivity spectra. The transmission coefficients were calculated and the distribution function of polaritons on the crystal boundary has been determined using these boundary conditions.

The analysis of the frequency dependence of the LPB distribution function performed in the frame of the diffusion approximation, as well as the determined boundary conditions indicate that the non-radiative surface polariton annihilation does not play an important role in the investigated crystals. As concerns the surface polariton states which can also influence the resonance luminescence of crystals [10], they are not formed in the investigated case, since the emitting surface is parallel to the optical axis of the crystal, and the exciton transition is polarized along this axis [25].

The spatial-energetic relaxation of polaritons has a diffusion character in the investigated samples, and the effective depth of the spatial distribution of polaritons determined from the ratio of intensities of LPB and UPB emission is around 5 – 6 µm at $T=2$ K. The probability of radiative polariton exition from a small internal solid angle is insignificant in the region $\omega>\omega_0$. Taking into account that the surface non-radiative annihilation is also insignificant one can consider that the diffusion polariton flux on the crystal boundary at $\omega>\omega_0$ is practically absent. The PL spectrum in the spectral region $\omega<\omega_o$ is basically influenced by the radiative exit of excitons through the boundary, and by the intensive diffusion in-depth of the sample.

In spite of the fact that the main results discussed here are obtained for a specific object (ZnP_2 crystals), the conclusions made from these results have a more general importance, and they can be used for the interpretation of PL spectra in other crystals.

3. BREWSTER EFFECTS IN EXCITON REFLECTIVITY SPECTRA

The Brewster effect in exciton spectra at oblique incidence has been discussed previously [17]. It was shown that in a model of isolated exciton resonance taking into account the spatial dispersion and the presence of a dead layer at the crystal surface it is possible to have a zero value of the reflection coefficient either in the p-polariztion component, or in the s-component in the limits of the exciton reflection band. The amplitude-phase spectral dependencies of the reflection coefficients of ZnTe, ZnP_2 and CdS crystals have been investigated with oblique incident light in the region of the lowest exciton resonances. The exciton Brewster effect has been observed in all these crystals (in the case of ZnP_2 it was observed in the s-polarization component). The experimental data are well explained by the theoretical model. At oblique incidence of the light on the flat boundary of a dielectric medium one can find an incidence angle φ=φ_BR, called Brewster angle, for which the reflection coefficient R_p becomes equal to zero ([76, § 1.5]; [77, § 4.2]). Then, the dependence of the phase difference Δ=ΔP-Δs between the p- and s- components of the reflected wave upon the φ shows a saltus in the point φ_BR equal to ±π. The absorption in the medium leads to the increase of the R_p (R_p≠0) coefficient at the φ_BR angle, and to the smearing of the sharp saltus in the (φ) function. In such cases one can speak only conventionally about the Brewster angle, i. e. φ_BR is a pseudo- Brewster incident angle ([77, § 4.2]). In this case, with changing φ, the reflection coefficient takes a minimum value. However, if a transparent dielectric film is deposited on the surface of the absorbing substrate, it is possible to have conditions when the exact Brewster effect R_p=0 is achieved. Moreover, the reflection coefficient can take zero value in the s-polarization component R_p=0 with certain parameters of the film and the substrate [78,79].

In this respect, it is of especial interest to study the Brewster effect in the spectra of oblique reflection in the region of exciton resonances of semiconductor crystals, for which the surface dead layer [3] plays the role of the transparent film. In this case, there are new possibilities for the experimental and theoretical investigation of the surface region of the crystal, the spatial dispersion (SD), the polariton effects, and other optical properties of the semiconductor caused by excitons [1-4, 78-81]. The adequate choice of the boundary conditions is of especial importance. It was shown [17] that Brewster phenomenon in the reflection of light can be observed also in the region of strong exciton absorption bands. The conditions for the observation of the Brewster effect are determined by the bulk parameters of the exciton resonance as well as by the properties of the surface transition layer.

3.1. Brewster Effect in the Model of the "Dead" Layer

The role of the surface and the spatial dispersion in the formation of optical spectra in the region of exciton resonances was widely discussed in the literature [4]. Different variants of the boundary conditions have been proposed for the explanation of experimental data. The details of the polariton behavior near the boundary of the crystal have been discussed. The model of boundary conditions based on the dead layer proposed by Hopfield and Thomas in 1963 is of a special importance in this sense. This is due to a general phenomenon characteristic for the most perfect crystals which is related to fast increase of the potential energy of the mechanical exciton with approaching the crystal surface at the distance L of the order of the exciton radius. The analysis of the Brewster effect in the exciton region of the spectrum was performed [17] in the frame o a theory taking into account the spatial dispersion. It was suggested that a dead (without excitons) layer exists on the crystal boundary, on the internal surface of which the Pekar additional boundary conditions are valid [2]. The dielectric function in the vicinity of the dipole-active exciton resonance was in-depth analyzed [1-7]

The considered system actually represents the absorbing crystal-substrate structure with a transparent dielectric film on the surface. Therefore, the amplitude reflection coefficient can be presented as [1-7]

$$r = (r_0 + r_1 e^{2i\theta})/(1 + r_0 r_1 e^{2i\theta}) \qquad (8)$$

where r_0 is the amplitude reflection coefficient of the light, incident from the vacuum on the surface of a semi-infinite dielectric medium with the dielectric constant ε_0 under a certain angle φ relative to the normal. r_1 is the amplitude reflection coefficient of the light incident from a medium with the dielectric constant ε_0 on the surface of a semi-infinite dielectric medium with the dielectric permittivity $\varepsilon(\omega, K)$ under the angle $\varphi = \arcsin(\sin\varphi/\sqrt{\varepsilon_0})$,

$$\theta = (\omega/c) l (\varepsilon_0 - \sin^2\varphi)^{1/2}$$

l is the phase thickness.

The formula (8) is valid for any angle of incidence φ for both p- and s- components of the polarization with taking into account the spatial dispersion in the coefficient r_1. The value of r_0 is calculated according to the usual Fresnel formula, while the calculations of r_1 need to use the SD theory. The procedure for the calculation of r_1 has been proposed in ref. [82]. From the formula (8) one can find the "Brewster condition" (r=0)

$$i \, tg \, \theta = (r_0 + r_1)/(r_0 - r_1), \quad (9)$$

which can de treated as a system of two non-linear equations with respect to ω and Γ. The solutions of this system depend upon the values of the angle of incidence φ and the parameters of the exciton resonance. The angle φ satisfying the condition (9) has the meaning of Brewster angle φ_{BR}.

3.2. Approximation of Normal Incidence at the Layer-Crystal Interface

The equation (9) can be presented in a convenient for the analysis form by using reasonable parameters of the problem. Usually one can consider that

$$\sin^2\varphi << \varepsilon_0, \quad q_0^2 \equiv \hbar\omega_0^2 \varepsilon_0 / 2Mc^2 \leq \omega_{LT}. \quad (10)$$

For instance, for CdS in the region of the exciton resonance $A_{n=1}$ [71,83-85] the parameters are the following: $\varepsilon_0 = 10$, $\hbar q_0^2 = 7 \cdot 10^{-2}$ meV, $\hbar\omega_{LT} = 2$ meV.

The corrections of the order of $\sin^2\varphi/\varepsilon_0$ have been neglected as compared with 1 when calculating r in ref. [17]. It was shown that r does not depend upon φ, i. e. $r_1(\varphi) \approx r_1(0°)$. The right equation (10) determines the conditions for which the contribution of the longitudinal exciton to the amplitude of the reflected wave can de neglected. From the equation (9) one can deduce the dependence of the frequency ω and the parameter Γ upon the angles θ and φ, corresponding to the conditions $r = 0$:

$$\Omega_{BR} + i\Gamma_{BR}/2 = \omega_L + [q_0 - q_\alpha(\varphi_{BR})\Phi_\alpha(\theta, \varphi_{BR})]^2 \quad \alpha = 1,2,3 \quad (11)$$

Where $\omega_L = \omega_0 + \omega_{LT}$ is the longitudinal exciton frequency, the values of the index $\alpha = 1, 2$ correspond to the p-component of the reflection, $\alpha = 1$, if $\varepsilon_0 \cos^2\varphi_{BR} > 1$, and $\alpha = 2$, if $\varepsilon_0 \cos^2\varphi_{BR} < 1$; the value of $\alpha = 3$ is for the s-component:

$$q_1 = q_2 = (\omega_{LT}/|\varepsilon_0 \cos^2 \varphi_{BR}-1|)^{1/2}, \quad q_3 = [\omega_{LT}/(\varepsilon_0-1)]^{1/2}$$

$$\Phi_1 = |\cos\theta| + i\varepsilon_0^{1/2}|\sin\theta|\cos\varphi_{BR} \qquad (12)$$

$$\pi\kappa \leq \theta \leq \arccos(q_0/q_1)+\pi\kappa, \quad \varepsilon_0^{-1/2} \leq \cos\varphi_{BR} \leq [(1+\omega_{LT}/q_0^2)/\varepsilon_0]^{1/2}$$

$$\Phi_2 = i\Phi_1^*, \quad \pi(\kappa+1/2) \leq \theta \leq -\arcsin[q_0/q_2\varepsilon_0^{1/2}\cos\varphi_{BR}])+\pi(\kappa+1) \qquad (13)$$
$$[(1+\omega_{LT}/q_0^2)/\varepsilon_0]^{-1/2} \leq \cos\varphi_{BR} \leq \varepsilon_0^{-1}$$
$$\Phi_3 = |\cos\theta|\cos\varphi_{BR} + i\varepsilon_0^{1/2}|\sin\theta|,$$

$$\pi\kappa \leq \theta \leq \arccos[q_0/q_3\cos\varphi_{BR}] + \pi\kappa, \quad q_0/q_3 \leq \cos\varphi_{BR} \leq 1. \qquad (14)$$

The inequalities (12), (13) and (14) which determine the region of allowed values of θ и φ_{BR} are obtained for the condition $\Gamma_{BR} \geq 0$. The index k takes integer values κ =0, 1, The equation (11) represents a simple analytical dependence between the parameters of the crystal in the region of exciton resonance, the values of the frequency ω_{BR} and the damping parameter Γ_{BR} for which the Brewster effect is realized. The inequality (10) determines the precision of calculations, which is in the limit of experimental errors of the determination of exciton resonance parameters. The relations obtained in ref. [17] are used for the analysis of experimental data. The following interesting observations can be deduced from the obtained formulae:

1. One can see that depending on specific values of parameters, the Brewster angle φ_{BR} in the region of exciton resonance can be changed in a wide interval (see. (12) — (14)), including the value of $\varphi_{BR}=0°$.
2. The possibility of the realization of the Brewster effect is also determined by the definite parameters. The Brewster effect is not realized for some values of parameters. Particularly, the SD ($q_0 \neq 0$) restricts the region of the allowed values of φ and l.
3. One can see from (11) and (14) that the reflection coefficient can become equal to zero also for the s-polarization component.
4. One can deduce from (8) that the frequency ω_{BR} and the damping parameter Γ_{BR} are periodical functions (with the period π) of the phase thickness θ.
5. In the dependences of ω_{BR} and Γ_{BR} upon φ_{BR} for the p-polarization component there are poles at $\varphi_{BR}=a\!\mathit{zccos}(\varepsilon_0^{-1/2})$. These poles in the approximation (10) corresponds to the "background" Brewster angle $\varphi_0=a\mathit{zctg}(\varepsilon_0^{1/2})$.
6. The equation (11) is exact for $\varphi_{BR}=0°$. In this case, $q_1=q_3$ and $\Phi_1=\Phi_3$), i. e. one can not distinguish between the s- and p- polarization components for the normal incidence of the light.

The case of $\varphi_{BR}=0°$ is analyzed in more details in ref. [17], since the reflectivity spectra are usually investigated with angles of incidence close to the normal incidence. By separating the real and imaginary parts of the equation (11) one can deduce for $\varphi_{BR}=0°$

$$\omega_{BR}=\omega_L+(q_3|\cos\theta|-q_0)^2 - \varepsilon_0 q_3^2 \sin^2\theta$$

$$\Gamma_{BR}=4\varepsilon_0^{1/2}q_3|\sin\theta|(q_3|\cos\theta|-q_0) \qquad (15)$$

Where the region of allowed values of θ is determined by the inequalities (12) and (14) for $\varphi_{BR}=0$, and $q_0/q_3 \leq 1$. In the absence of SD ($q_0=0$), the last relation is always accomplished, and θ (see (14)) can take values from the interval $\pi\kappa \leq \theta \leq \pi(\kappa+1/2)$ ($\kappa=0,1,...$), while in the interval $\pi(\kappa+1/2)<\theta<\pi(\kappa+1)$ the Brewster effect is impossible at normal incidence. With increasing the SD (q_0) parameter, the region of the allowed values of θ (as well as l) is narrowed, and at $q_0>q_3$ the reflection coefficient can not become equal to zero for any thicknesses of the dead layer l.

3.3. Brewster Effect in the Amplitude and Phase Spectra of Exciton Reflectivity in ZnP$_2$ Crystals

A very intensive reflectivity line was observed in the black modification of the ZnP$_2$ crystals in the spectral range of 1,560—1,568 eV corresponding to the dipole-active exciton transition in the n=1 state [23,25-28], the optical transition being allowed only in the polarization E∥C$_{2z}$ where C$_{2z}$ is one of second order axis. By using the crystal face parallel to the C$_{2z}$ axis in the investigation of reflection, and by orienting this axis perpendicularly to the plane of incidence, the possibility to investigate the Brewster effect emerging only due to s-polarization was realized. The exciton contribution in the p-polarization was absent. The phase spectra $\Delta=\Delta_p-\Delta$ of ZnP$_2$ are illustrated in Figure 17 for the angles of incidence $\varphi=10°$ (1 – full circles), 65°, 79° (2, 3 – triangles), 85° (4 – open circles). The comparison of 2, 3, 1and 4 curves shows that significant changes of the phase spectrum occur in the interval of angles of incidence $65°<\varphi<85°$. The shift of the curve 4 by π as compared to the curve 2 at the ends of this spectral interval indicates on the occurrence of the usual (non-excitonic) Brewster effect in p-polarization component. On the other hand, a dramatic change of the form of the phase spectrum with the transition from curves 2,3 to the curve 4 results from the excitonic Brewster effect.

The exciton Brewster angle for the s-component $\varphi_{BR}^{(s)}=81,9°$ and the usual Brewster angle of the p-component $\varphi_0^{(p)}=72,5°$ where found in ref. [17] by using the parameters of the n=1 exciton resonance in ZnP$_2$ obtained from the theoretical analysis of amplitude-phase reflectivity spectra. Therefore, with increasing the angle of incidence, initially the usual Brewster effect in the p-component occurs, which leads to the shift of the phase curve of the type 2 by 180° on the scale Δ (see. curve 3 for $\varphi=79°$). Later-on, the excitonic Brewster effect occurs, which means the transition from the curve 3 to the curve 4. The inset of Figure 17 illustrates the measured phase godographs corresponding to the spectral dependences 2 and 4. The arrows on the godograph circles indicate the direction of the frequency increasing, and the disruptions correspond to the ends of the spectral interval. Since the origin of coordinates is situated inside the circle 2, the total phase shift is equal to $\approx 2\pi$ (compare with the curve 2 of the spectral dependence $\Delta(\omega)$). Since the curve 4 does not contain the origin of coordinates, the phase returns to the initial value with passing the frequency though the resonance (compare with the curve 2 of the spectral dependence $\Delta(\omega)$). The circuition of the godograph circle is curried out clockwise in the case 1 – 3. In the case 4, the direction of the circuition is inverted. Therefore, the excitonic Brewster effect in the s-polarization component is accompanied by the change of the circuition direction of the phase godograph circle, which is a characteristic difference from the case of excitonic Brewster effect in the p-component (see [3]). Apart from that, at an exact angle of incidence $\varphi_{BR}^{(s)}$ at frequency $\omega_{BR}^{(s)}$ the phase godograph curve goes to the infinite, which corresponds to the reduction of the reflection coefficient R to zero in the ratio $\rho=(R_p/R_s)^{1/2}$.

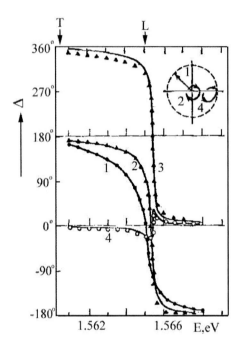

Figure 17. The spectral dependence of the phase shift $\Delta=\Delta_p-\Delta_s$ of the reflection coefficients r_p и r_s in the region of the lowest dipole-active exciton state of ZnP$_2$ (T=2K) crystals at different angles of incidence of the light: φ—10° (curve 1), 65° (2), 79° (3), 85° (4); circles - experiment, curves - theory. The inset illustrates the measured phase godographs (the dependence of the $\rho=|r_p/r_s|$ upon the Δ in polar coordinates) for the angles of incidence φ—65° (2), 85° (4). The crystal is oriented in such a way that the optical transition in the exciton state is allowed only in the s-polarization.

4. EXCITON SPECTRA IN CRYSTALS WITH WEAK POLARITON EFFECT ($\omega_{LT} \leq \Gamma$)

The exciton spectra with a weak polariton effect are observed in many crystals [86-92]. Most interesting effects have been observed in CuGaS$_2$, CuAlS$_2$, CuAlSe$_2$, and CuGaSe$_2$, chalcopyrite compounds which crystallize in structures with the D_{2d}^{12} space group. Three exciton series A, B, C have been observed in these crystals [86-92]. The optical transitions in these crystals are polarized. If one not takes into account the spin-orbital splitting, the transitions from the Γ_1 valence band to the Γ_1 conduction band are allowed in the E∥c polarization, while the transitions from the Γ_5 valence band to the Γ_1 conduction band are allowed in the E⊥c polarization. The spin-orbital splitting leads to the modification of the selection rules.

The energy band structure of chalcopyrites is theoretically calculated by analogy with the zincblende structure. The lower conduction band is described by the Γ_1 representation for the zincblende structure (T_d symmetry). The upper valence band is described by the Γ_{15} representation. With the transition from the T_d crystal symmetry (zincblende) to the D_{2d} symmetry, the crystal field (Δ_{cr}) without taking into account the spin-orbital interaction (Δ_{so}) leads to the splitting of the p-type Γ_{15} valence band into two sub-band Γ_5 and Γ_1. The representation of the lower Γ_1 conduction band transforms to the Γ_1 representation. The combined interaction of the crystal field and the spin-orbital interaction with the transition to the chalcopyrite structure splits the state of the p-type Γ_{15} valence band into the Γ_7 (A), Γ_6 (B) и Γ_6 (C) levels. The lower conduction band in the chalcopyrite structure is described by the Γ_6 symmetry [93-97].

The interaction of electrons of the conduction band Γ_6 and holes of band Γ_7 is determined by product of irreducible representations $\Gamma_1 \times \Gamma_6 \times \Gamma_7 = \Gamma_3 + \Gamma_4 + \Gamma_5$. The Γ_4 exciton allowed in E∥c polarization, Γ_5 allowed in E⊥c polarization and Γ_3 forbidden in both polarizations are formed as a result of this interaction. In the short-wavelength region, the $\Gamma_1 \times \Gamma_6 \times \Gamma_6 = \Gamma_1 + \Gamma_2 + \Gamma_5$ interaction leads to the formation of the Γ_5 exciton series (B-exciton series) which is allowed in the E⊥c polarization, and Γ_1 and Γ_2 excitons which are forbidden in both polarizations.

4.1. Free Exciton Spectra in CuGaS$_2$ Crystals

The Γ_4 exciton series allowed in the E∥c polarization in CuGaS$_2$ crystals has a big oscillator strength. Three lines of the Γ_4 exciton hydrogen-like series have been revealed in the reflectivity spectra of CuGaS$_2$ crystals measured in E∥c, k⊥c polarization at 10 K as follows: n = 1 (ω_t = 2.5011 eV, ω_L = 2.5045 eV), n = 2 (2.5337 eV), n = 3 (2.5377 eV) (Figure 18) [98-105].

The reflectivity spectra in the vicinity of line $n = 1$ has a usual shape with a maximum and a minimum. These peculiarities are due to the presence of longitudinal and transversal excitons. These data were used to estimate the energy of the longitudinal-transversal splitting of Γ_4 excitons which is equal to 3.9– 4.4 meV. The luminescence bands related to the emissioin from the upper (ω_{n1}^L) and lower (ω_{n1}^T) exciton polariton branches are observed in the luminescence spectra excited by the λ = 4880 Å line of an Ar$^+$ laser. These results agree with the data presented in [98-105]. In our spectra, the emission maxima are shifted by 1.8 Å toward shorter wavelengths with respect to R_{min} and R_{max} in the reflectivity spectra. In the range of higher energies, the reflectivity spectra show line $n = 2$ with the maximum at 2.5337 eV and the minimum at 2.5347 eV and line $n = 3$ with the maximum at 2.5377 eV. In the vicinity of $n = 2$ in the luminescence spectra, the peaks denoted as ω_{n2}^L and ω_{n2}^T are formed, which, most likely, are caused by transverse–longitudinal splitting ω_{LT} of state $n = 2$ of the A-exciton series (Γ_4). Splitting ω_{LT} for $n = 2$ equals 1.1–1.3 meV. The emission peaks at 2.5337 and 2.5347 eV cannot be attributed to the lines of the resonance Raman scattering because, on the energy scale, they are separated from the laser emission line 488 nm less than the energy of optical phonons in a CuGaS$_2$ crystal. In addition, under the Hg-lamp excitation (2.82 eV, [100]), the luminescence spectra have a narrow band (ω_{n1A}^T = 2.4996 eV) caused by emission of exciton-polaritons and, in the short-wavelength range, also lines $n = 2$ (2.5332 eV) and $n = 3$ (2.5382 eV). These lines cannot be interpreted as the Raman lines excited by a Hg-lamp. Taking into account the energy position of lines $n = 2$ and $n = 3$, one can determine the Rydberg constant for the A-exciton series, which is equal to 0.0324 eV. The energy of the continuum edge (E_g, $n = \infty$) is 2.5413 eV.

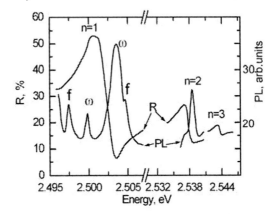

Figure 18. Reflectivity and luminescence spectra of CuGaS$_2$ crystals measured in the E$\|$c, k$\perp$c polarization. Reprinted with permission from [126], N.N. Syrbu, L.L.Nemerenco, I.G.Stamov, V.N. Bejan, V.E. Tezlevan (2007) *Interference of resonance luminescence of exciton*. Optics Communications **272**, 124-130. Copyright (2007), with permission from Elsevier.

From the reflectivity measurements in the IR (400 cm^{-1}) and the near IR (12000 cm^{-1}) range, we also determined the background dielectric constant

$$\varepsilon_b = \left| \frac{1+\sqrt{R}}{1-\sqrt{R}} \right|^2.$$

In the wavelength region of 2.5 μm we have $R = 0.21$ and $\varepsilon_b = 7$. The reflectivity thus obtained [98-101] agrees fairly well with the value of 0.2 obtained in [106]. At $\varepsilon_b = 7$, the reduced effective mass of $A(\Gamma_4)$ excitons is

$$\mu = \frac{\varepsilon_b^2 R}{R_{H_2}} = 0.117 m_0,$$

where R (0.03247 eV) is the Rydberg constant of A exciton and R_{H_2} is the Rydberg energy of a hydrogen atom (13.6 eV). The Bohr radius (a_B) of the S-state of a Γ_4-exciton equals 0.32×10^{-6} cm.

In CuGaS$_2$ crystals as well as in A^{2}B^6 compounds [12-19], the light (I_0) incident onto a crystal generates the waves in the crystal bulk which are described by the equation [4,7,22]

$$n_{1,2}(\omega) = \left\{\varepsilon_0 + \frac{1}{2}\left[a \pm (a^2+b)^{1/2}\right]\right\}^{1/2}$$

where

$$a = \frac{Mc^2}{\hbar\omega_0\omega^2}(\omega^2 - \omega_0^2 + i\omega\gamma) - \varepsilon_\infty, \quad b = 8\frac{Mc^2\varepsilon_0\omega_{LT}}{\hbar\omega^2}$$

This result corresponds to two polariton branches with the upper one denoted as ω^L and the lower one, as ω^T. A "dead" exciton-free layer with a thickness of l is formed on the surface. The light beams reflected from the two interfaces interfere and, thus, give rise to the rotation of the profile of the excitonic reflectivity spectrum. The light reflectivity R for a crystal with a dead layer is described as

$$R = \left|\frac{r_{12} + r_{23}e^{i\varphi}}{1 + r_{12}r_{23}e^{i\varphi}}\right|^2$$

where r_{12} is a constant, r_{23} is the function characterized by a resonant profile, which determines the reflectivity profile of the excitonic spectra, and

$$\varphi = \frac{4\pi n_\infty l}{\lambda},$$

where λ is the light wavelength.

As is evident from these expressions, the quantity R varies periodically with l with a period of $\lambda/2n_\infty$. In the vicinity of the transition with the frequency ω_0, the permittivity is given by

$$\varepsilon(\omega,k) = \varepsilon_{b\perp}\left(1 + \frac{\omega_{LT}}{\frac{\hbar^2 k^2}{2M} - \omega + \omega_0}\right).$$

The contours of exciton reflectivity spectra were calculated in the frame of classical optics by taking into account the spatial dispersion and the presence of the dead layer. The following parameters have been deduced from these calculations for the n = 1 line of the Γ_4

exciton: the longitudinal-transversal splitting ω_{LT} = 4.0 meV, the damping parameter γ = 1.3 meV, and the translation mass $M = 2.0m_0$. The dispersion of the upper and lower polariton branches has been determined from these calculations for the Γ_4 exciton (inset in Figure 19).

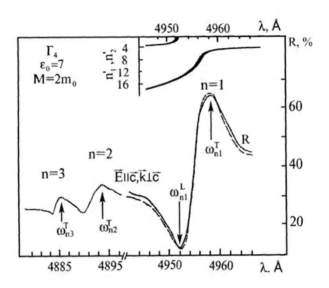

Figure 19. Reflectivity spectra of CuGaS$_2$ crystals measured at 8.6 K in the E∥c, k⊥c polarization (full line), and the results of calculations (dashed line) for the n=1 line of the Γ_4 exciton. The calculated dispersion of the LPB and UPB is inset. Reprinted with permission from [99] N.N. Syrbu, V.V. Ursaki, I.M. Tiginyanu, V.E. Tezlevan, M.A. Blaje (2003) *The interference*, Journal of Physics and Chemistry of Solids 64, 1967-1971. Copyright (2003), with permission from Elsevier.

The effective masses of light and heavy holes have been determined from the relations $M = m_V^* + m_C^*$ and $\frac{1}{\mu} = \frac{1}{m_V^*} + \frac{1}{m_C^*}$. At the exciton mass $M = 2m_0$, we have $m_C^* = 0.124m_0$; at $M = 2.5m_0$, we have $m_C^* = 0.122m_0$ and at $M = 3.5m_0$ we have $m_C^* = 0.121m_0$. These data show that the mass varies insignificantly even if the M-value is determined with a considerable error. The translational mass M for the Γ_4 excitons (the A series) was determined from the calculated profiles of the reflectivity spectra within an accuracy of $\pm 0.2m_0$. Thus, if $M = 2m_0$, the electron mass is $m_C^* = 0.12m_0$, and the hole mass is $m_V^* = 1.87m_0$.

The lines at 2.6217 eV and 2.6536 eV, which are the lines n =1 and n =2 of the B-exciton series, as well as the lines at 2.6323 eV and 2.6611 eV corresponding to the n =1 and n =2 of the C-exciton series are observed in the short-wavelength region (Figure 20). The position of the absorption lines corresponding to the ground states of the A, B, and C excitons agree well with the literature data [98-101]. A series of peaks at 2.5016, 2.5337 and 2.5375 eV which correspond to the n = 1, 2 and 3 lines of the A-exciton series (Γ_4-exciton) is observed in the λ-modulated reflectivity spectra of CuGaS$_2$ crystals measured in the region of the absorption edge at 10 K in E∥c, k⊥c polarization (Figure 21).

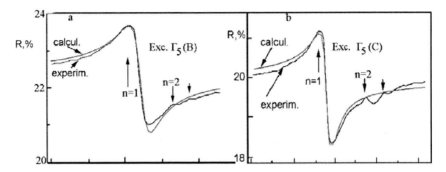

Figure 20. Reflectivity spectra of CuGaS$_2$ crystals measured at 8.6 K in the E⊥c polarization (full line), and the results of calculations (dashed line) for the n=1 line of the Γ$_5$(B) and Γ$_5$(C) exciton.

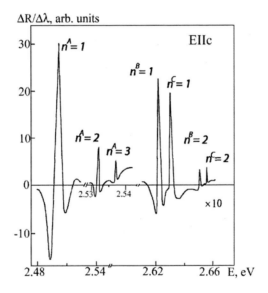

Figure 21. The λ-modulated reflectivity spectra of CuGaS$_2$ crystals measured at 10 K in the E∥c, k⊥c polarization. Reprinted with permission from [126], N.N. Syrbu, L.L.Nemerenco, I.G.Stamov, V.N. Bejan, V.E. Tezlevan (2007) *Interference of resonance luminescence*. Optics Communications **272**, 124-130. Copyright (2007), with permission from Elsevier.

The λ-modulated reflectivity spectra were calculated for lines $n = 1$ of A-, B-, and C-excitons measured at 10 K. For line $n = 1$ of A-excitons, the best agreement between the experimental and calculated curve was attained at the following parameters: $\varepsilon_b = 7$, $\gamma = 0.5$ meV, $\omega_0 = 2.5006$ eV, $\omega_{LT} = 4$ meV, $l = 15$ Å, and $M = 2m_0$; for the B-exciton line, the parameters are $\varepsilon_b = 7$, $\gamma = 0.5$ meV, $\omega_0 = 2.6217$ eV, $\omega_{LT} = 3.5$ meV, $M = 2m_0$, and $l = 15$ Å, and for the C-exciton line, $\varepsilon_b = 7$, $\gamma = 0.7$ meV, $\omega_0 = 2.6323$ eV, $\omega_{LT} = 2.5$ meV, $l = 15$ Å, and $M = 2m_0$.

At the same time it should be mentioned that, for the B series with $\mu = 0.148m_0$ we have $m_V^* = 0.74m_0$, while, as follows from the data on the C series, at $\mu = 0.144m_0$ we have m_{V3}^*

= 0.89m_0. When determining the effective masses m_{V1}, m_{V2} and m_{V3}, we took into account the effective mass M calculated for an A (Γ_4) exciton and the μ values for the B and C series.

4.2. Free Exciton Spectra in CuAlS$_2$ Crystals

The n = 1 (ω_t = 3.543 eV, ω_L = 3.546 eV) and n = 2 (3.565 eV) lines of the Γ_4 exciton hydrogen-like series are observed in the reflectivity spectra of CuAlS$_2$ crystals measured at 10 K in the E∥c polarization (Figure 22). The reflectivity spectra in the region of the n = 1 line are of a usual excitonic shape with a maximum at 3.543 eV and a minimum at 3.546 eV. These peculiarities are due to presence of the transversal and longitudinal excitons. A longitudinal-transversal exciton splitting of 3 meV[107] is estimated for the Γ_4 excitons from these data. A Rydberg constant of 32 meV is determined for the Γ_4 exciton series from the position of n = 1 and n = 2 lines (Figure 22). The energy of the continuum (E_g, n = ∞) is equal to 3.575 eV. These energy values of the ground (n= 1) and excited (n= 2) states of excitons are in accordance with previously reported values of 3.534 and 3.665 eV, measured at 77 K [108].

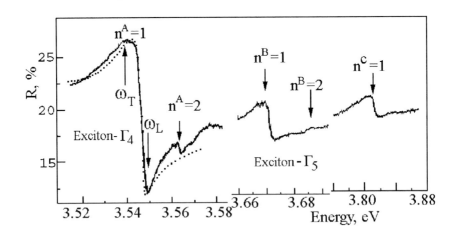

Figure 22. Optical reflectivity spectra of CuAlS$_2$ crystals. The dot-line is the result of calculation of the reflectivity contour of the n=1 line of the Γ_4 exciton taking into account the spatial dispersion, i. e. limited value of the exciton mass M, and the boundary conditions. Reprinted with permission from [107], N.N. Syrbu, B.V. Korzun, A.A. Fadzeyeva, R.R. Mianzelen, V.V. Ursaki, I. Galbic (2010) *Exciton*. Physica B: Condensed Matter, in Press. Copyright (2010), with permission from Elsevier.

The reported value of ε_b in CuAlS$_2$ crystals equals 7.05 in the (E∥c) polarization and 8.14 in the (E⊥c) polarization far from the exciton resonances (ν = 4000-3000 cm^{-1}) [106]. According to our data, the coefficient of reflection equals 0.20-0.22, and the value of ε_b is 7.1-7.3 in the long-wavelength region of exciton resonances. The mean value of the background dielectric constant of 7.26 near the exciton resonance was used in calculations. With ε_b = 7.26 and Rydberg constant R =0.032 eV, the Γ_4-excitons reduced mass equals to $\mu = \varepsilon_b^2 R/R_H$ = 0.11m_0, where R_{H_2} is the Rydberg energy of the hydrogen atom (13.6 eV). The Bohr radius

(a_B) of the S-state of the Γ_4-exciton equals 0.3×10^{-6} cm. A maximum at 3.668 eV (transversal exciton) and a minimum at 3.670 eV (longitudinal exciton) are observed in the E⊥c polarization for the Γ_5 exciton series (Figure 22). The longitudinal-transversal splitting of the Γ_5 exciton equals 2.0 meV. The n=2 excited exciton state is observed at 3.687 eV. The binding energy of the Γ_5 exciton equals 25 meV, and the energy of the continuum equals 3.693 eV. The C-exciton is observed at 3.813 eV(n = 1) in the same polarization. A band at 4.39 eV was observed at 77K in the energy interval of 3.6 - 4.98 eV. Since no other lines were observed between the 3.665 eV and 4.39 eV at 77K, the 4.39 eV band has been associated with C-exciton series [107,108]. As mentioned above, a line at 3.813 eV is observed in the reflectivity spectrum (Figure 22). Taking into account these data and the exciton spectra of $CuGaS_2$ crystals [98-101], we assume that the 3.813 eV line is related to the ground state of the C-exciton.

For the B-exciton series, the coefficient of reflection equals 21 % at 3.6 eV, and the dielectric constant ε_b equals 7.2. The reduced mass equals $0.09m_0$ with the binding energy of 25 meV.

A weak luminescence is observed at 200 K in $CuAlS_2$ crystals under the excitation by the 325 nm line of a He-Cd laser. The luminescence intensifies with decreasing temperature to 100 K, and further to 10 K (Figure 23). A broad photoluminescence band is observed at 10 K around 3 eV, along with two narrow bands around 3.3 eV and a band at 3.54 eV. The band at 3.54 eV is assumed to be related to the recombination of Γ_4 exciton polaritons, while the narrow lines are due to the recombination of donor-acceptor pairs (DAP).

The contour of the reflectivity spectrum was calculated for the n = 1 line of the Γ_4 exciton (the A-series). The coefficient of reflection for the B and C exciton series changes in a small limit of around 4 – 5 % from the minimum to the maximum. In this case, the errors in the calculation of the value of the exciton translation mass are bigger. A comparison of the calculated and the measured contours of the reflectivity spectra for the ground state of the Γ_4 exciton is presented in Figure 22. The translation mass M of the Γ_4 excitons determined from these calculations varies in the limits of $(1.0 – 1.3)m_0$ for different crystals.

The values of the reduced mass of excitons, as in the case of $CuGaS_2$ crystals are determined from the relation

$$\mu = \frac{\varepsilon_b^{\text{II}} \varepsilon_b^{\perp} R_j}{R_{H_2}}$$

where j corresponds to Γ_4 and Γ_5 excitons, R_j is the binding energy of the j-exciton, R_{H_2} is the Ridberg constant of the hydrogen atom, $\varepsilon_b^{\text{II}}$ and $\varepsilon_b^{\perp}$ are the background dielectric constants.

The effective masses in the conduction and three valence bands have been determined on the basis of these data, and taking into account that $M = m_v^* + m_c^*$ and $1/\mu = 1/m_v^* + 1/m_c^*$, where m_c^*, m_{vl-3}^* are the effective masses in the conduction band and in the $\Gamma_7(V_1)$, $\Gamma_6(V_2)$, $\Gamma_7(V_3)$ valence bands. With the $M=1.2m_0$ and $\mu=0.11m_0$ the electron effective mass m_c^* equals $0.12m_0$, and the effective mass of holes m_{V1}^* equals $1.1m_0$. The obtained value $m_v^* = 1.1\ m_o^*$ coincides in the limits of experimental errors with the value of $1.2\ m_o^*$ obtained previously for the $CuGaS_2$ crystals [98-101]. These data demonstrate that the value of m_c^*

changes insignificantly in spite of significant errors in the determination of M. The translation mass M of the Γ_4 excitons (A-series) has been determined from the calculation of reflectivity spectra contours with the $\pm 0.2 m_0$ errors. Therefore, the electron mass $m_c^* = 0.12 m_0$, and the heavy hole mass $m_{V1}^* = 1.10 m_0$ have been deduced with the value of $M = (1.0-1.3) m_0$.

For the Γ_5 excitons, the coefficient of reflection in the long-wavelength region of the exciton resonance (E = 3.6 eV) equals 21%, and the dielectric constant equals 6.8-7.0. The reduced mass μ equal to $0.09 m_0$ have been calculated assuming the binding energy of the B-exciton equal to 25 meV. The parameters of the V_2 valence band can be calculated using the parameters of the Γ_5 exciton series. One should take into account that the Γ_5 exciton series is formed by the interaction of electrons from the C_1 ($m_c^* = 0.12 m_0$) conduction band and the holes from the V_2 valence band. The value of light hole mass $m_{V2}^* = 0.36 m_0$ has been estimated using the value of $m_c^* = 0.12 m_0$ and the reduced mass of the B-series equal to $\mu = 0.09 m_0$.

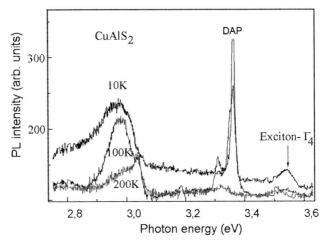

Figure 23. The luminescence spectra of CuAlS$_2$ crystals excited by the 325 nm line of a He-Cd laser measured at different temperatures. Reprinted with permission from [107], N.N. Syrbu, B.V. Korzun, A.A. Fadzeyeva, R.R. Mianzelen, V.V. Ursaki, I. Galbic (2010) *Exciton*. Physica B: Condensed Matter, in Press. Copyright (2010), with permission from Elsevier.

The parameters of the exciton series and the energy of the continuum (E_g) determined in this work allows one to reliable determine the splitting of the upper valence bands in the center of the Brillouin zone due to the crystal field (Δ_{cf}) and spin-orbit interaction (Δ_{so}).

4.3. Free Exciton Spectra in CuGaSe$_2$ Crystals

The contours of exciton reflectivity spectra of CuGaSe$_2$ crystals measured in E‖c polarization at 10K (full curve) as well as the calculated data according to the dispersion relations (dotted curve) are presented in Figure 24. The calculations shown that the longitudinal transversal splitting of the ground state of the Γ_4 exciton ω_{LT} equals 0.8-3.0, while the damping parameter changes in the limits of 1.4 – 1.8 meV. The following parameters have been deduced from the analysis: the energy of the longitudinal exciton $E_L =$

1.7230 eV, the energy of the transversal exciton E₀ = 1.7205 eV (ω_LT= 2,5 meV), the background dielectric constant ε_b= 6.2, the damping parameter γ = 0.8 meV, the exciton mass M = 0.25m₀, the dead layer thickness l = 54Å

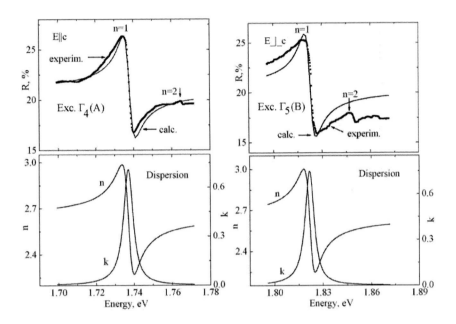

Figure 24. The reflectivity spectra of CuGaSe₂ crystals in the region of A and B excitons measured at 10K, and the calculated values of n and k. Reprinted with permission from [109], S. Levcenko, N. N. Syrbu, V. E. Tezlevan, E. Arushanov, J. M. Merino, and M. León (2008) Exciton spectra and energy band structure of CuGaSe₂ single crystals. J. Phys. D: Appl.Phys. **41**, 055403. Copyright (2008), with permission from Institute of Physics.

The best coincidence of the calculated and experimental reflectivity spectra are realized when the dielectric constant has a small gradient. The value of the gradient is determined by ε_F^{max} and ε_F^{min}. The minimum value of ε_F^{min} (the background value of the dielectric constant) corresponds to the energy at which the exciton resonance starts to contribute to the reflectivity spectra, and ε_F^{max} corresponds to the energy at which the s-exciton state does not anymore influence the reflectivity spectra (does not contribute to the ε).The gradient Δε does not exceed 0.15 for any exciton series. This gradient is due to the fact that, in the frequency interval of transitions in the S-state of A-excitons, there is a contribution to the dielectric constant of the B-exciton series, which increases with the increase of frequency. There is a contribution to the dielectric constant from the neighboring exciton series in the frequency region of n=1 lines of each exciton series. The investigation of exciton spectra of CuGaSe₂ crystals [109] shown that there are crystals with different form of the reflectivity spectra in the region of long-wavelength excitons. In these crystals, the dependence of the reflectivity spectra is sharper in the region ω_T of the long-wavelength excitons. This is an indicative of a bigger exciton mass [109].

The measured and calculated reflectivity spectra of the A-exciton series are shown in Figure 24. The following parameters have been deduced from the analysis: $\varepsilon_b = 9.4$, $\omega_0 = 1.7376$ eV, M = 1.1m$_0$, $\gamma = 0.5$ meV, $\omega_T = 1.7369$ eV and $\omega_{LT} = 0.7$ meV. The reflectivity spectra of the B-exciton series (n =1 and n = 2) as well the C-exciton series (n=1) are also shown in Figures 24 and 25. This series has nearly the same oscillator strength as the A-exciton series. The value of dielectric constant ε_b used in the calculations of the reduced exciton mass was estimated from the ε_∞ value determined from the IR reflectivity spectra. Far from the exciton resonances (ν = 4000-3000 cm^{-1}), ε_∞ equals 7.2 and 8.2 for the (E∥c) and (E⊥c) polarization, respectively [110].

The λ-modulated reflectivity spectra of CuGaSe$_2$ crystals measured in E∥c polarization at 10 K are shown in Figure 26. The ground n=1 and the excited n=2 exciton states are observed at 1.736 eV and 1.772 eV, respectively. The minima in the λ-modulated reflectivity spectra are close to the mean value of ω_T and ω_L in the exciton spectra.

The measurements of polarized reflectivity and λ-modulated reflectivity spectra demonstrate that the n=1 line of the A-exciton series is allowed in the E∥c polarization, and is forbidden in the E⊥c polarization.

On the basis of these data, and taking into account that M = $m_v^* + m_c^*$ and $1/\mu = 1/m_v^* + 1/m_c^*$, where m_c^*, m_{v1-3}^* are the effective masses in the conduction band and in the $\Gamma_7(V_1), \Gamma_6(V_2), \Gamma_7(V_3)$ valence bands, the effective masses of the conduction band and the three valence bands have been determined. The obtained value of $m_{v_1}^* = 1,23 m_0$ coincides in the limits of experimental errors with the reported value of 1.2m$_0$ [109,110].

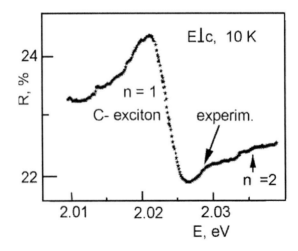

Figure 25. Contour of reflectivity spectra of the –excitons in CuGaSe$_2$ crystals measured in E⊥c polarization at 10 K. Reprinted with permission from [109], S. Levcenko, N. N. Syrbu, V. E. Tezlevan, E. Arushanov, J. M. Merino, and M. León (2008) Exciton spectra and energy band structure of CuGaSe$_2$ single crystals. J. Phys. D: Appl.Phys. 41, 055403. Copyright (2008), with permission from Institute of Physics.

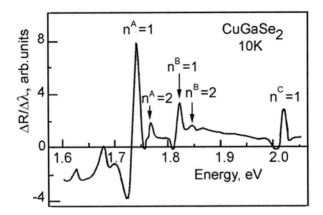

Figure 26. Wavelength modulated reflectivity spectra of CuGaSe$_2$ crystals. Reprinted with permission from [109], S. Levcenko, N. N. Syrbu, V. E. Tezlevan, E. Arushanov, J. M. Merino, and M. León (2008) Exciton spectra and energy band structure of CuGaSe$_2$ single crystals. J. Phys. D: Appl.Phys. **41**, 055403. Copyright (2008), with permission from Institute of Physics.

4.4. Free Exciton Spectra in CuAlSe$_2$ Crystals

The n = 1 (ω_t = 2.8212 eV, ω_L = 2.8237eV), n = 2 (2.8390 eV) and a weak line at 2.8442 eV of the hydrogen-like series of the Γ_4 exciton are observed in the reflectivity spectra of the CuAlSe$_2$ crystals measured at 10 K in the E∥c, κ⊥c polarization (Figure 27). On the basis of these data, a longitudinal-transversal splitting energy of 2.5 meV was estimated for the Γ_4 excitons. The Rydberg constant of 24 meV has been determined from the energy position of the n = 1 and n = 2 lines. The energy of the continuum (E_g, n = ∞) is 2.845 eV. The values of energies for the ground (n = 1) exciton states fairly well agree with the previously reported values of 2.737, 2.851 и 3.012 eV deduced at 77 K for the A, B, and C excitons [112,113].

Far from the exciton resonances (v = 4000-3000 см$^{-1}$), the background dielectric constant ε_b equals 6.67 and 8.28 in the (E∥c) and (E⊥c) polarization, respectively, in CuAlSe$_2$ crystals [106,110]. The reflection coefficient in the region of exciton resonances equals 0.24-0.25, and the ε_b value varies in the interval of 7.4-8.2. The value of the background dielectric constant near the exciton resonance has been used in calculations.

With ε_b = 7.6 and Rydberg constant R =0.024 eV, the Γ_4-excitons reduced mass equals to $\mu = \varepsilon_b^2 R/R_H = 0.1 m_0$, where R_{H_2} is the Rydberg energy of the hydrogen atom (13.6 eV). The Bohr radius (a$_B$) of the S-state of the Γ_4-exciton equals 0.3×10^{-6} cm. A maximum at 2.851 eV (transversal exciton) and a minimum at 2.853 eV (longitudinal exciton) are observed in the E⊥c polarization for the Γ_5 exciton series (Figure 27). The longitudinal-transversal splitting of the Γ_5 exciton equals 2.0 meV. The n=2 excited exciton state is observed at 2.868 eV. The binding energy of the Γ_5 exciton equals 22 meV, and the energy of the continuum equals 2.873 eV. The C-exciton is observed at 3.023 eV (n=1) and 3.039 eV (n=2) in the same polarization. The form of line for the ground exciton state is shown in Figure 27. The Rydberg constant is 18 meV and the energy of the continuum is 3.038 eV.

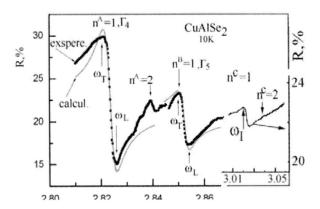

Figure 27. The experimental and calculated contours of the reflectivity spectra of CuAlSe₂ crystals at 10K for Γ_4 and Γ_5 excitons. Reprinted with permission from [171], N.N. Syrbu, A.V. Dorogan, V.V. Ursaki, A. Masnik (2010) Wavelength modulated optical reflectivity spectra of CuAl$_{1-x}$Ga$_x$Se$_2$ crystals. J. Opt. accepted. Copyright (2010), with permission from Institute of Physics.

The measured and calculated reflectivity contours are shown in Figure 27. The following parameters were deduced from this analysis ε_b =7.6, ω_T =2.8220 eV, ω_{LT} =2.5 meV, γ =3.5 meV. For the Γ_5 excitons the parameters are as follows: ε_b =7.0, ω_T =2.8520 eV, ω_{LT} =1.0 meV, γ =2.5 meV. The translation mass M of (1.0-1.3)m₀ and 0.5-0.8 m₀ was determined for the Γ_4 and Γ_5 excitons in CuAlSe₂ crystals. With the M=1.3m₀ and μ=0.1m₀ the electron effective mass m_c^* equals 0.11m₀, and the effective mass of holes m_{V1}^* equals 1.2m₀. The obtained value of electron (m_c^*) and hole (m_v^*) masses for CuAlSe₂ do not significantly differ from the masses obtained previously for the CuGaSe₂ crystals [109-113]. The parameters of Γ_5 excitons do not significantly differ from the parameters of the Γ_4 excitons.

These data demonstrate that the value of m_c^* changes insignificantly in spite of significant errors in the determination of M. The translation mass M of the Γ_4 excitons (A-series) has been determined from the calculation of reflectivity spectra contours with the ±0.2m₀ errors. Therefore, the electron mass m_c^* = 0.12m₀, and the heavy hole mass m_{V1}^* = 1.10m₀ have been deduced with the value of M = (1.0-1.3)m₀.

With the translation mass M=(0.5-0.8)m₀ and the Rydberg constant of the B-exciton R= 22 meV, the effective electron mass of 0.1m₀, and the light hole mass m_{V2}^* equal to (0.4-0.7)m₀ were obtained. For the C-exciton series the reduced effective mass is μ=0.076m₀ and the hole mass m*$_{V3}$ is equal to 0.25 m₀. The parameters of bands are shown in Figure 28.

The energy band structure of I-III-VI₂ crystals is calculated as for the nearest analogs of the zincblende [93-97]. A stronger decrease of the bandgap and the spin-orbit splitting is found as compared to the zincblende analogs. In most of crystals from this group, including CuAlSe₂, the bandgap is decreased by 1 eV as compared to the ZnSe analog, and the spin-orbit splitting decreases from 0.45 eV to 0.23 eV. These effects are explained by the hybridization of the p- and d-states, which determine the upper valence bands in the center of the Brillouin zone [86,93,94].

The interval between the levels $\Gamma_7(V_1) - \Gamma_6(V_2)$ in I-III-VI₂ structures is assigned as E_1, and the gap between the levels $\Gamma_6(V_2) - \Gamma_7(V_3)$ is assigned as E_2 provided that $\Delta_{cf} < E_g$. These values are deduced from the Hamiltonian matrix and are determined by the following relation:

$$E_{1(2)} = \frac{1}{2}(\Delta_{so} + \Delta_{cf}) \pm \left[\frac{1}{4}(\Delta_{so} + \Delta_{cf})^2 - \frac{2}{3}\Delta_{so}\Delta_{cf}\right]^{1/2}$$

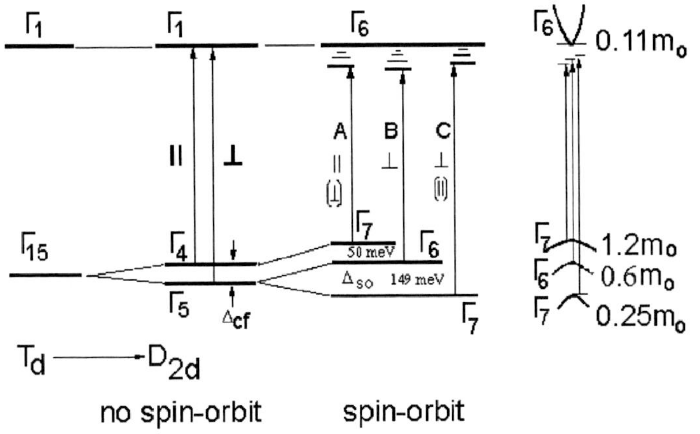

Figure 28. Band structure at the Γ point showing the transition from zincblende (T$_d$) to chalcopyrite (D$_{2d}$) structure. Reprinted with permission from [171], N.N. Syrbu, A.V. Dorogan, V.V. Ursaki, A. Masnik (2010) Wavelength modulated optical reflectivity spectra of CuAl$_{1-x}$Ga$_x$Se$_2$ crystals. J. Opt. accepted. Copyright (2010), with permission from Institute of Physics.

Table 1. Band and Exciton parameters of CuAlS$_2$, CuAlSe$_2$, CuGaS$_2$ and CuGaSe$_2$ crystals

Crystals	Exciton state	A(eV)	B(eV)	C(eV)	Δ_{cf}(eV)	Δ_{so}(eV)	Effective masses
CuAlS$_2$	n=1 n=2 R	3.543 3.565 0.032	3.668 3.687 0.025	3.813	-0.04	0.23	m$_c^*$ =0.12m$_0$ m$_{v1}^*$ = 1.10 m$_o$ m$_{v2}^*$ =0.36m$_0$ m$_{v3}^*$
	E$_g$ (n=∞)	3.575	3.693				
CuAlSe$_2$ [39]	n=1 n=2 R	2.821 2.839 0.024	2.851 2.868 0.022	3.023 3.039 0.018	-0.02	0.15	m$_c^*$ =0.11m$_0$ m$_{v1}^*$ = 1.20 m$_o$ m$_{v2}^*$ =1.20m$_0$ m$_{v3}^*$ =0.25m$_o$
	E$_g$ (n=∞)	2.845	2.873	3.041			
CuGaS$_2$	n=1 n=2 n=3 R	2.5011 2.5303 2.5357 0.0392	2.6217 2.6536	2.6323 2.6611	-0.014[29] -0.017[38] -0.016[28]	0.125[29] 0.126 [38] 0.119[28]	m$_c^*$ =0.12m$_0$ m$_{v1}^*$ = 1.87 m$_o$ m$_{v2}^*$ =0.74m$_0$ m$_{v3}^*$ =0.89m$_o$
	E$_g$ (n=∞)						
CuGaSe$_2$	n=1 n=2 R	1.7380 1.7650 0.036	1.8235 1.8483 0.0357	2.022	-0.113[37] -0.112[38] -0.093[28]	0.23[37] 0.231[38] 0.227[28]	m$_c^*$ =0.12m$_0$ (m$_c^*$ =0.14m$_0$[37]) m$_{v1}^*$ = 1.31 m$_o$ (m$_{v1}^*$ = 1.26 m$_o$[37]) m$_{v2}^*$ =1.31m$_0$
	E$_g$ (n=∞)	1.7745	1.8592				

By using this relation and taking into account the energy position of the n = 1 lines of the A, B, and C excitons one can calculate the value of the crystal field and spin-orbit splitting. Table 1 summarizes the calculated values of the Δ_{cf} and Δ_{SO} from the position of ground (n = 1) states of the A, B and C excitons in CuAlS$_2$, CuAlSe$_2$, CuGaS$_2$ and CuGaSe$_2$ crystals. The excitons spectra of CuGaS$_2$, CuGaSe$_2$ and CuAlSe$_2$ crystals have been previously published [93-113]. The results presented in this table demonstrate that the crystal field splitting is nearly the same in three of these crystals, and is slightly different in CuGaSe$_2$ crystals. The valence band splitting in CuAlS$_2$ and CuGaSe$_2$ crystals is equal to 0.23 eV, while for CuAlSe$_2$ and CuGaS$_2$ crystals it is 0.15 eV and 0.12 eV, respectively. The upper valence band splitting in I-III-VI$_2$ crystals is determined by the wave functions of the ions (atoms) responsible for the origin of these bands. It was theoretically shown [10-14] that the valence bands in I-III-VI$_2$ crystals originate from 3d, 4s, 4p orbitals of the Cu atoms, 5s and 5p orbitals of the In, Ga, or Al atoms, and 3s, 3p orbitals of the S or Se atoms. In the region of the upper valence band, the 3p states of the sulfur atoms and the 3d states of the copper atoms have the highest density. One should mention at the same time that the ion charges in these crystals are anisotropic [106,110]. The ion charges are determined from the calculations of the anisotropy of the phonon spectra and are certainly related to the form of the electron orbitals of ions in the crystal lattice. The ionic charges in these crystals are rotating ellipsoids oriented along some directions in the crystal lattice. The effective ionic charges for Cu, Ga, Al, Se, and S have been determined from the analysis of infrared reflectivity spectra in CuGaSe$_2$, CuAlSe$_2$, and CuAlS$_2$ crystals measured in E$\|$c and E$\perp$c polarizations [106,110]. The calculations demonstrated that the effective ionic charges for each of ions are different for E$\|$c and E$\perp$c polarizations. The values of $\Delta Z = Z_{eff}(E\|c) - Z_{eff}(E\perp c)$ have been calculated for each of ions in CuGaSe$_2$, CuAlSe$_2$, and CuAlS$_2$ crystals. It was found that the ellipsoids of the ionic charges (electron orbitals) of all the three ions preserve their orientation with the transition from CuGaSe$_2$ to CuAlSe$_2$. With the transition from CuAlSe$_2$ to CuAlS$_2$, the ellipsoid of the sulfur ion charge preserves its orientation with respect to Se ion, while the aluminum and copper ions change the configuration of their electron orbitals [106,110]. Since the upper valence band is predominantly determined by the d-states of the copper ions, it is reasonable to assume that these changes will influence the crystal field and spin-orbit splitting of the valence bands, and may be the values of the heavy and light holes effective masses. The effective mass of holes $m_{V2}*=0.36m_0$ in CuAlS$_2$ crystals, and $m_{V2}*=1.20m_0$, $m_{V2}*=1.31m_0$ in CuAlSe$_2$ and CuGaSe$_2$ crystals, respectively (table 1).

Previously [86,93,94], the crystal field splitting has been estimated from the following relation:

$$\Delta_{kp} = -3/2b(2-c/a)$$

where a and c are the crystal lattice constants, b is the deformation potential, which equals 1.0 for the I-III-VI$_2$ chalcopyrite compounds. The following relation is used for the estimation of the influence of the p-d hybridization on the spin-orbit splitting:

$$\Delta_{SO} = \beta\Delta_p + (1-\beta)\Delta_d$$

where the spin-orbit splitting of the p-states equals Δ_p = 0.43 eV for the Cu atoms, and Δ_d = -0.13 eV is the negative spin-orbit splitting of the d-levels, β is the content of the p-states in the upper bands in percents [86,93,94]. By using these relations, the income of the p- and d-states in the upper valence bands of CuAlS$_2$ crystals has been estimated to be around 20-28% and 80-72 %, respectively.

5. POLARITON MODES AT OBLIQUE INCIDENCE OF THE LIGHT ON THE CRYSTAL SURFACE

Methods for the calculation of contours of exciton polariton reflectivity spectra are reported in many papers. They are based on a simple Pekar's model [1-4, 80-82]. The reflectivity spectra at oblique incidence on CdS surface have been discussed by Permogorov et al [114, 115]. At oblique incidence of the light, it is necessary to take into account the spatial dispersion, damping parameter, and the participation of the surface dead layer (without excitons) simultaneously at different orientation of the wave vector and the polarization of light with respect to the crystal axis. The influence of all these parameters can change the shape of the reflectivity spectra.

Broser et al. performed a careful quantitative study of reflectivity spectra in the region of exciton resonances without neglecting any important parameter. Pekar's model has been used for calculations. It was shown that the calculated and measured spectra agree quite well.

In order to calculate reflectivity spectra of a solid, it is necessary to know the frequency dependence of the refractive indices $n_p = c|k|/\omega$ of each electromagnetic wave. The n_p are obtained from the classical dispersion relation, which connects frequency ω and wave vector k of the propagating wave and which contains as a material property the dielectric tensor ($\varepsilon_{ij}(\omega,\kappa)$). In the case of excitonic polaritons it has been shown that the tensor can de described by a Lorentzian oscillator, in which the resonance frequency is replaced by the k-dependent exciton energy [1-4].

The common solution of the dispersion relation involves complicated tensor operations. Two special cases are often discussed in literature:

1. The normal of the reflecting surface is not parallel to any of the principal axes of ε_{ij}, but normal incidence of light is assumed [116].
2. Oblique incidence is allowed, but the reflecting surface and the plane of incidence are parallel to the axes of ε_{ij} [116-120].

Broser et al treated the second case with the restriction to the spectral range of the lowest-energy exciton of CdS (A-exciton), being active only for E⊥c. In this range the two components of ε_{ij} perpendicular to the c-axis ($\varepsilon_\perp$) contain the excitonic resonance while the third is a constant ($\varepsilon_{0\parallel}$). (For simplicity it was assumed that $\varepsilon_{0\parallel} = \varepsilon_{0\perp} = \varepsilon_0$, ε_0 being the background dielectric constant). The possible orientations of $\varepsilon_\perp$, and therefore of the c-axis, towards the normal of incidence result in six different dispersion relations for which the n_p^2 are listed in Table 2.

ω_0 is the exciton energy for $\kappa = 0$, $4\pi\alpha_0$ the polarizability, which is proportional to the longitudinal-transverse splitting ω_{LT}, $M_\perp$ and $M_\parallel$ are the effective masses perpendicular and parallel, respectively, to the c-axis. Γ describes a damping constant and θ the angle of incidence. n_0 is the refractive index of the medium in which the crystal is embedded. The six different geometrical arrangements in Table 2 are named: the non-polariton (σz), the transverse (σx), (σy), the isotropic (πz), the mixed-mode (π,x), and the quasi-transverse (πy) case. The later has not yet been treated in the literature; it implies that the plane of reflectance is the basic plane of the hexagonal prism. For normal incidence two pure transverse modes occurs; with increasing angle of incidence a longitudinal component is mixed in.

Table 2. Index of refraction of the different polariton modes for the different orientations of the c–axis

σz		$n^2 = \varepsilon$
σx		$n_{1,2}^2 = \frac{1}{2}\left\{\varepsilon_0 - \alpha M_\perp - n_0^2 \sin^2\theta\left(\frac{M_\perp}{M_\parallel} - 1\right) \pm \sqrt{\left[\varepsilon_0 + \alpha M_\perp + n_0^2 \sin^2\theta\left(\frac{M_\perp}{M_\parallel} - 1\right)\right]^2 + 4\varepsilon_0 b M_\perp}\right\}$
σy		$n_{1,2}^2 = \frac{1}{2}\left\{\varepsilon_0 - \alpha M_\parallel - n_0^2 \sin^2\theta\left(\frac{M_\parallel}{M_\perp} - 1\right) \pm \sqrt{\left[\varepsilon_0 + \alpha M_\parallel + n_0^2 \sin^2\theta\left(\frac{M_\parallel}{M_\perp} - 1\right)\right]^2 + 4\varepsilon_0 b M_\parallel}\right\}$
πz		$n_{1,2}^2 = \frac{1}{2}\left[\varepsilon_0 - \alpha M_\perp \pm \sqrt{(\varepsilon_0 + \alpha M_\perp)^2 + 4\varepsilon_0 b M_\perp}\right]$ $n_3^2 = -(a+b)M_\perp = \frac{n_1^2 n_2^2}{\varepsilon_0}$
πx		$n_{1,2}^2 = \frac{1}{2}\left\{\varepsilon_0 - (a+b)M_\perp - n_0^2 \sin^2\theta\left(\frac{M_\perp}{M_\parallel} - 1\right) \pm \sqrt{\left[\varepsilon_0 + (a+b)M_\perp + n_0^2 \sin^2\theta\left(\frac{M_\perp}{M_\parallel} - 1\right)\right]^2 + 4bM_\perp n_0^2}\right\}$
πy		$n_{1,2}^2 = \frac{1}{2}\left\{\varepsilon_0 - \alpha M_\parallel - n_0^2 \sin^2\theta\left(\frac{M_\parallel}{M_\perp} - 1\right) \pm \sqrt{\left[\varepsilon_0 + \alpha M_\parallel + n_0^2 \sin^2\theta\left(\frac{M_\parallel}{M_\perp} - 1\right)\right]^2 + 4bM_\parallel(\varepsilon_0 -}\right.$
		$\alpha = \frac{(\omega_0^2 - \omega^2 - i\Gamma\omega)c^2}{\hbar\omega_0\omega^2} \quad b = \frac{4\pi\alpha_0\omega_0 c^2}{\hbar\varepsilon_0\omega^2}$

In Figure 29 the real and the imaginary parts of the dispersion relations $\omega = f(n^2)$ for the isotropic case are shown for different damping constants, using the parameters of the A-exciton of CdS. The real parts of the three branches do not change for damping constants up to 10^{-4} eV, and even for $\Gamma = 10^{-3}$ eV they are only slightly altered. As the usual experimental damping constants have values around $\Gamma \approx 10^{-4}$ eV, this implies that in most cases the dispersion relations for $\Gamma = 0$ can be used without difficulty for discussions. The imaginary part of the dispersion relations, however, is very sensitive to damping; its shape shows mirror symmetry for n_1^2 and n_2^2. The influence of damping to the imaginary part of n_p^2 is always strong if the excitonic part of the polariton is dominant, and weak if the photonic part of the poariton is dominant. The imaginary part of n_3^2 is constant for all ω in accordance with the fact that the longitudinal mode is a pure excitonic mode.

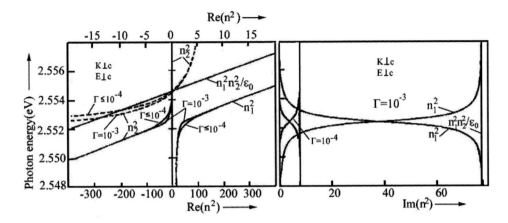

Figure 29. Dispersion relations for the isotropic case. $4\pi\alpha_0=0,01397, \omega_0=2,5524 eV, M_{ex}^{\perp}=0,9m_o$. Reprinted with permission from [82], I. Broser, M. Rosenzweig, R. Broser, M. Richard, E. Birkicht (1978) A quantitative study of excitonic polariton reflectance in CdS. Physica Status Solidi (b) **90**, 77-91. Copyright (1978), with permission from Wiley-VCH Verlag GmbH & Co.KGaA.

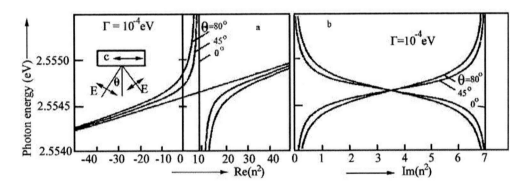

Figure 30. Dispersion relations for the mixed- mode case $4\pi\alpha_0=0,01397, \omega_0=2,5524 eV, M_{ex}^{\perp}=0,9m_o$, $M_{ex}^{\parallel}=5m_o$. Reprinted with permission from [82], I. Broser, M. Rosenzweig, R. Broser, M. Richard, E. Birkicht (1978) A quantitative study of excitonic polariton reflectance in CdS. Physica Status Solidi (b) **90**, 77-91. Copyright (1978), with permission from Wiley-VCH Verlag GmbH & Co.KGaA.

In the mixed-mode case (πx), only two modes of propagation exist. Again the two branches in the real part remain unaltered as long as the damping constant is limited to values bellow $\Gamma = 10^{-4}$ eV. However, the relation $\omega = f(n^2)$ is now a function of the angle of incidence, as shown in Figure 30. For $\theta = 0°$ the exciton and the photon exist without any interaction. With increasing θ the oscillator strength and the longitudinal-transverse splitting factor increase. Thus, we have a case for which an important parameter of the polariton dispersion can be altered by orders of magnitude, simply by tuning the angle of incidence. The imaginary part behaves similarly to the isotropic case; it varies, however, with θ. For $\theta = 0°$ it is zero for the photon-like branch, but constant for the exciton-like one.

5.1. The Effective Index of Refraction

The knowledge of the frequency dependence of the n_p does not allow the calculation of the reflectivity spectra without any further assumptions. As a result, of "spatial dispersion" the Maxwellian boundary conditions, the continuity of the tangential components of *E* and *H*, have to be extended. The Pekar's additional boundary conditions have been used, which assume the vanishing of the excitonic polarization at the surface. Thus an effective index of refraction, n*, results. It describes the reflectivity by r = (n *-1) / (n* +1). For the six cases discussed, the n* are listed in Table 3.

The relations πz, πx, and σx were published by Permogorov et al. [114, 115] and Skettrup [120]. Again the expressions for σ-polarization are especially simple, as here only two transverse modes exist. Two waves propagate also for the mixed-mode (πx) and the quasi-transverse (πy) case. The isotropic case (πz) needs the consideration of three different refractive indices, representing two transverse and one longitudinal waves.

Figure 31 shows the frequency dependence of n* for different angles of incidence for the isotropic case of CdS. The real part of n* (Figure 31) shows a minimum at ω_L ($n_2 = n_3 = 0$), followed by a strong increase which ends in a sharp maximum. The energy of the maximum shifts slightly with the angle of incidence.

Table 3. Effective index of refraction for the different orientations of the c-axis

σZ		$n^* = \dfrac{n_0 \cos\theta}{\sqrt{\varepsilon_0 - s}}$
σx, σy		$n^* = n_0 \cos\theta \dfrac{\omega_1 + \omega_2}{\varepsilon_0 - s + \omega_1 \omega_2}$
πZ		$n^* = \dfrac{\varepsilon_0 \cos\theta}{n_0} \dfrac{\varepsilon_0 s(n_1^2 - n_2^2) + \omega_3 [n_2^2(n_1^2 - \varepsilon_0)\omega_1 - n_1^2(n_2^2 - \varepsilon_0)\omega_2]}{\varepsilon_0 (n_1^2 - n_2^2)\omega_1 \omega_2 \omega_3 + s[n_2^2(n_1^2 - \varepsilon_0)\omega_2 - n_1^2(n_2^2 - \varepsilon_0)\omega_1]}$
πx		$n^* = \dfrac{\varepsilon_0 \cos\theta}{n_0} \dfrac{\omega_1 + \omega_2}{\varepsilon_0 - s + \omega_1 \omega_2}$
πy		$n^* = \dfrac{\varepsilon_0 \cos\theta}{n_0} \dfrac{\varepsilon_0 - s + \omega_1 \omega_2}{(\varepsilon_0 - s)(\omega_1 + \omega_2)}$
$n_0 \mid n_s \mid n^*$		$n^* = \dfrac{1}{\delta} \dfrac{1 + \delta n^* + (1 - \delta n^*) e^{i\Phi}}{1 + \delta n^{**} - (1 - \delta n^*) e^{i\Phi}}$ $\delta_\sigma = \dfrac{\sqrt{n_s^2 - s}}{n_0 \cos\theta}$
		$\delta_\pi = \dfrac{\sqrt{n_s^2 - s}}{\cos\theta} \dfrac{n_0}{n_s^2}$
		$s = n_0^2 \sin^2\theta, \quad \omega_p = \sqrt{n_p^2 - s} \quad \Phi = 2l\dfrac{\omega}{c}\sqrt{n_s^2 - s}$

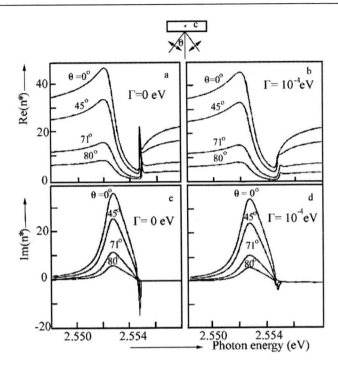

Figure 31. Effective index of refraction (n*) in the isotropic case $4\pi\alpha_0=0,01397, \omega_0=2,5524 eV$, $M_{ex}^{\perp}=0,9 m_0$. Reprinted with permission from [82], I. Broser, M. Rosenzweig, R. Broser, M. Richard, E. Birkicht (1978) A quantitative study of excitonic polariton reflectance in CdS. Physica Status Solidi (b) **90**, 77-91. Copyright (1978), with permission from Wiley-VCH Verlag GmbH & Co.KgaA.

The imaginary part of n* (Figure 31) crosses the zero-line for all θ> 0 °, having a sharp minimum exactly at the energy of the maximum of Re(n*). The energetic position of both these extrema si that of $n_2=n_0 \sin\theta$, which explains the shift with θ. The fact that Im(n*) becomes negative, does not means that the absorption becomes negative, as would be assumed by the usual assumption α≈Im (n*). The effective index of refraction n* describes only the reflectivity, but not the absorption. In Figure 31b and 31d the influence of damping is shown. The sharp singularities are smaller than before, but still visible for values around 10^{-4} eV. They disappear totally for some 10^{-4} eV. It is worth noting that now the effect at θ=71° и θ= 80° is more pronounced than at θ= 45°.

The influence of an exciton-free surface layer can be described by the well-known expressions for a classical optical system consisting of a thin layer of thickness *l* and refractive index $n_s= \sqrt{\varepsilon_o}$ embedded into two media, one with index n_o (vacuum, air, liquid He) and a second one with index n* (bulk) [29]. The total reflectivity, which results from the interference of the electromagnetic waves reflected from the first and the second boundary, can be expressed again by an effective index of refraction n^+, also listed in Table 3.

5.2. Calculated Reflectivity Spectra of CdS

Knowledge of the effective indices of refraction enables us to discuss the expected reflectivity spectra and their dependences on the angle of incidence, damping, and thickness of an exciton-free surface layer. Two typical orientations of the crystal, namely the isotropic (πz) and the mixed-mode (πx) cases, are treated in the following. The use of π-polarization has several advantages: First, the anomalies near ω_L are most pronounced; second, the reflectivity outside the resonance changes with θ from a small value (20 % at $\theta= 0°$) over zero (at Brewster angle at $\theta= 71°$) to unit (at $\theta= 90°$). Especially important, however, is the fact that the influence of a exciton-free layer with a refractive index $n_s = \sqrt{\varepsilon_0}$ disappears at the Brewster angle.

Figure 32a shows, for the isotropic case, calculated reflectivity spectra for four angles of incidence, neglecting damping and a surface layer. One observes indeed a sharp positive spike at $\theta= 45°$ and sharp negative ones for higher angles. The energetic position of these spikes varies with the angle of incidence in the same way as the sharp extrema in the curves of n* (see Figure 31). It was checked a great number of crystals, but never was found one with a damping constant smaller than several 10^{-5} eV. For $\Gamma= 5 \times 10^{-5}$ eV (see Figure 32b) the positive spike for $\theta= 45°$ disappears totally. The negative spike for $\theta=71°$ and $\theta=80°$ still exists up to a damping constant of some 10^{-4} eV. Therefore it seams that the experimental evidence of spike for $\theta= 45°$ without any surface layer is still missing.

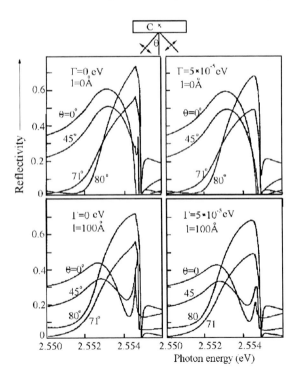

Figure 32 Calculated reflectivity spectra: isotropic case $4\pi\alpha_0=0,01397$, $\omega_0=2,5524$eV, $M_{ex}^\perp=0,9m_0$. Reprinted with permission from [82], I. Broser, M. Rosenzweig, R. Broser, M. Richard, E. Birkicht (1978) A quantitative study of excitonic polariton reflectance in CdS. Physica Status Solidi (b) **90**, 77-91. Copyright (1978), with permission from Wiley-VCH Verlag GmbH & Co.KGaA.

The angular dependence in the mixed-mode arrangement (Figure 33) shows much narrower structures. There is, as expected no effect for θ= 0, and for θ= 45°, Very pronounced peak with a halfwidth of about 0.2 meV appears, while at θ= 80° a reflectivity minimum with R=0 occurs.

The calculations show clearly that damping lowers the reflectivity, but does not change the shape of the curves as drastically as in the isotropic case. It can be seen that even for Γ = 5x10^{-5} eV (Figure 33b) a narrow peak still exists, which allows an exact determination of energy parameters. For example, the position of the small peak at θ=10° differs from ω$_L$ less than 0.05 meV. Measurements in the mixed-mode case are in this aspect more effective that in all other geometric arrangements.

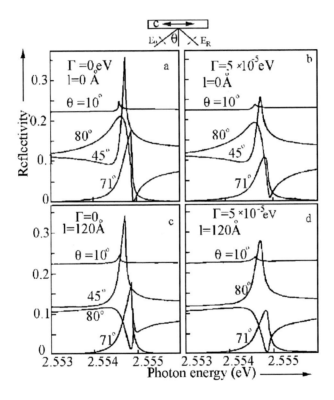

Figure 33. Calculated reflectivity spectra: mixed mode case 4πα$_0$=0,01397, ω$_0$=2,5524eV, M$_{ex}^⊥$=0,9m$_0$, M$_{ex}^∥$=5m$_0$. Reprinted with permission from [82], I. Broser, M. Rosenzweig, R. Broser, M. Richard, E. Birkicht (1978) A quantitative study of excitonic polariton reflectance in CdS. Physica Status Solidi (b) **90**, 77-91. Copyright (1978), with permission from Wiley-VCH Verlag GmbH & Co.KGaA.

The determination of the damping constant is an efficient tool for defining the perfectness of a crystal: impurities, defects, and phonons reduce the lifetime of excitons, which results in an increase of Γ. As shown in Figure 33a, the isotropic case allows one to test sensitively a range of Γ between 5x10^{-5} and 5x10^{-4} eV, since in this range spikes disappear. This is only valid, however, if we can exclude the existence of an exciton-free surface layer. Realistic values for Γ between 5x10^{-5} and 5x10^{-4} eV can be better determined from measurements in the mixed-mode case (Figure 34b), especially because layer effects are less important (Figure 34d).

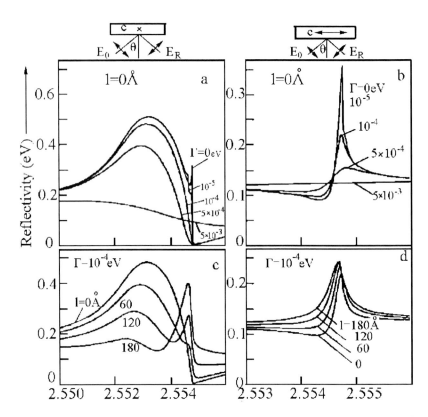

Figure 34. Calculated reflectivity spectra. Influence of the damping and layer thickness for θ=45°, $4\pi\alpha_0$=0,01397, ω_0=2,5524eV, $M_{ex}^{\perp}$=0,9m_0, $M_{ex}^{\parallel}$=5m_0. Reprinted with permission from [82], I. Broser, M. Rosenzweig, R. Broser, M. Richard, E. Birkicht (1978) A quantitative study of excitonic polariton reflectance in CdS. Physica Status Solidi (b) **90**, 77-91. Copyright (1978), with permission from Wiley-VCH Verlag GmbH & Co.KGaA.

The shape of the reflection spectra is completely altered if an exciton-free surface layer of more than about 50 Å thickness exists. In the isotropic case additional spikes occurs (see Figure 32c) at Γ = 0 even for low angles of incidence. In contrast to the behavior of crystals without exciton-free layers, these spikes exist also for realistic damping constants (see Figure 32d). They are most pronounced around θ=45°, so that it might happen that a crystal with almost no fine structure at θ=0° shows a peak at θ=45°. The experiments of Permogorov et al. [114] may be explained in this way, especially since exciton-free layers have been mentioned by the authors.

In the mixed-mode geometry, exciton-free layers of about 100 Å have marked influence only at higher angles of incidence (Figure 33c). Figure 35a shows for θ=80° the variation of the reflectivity curves as a function of layer thickness *l*. With increasing thickness the pronounced reflection minimum at *l* = 0 changes to a very sharp maximum at *l* ≥ 100 Å. However, such narrow peaks are very sensitive to damping. This is evident from Figure 35b where for Γ between 10^{-5} and 10^{-4} eV the structure disappears entirely.

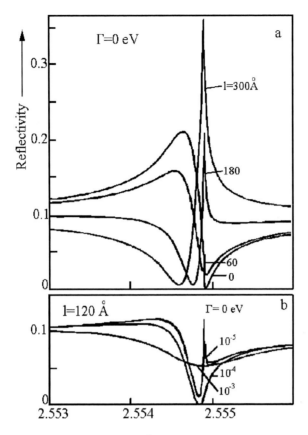

Figure 35. Calculated reflectivity spectra for θ =80°. Influence of damping and layer thickness. $4\pi\alpha_o$ = 0.01397, ω_o =2.5524 eV, $M^{\perp}_{EX}$ =0.9m_o, $M^{\parallel}_{EX}$ =5m_o. Reprinted with permission from [82], I. Broser, M. Rosenzweig, R. Broser, M. Richard, E. Birkicht (1978) A quantitative study of excitonic polariton reflectance in CdS. Physica Status Solidi (b) **90**, 77-91. Copyright (1978), with permission from Wiley-VCH Verlag GmbH & Co.KGaA.

In the mixed-mode geometry, exciton-free layers of about 100 Å have marked influence only at higher angles of incidence (Figure 33c). Figure 35a shows for θ=80° the variation of the reflectivity curves as a function of layer thickness l. With increasing thickness the pronounced reflection minimum at l = 0 changes to a very sharp maximum at $l \geq$ 100 Å. However, such narrow peaks are very sensitive to damping. This is evident from Figure 35b where for Γ between 10^{-5} and 10^{-4} eV the structure disappears entirely.

In the mixed-mode case at low angles of incidence only smaller effects of surface layers occur (see Figure 33d), so that the determination of damping by means of lineshape analysis should be done preferentially at medium θ (see Figure 34d. This small influence of l is in sharp contrast to the isotropic case (see Figure 34c). It is important to note that for all π-polarized spectra at the Brewster angle (θ= 71°) no influence of the layer can be expected (see Figure 32). Here the difficulty arises that no background reflectivity exists outside the resonance region.

5.3. Spectra of Oblique Reflectance in Crystals with Weak Polariton Effect ($\omega_{LT} < \Gamma$)

In the single-axis crystals, for instance CdS, the exciton mass M entering the expression for the dielectric function is anisotropic [12-22]:

$$\varepsilon_{\text{II}}(\omega, k) = \varepsilon_{\infty \text{II}} \left(1 + \frac{\omega_{LT}}{\frac{\hbar k_x^2}{2M_{\text{II}}} + \frac{\hbar k_z^2}{2M_\perp} - \omega + \omega_0} \right)$$

were k_x, k_z are the projections of the wave vector **k** on the x and z axis, M_{II} is the exciton translation mass at **k**||c, i. e. **k**||z, and $M_\perp$ is the exciton translation mass at **k**⊥c.

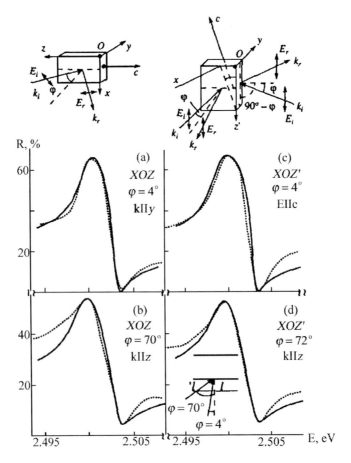

Figure 36. Calculation of the reflectivity spectra at different angles of incidence on the surface of CuGaS₂ crystals (dot line is for experiment, full line is for calculation). The schemes of excitation and registration are shown it the upper part of the Figure.

Figure 36a presents the reflectivity spectra of CuGaS$_2$ crystals from the surface containing the z and x axis (i. e. the c-axis is in the plane of the drawing plane). The wave vector is practically parallel to the y-axis, and perpendicular to the z-axis at small angles of incidence, close to the normal incidence ($\varphi < 4°$). The oscillations of normal additional waves in the plane perpendicular to the plane of incidence (in the x-direction) are always perpendicular to the c-axis, i. e. to the z-axis, and transversal with respect to the wave vector **k**. Such oscillations are non-active, and are not observed in spectra. The oscillation waves parallel to the plane of incidence (in the y-direction), i. e. parallel to the c-axis, are active and transversal with respect to the wave vector **k**. A minimum is observed in reflectance at 2.5035 eV at $\varphi \approx 4°$. The exciton mass $M^{k\|y}(M^y)$, i. e. $M_\perp$ contributes to the reflectivity spectrum.

The increase of φ to 70° leads to the propagation of the light wave in the direction of the z-axis (close to the **k**∥z orientation, see Figure 36b). In this case, the contour of the reflectivity spectra (non-polarized light) is determined by the M_II exciton mass. The E⊥c component of the light does not influence the contour of reflectivity spectra, since the oscillator strength of the Γ_5 exciton allowed in this polarization is by a factor of 3500 lower as compared to the oscillator strength of the Γ_4 exciton (allowed in the E∥c polarization.) [100-104]. The length of the light trajectory through the "dead" surface layer should be bigger at angles of incidence $\varphi \cong 70°$. It is 29.4 Å at $\varphi = 70°$, while it is 10 Å at incidence close to the normal ($\varphi \approx 4°$). These data agree with the thickness of the "dead" layer deduced from the calculations of reflectivity spectra at $\varphi = 70°$ which is equal to 29 Å for the *XOZ* surface, and 32 Å for the *YOZ* surface. The calculations of reflectivity spectra at incidence close to the normal ($\varphi \approx 4°$) shown that the thickness of the exciton-free layer for these surfaces (*XOZ* and *YOZ*) is equal to 5 and 9 Å, respectively.

The reflectivity spectra for the surface (*YOZ'*), i. e. the surface (112) with the c-axis in the horizontal plane are shown in Figure 36b. At small angles of incidence, the reflectivity spectra with the orientation of the vector E∥c are analyzed. At big angles of incidence, the spectra are measured with non-polarized light. The oscillation in the plane perpendicular to the plane of incidence E⊥c, i. e. s-component is perpendicular to the c-axis, are not active. The oscillations of the **E** vector in the plane of incidence, i. e. p-component is parallel to the c-axis, are active at $\varphi \approx 4°$. The oscillations in the E∥c polarization are transverse with respect to the wave vector **k** and practically are nearly parallel to the c-axis. Such oscillations are active. The longitudinal oscillation, perpendicular to the c-axis, are not active. The contour of reflectivity spectra at $\varphi \approx 4°$ are formed by the transversal exciton mass $M_\perp$. The contour of reflectivity spectra at big angles φ are determined by the component of the Γ_4-exciton corresponding to the **k**∥c orientation. At big φ, the longitudinal oscillations have a dipole momentum parallel to the c-axis, which means that at $\varphi \to 90°$ the main contribution to E comes from the longitudinal waves and the $M_\mathrm{II}(\mathbf{k}\|c)$ exciton mass.

A minimum of the reflectivity spectra is observed at 2.5035 eV. Since the reflectivity spectra were measured form the same surface of the same sample for all geometries, it was supposed [98-100] that the main influence to the position of the minimum in reflectivity spectra comes predominantly from the *M* exciton mass.

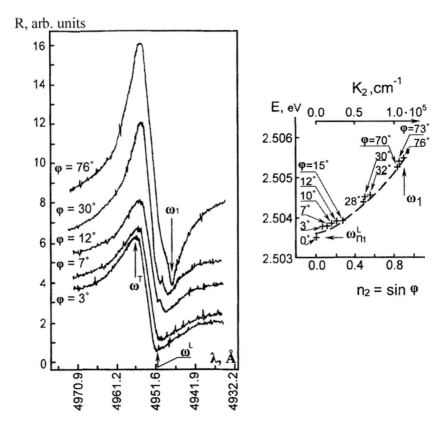

Figure 37. The angular dependence of the reflectivity spectra in the E‖c polarization, and the refraction coefficient n_2 of the Γ_4 excitons.

Figure 37 presents the reflectivity spectra of the XOZ plane of CuGaS$_2$ crystals for the E‖c, k⊥c polarization as a function of the angle of incidence (φ). The E‖c, k⊥c polarization is maintained when the angle of incidence change from 0 to 90° (the measurements are carried out from 3 to 76°). Therefore, the changes of the reflectivity spectra as a function of φ are due to the peculiarities of the propagation of Γ_4 polariton waves in the surface region of the crystal. The oscillation waves in the plane of incidence (the s-component) are transverse with respect to the wave vector **k**. Transverse oscillations are observed predominantly in the angular dependence. These waves are active and have big oscillator strength, since they correspond to the E‖c polarization (the Γ_4 exciton). The longitudinal waves correspond to the polarization perpendicular to the c-axis (E⊥c). In this polarization, the Γ_5 exciton is active which has oscillator strength by a factor of 2500 lower as compared to the Γ_4 exciton. Therefore, in this geometry and polarization the main contribution to the refraction coefficient (polariton branch) comes from the Γ_4 exciton. The angular dependence of the refractive index n_2 is shown in the right part of the figure.

The calculated profiles of reflectivity spectra at the angles $\varphi \approx 70°$ to the *XOZ* and *YOZ* surfaces showed better agreement with the experimental data at the following parameter values: $\varepsilon_b = 7$, $\omega_0 = 2.4985$ eV, $\omega_{LT} = 4.4$ meV, $\gamma = 0.3$ meV (*XOZ*), $\gamma = 0.35$ meV (*YOZ*), $l = 29$ Å (*XOZ*), $l = 32$ Å (*YOZ*), $M = 0.9m_0$ for the *XOZ* surface, and $M = 0.8m_0$ for the *YOZ* surface. These data show that the exciton mass is anisotropic ($M_{\text{II}} = 0.9 m_0$ and $M_\perp = 2 m_0$).

The effective mass $m^*_{V_1}$ of the upper valence band for the **k**II*c* and **k**⊥*c* directions was determined under the assumption that m^*_C is parabolic. For the **k**II*c* direction, the mass m_{V_1} equals $0.77 m_0$ and for the **k**⊥*c* direction, it equals $1.87 m_0$. It should be kept in mind that, at $\varphi = 72°$, the wave vector **k** is almost parallel to the *c*-axis, since it forms an angle of 35° with the *ZOY* surface.

In CdS crystals, the anisotropy in the exciton mass was detected in the luminescence measurements along different **k** directions with respect to the crystal *c* axis [3]. The anisotropy of the effective mass can be evaluated from the energy shift of the luminescence maximum with the variation of the wave vector direction and the relaxation of the excitons–polaritons via the upper branch. The energy of the longitudinal exciton is known to be determined by the formula [1–6]

$$E_{n(k)} = E_g - \frac{R}{n^2} + \frac{\hbar^2 k^2}{2M} = E_{nt} + \frac{\hbar^2 k^2}{2M}$$

where $E_{n(k)}$ is the energy of a transverse exciton, *M* is the exciton mass at the appropriate orientation of the wave vector

$$R = \frac{\mu e^4}{2(4\pi\varepsilon_b)^2 \hbar^2 \varepsilon^2} = \frac{\varepsilon^2}{2(4\pi\varepsilon_b)\varepsilon a_B}$$

is the Rydberg constant expressed in the SI units, and a_B is the Bohr radius. Thus, the exciton mass determines the energy position of the minimum of the reflectivity spectra of the Γ_4-exciton. The Γ_5 excitons are allowed for the polarization **E** ⊥ *c* and are characterized by an extremely low oscillator strength. For this reason, these excitons cannot be detected at the polarization **E** ∥ *c*, along with the Γ_3 excitons forbidden for both polarizations.

Figure 40 shows the emission spectra of CuGaS$_2$ crystals excited by the 476.5 nm line of an Ar+ laser for the orientations **k**II*x* ($\varphi = 4°$) and **k**II*z* ($\varphi = 85°$). The emission peak for the orientation **k**II*z* is observed at a shorter wavelength (2.5040 eV) than that (2.5033 eV) for **k**II*x* ($\varphi = 4°$). The dependence of the energy position of a longitudinal exciton given above shows that for a lower value of the exciton mass, the emission energy should shift towards shorter wavelengths. Indeed, it is really observed experimentally. The emission peak at the orientation **k**∥*z* lies in the short-wavelength range (Figure 40). Excitation by line 488 nm of an Ar$^+$ laser, causes the polaritons to be scattered mainly via the lower polariton branch. The luminescence peak is observed at an energy of 2.4995 eV (Figure 40). These results support the data about the anisotropy of the exciton mass obtained from the angular variations of the reflectivity spectra.

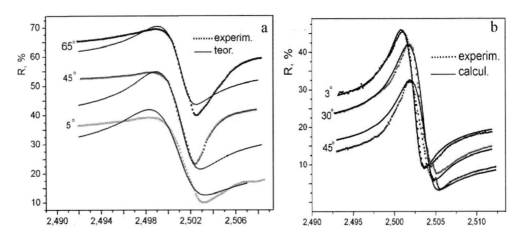

Figure 38. Angular dependences of the reflectivity spectra for the XOZ(a) and YOZ(b) surfaces. For the wave vector κ∥y the exciton mass M is smaller than the mass for the κ∥x direction.

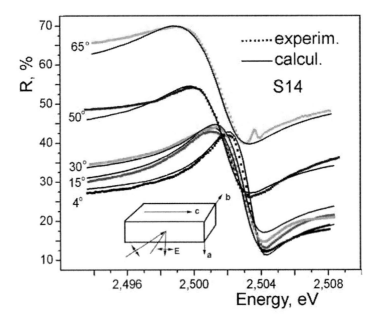

Figure 39. Mixed modes in CuGaS$_2$ crystals. The effective mass M decreases with the transition from the E∥c orientation to the E⊥c one.

5.4. Interference of Additional Waves in the Case of Weak Exciton Polaritons

The investigation of the additional wave interference caused by the polariton branches proved to be extremely interesting.

The interference of exciton polariton waves was observed in the reflectivity and absorption spectra of very thin (d <1 μm) CdS crystals [121-124].

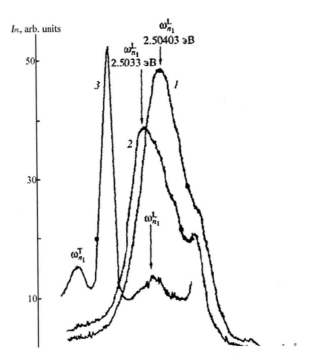

Figure 40. Luminescence spectra of CuGaS$_2$ crystals excited by the lines $\lambda = 476.5$ nm (*1, 2*) and 488 nm (*3*) of an Ar$^+$ laser at the directions of the wave vector **k** ∥ z, $\varphi=85°$ (*1*) and **k**⊥y, $\varphi = 4°$ (*2*).

Two types of interference oscillations were found to be superimposed in the short-wavelength side of the ω_L. The long period oscillations are due to the interference of upper branch exciton polaritons. On the background of these oscillations one can observe short period oscillations occurring as a result of lower and upper branch exciton polariton mutual interference. The interference spectra in the region of exciton resonances allow an accurate determination of the main exciton polariton parameters [121-124].

Thin single-crystalline platelets with parallel faces are necessary in order to realize such kind of oscillations. Apart from that, several additional conditions should be fulfilled: a low value of the damping parameter $\gamma < 1$ meV and enough free path length of the lower branch exciton polariton d $\cong L_1$, where $L_1 = (n_1'' \cdot k_0)^{-1}$, n_1'' is the imaginary part of the refraction index, $k_0 = \omega/c$ is the wavelength of the light in vacuum [121-124].

A weak maximum at 2.5001 eV and a minimum at 2.50175 eV are observed in the reflectivity spectra measured in E⊥c, **k**∥y polarization (Figure 41). They are due to to the transverse and longitudinal mode of the Γ_5 exciton, respectively. The reflection coefficient changes in the limits of 1-2% from the minimum to maximum. The increase of the reflection coefficient in E⊥c polarization starts at higher wavelengths as compared to E∥c polarization. In the E∥c polarization, the reflection coefficient starts to increase at 4948 Å, while in the E⊥c polarization the reflection coefficient decreases. One can assume that the observed peak of reflectivity observed in the E⊥c polarization does not come from the residual reflectivity of the Γ_4 exciton, but is due to the Γ_5 exciton which is allowed in this polarization.

The crystals with thickness less than 10 μm are transparent in the E⊥c polarization. Two maxima are observed at 2.50109 and 2.50012 eV in the absorption spectra of 3 μm thick crystals measured in E⊥c polarization (Figure 41). A similar doublet structure was observed previously [98-101]. The short-wavelength maximum corresponds to the ω_T frequency of the Γ_5 exciton, while the long-wavelength maximum at 2.50012 eV is due to the forbidden Γ_3 exciton. Absorption bands due to the bound excitons are observed in the long-wavelength region. A doublet band due to the $n = 2$ state of the Γ_3 exciton is observed in the short-wavelength (~ 4900 Å) region of the absorption spectra. The doublet observed at 2.5252 eV and 2.5242 eV results from the lifting of orbital degeneracy of higher Γ_5 exciton states under the action of crystal field. A weak absorption peak due to the $n = 3$ state of the Γ_5 exciton is observed at 2.5307 eV. The following parameters of the Γ_5 exciton are deduced from these data: $n = 1$ ($\omega_T = 2.5011$ eV, $\omega_L = 2.5017$ eV), $n = 2$ (2.5252 eV, 2.5242 eV), $n = 3$ (2.5307 eV), Rydberg constant 30.5 - 32.4 meV, $E_g = 2.5336$ eV, exciton mass $M = 0.6 m_0$.

The allowed Γ_4 exciton with a large oscillator strength is seen in E∥c polarization. The absorption coefficient for this exciton at the frequency ω_t is higher than 10^4 cm^{-1}, so that the crystals with ~1.5 μm thickness are opaque for the Γ_4 exciton waves. These observations are consistent with the data reported in Ref. [86,87]. The Γ_5 exciton allowed in E⊥c polarization displays a much smaller oscillator strength. The absorption coefficient at $\omega_t(\Gamma_5)$ frequency does not exceed 600 cm^{-1} (Figure 41). As mentioned above, according to the data reported in Ref. [98-101], the Γ_4 exciton oscillator strength is 2500 times greater than that of the Γ_5 exciton. According to our data, this difference is less (about one order of magnitude), in accordance with the evaluations of ref [98-101]. Nevertheless, the 1.5 μm thick crystals are transparent in E⊥c polarization and therefore we were able to measure the absorption coefficient of the Γ_5 exciton up to the region of continuum in these crystals. As a result we have observed the n = 1, 2 and 3 lines of the Γ_5 exciton (Figure 41) [125,126].

In the reflectivity spectra of thin platelets (d ~ 1.5 ± 0.2 μm) in E⊥c polarization we have observed Fabry-Perot interference in the spectral range from 4900 A to 5000 A (Figure 42, curve a). Since the crystal with d = 1.5 μm is opaque for the Γ_4 (E∥c) excitons, we actually observed the interference of Γ_5 exciton polariton waves. The arrow in Figure 42 (curve a) marks the ω_L (Γ_5) frequency, on the left and right sides of which one can observe the interference of exciton waves. The fine structure of the spectrum corresponds to the condition of Fabry-Perot interference of the lower branch exciton polaritons. The minima of these interference bands are labeled with N (6 to 38) numbers. The structure of the spectrum on the short-wavelength side of $\omega_L(\Gamma_5)$ is rather complicated. One can observe two series of interference oscillations in this spectral range. The long-period oscillations (the interference bands labeled 2 to 12) are due to the interference of the upper branch polariton waves. On the background of these oscillations we can see short-period oscillations related to the mutual interference of the upper and lower branch polaritons.

The interference bands in Figure 42 follow the condition:

$$2n'_r d = \lambda_0 N$$

where d is the crystal thickness, n'_r is the real part of the exciton polariton refractive index n_r (r = 1,2 for the lower and upper polariton branches, respectively), $\lambda_0 = 2\pi c/\omega$ is the light wavelength in vacuum, N is the number of interference band.

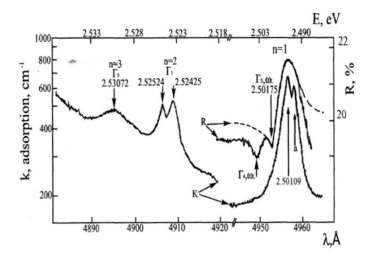

Figure 41. Absorptions and reflectivity spectra of CuGaS$_2$ crystals measured in the E⊥c, k⊥c polarization. Full line is for experiment, dashed line is for calculations. Reprinted with permission from [99] N.N. Syrbu, V.V. Ursaki, I.M. Tiginyanu, V.E. Tezlevan, M.A. Blaje (2003) *The interference*, Journal of Physics and Chemistry of Solids **64**, 1967-1971. Copyright (2003), with permission from Elsevier.

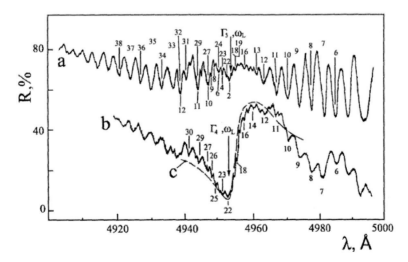

Figure 42. The interference in the reflectivity spectra of thin CuGaS$_2$ crystals in E⊥c polarizations (curve a), in E∥c polarizations (curve b), and reflectivity spectra of thick samples in E∥c polarization (curve c) at 9 K. Reprinted with permission from [99] N.N. Syrbu, V.V. Ursaki, I.M. Tiginyanu, V.E. Tezlevan, M.A. Blaje (2003) *The interference of additional waves of forbidden polaritons excited by allowed polaritons in CuGaS$_2$*, Journal of Physics and Chemistry of Solids **64**, 1967-1971. Copyright (2003), with permission from Elsevier.

The number of interference band (indicated for each line) was determined from the refractive index away from excitons (in the frequency range ~12000 cm^{-1}). Two series of interference oscillations are observed in the short-wavelength side of the ω_L (Γ_5).

On the basis of calculations of the Fabry-Perot and additional wave interference one can plot the exciton polariton branches (the squares in Figure 43). Each experimental point on the curve corresponds to the indicated number of the interference band. Figure 43 shows also the polariton branches deduced from the reflectivity spectra calculations according to the above presented formula for the Γ_4 and Γ_5 excitons using different ε_0 and M parameters.

Figure 42 presents also the reflectivity spectra of thin (d ~ 1.8 μm, curve c) and thick (curve b) CuGaS$_2$ crystals measured in E||c polarization. The reflectivity spectrum contour of the thin sample contains interference bands, which practically coincide with those exhibited in E⊥c polarization. However, the overall reflectivity spectrum contour corresponds to the n =1 line of the large oscillator strength Γ_4 exciton allowed in E||c polarization. The Γ_4 exciton additional waves can not interfere since they are totally absorbed in crystal (the crystal is opaque for these waves). Evidently, the interference pattern in E||c polarization is due again to the Γ_5 exciton additional polariton waves as the crystal of this thickness is transparent only for these waves. In this connection, one can suppose that additional polariton waves of the small oscillator strength Γ_5 exciton are excited by the big oscillator strength Γ_4 exciton.

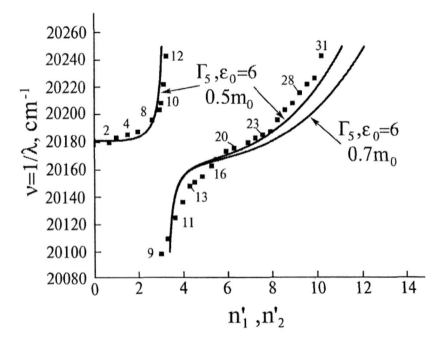

Figure 43. The dispersion curves for the upper and lower branches of the Γ_5 exciton polariton calculated from the reflectivity spectra of n=1 line with ε_0=6, M=0.7m$_0$ and M=0.5m$_0$ parameters (solid curves) and from the interference in reflectivity spectra (square symbols). Calculation data are presented in Table 1. Reprinted with permission from [99] N.N. Syrbu, V.V. Ursaki, I.M. Tiginyanu, V.E. Tezlevan, M.A. Blaje (2003) *The interference of additional waves of forbidden polaritons excited by allowed polaritons in CuGaS$_2$*, Journal of Physics and Chemistry of Solids **64**, 1967-1971. Copyright (2003), with permission from Elsevier.

The repumping of oscillations between the interacting exciton states was previously investigated in $A^{II}B^{VI}$ crystals [121-124]. In CdS crystals with wurtzite structure, the linear, with respect to the wave vector, members in the conduction band and the second valence band change the dispersion curve of the n=1 B exciton in way that it becomes non-parabolic for wave vectors $k \perp c$.

$$\hbar\omega_{ex}^{\pm}(k) = \hbar\omega_T^B + \frac{\hbar^2 k^2}{2m_{\perp}^*} + \frac{\Delta_i}{2} \pm \left[\left(\frac{\Delta_i}{2}\right)^2 + (\varphi k)^2\right]^{1/2} \quad (16)$$

where φk is a linear with respect to k member in the expression of exciton energy; $m_{\perp}^* = m_{e\perp}^* + m_{h\perp}^*$ is the transverse effective mass of the exciton in the $k \perp c$ direction, $\hbar\omega_T^B$ is the energy of the B exciton at $k = 0$. The linear with respect to the wave vector member φk describes the mixing of ($n = 1$) states of B-exciton with a symmetry Γ_1, Γ_2, Γ_{5T} and Γ_{5L}. The Γ_1 (E‖c) and Γ_{5T} (E⊥c) in CdS are long-wavelength excitons, $\Delta_i = 1, 2$ is the splitting value of interacting states at $\vec{k} = 0$, $\Delta_1 = 0$ (for Γ_2 and Γ_{5T}) and $\Delta_2 > 0$ (for Γ_1 and Γ_{5L}). A most strong repumping of the oscillator strength between the interacting states occurs when they are degenerated at k = 0. In this case, for E⊥c both hybrid states will have the same oscillator strength independent on $\vec{k}$. Each of them interacts with the light. Since $\Delta_1 = 0$, the equation 16 becomes:

$$\hbar\omega_{ex}^{\pm}(k) = \hbar\omega_T^B + \frac{\hbar^2 k^2}{2m_{\perp}^*} + \varphi k$$

In the E‖c polarization in CdS crystals $\Delta_2 > 0$, and weak mixing of interacting states takes place. In these crystals, almost all the oscillator strength is concentrated in a hybrid sate predominantly of the Γ_1 character, and it strongly interacts with the light. In CuGaS$_2$ crystals there is a little another situation. As mentioned above, excitons with Γ_4, Γ_5 and Γ_3 symmetry are observed in the long-wavelength region. The Γ_4 exciton is dipole-active in the E‖c polarization, while the Γ_5 exciton is active E⊥c pol;arization, and Γ_3 exciton is not active in both poilarization. Practically all the oscillator strength in CuGaS$_2$ crystals is concentrated in the Γ_4 exciton. The oscillator strength of the Γ_4 exciton is by a factor of 2500 higher as compared to the Γ_5 exciton. Therefore, the light in E‖c polarization excites the Γ_4 excitons with a "giant" oscillator strength, which consequently excite the Γ_5 exciton waves due to their close resonance frequency. The Γ_5 exciton waves excited by the waves of the Γ_4 excitons interfere in the crystal determining the fine structure of reflectivity spectra in the E‖c polarization. In this polarization, the waves of the Γ_4 excitons can not interfere, since the crystal with the thickness of ~1.5 μm is practically absolutely non-transparent. Therefore, the interference observed in reflectivity spectra in the E‖c polarization (Figure 42b) is the interference of additional waves of the Γ_5 exciton excited by the Γ_4 exciton.

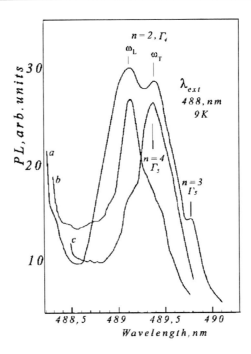

Figure 44. Luminescence spectra of CuGaS$_2$ crystals ($d \sim 320$ μm) for the polarization E||c, k⊥c (b), E||c, k||c (c) and nonpolarized (a) excited by the lines 488 nm of Ar$^+$ laser. Reprinted with permission from [126], N.N. Syrbu, L.L.Nemerenco, I.G.Stamov, V.N. Bejan, V.E. Tezlevan (2007) *Interference of resonance luminescence*. Optics Communications **272**, 124-130. Copyright (2007), with permission from Elsevier.

5.5. Interference in Luminescence Spectra of Exciton Polaritons

As shown above, the excited states of the Γ_4 and Γ_5 in the short-wavelength range are well established in CuGaS$_2$ crystals. The Γ_4 exciton is observed in reflectivity spectra, while the Γ_5 exciton is observed in absorption spectra.

Radiation bands at 2,5307 eV, 2,5332 eV and 2,5347 eV are observed under the excitation with the wavelength λ=4880 Å of CuGaS$_2$ crystals at 9 K (Figure 44). In the polarization E||c, k⊥c the short-wavelengh lines 2,5347 eV is the most intense, and in the polarization E||c (k is almost parallel to the axis c) the maximum 2,5332 eV is found. The radiation band 2,5307 eV may be explained by the excited state n=3 of the exciton Γ_5. The line n=4 of the exciton Γ_5 is found at the energy 2.5332 eV. At the same energy the luminescence line is found (Figure 44). The excited state lines n=2 of the exciton Γ_4 revealed at the energies 2,5337 (ω_t) and 2,5347 (ω_L) coincide with the radiation lines shown in Figure 4. Thus, in the radiation spectra the lines n=3 (Γ_5) and n=4 (Γ_5) and/or n=2 (Γ_4) are revealed. In the polarized light all three peaks of radiation are found (curve a, Figure 44). In the samples of small thickness in the radiation spectra in the region of n=3 and n=4 of the exciton Γ_5 fine structure is found (Figure 44). The distance between the bands is much less than the energy of the Γ phonons or the difference between the energies of Γ phonons. This structure of the bands is explained by the luminescence interference in thin plate CuGaS$_2$.

Earlier in work [100,126] we have reported on interference in the spectra of absorption of the Γ_5 exciton wave observed up to the states $n=3$ and $n=4$ of the exciton Γ_5.

Interference in the luminescence spectra is observed in the samples of the thickness 0,5-1,5 µm (Figures 45, 46). At excitation by the line 4765 Å of the Ar$^+$ laser the radiation bands are not practically found, i.e. their intensity is very weak. We consider that the observed radiation possesses a resonance character, because the excitation energy of 2,5402 eV is close to the lines $n=2$ of the exciton Γ_4. In these luminescence spectra another peculiarity is observed too. The intensity of the interference bands (bands 3, 2, 1) decreases as the wavelength increases relative to the central band of radiation of 2,5307 eV. From the short-wavelengh side, the intensity of the interference bands (bands 19, 20) also decreases as the wavelength decreases relative to the band 2,5332 eV. Interference takes place at three frequencies where the energies $n=4$ of the exciton Γ_5 and the energies $n=2$ of the exciton Γ_4 coincide. We consider that at excitation of these states there occurs an exchange between two polariton branches, i.e. the exchange between ω_t of the state $n = 2$ of the exciton Γ_4 and ω_t (or ω_L) $n=4$ of the exciton Γ_5. Taking into account the absorption coefficient for excitons Γ_4 and Γ_5 it may be surely considered that the Fabry-Perot interference at the thickness $d\sim0,5$-1,5 µm takes place only for the exciton Γ_5 wave. For the waves of the excitons Γ_4 these thicknesses are opaque. These waves appearing in the surface region attenuate before reaching the second face of the crystal.

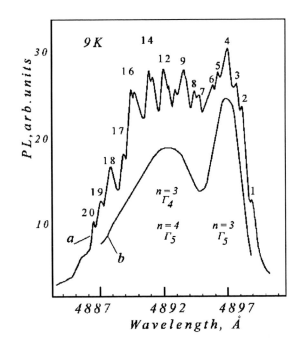

Figure 45. The interference in the luminescence spectra of thin ($d \sim 0.6$ µm) CuGaS$_2$ crystals excited by the 488 nm line of Ar$^+$ laser. Reprinted with permission from [126], N.N. Syrbu, L.L.Nemerenco, I.G.Stamov, V.N. Bejan, V.E. Tezlevan (2007) *Interference of resonance luminescence of exciton*. Optics Communications **272**, 124-130. Copyright (2007), with permission from Elsevier.

It should be noted that the oscillator strength of the exciton Γ_4 is much lager than that of the exciton Γ_5 [98-101,126,127]. Therefore, the state $n=2$ Γ_4 of the exciton polariton is excited, it causes an excited polariton wave of the exciton Γ_5 (in the region $n=3$ and $n=4$) by the energy exchange. The Fabry-Perot interference characterizes polariton branches (the refractive index) of the exciton Γ_5. In CuGaS$_2$ crystals polariton branches are restored in the region of "bottle neck" for the states $n=1$ of the exciton polaritons Γ_4 and Γ_5 [98-101,126]. For the upper polariton branch ω_L ($n=1$) of the exciton Γ_5 the refractive index is equal to 3,45 at the wavelength 4840 Å. We suppose that the refractive index value will be equal to 4 in the region of the polariton branch ω_t ($n=2$) of the polariton Γ_5, and it will be equal to 4,5 in the region ω_L ($n=3$) of the polariton Γ_5. Proceeding from these approximations, we estimate the order of the interference bands and dispersion of the polariton branch ω_L ($n=3$) of the exciton Γ_5 (Figure 47, Table 4).

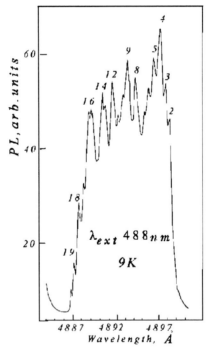

Figure 46. The interference in the luminescence spectra of thin ($d \sim 1.2$ μm) CuGaS$_2$ crystals excited by the 488 nm line of Ar$^+$ laser. Reprinted with permission from [126], N.N. Syrbu, L.L.Nemerenco, I.G.Stamov, V.N. Bejan, V.E. Tezlevan (2007) *Interference of resonance luminescence of exciton*. Optics Communications **272**, 124-130. Copyright (2007), with permission from Elsevier.

The upper polariton branch ω_L ($n=1$) transits into the transverse (lower) polariton branch ω_t of the state $n=2$ (Figure 47). The upper polariton branch ω_L of the state $n=2$ transits into the transverse polariton branch ω_t of the state $n=3$. Considering that in the region of "bottle neck" of the states $n=3$ the refractive index is equal to 4, the samples thickness is equal to 0,8 μm, from the Fabry-Perot interference condition $2nd = \lambda N$ the order of the interference bands (N) for each found band (m) is determined (Table 4). Proceeding from these data, i.e. taking into account these suppositions, the change of the refractive index of the polariton branch ω_t for the state $n=3$ of the exciton Γ_5 is calculated (Table 4, circles in Figure 47). At higher frequencies the interference bands ($m=9-20$) are caused by the change of the refractive index of the polariton branches of the states $n=4$ of the exciton polaritons Γ_5.

Table 4. Position of the luminescence interference bands in the region of polariton branches ω_t and ω_L of the exciton polaritons Γ_5 in CuGaS$_2$ crystals at 10 K at excitation of 488 nm of Ar$^+$ laser

Band numbers, m	Wavelength, λ, Å	Interference maximum number, N	Refractive index	Band numbers, m	Wavelength, λ, Å
1	4900	14	4,287	11	4893,5
2	4899,5	16	4,899	12	4893
3	4899,3	18	5,512	13	4879,2
4	4898,9	20	6,124	14	4891,6
5	4898,0	22	6,735	15	4890,6
6	4897,2	24	7,356	16	4889,7
7	4896,4	26	7,957	17	4875,2
8	4896	28	8,568	18	4888,2
9	4895,5	30	9,179	19	4887,2
10	4894,7	32	9,7894	20	4886,2

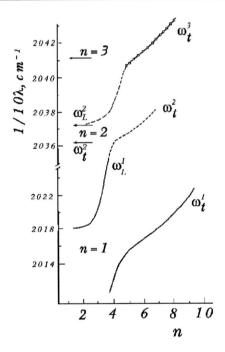

Figure 47. The dispersion curves for the upper and lower branches of the Γ_5 exciton polariton calculated from the absorption spectra of $n=1$ line with $\varepsilon_0 =6$, M=0.7m_0. Reprinted with permission from [126], N.N. Syrbu, L.L.Nemerenco, I.G.Stamov, V.N. Bejan, V.E. Tezlevan (2007) *Interference of resonance luminescence*. Optics Communications **272**, 124-130. Copyright (2007), with permission from Elsevier.

6. EXCITON-PHONON LUMINESCENCE SPECTRA WITH EXCITON-POLARITON ANNIHILATION

The free excitons in semiconductors are characterized by the wave vector **k**, effective translation mass $M = m_e + m_h$, and kinetic energy E. The relations between the exciton energy and the wave vector in the approximation of effective mass is expressed as

$$E_n = E_{0n} + E = E_{0n} + \frac{\hbar^2 k^2}{2M},$$

were E_{0n} is the energy of the n quantum state.

After optical excitation of a crystal and relaxation processes, a steady-state concentration of excitons and a steady-state energy distribution is established. If the exciton lifetime is long enough for establishing the thermal equilibrium, and the exciton band is characterized by parabolic dispersion, the distribution of excitons over the kinetic energy E is described by the Maxwell distribution

$$\frac{dN}{N}(E) = \frac{2}{\pi^2}\left(\frac{\pi}{kT}\right)^{3/2} E^{1/2} \exp(-E/kT) dE \qquad (17)$$

were N is the total number of excitons, k is the Boltzmann constant, T is the crystal temperature.

For processes of radiative annihilation of excitons with simultaneous excitation of optical phonons, the energy and momentum conservation laws are following:

$$E_i - \hbar\omega_s = \sum_{n=1}^{N} \hbar\Omega_n \qquad (18)$$

$$\mathbf{k}_i - \mathbf{k}_s = \sum_{n=1}^{N} q_n \qquad (19)$$

Were E_i and $\mathbf{k}_i$ are the energy and the wave vector of the initial state in the exciton band, $\hbar\omega_s$ and $\mathbf{k}_s \approx 0$ are the energy and the wave vector of the emitted light, $\hbar\Omega_n$ and q_n are the energy and the wave vectors of the excited optical phonons, N is the number of emitted phonons, which will be referred as the order of exciton-phonon luminescence.

As one can see from the conservation law, excitons with any values of energy and wave vector can participate in the exciton-phonon annihilation processes. The wave vector on exciton is transmitted to the phonons during the optical transition. Particularly, in the case of first-order process schematically illustrated on the energy diagram (Figure 48), the wave vector of the excited phonon equals to the wave vector of the annihilated exciton (with the precision of the wave vector of the light).

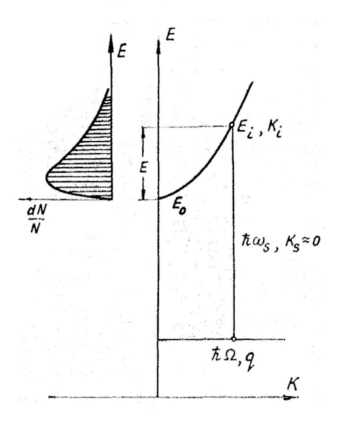

Figure 48. Schematic presentation of the exciton annihilation process with a concomitant formation of one optical phonon. The quasi-equilibrium distribution of excitons over the kinetic energy E is shown in the left part of the Figure.

Taking into account the weak dependence of the optical phonons upon the wave vector near the center of the Brilloin zone, one can consider the values of $\hbar\Omega_n$ in (18) as being constant and equal to the highest value of the phonon energy. Then, according to the energy conservation law one can expect that the spectrum of the exciton-phonon annihilation will contain lines with the long-wavelength limit shifted by (1, 2, 3, etc.) $\hbar\Omega$ to lower energies as compared to E_0. The form of these lines should reflect the distribution of excitons over the kinetic energy , and, in the case of thermal equilibrium , can be expressed as

$$I_n \propto E^{1/2} \exp\left(-\frac{E}{kT}\right) W_N(E) \qquad (20)$$

where E is the exciton kinetic energy, which is measured from the long-wavelength limit of the line in the spectrum. The factor $W_N(E)$ describes the dependence of the probability of exciton-phonon annihilation of the N-order upon the exciton kinetic energy.

One should mention that differences in the $W_N(E)$ functions for processes of different order are a qualitative criterion of the fulfillment of the momentum conservation law for the exciton-phonon annihilation. The processes of different order are primarily different from the point of view of formulation of the momentum conservation law, which can lead to different spectral form of lines only when this law is fulfilled. The violation of the momentum conservation law in defected crystals was deduced from the change of the $W_1(E)$ function for ZnO crystals.

One should mention that, in the case of equilibrium distribution of excitons over the kinetic energy, the form of the exciton-phonon annihilation line is universal, i. e. it depends only on the type of the $W_N(E)$, but does not depend on other parameters of the crystal, such as the bandgap, effective masses, etc. Due to this fact, the exciton-phonon annihilation spectra can be used for the determination of the "exciton" temperature of the crystal, which can exceed the lattice temperature in the case of intense excitation.

Among other materials, exciton-phonon luminescence spectra of CdS crystals are most thorough investigated. This is due to the availability of perfect CdS crystals as well as due to the convenience of the spectral range (485 – 520 nm) of the intrinsic luminescence. At the same time, the main regularities of exciton-phonon luminescence processes are revealed in spectra of CdS crystals.

The parameters of the $CuGaS_2$ crystals are very close to those of CdS crystals. The intrinsic luminescence of the chalcopyrite is situated in a convenient spectral region. The luminescence spectra of the chalcopyrite single crystals in the region of the absorption edge at liquid helium temperature contain luminescence lines due to the free and bound excitons. The optical transitions mainly interact with longitudinal (LO) phonons. The resonant Raman scattering lines (P_5, P_6, P_7 lines in Figure 49) are much narrow as compared to the phonon luminescence lines (E_{LO}).

The emission due to bound excitons masks the emission of free excitons in some samples. Samples with a weak luminescence of bound excitons have been chosen for experiments. The form of the exciton luminescence spectra practically does not depend upon the frequency of the excitation light. The luminescence spectra were measured under the excitation of Ar^+ laser lines as well as under the excitation of lines from xenon or mercury lamp. Resonance Raman scattering lines, emission from the upper and lower exciton polariton branches, emission due to bound excitons D, as well as the exciton-phonon luminescence (E_{LO}) is observed under the excitation of 4880Å Ar^+ laser line.

Probably, thermal equilibrium is set in the investigated samples, and the density of states in exciton bands changes as $E^{1/2}$, i. e. one can consider that the bands are parabolic. The width of exciton-phonon luminescence lines at low temperatures is associated with the energy distribution of excitons in the region of the polariton resonance. These regularities have been investigated in details in many crystals A^2B^6 [128-135].

The bottom of the exciton band shifts to lower energies with changing the composition of the solid solution near the $CuGaS_2$ composition as shown in Figure 50. In these crystals, the energy of the 5145 Å Ar^+ laser line is a little higher than the ground state of the Γ_4 exciton.

The reflectivity spectra of crystals with this composition in the E∥c polarization are shown in Figure 50. A maximum at 2.38 eV and a minimum corresponding to the longitudinal exciton are observed in the reflectivity spectra. A weak maximum is observed in the region of the minimum in the spectrum. One can suggest that this maximum is due to a "spike" which determines the longitudinal frequency of the exciton polaritons. Exciton phonon luminescence is observed in these crystals under the excitation by 5145 Å laser line (bands 1-5 in Figure 50).

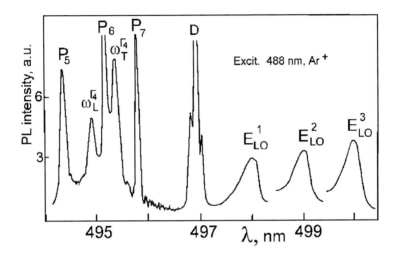

Figure 49. Luminescence spectra of $CuGaS_2$ crystals due to the annihilation of free excitons.

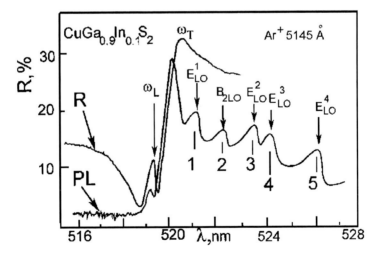

Figure 50. Reflectivity and exciton-phonon luminescence spectra of $CuGa_{0.9}In_{0.1}S_2$ solid solutions.

6.1. Energy Exchange between the Upper and Lower Modes of Polariton Branches

The investigation of resonance Raman scattering by using a tunable frequency interval obtained by passing the radiation of an incandescence lamp through a monochromator has been previously performed in CdS crystals [129-137]. It was shown that secondary emission spectra related to the 2 LO scattering reproduce the stationary energy distribution of the initial polariton states in the investigated sample. In II-VI crystals (CdS, CdSe, ZnTe, etc) under the excitation by the light with the frequency ω_0 higher than the resonance exciton frequency ω_{exc}, an excitonic luminescence band related to the recombination of excitons in the ground or excited states with $\mathbf{k} = 0$ is observed in the region of ω_{exc}. Simultaneously, a series of narrow lines is observed at higher frequencies, shifted from the excitation ω_0 line by a frequency multiple to the frequency of the longitudinal optical phonon [129-131]. The emission line at the frequency $\omega_0-2\omega_{LO}$ was attributed to the resonance Raman scattering [132], or to the luminescence of hot excitons [133], which emit longitudinal optical phonons as a result of the cooling process. In the second case, the exciton is formed as a result of indirect excitation with the participation of a LO-phonon, and it radiates also with the emission of one LO-phonon. As a result, the emission line shifted by $(N + i)\omega_{LO}$ from the excitation line corresponds to N intermediate exciton states. It was shown that the resonant Raman scattering is equivalent to the hot luminescence if only the ground state of the exciton is considered as intermediate state in the resonance scattering process [134]. In this case, the contour of the emission line is more gradual on the short wavelength side, and more sharp on the long wavelength side [130,131]. The more detailed analysis shows that the emission for the $\omega_0 - 2\omega_{LO}$ line has an additional component, which is comparable to the main one. This component can not be explained by a simple kinetic treatment in the frame of the hot luminescence theory [134].

In contrast to the II-VI compounds, the chalcopyrite crystals are characterized by the presence of several polar modes. The unit cell of these crystals contains two formula units with 21 optical vibration modes. The vibration modes at the center of the Brillouin zone are described by the following irreducible representation:

$$\Gamma = A_1(\Gamma_1) \oplus 2A_2(\Gamma_2) \oplus 3B_1(\Gamma_3) \oplus 4B_2(\Gamma_4) \oplus 7E(\Gamma_5).$$

One mode of B_2 symmetry and one mode of E symmetry are acoustical. The A_2 modes are forbidden in the IR and Raman spectra. Three vibration modes of B_2 symmetry are active in the IR spectra in the E∥c polarization, while six vibration modes of E symmetry are active in the E⊥c polarization [135-137].

In the past few years, the resonant Raman scattering in $CuGaS_2$ has been intensively examined [100,101,138-143]. The polarization measurements in various resonant Raman scattering geometries under excitation by different wavelengths have been performed [24-29].

The emission lines of argon or He-Cd lasers were used as excitation source to achieve the resonance scattering conditions [138-140]. Symmetry considerations for (100) - surface result in the following allowed modes for a given back-scattering configuration: A for x(z,z)x; E(LO) for x(z,y)x and x(y,z)x; and A + B for x(y,y)x, respectively [100,101,136].

Photoluminescence and resonance Raman scattering spectra of $CuGaS_2$ crystals were investigated at low temperature (10 K) under the excitation with the radiation from a spectral interval obtained by passing the radiation of an incandescent or xenon lamp through a monochromator. The thermalized luminescence of the Γ_4 and Γ_5 excitons was revealed under the excitation by an interval of photon energies higher than the energy of the ground state of long-wavelength Γ_4 excitons ($\omega_T(\Gamma_4)+1E^1_{LO}$). The energy conversion between the polariton modes in luminescence and resonance Raman scattering spectra was considered.

A 100 W incandescent lamp or a 500 W xenon lamp were used as light source (L). An MDR-23 monochromator was used for the selection of the frequency interval of the excitation light. The selected frequency interval was directed by means of the mirror M, condenser C, polarizer P, and filter F to the sample installed on the cold station of a LTS-22C330 cryogenic system assuring a sample temperature of 10 K. The emission from the sample is focused by the condenser C through the analyzer A, and filter F on the entrance slit of a double DFS-32 Raman spectrometer. The signal is measured by a photomultiplayer tube PMT and the registration system shown in Figure 51. Therefore, the first monochromator is used for the selection of a narrow spectral interval of the excitation light, while the second monochromator serves for the registration of the secondary emission spectrum. The intensity of the emission from the crystal is rather high near the resonance. This allows one to measure the emission spectra by using a narrow excitation interval of less than 10 cm^{-1}, therefore excluding the employment of an expensive tunable laser in experiments with resonance Raman scattering.

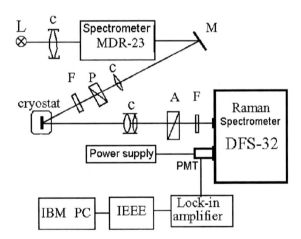

Figure 51. Schematics of the experimental set-up for measuring luminescence and Raman scattering spectra. Reprinted with permission from [127], N.N. Syrbu, V.E. Tezlevan, I.Galbic, L. Nemerenco, V.V. Ursaki (2009) *Exciton*. Optics Communications 282, 4562-4566. Copyright (2009), with permission from Elsevier.

The first investigations of the secondary emission by using a tunable frequency interval obtained by passing the radiation of an incandescence lamp through a monochromator performed in CdS crystals [130-134] demonstrated the necessity of high emission intensity. The optical properties and the emission intensity of $CuGaS_2$ crystals are similar to those of CdS crystals [100,101,125,126]. It was shown that the secondary emission spectra due to 2

LO scattering reproduce the stationary energy distribution of the initial polariton states [144-147]. "Photon-like" polaritons of the low polariton branch are formed under the excitation of the crystal in the region of the exciton resonance [1-5]. The scattering of these polaritons on LO phonons represents the pre-resonance Raman scattering. The scattering intensity on 1 LO phonons in the pre-resonance conditions is lower as compared to the intensity of 2 LO phonon scattering.

The lowest frequency (energy) of the optical phonon at T = 10 K in CuGaS$_2$ crystals equals 75 cm^{-1}. B$_2^1$$_{LO}$ (89 см$^{-1}$), B$_1^1$ (115 см$^{-1}$), E^2$_{LO}$ (141 см$^{-1}$), E^3$_{LO}$ (177 см$^{-1}$) and other high energy phonons are observed at T = 10 K. All the selection rule allowed vibration modes have been observed in CuGaS$_2$ crystals at 300 K and 9 K [100,101,135-137].

The emission spectra in the region of the "bottle neck" are presented in Figure 52. The spectra were measured with different values of the excitation energy of the light from a xenon lamp passed through a MDR-23 monochromator. An emission maximum at 2.500 eV is observed under the excitation by 2.507 eV (494.4 nm). This value is close to the ω_T energy of the transversal Γ_4 exciton. The reflectivity at the ω_T frequency of the n = 1 band of the Γ_4 exciton reaches the value of 50 %. It could be that the emission band at 2.500 eV which is related to the emission of the thermalized exciton is partially affected by the scattered light. An emission band separated by 75 cm^{-1} from the excitation band emerges in the long wavelength region under the excitation by the 2.506 eV (494.7 nm) which is close the resonance energy of the longitudinal exciton. This value is nearly equal to the energy of the E$_{LO}^1$ phonon [100,101,127]. By changing the excitation energy to 2.502 eV (495.4 nm), the E$_{LO}^1$ band is shifted to the infrared region by nearly the same energy, i. e. E$_{LO}^1$ is separated from the E$_{exc}$ by 75 cm^{-1}. This shift is accompanied by the intensification of the long-wavelength luminescence wing (Figure 52). The intensification of the luminescence wing may be due to the increasing probability of the scattering by acoustical phonons as compared to optical phonons for the polariton excited by the light in the region of ω_{LT}. The participation of higher energy LO phonons is not excluded.

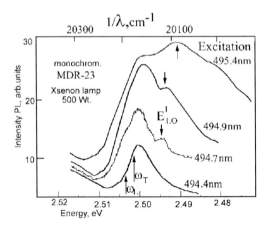

Figure 52. The emission spectra of CuGaS$_2$ crystals under the excitation with the radiation from a spectral interval of a 500 W xenon lamp. Reprinted with permission from [127], N.N. Syrbu, V.E. Tezlevan, I.Galbic, L. Nemerenco, V.V. Ursaki (2009) *Exciton*. Optics Communications 282, 4562-4566. Copyright (2009), with permission from Elsevier.

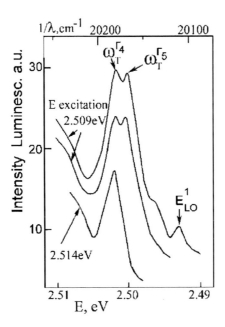

Figure 53. The emission spectra of CuGaS$_2$ crystals measured in E∥c polarization under the excitation with the radiation from a spectral interval of a 100 incandescent lamp. Reprinted with permission from [127], N.N. Syrbu, V.E. Tezlevan, I.Galbic, L. Nemerenco, V.V. Ursaki (2009). *Exciton*. Optics Communications 282, 4562-4566. Copyright (2009), with permission from Elsevier.

As a result, some stationary energy distribution of polaritons is established at the moment of LO-phonon emission. The excitation to the ground state of the upper polatiton branch occurs under the excitation by the light with the energy near ω_L or with higher energy. The luminescence is due to the scattering of these states by acoustical and optical phonons to the states of the lower polariton branch. This is confirmed by the analysis of the character of the luminescence curve presented in Figure 52. These spectra represent the Raman scattering excited by a tunable monochromatic radiation source, and demonstrate the energy exchange between the lower and upper exciton polariton branches.

Figure 53 presents the emission spectra of CuGaS$_2$ crystals excited by the radiation from a spectral interval obtained by passing the radiation of a 100 W incandescent lamp through a MDR-23 monochromator. The spectral slit width is taken small enough in order to minimize the scattered light in the region ω_T. The sample was replaced by a mirror to control the amount of scattered light in the region ω_T, and the spectral slit width was adjusted to set the scattered light in the region ω_T equal to zero. A luminescence band in the region of 2.501 eV is observed under the excitation by the energy of 2.514 eV (20280 cm^{-1}), i. e. by 75-80 cm^{-1} higher than the energy of the transversal exciton (Figure 53). This emission band is due to the luminescence of the transversal excitons ($\omega_T \sim 2.5011$ eV). A doublet is observed in the luminescence spectra under the excitation by (E∥c) polarized light with the energy of 2.509 eV, i. e. a line in the long wavelength region (2.500 eV) appears in addition to the ω_T line. The energy of this line coincides with the energy of the transversal Γ_5 exciton which is active in the E⊥c polarization [100,101,125-127]. Therefore, by adjusting the energy of the (E∥c) polarized excitation light equal to $E(\omega_T,\Gamma_5)+E_{phonon}(E^1_{LO}) = 2.509$ eV, one can observe the

$\omega_T(\Gamma_4)$ and the $\omega_T(\Gamma_5)$ bands as well as the one-phonon E^1_{LO} (2.495 eV) band. The emergence of the 2.500 eV emission band corresponding to the energy of the transversal Γ_5 exciton indicates on the excitation of the Γ_4 exciton which is active in the (E||c) polarization with a consequent transmission of the energy to the Γ_5 exciton, i. e. the energy conversion between the branches of different excitons occurs. The position of the E^1_{LO} (2.495 eV) band remains practically constant when the energy of excitation changes. The E^1_{LO} (2.495 eV) band is separated by 75 cm^{-1} from the resonant value of the ω_T exciton. One can suggest that this band corresponds to a phonon replica of the luminescence band related to the annihilation of the exciton, and it does not correspond to the resonant Raman scattering.

Two maxima at 2.530 eV and 2.504 eV emerge in the emission spectra of CuGaS$_2$ crystals under the excitation in the region of the continuum with the energy of 2.541 eV (487.8 nm). These maxima are due to the luminescence of the n = 2 and n =1 states of the Γ_4 excitons, respectively (curve (a) in Figure 54). At the same time, broad bands at 2.525 eV and 2.520 eV are observed. The luminescence peaks related to n = 1 and n = 2 do not change their energy position under the excitation by 2.540 eV (488.0 nm) excitation, while the bands at 2.525 eV and 2.520 eV are shifted to the long wavelength region by the energy equal to the change of the excitation energy (curve (b) in Figure 54). The same behavior is observed with decreasing the excitation energy to 2.539 eV (488.3 nm). The 2.525 eV and 2.520 eV bands are shifted, while the luminescence peaks related to n = 1 and n = 2 do not change their position (curve (c) in Figure 54). The shift of the 2.525 eV and 2.520 eV bands indicates that they are Raman scattering lines (Figure 54). These bands are the most distant bands from the luminescence n = 1 line.

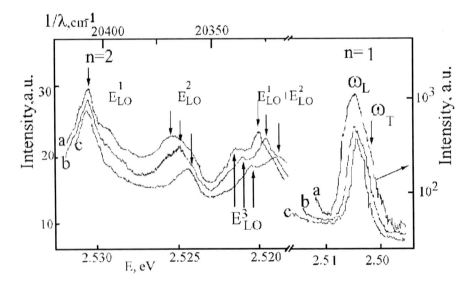

Figure 54. The photoluminescence and Raman scattering spectra of CuGaS$_2$ crystals under the excitation with the radiation from a spectral interval of a 500 W xenon lamp: (a) 487.8 nm (2.541) eV; (b) 488 nm (2.540 eV); (c) 488.3 nm (2.539 eV). Reprinted with permission from [127], N.N. Syrbu, V.E. Tezlevan, I.Galbic, L. Nemerenco, V.V. Ursaki (2009) *Exciton*. Optics Communications 282, 4562-4566. Copyright (2009), with permission from Elsevier.

The intensity of the emission associated with the ω_T (2.5011 eV) n = 1 of the Γ_4 excitons is much higher than the intensity of short-wavelength bands. The intensity of the emission at the frequency of the transversal exciton increases from 100 to 1000 arbitrary units with increasing the excitation energy from 2.539 eV to 2.541 eV. This indicates on the resonance character of the process and to the correspondence of the resonance frequency to the frequency of the continuum. In a previous work [129,131], the region of the 2LO scattering has been chosen for the investigation of the secondary emission in CdS, since the probability of this scattering does not depend upon the wave vector of the initial state. A bigger number of phonons emerge in the resonant Raman scattering spectra of $CuGaS_2$ crystals [101,125-127,140-143], since the unit cell contains a bigger number of atoms. We selected a region of 2LO scattering with the participation of low energy optical phonons (72-75 and 89-90 cm^{-1}) for the investigation of resonant Raman scattering.

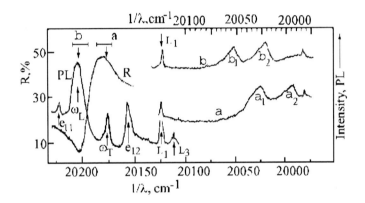

Figure 55. The emission spectra of $CuGaS_2$ crystals under the excitation by the 4765 Å Ar laser line (curve PL), the reflectivity spectrum (curve R), and the emission spectra under the excitation by monochromatic light in the region of the "bottle neck". Reprinted with permission from [127], N.N. Syrbu, V.E. Tezlevan, I.Galbic, L. Nemerenco, V.V. Ursaki (2009) *Exciton*. Optics Communications 282, 4562-4566. Copyright (2009), with permission from Elsevier.

It is known that photon-like polaritons of the lower polariton branch are created under the excitation in the region of the low exciton resonance [1-5]. The scattering of these polaritons on LO phonons represents pre-resonance Raman scattering. The intensity of this scattering is high enough near the resonance. As mentioned above, a spectral interval of the incandescence lamp has been used for the excitation.

The contours of reflectivity (R) and photoluminescence (PL) spectra of the ground state of the Γ_4 exciton under the excitation by the 4765 Å Ar laser line are shown on the left part of the Figure 55. Narrow e_{11}, e_{12}, ω_L, ω_T lines as well as the lines L_1 (20125 cm^{-1}) and L_3 (20113 cm^{-1}) are observed in the emission spectra of $CuGaS_2$ crystals. The e_{11} and e_{12} lines are related to the resonance Raman scattering. The ω_L, and ω_T lines come from the emission of exciton polaritons, while the L_1 and L_3 lines are due to the emission of bound excitons [147-150]. The L_1 line is the narrowest and the most intensive one among these lines.

The emission spectra measured under the excitation by the radiation passed through a MDR-23 monochromator are shown in the right part of Figure 55. The excitation has been performed with energies 2.501 eV (20175 cm^{-1}) (a), and 2.504 eV (20203 cm^{-1}) (b). The position of the excitation energies are indicated by the arrows "a" and "b" in the left part of the Figure. The width of the excitation band indicated by bars is around 7 cm^{-1}. The L$_1$ (20125 cm^{-1}) line is observed in these spectra. The form and the position of this line are independent on the frequency of the excitation light, i. e. it is due to the luminescence of bound excitons [147-150]. Apart from that, emission maxima labeled as "a$_1$, a$_2$" and "b$_1$, b$_2$" are observed in the emission spectra. These lines are separated from the excitation line by 2E^1 and 2B$_1^1$ LO phonon energy, and their position changes with changing the energy of the excitation photons. Therefore, one can suggest that these lines are due to resonance Raman scattering.

A further increase of the frequency of the excitation light (20280 cm^{-1}) leads to broadening of the 2E^1 and 2B$_1^1$ LO phonon bands and their mergence in one broad band. This behavior can be explained by the emergence of a maximum with the energy by 2E$^1_{LO}$ lower than the energy of the ground Γ_4 exciton state. This secondary emission can be attributed to the exciton-phonon luminescence. Therefore, a gradual transition of resonance Raman scattering line to luminescence line occurs with increasing the frequency of the excitation light in the region of photon-exciton resonance. A similar behavior has been observed and investigated in CdS crystals [1-7,144-146]. The emergence of the luminescence in CdS crystals was explained by the higher probability of scattering by acoustical phonons as compared to optical phonons for the polariton excited by the light in the region of longitudinal-transversal splitting. As a result, some stationary energy distribution of polaritons is established at the moment of LO-phonon emission, the degree of quasi-equilibrium being determined by the lifetime of polaritons. The light with the frequency higher than the frequency of the transversal exciton excites mainly states of the upper polariton branches. The luminescence is due to the scattering of these states by acoustical phonons to the states of the lower polariton branch. A similar dependence of the secondary emission spectrum upon the frequency of the excitation light is observed in the region of the 1-LO scattering line. One of the most important observations is the gradual increase of the intensity ratio of the 1LO line to the 2LO line with the transition from the Raman scattering to the luminescence [2-5].

Two maxima at 1,7198 eV (ω_T) and 1,7226 eV (ω_L) are observed in the luminescence spectra of CuGaSe$_2$ crystals measured at 10 K under the excitation by the 5145 Å line of an Ar$^+$ laser (Figure 56). The maximum at 1,7198 eV differs by 0.7 meV from the energy of the transverse exctiton ω_0 (1,7205 eV) determined from the calculation of reflectivity spectra. The maximum at 1,7226 ascribed to ω_L differs by 0.3 meV from the energy of the longitudinal exciton ω_L (1,7230 eV) determined from the calculation of reflectivity spectra. One can conclude that the spectra presented in Figure 52 are determined by the emission from the upper (ω_L) and lower (ω_T) exciton polariton branches.

Several peaks of lower intensity 1-18 (Figure 56 and 57), are observed in the long-wavelength region of the spectrum. The energy position and form of these spectra suggest that they are due to the annihilation of exciton polaritons with the emission of phonons. The peaks 1 - 4 situated in the vicinity of the ω_T band are weaker that the 6, 7, 8, 10, 13 peaks. The peaks 1, 2, and 3 are separated from the ω_T by 44,8 cm^{-1}, 62,9 cm^{-1} and 68,6 cm^{-1}, respectively. These energies agree with the energy of one-phonon and two-phonon vaibartion modes at the center of Brillouin zone deduced from the IR reflectivity spectra [106,110,151]. One of the most probable interpretations of peaks is presented in Figures 56 and 57. However, some other vibration modes could be responsible for the bands, taking into account a big variety of phonons in CuGaSe$_2$ crystals. Undoubtedly, the annihilation of exciton polaritons occurs via the increase of the population density of the lower polariton branch.

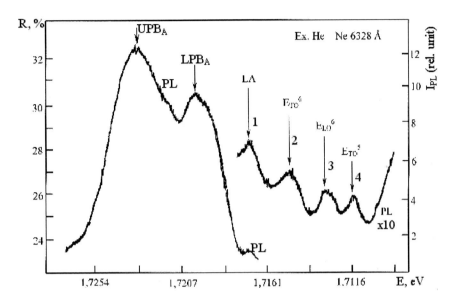

Figure 56. Luminescence spectra of CuGaSe$_2$ crystals measured at 9K. Reprinted with permission from [150], N.N. Syrbu, I.M. Tighinyanu, V.V. Ursaki, V.E. Tezlevan, V.E. Zalamai, L.Nemerenco (2005). *Polariton emission*. Physica B 365, 43-46. Copyright (2005), with permission from Elsevier.

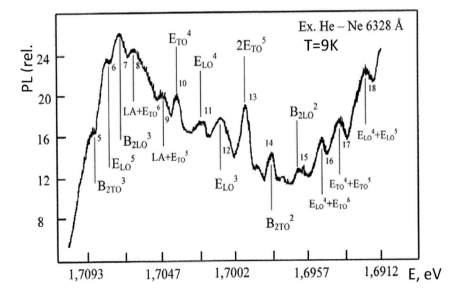

Figure 57. Luminescence spectra measured in the long-wavelength region of the spectrum with respect to the exciton polaritons. Reprinted with permission from [150], N.N. Syrbu, I.M. Tighinyanu, V.V. Ursaki, V.E. Tezlevan, V.E. Zalamai, L.Nemerenco (2005). *Polariton emission*. Physica B **365**, 43-46. Copyright (2005), with permission from Elsevier.

7. BOUND EXCITONS IN MULTINARY COMPOUNDS

As mentioned above, the energy of the free transverse exciton Γ_4 is equal to 2,5011 eV, and the energy of the long-wavelength exciton Γ_5 is 2,5001 eV [99,100,149]. The energy of the forbidden exciton Γ_3 according to Ref. [14] is equal to 2,4995 eV. These data were taken into account for identification of bands of bound excitons.

Figure 58 shows luminescence spectra of $CuGaS_2$ crystals at 9 K excited by the line 4765Å of Ar^+ laser. In the spectra three wide bands of radiation (A, B and C) at the energies 2,3982 eV, 2,3477 eV and 2,3055 eV, respectively are found. In the region of shorter wavelength other lines of radiation determined by the excitons bound on neutral donor are found (they are not discussed here). On the top of the most intense band B a narrow line of radiation F_0 (2,3964 eV) is found. The radiation bands B, C and D have the half-width being an order higher than the radiation line F_0 (Figure 58). The narrow line of radiation has the half-width of 1-2 meV and it is differently shown in different samples. From different technological lots the samples were found wherein the band B had higher intensity than that of the line F_0. In these samples the line F_0 was found as an absorption band (fragment of Figure 59).

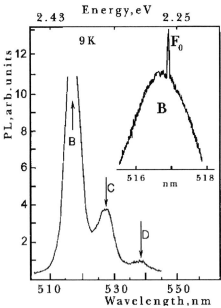

Figure 58. Luminescence spectra of $CuGaS_2$ crystals excited by the line 4765 Å of argon laser at 9 K and fragment of the top of the short-wave maximum of luminescence B. Reprinted with permission from [149], N.N.Syrbu, L.L. Nemerenco, V.N. Bejan, V.E. Tezlevan (2007). *Bound exciton*. Optics Communications 280, 387-392. Copyright (2007), with permission from Elsevier.

One can consider that the radiation bands B, C and D have a nature of origin different from the radiation line F_0. At strong luminescence in the region of the band B a part of the radiation energy is absorbed in result of electron transitions to the level of the bound exciton F_0. In the region of the radiation bands C and D narrow lines of radiation are not observed. In the absorption spectra the absorption band F_0 strictly at the energy 2,3964 eV and the absorption bands f_1(2,4013 eV), f_2 (2,4040 eV) and f_3 (2,4071 eV) are found (Figure 59).

These lines are very close and are similar to the hydrogen-like series of the bound exciton. The distance between the lines decreases ($F_0 - f_1 = 4,9$ meV, $f_1 - f_2 = 3,2$ meV and $f_2 - f_3 = 1,8$ meV) as the transition energies increase. In the region of shorter waves the absorption lines k_1 (2,4204 eV) and k_2 (2,4167 eV) are found. These lines have a nature different from the absorption lines $F_0, f_1 - f_3$, and different from the radiation bands B, C and D.

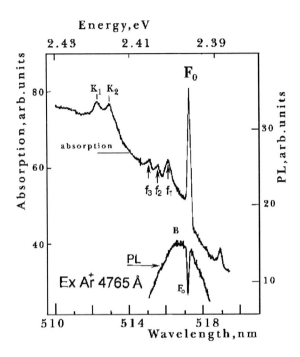

Figure 59. Absorption spectra of $CuGaS_2$ crystals at 9 K and fragment of the radiation band B for the samples with intense luminescence. Reprinted with permission from [149], N.N.Syrbu, L.L. Nemerenco, V.N. Bejan, V.E. Tezlevan (2007). *Bound exciton*. Optics Communications 280, 387-392. Copyright (2007), with permission from Elsevier.

At excitation of $CuGaS_2$ crystals by the line 5145 Å of Ar^+ laser at 9 K there are found the radiation bands B and C, which are overlapped by series of narrow bands of radiation F_0, f_1, f_2, f_3 and series of the radiation lines x_1 (2,3919 eV), x_2 (2,3868 eV), x_3 (2,3843 eV), x_4 (2,3830 eV) and x_5 (2,3821 eV) (Figure .60). In the region of longer waves there are found the radiation lines y_1 (2,3616 eV), y_2 (2,2572 eV), y_3 (2,3524 eV), y_4 (2,3447 eV), y_5 (2,3429 eV) and y_6 (2,3382 eV), which overlap the wider band of radiation C. The radiation bands $F_0, f_1 - f_3$ with the accuracy up to ±0,1 meV coincide with the synonymous absorption bands. This testifies to the fact that these lines are not lines of the Raman scattering, but they are determined by the bound exciton transitions. The transitions into the ground state of the bound exciton ($n = 1$) determine the line of radiation and absorption F_0 (2,3964 eV). From the short-wave side of the radiation line F_0 a narrow line of radiation R(2,3976 eV) is observed, which is not found in absorption and therefore it is the Raman phonon scattering B_{2LO}^1 (89 cm^{-1}). The radiation lines $x_1 - x_5$ are located from the long-wave side of the line F_0, they gather into the long-wave side. These lines are similar to the hydrogen-like series, with continuum in the region of 2,38 eV, in the absorption spectra they are not found. Taking into account the above stated we may suppose that groups of lines $F_0 - f_3$ are determined by excited states of bound excitons with continuum from the short-wave side in the region ≈2,4080 eV. The radiation lines $x_1 - x_5$ are determined by excitons bound on acceptor with continuum from the short-wave side of the series $x_1 - x_5$. It is quite possible that the levels responsible for the lines $x_1 - x_5$ are located near the valence band, and the levels $F_0 - f_3$ - below the conduction band (Figure 61). In this model optic transitions from the levels $x_1 - x_5$ to the levels $F_0 - f_3$ and continuum are not found in absorption since they are unfilled levels of the acceptor. In the radiation spectra, transitions to the acceptor levels are possible, hence they are observed in luminescence. The radiation lines $y_1 - y_6$ are possibly determined by electron transitions between the bound exciton levels ($F_0 - f_3$) and the acceptor levels ($x_1 - x_5$).

The discussed luminescence spectra are resonant, because the excitation energy (2,4094 eV) is close to the continuum energy ($E_\infty = 2,4080$ eV). In Ref. [141] the radiation bands 2,397 eV (a_0), 2,404 eV (δ_0) and 2,417 eV are found. These radiation lines practically coincide with the lines of absorption and radiation found in our samples F_0 (2,3964), f_2 (2,4040 eV) and 2,4167 eV (k_2). The lines of radiation and absorption are found in samples of different technological lots, obtained in different laboratories. Therefore, we suppose that the center whereon the exciton is bound is determined by intrinsic defects. In our samples at excitation by short-wave lines (4880 Å) of Ar^+ laser, the radiation lines k_1 and k_2 are found, being observed in absorption. This testifies to the fact that these lines do not belong to the series $F_0 - f_3$.

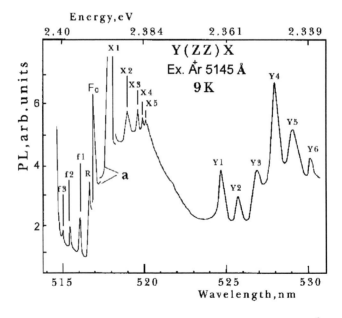

Figure 60. Luminescence spectra of $CuGaS_2$ crystals at 9 K excited by line 5145 Å of argon laser. Reprinted with permission from [149], N.N.Syrbu, L.L. Nemerenco, V.N. Bejan, V.E. Tezlevan (2007). *Bound exciton*. Optics Communications 280, 387-392. Copyright (2007), with permission from Elsevier.

Exciton Polariton Dispersion in Multinary Compounds 85

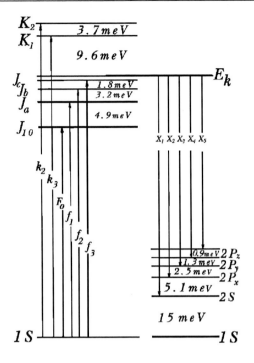

Figure 61. Energy diagram of the levels of the exciton bound on neutral acceptor. Reprinted with permission from [149], N.N.Syrbu, L.L. Nemerenco, V.N. Bejan, V.E. Tezlevan (2007). *Bound exciton in CuGaS$_2$*. Optics Communications 280, 387-392. Copyright (2007), with permission from Elsevier.

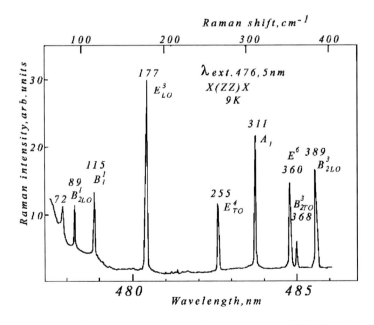

Figure 62. Raman scattering of CuGaS$_2$ crystals at 9 K in the geometry X(ZZ)X excited by the line 4765 Å of argon laser. Reprinted with permission from [149], N.N.Syrbu, L.L. Nemerenco, V.N. Bejan, V.E. Tezlevan (2007). *Bound exciton*. Optics Communications 280, 387-392. Copyright (2007), with permission from Elsevier.

Table 5. Raman scattering lines at 9 and 300 K

Symmetry of modes	Results of work [20]	Results of work [21]	Results of work [22]	Results of work [19]	Results of work [149]. Excitation by Ar⁺ laser		
					$\lambda =$ 5145Å T = 300 K	$\lambda =$ 4965Å T = 9 K	$\lambda =$ 5145Å T = 9 K
E^1_{TO}	75			75	75.0	74.6	75.0
E^1_{LO}	76			75		84.7	75.0
E^2_{TO}	95			147	137.0	137.1	137.0
E^2_{LO}	98			147	147.0		141.2
E^3_{TO}	147	156	165	167	167.0	167.8	
E^3_{LO}	167	160		167			182.3
E^4_{TO}	260	262	258	273		253.0	
E^4_{LO}	278	276		283	260.0	262.0	
E^5_{TO}	335	332	332	332	327.0	328.0	
E^5_{LO}	352	352	349	347	348.0	348.0	
E^6_{TO}	365	363	364	365	362.0		
E^6_{LO}	387	384	366	367		370.0	372.7
B^1_{2TO}	259	95	94	95			
B^1_{2LO}	284	95		95	95.0	95.0	95.2
B^2_{2TO}	339	262	277	286	277.0	277.0	
B^2_{2LO}	369	281		286	294.0	298.0	
B^3_{2TO}	371	368	349	367		366.0	
B^3_{2LO}	402	401	366	393	385.0	386.0	383.2
B^1_1	138	97	97	116	112.0	113.7	
B^2_1	203	203		238		197.0	
B^3_1	243	358	358	401	394.0		
A_1	312	312	312	312	312.0	312.0	

Figure 62 shows the Raman scattering spectra of the same crystals at 9 K and excitation by the line 4765 Å of Ar⁺ laser. The Raman scattering is registered in the geometry $x(zz)\overline{x}$ (back scattering). In this case the excitation energy got into the region of continuum of the long-wave exciton Γ_4 and Γ_5. In the scattering spectra the oscillation modes E^1_{TO} (72 cm⁻¹),

B^1_{2LO} (89 cm^{-1}), E^3_{LO} (177 cm^{-1}), E^4_{TO} (255 cm^{-1}), A_1 (311 cm^{-1}), E^6_{LO} (360 cm^{-1}), B^3_{2TO} (368 cm^{-1}) and B^3_{2LO} (389 cm^{-1}) are found. The Raman scattering at room temperature and at the geometries x(xy)z, x(xz) y is shown in Figure 63. In the spectra the intense mode A_1 is distinguished, practically all the rest modes are weaker. The obtained oscillation modes at 300 K, 9 K are comparable with the literature data [135-137,141,152], they are given in Table 5. These results show that the crystal quality was good, and the oscillation modes correspond to the ones known in literature.

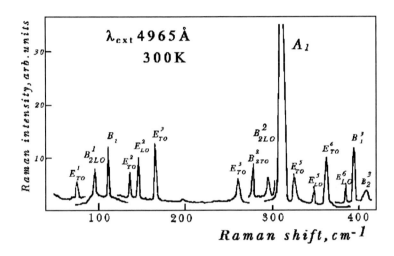

Figure 63. Raman scattering of CuGaS$_2$ crystals at 300 K excited by the line 4965 Å of argon laser. Reprinted with permission from [149], N.N.Syrbu, L.L. Nemerenco, V.N. Bejan, V.E. Tezlevan (2007). *Bound exciton*. Optics Communications 280, 387-392. Copyright (2007), with permission from Elsevier.

It should be noted that in the Raman scattering at 9 K (Figure 62) the intensity of the oscillation modes differs from the intensity of the scattering lines registered at 300 K (Figure 63). At 300 K the mode A_1 (312 cm^{-1}) is the most intense, and at 9 K - E^3_{LO} (177 cm^{-1}). At the same time we can state that at the temperature 9 K we have not found any change in the structure of spectra of oscillation modes testifying to phase transitions.

7.1. Resonance Impurity Emission in CuGaS$_2$ Crystals

Figure 64 shows the radiation spectra of CuGaS$_2$ crystals at 9 K in the geometry X(ZZ)X and excitation by line 4965 Å of Ar$^+$ laser [147]. The excitation energy 2.4968 eV is less than the energy of the transverse long-wave excitons Γ_4, Γ_5 and Γ_3. Under these conditions of excitation, as a rule, Raman dispersion and not luminescence is observed, since the excitation energy gets into the region of the crystal transparency. In the radiation spectra (Figure 64) intense narrow line L$_1$ (2.4948 eV), line L$_2$ (2.4938 eV), L$_3$ (2.4933 eV) and L$_4$ (2.4934 eV) are observed. The intense radiation line L1 is 2.1 meV apart the laser excitation line. The radiation line L$_4$ is 3.8 meV apart the laser radiation line. Therefore, the whole group of lines

L_1-L_4 cannot be explained by the Raman scattering. Numerous studies of the Raman scattering in these crystals show that in the spectra the intense mode A_1 at 312 cm^{-1} is observed, and other oscillation modes have weak intensity [135, 136]. Minimal energy of the optic phonon in these crystals is equal to 8 meV.

Hence, the lines found lines are the luminescence spectra. Excitation of the luminescence spectra is possible due to the excitation energy being very close to the optic transition energy. The found radiation spectra are caused by excitons bound on neutral donor (acceptor). In work [141] the lines 2.495 eV (I_b), 2.493 eV (I_c) are found, being apparently analogues of lines L_1 and L_2 shown in Figure 64. Bands C_1 (2.4584 eV) and C_2 (2.4561 eV) presented in Figure 64 may be analogues of bands H (2.457 eV), H*(2.460 eV) revealed by the authors of work [141]. Band I_a (2.501 eV) in work [141] is attributed to bound excitons. According to the above given data and the results reported in works [99-101] the band at the energy 2.501 eV is determined by free excitons, namely by the transverse mode of exciton polariton Γ_4. The authors of [141,142] attribute short-wave radiation lines (2.495-2.493 eV) to excitons bound on neutral donor (acceptor). Studies of splitting of the lines 2.4939 eV and 2.4920 eV in the magnetic field carried out in work [142] have confirmed this interpretation.

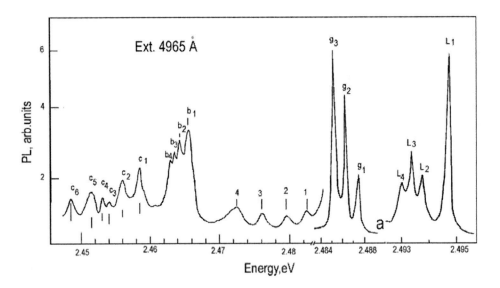

Figure 64. Luminescence spectra of CuGaS2 crystals at 10 K and at excitation by 4965 Å line of Ar+ laser. Reprinted with permission from [147], N.N. Syrbu, L.L. Nemerenco, V.E. Tezlevan (2007). Resonance impurity radiation. Optical Materials 30, 451-456. Copyright (2007), with permission from Elsevier.

In the samples investigated by us in the region of longer waves from lines L_1-L_4 a narrow group of lines g_1(2.4875 eV), g_2(2.4861 eV), g_3(2.4851 eV) and group b_1(2.4655 eV), b_2(2.4643 eV), b_3(2.4636 eV), b_4(2.4631 eV) are observed (Figure 64). Between these groups of lines wider radiation lines (1-4) are observed, having apparently another nature, i.e. they are not caused by bound excitons of the center under discussion. In the long-wave region from band b_4 narrower radiation lines C_1(2.4584 eV), C_2(2.4561 eV), C_3(2.4542 eV), C_4(2.4528 eV) and wider bands at the energy of 2.4514 eV(C_5) and 2.4485(C_6) eV are found.

Figure 65 shows luminescence spectra of CuGaS$_2$ crystals excited by line 4880 Å of Ar$^+$ laser. In the region of energies exceeding n=1 of free exciton Γ$_4$ resonance Raman scattering of 1LO, 2LO exciton polaritons is found. These spectra have been discussed by us in work [99-101], so they are not considered in the present paper. Figure 65 shows radiation spectra in the energy range E < E$_n$ =1 (of free exciton 2.5001 eV). In the radiation spectra intense lines at 2.49478 eV (L$_1$), 2.4938 eV (L$_2$), 2.4933 eV (L$_3$), 2.4930 eV (L$_4$) 2.4875 eV(g$_1$) and 2.4861 eV (g$_2$) are found. Dashed line indicates position of line g$_3$, which is found at excitation by line 496,5 nm and is shown in Figure 64. In the long-wave region lines being weaker in intensity in the interval 2.4779 - 2.4318 eV are observed (Figure 65). The radiation lines L$_1$-L$_4$, g$_1$, g$_2$ shown in Figure 65 coincide in energy with the energy position of short-wave lines presented in Figure 64. This confirms the fact that they are caused by luminescence and they are not Raman scattering lines, except for line g$_3$. Line g$_3$ is 94 cm^{-1} apart the frequency of laser line (4965 Å). This value coincides with the frequency (94 cm^{-1}) of the symmetry phonon B$_2$ [135-137].

Thus, in CuGaS$_2$ crystals at excitation by lines 4880 and 4965 Å luminescence from the bound exciton levels is observed. The excitons are bound on neutral donor (acceptor) [141,142]. Analogous spectra are investigated in detail in CdS crystals [153]. In the radiation spectra transitions from the bound excitons levels to the ground and excited states of donor (acceptor) are possible.

It is known that the hydrogen-like donor (acceptor) binding energy may be estimated from the ratio

$$\frac{G}{E_D} = \left(1 + \frac{m_e^*}{m_h^*}\right)^{-1},$$

where G is the exciton binding energy, E_D is the donor (acceptor) binding energy, m_e^*, m_h^* is the effective mass of electrons and holes, correspondingly [8]. In CdS the ratio $m_h^*/m_e^* >> 4$. Hapfeld has shown that excitons may be bound on donor (acceptor), when the condition $m_h^*/m_e^* > 1,4$ is realized. For CuGaS$_2$ crystals the effective mass of holes m_h^* is equal to 1.87m$_0$, and the effective mass of electrons m_e^* is equal to 0.12 m$_0$ [99-101,125,126]. The binding energy of free exciton Γ$_4$ is equal to 39.2 meV, that of exciton Γ$_5$ - 39,4 meV. Therefore, the ratio of the effective masses of electrons and holes in CuGaS$_2$ crystals is equal to 15,6. Estimating by the above given dependence the binding energy of donor (acceptor) E$_D$(E$_A$) in CuGaS$_2$ crystals, taking into account parameters of excitons Γ$_4$ we will obtain that it may achieve (41-42) meV. Proceeding from these data and the luminescence spectra (Figure 64, 65) one can suppose that excitons bound on neutral donor (acceptor) have the energy diagram of the levels shown in Figure 66. The short-wave radiation lines L$_1$-L$_4$ are determined by transitions from the bound exciton level to the donor (acceptor) ground state level. The second group of narrow lines b$_1$-b$_4$ is caused by transitions from the bound exciton levels to the donor (acceptor) excited levels.

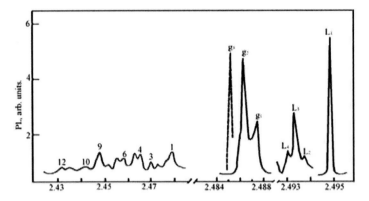

Figure 65. Luminescence spectra of CuGaS$_2$ crystals at 10 K and at excitation by 488 nm line of Ar+ laser. Reprinted with permission from [147], N.N. Syrbu, L.L. Nemerenco, V.E. Tezlevan (2007). Resonance impurity radiation. Optical Materials 30, 451-456. Copyright (2007), with permission from Elsevier.

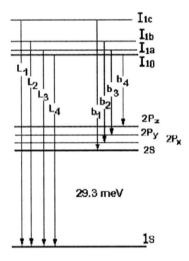

Figure 66. Energy levels and electron transitions of exciton bound on neutral donor (acceptor). Reprinted with permission from [147], N.N. Syrbu, L.L. Nemerenco, V.E. Tezlevan (2007). Resonance impurity radiation. Optical Materials 30, 451-456. Copyright (2007), with permission from Elsevier.

Thus, if the radiation line L$_1$ is caused by transitions I$_{1c}$ - 1S, line L$_2$ – by transitions I$_{1b}$ - 1S, line L$_3$ – by transitions I$_{1a}$ -1S and line L$_4$ – by transitions I$_{1o}$ - 1S, then the energy intervals I1o - I1a, I1a - I1b and I1b - I1c are equal to 1.0, 0.5 and 0.3 meV, correspondingly. The radiation maxima b$_1$(2.4655 eV), b$_2$(2.4643 eV), b$_3$(2.4636 eV) and b$_4$(2.4631 eV) are apparently caused by transitions from the bound exciton levels to the donor (acceptor) excited levels. We consider that the long-wave radiation line (b$_4$) is connected with transitions from I1o to the level 2P$_z$. The short-wave radiation line b$_1$ is caused by transitions I1c to the levels 2S. The radiation line b$_2$ is apart the line b$_1$ by the energy distance 1.2 meV. If we consider that b$_2$ is caused by transitions I1$_b$ -2P$_x$, and the radiation line b$_3$ is caused by transitions I1a - 2P$_y$, we will obtain that the distance between the excited levels of donor (acceptor) is equal to

0.2 meV (Figure 66). In this case the energy distance between the levels 1S - 2S is equal to 29.3 meV.

The theory of the energy levels of an exciton in a slightly anisotropic crystals was worked out by Hopfield and Thomas [3] and by Wheeler and Dimmock [154]. These theories apply equally well to a donor in semiconductors.

The effective mass Hamiltonian for the donor electron at the zone center in a uniaxial crystal, with the c axis along the z direction, is given by [151]

$$H = \frac{[\vec{p}_\perp(e/c)\vec{A}_\perp]^2}{2m_\perp m_e} + \frac{[p_z(e/c)A_z]^2}{2m_\parallel m_e} - \frac{e^2}{\varepsilon_0(x^2+y^2+z^2\varepsilon_\perp/\varepsilon_\parallel)} + g_\perp \beta \vec{H}_\perp \cdot \vec{S}_\perp + g_\parallel \beta H_z S_z \quad (21)$$

where

$$\vec{A} = \frac{1}{2}(\vec{H} \times \vec{r}), \quad \beta = e\hbar/2m_e c, \quad \varepsilon_0(\varepsilon_\perp \varepsilon_\parallel) \quad (22)$$

Wheeler and Dimmock [154] make the coordinate transformation

$$(x, y, z) = [x', y', z'(m_\perp/m_\parallel)^{1/2}] \quad (23)$$

In this coordinate system the nonmagnetic part of the Hamiltonian simplifies to

$$H = \frac{\vec{p}'^2}{2m_e m_\perp} - \frac{e^2}{\varepsilon_0 r'} + H_\alpha \quad (24)$$

where

$$H_\alpha = -(e^2/\varepsilon_0)\left\{[x'^2 + y'^2 + z'^2(1-\alpha)]^{-1/2} - r'^{-1}\right\},$$

$$\alpha = 1 - \varepsilon_\perp m_\perp / \varepsilon_\parallel m_\parallel \quad (25)$$

H_α is a small perturbation, when the anisotropy factor α is small. For $\alpha = 0$, the donor energy levels in zero magnetic field are hydrogenic. The donor binding energy for $\alpha = 0$ is given by

$$E_0 = E_H m_\perp / \varepsilon_{0\perp} \varepsilon_{0\parallel} \quad (26)$$

where E_H is the hydrogen atom binding energy.

Using first-order perturbation theory, Wheeler and Dimmock calculate the donor energy levels, for a magnetic field in the z direction, to be

$$E_{1s} = -E_0\left(1+\frac{1}{3}\alpha+\frac{1}{20}\alpha^2\right)+\sigma H_z^2 \pm \frac{1}{2}g_{\text{II}}\beta H_z$$

$$E_{2s} = -\frac{1}{4}E_0\left(1+\frac{1}{3}\alpha+\frac{1}{20}\alpha^2\right)+14\sigma H_z^2 \pm \frac{1}{2}g_{\text{II}}\beta H_z$$

$$E_{2p_z} = -\frac{1}{4}E_0\left(1+\frac{3}{5}\alpha+\frac{9}{28}\alpha^2\right)+6\sigma_z^2 \pm \frac{1}{2}g_{\text{II}}\beta H_z$$

$$E_{2p_x} = E_{2p_y} = -\frac{1}{4}E_0\left(1+\frac{1}{5}\alpha+\frac{9}{140}\alpha^2\right)+12\sigma H_z^2 \pm \frac{1}{2}g_{\text{II}}\beta H_z \pm \beta H/m_\perp \quad (27)$$

where σ, the constant determining the diamagnetic shift, is given by

$$\sigma = \beta^2/2m_\perp^2 E_0 \quad (28)$$

One can estimate the binding energy of the donor acceptor by adding the average binding energy of the $2p$ states to our measured average energy separation between $2p$ and $1s$ states. According to Eqs. (26) and (27), the average binding energy of the $2p$ states is

$$E_B(2p) = \frac{1}{4}\left(E_H m_\perp / \varepsilon_{0\text{II}} \varepsilon_{0\perp}\right)\left(1+\frac{1}{3}\alpha+\frac{3}{20}\alpha^2\right) \quad (29)$$

The average energy separation of the $2p$ and $1s$ levels is 29.30±0.02 meV (Figure 5). To evaluate Eq. (29) we can use our measured values of $m_\perp$ and α.

The determined value of the electron effective mass is equal to $0.14m_0$, and the dielectric constants $\varepsilon_{0\perp}$ and $\varepsilon_{0\text{II}}$ equal to 26 and 27, respectively, for polarization E⊥c and E∥c. Considering the anisotropy of the electron effective mass being absent, one obtains the anisotropy factor.

$$\alpha = 1 - \frac{\varepsilon_\perp \cdot m_\perp}{\varepsilon_{II} \cdot m_{II}} \approx 0.14$$

Thus, using the obtained value of α and the above given parameters we have determined the state binding energy $E_D(2p) = 11.9\,\text{meV}$ and the donor (acceptor) binding energy equal to 41.2 meV (29.3+11.9).

In the luminescence spectra one more group of intense lines gl-g4 is observed (Figure 64, 66). Apparently, these lines are also determined by excitons bound on donor or acceptor, but this is another center. If this supposition is correct, these lines take place between the bound exciton levels on the donor ground state.

8. WAVELENGTH MODULATED EXCITON REFLECTIVITY SPECTRA IN CuAlSe$_2$ CRYSTALS

The relative increment of the reflection coefficient $\Delta R/R$ is measured in the wavelength modulated spectra. At the same time, the expressions for the $\Delta\varepsilon_1$ and $\Delta\varepsilon_2$ changes of the real (ε_1) and imaginary (ε_2) components of the dielectric function ε are obtained in the theoretical considerations. The expression of the $\Delta R/R$ as a function of $\Delta\varepsilon_1$ and $\Delta\varepsilon_2$ can be written as [155]:

$$\Delta R/R = \alpha(\varepsilon_1, \varepsilon_2)\Delta\varepsilon_1 + \beta(\varepsilon_1, \varepsilon_2)\Delta\varepsilon_2, \qquad (30)$$

where
$$\alpha = C_1[(\varepsilon_1 - 1)A_+ + \varepsilon_2 A_-];$$

$$\beta = C_2[(\varepsilon_1 - 1)A_+^{-1} - \varepsilon_2 A_-^{-1}];$$

$$A_\pm = \pm \frac{\sqrt{2}\left[(\varepsilon_1^2 + \varepsilon_2^2)^{1/2} \pm \varepsilon_1\right]^{1/2}}{(\varepsilon_1^2 + \varepsilon_2^2)^{1/2}}, \quad C_1 = [(\varepsilon_1 - 1)^2 + \varepsilon_2^2]^{-1}, \qquad (31)$$

$$C_2 = 2\varepsilon_2/[(\varepsilon_1 - 1)^2 + \varepsilon_2^2](\varepsilon_1^2 + \varepsilon_2^2)$$

As one can see, the $\Delta R/R$ depends on both $\Delta\varepsilon_1$ and $\Delta\varepsilon_2$. One of the α or β coefficients is much bigger than another one in the region of the fundamental absorption edge. The contours of the wavelength modulated reflectivity spectra for the n = 1 line of the A, B. and C excitons have been calculated taking into account the spatial dispersion and the presence of the dead layer. The calculated spectra better coincide with the experimental ones when the dielectric constant has a small gradient. This gradient is determined by ε_b^{max} and ε_b^{min}. The minimum value of the background dielectric constant ε_b^{min} corresponds to the energy at which an input emerges to the exciton resonance in the reflectivity spectra, while the maximum value ε_b^{max} corresponds to the energy at which the s-state of the exciton does not influence anymore the reflectivity spectra (there is no input to the ε). The gradient $\Delta\varepsilon$ is smaller than 0.15 for any of the exciton series. This gradient is due to the fact that, in the frequency interval corresponding to the transitions in the s-state of the A excitons, there is an input to the dielectric constant from the B-exciton series, which increases insignificantly with the increase of the frequency of the light.

The wavelength modulated reflectivity spectra of CuAlSe$_2$ and CuGaSe$_2$ crystals are presented in Figure 67 and Figure 68, respectively. The n = 1 and n = 2 lines of the A, B, and C extcitons are revealed in the spectra for both the compositions. Apart from that, the n = 3 line of the A exciton is observed in the spectrum of the CuAlSe$_2$ crystals. The maxima of the

A exciton series (Γ_4 excitons) in the CuGaSe$_2$ crystals are observed at 1.740 eV (n = 1) and 1.766 eV (n = 2), while the B exciton series (Γ_5 excitons) is observed at 1.822 eV (n = 1) and 1.848 eV (n = 2). Additionally, the s-state of the C excitons is observed at 2.030 eV. As expected, the maxima of the first derivative of the reflectivity spectra for the exciton states correspond to an intermediate energy between the energy of the longitudinal and transversal exciton.

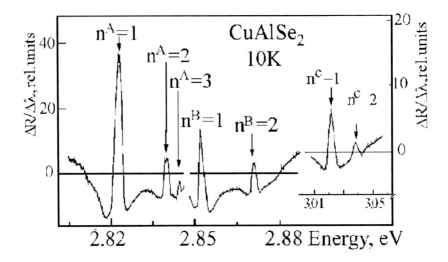

Figure 67. Wavelength modulated reflectivity spectra of CuAlSe$_2$ crystals measured at T = 10K. Reprinted with permission from [171], N.N. Syrbu, A.V. Dorogan, V.V. Ursaki, A. Masnik (2010) Wavelength modulated optical reflectivity spectra of CuAl$_{1-x}$Ga$_x$Se$_2$ crystals. J. Opt. accepted. Copyright (2010), with permission from Institute of Physics.

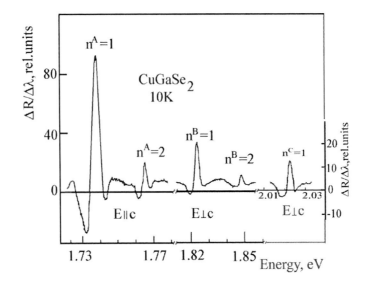

Figure 68. Wavelength modulated reflectivity spectra of CuGaSe$_2$ crystals measured at T = 10K. Reprinted with permission from [171], N.N. Syrbu, A.V. Dorogan, V.V. Ursaki, A. Masnik (2010) Wavelength modulated optical reflectivity spectra of CuAl$_{1-x}$Ga$_x$Se$_2$ crystals. J. Opt. accepted. Copyright (2010), with permission from Institute of Physics.

Table 6. Parameters of exciton spectra in CuAl$_X$Ga$_{1-X}$Se$_2$ crystals

CuAl$_X$Ga$_{1-X}$Se$_2$ composition	Exciton states	A –exciton energy, eV	B-exciton energy, eV	C-exciton energy, eV
X=0.0 (CuGaSe$_2$) According to [17] and [171]	n=1	1.7380	1.8235	2.022
	n=2	1.7650	1.8483	
	R	0.036	0.0357	
	E$_g$	1.7745	1.8592	
X=0.2	n=1	1.920	1.980	2.200
	n=2	1.943	2.002	2.218
	R	0.030	0.029	0.024
	E$_g$	1.950	2.009	2.224
X=0.3	n=1	1.980	2.08	2.280
	n=2	2.000	-	-
	R	0.027	-	-
	E$_g$	2.007	-	-
X=0.4	n=1	2.080	2.190	2.400
	n=2	2.102	2.211	2.418
	R	0.029	0.028	0.024
	E$_g$	2.209	2.218	2.424
X=0.5	n=1	2.180	2.280	2.500
	n=2	2.202	2.299	2.518
	R	0.029	0.027	0.024
	E$_g$	2.209	2.307	2.524
X=0.7	n=1	2.480	2.522	2.680
	n=2	2.500	2.540	2.697
	R	0.026	0.024	0.023
	E$_g$	2.506	2.546	2.703
X=0.8	n=1	2.483	2.575	2.770
	n=2	2.501	2.592	2.786
	R	0.024	0.023	0.021
	E$_g$	2.507	2.598	2.791
X=1.0 CuAlSe$_2$	n=1	2.8212	2.851	3.023
	n=2	2.8390	2.868	3.039
	n=3	2.8442	-	-
	R	0.024	0.022	0.018
	E$_g$	2.8450	2.873	3.041

Note that the wavelength modulated reflectivity spectra allow one a better determination of the ground and excited exciton states (Figure 69). At the same time, such parameters as the exciton mass, the frequency of the longitudinal (ω_L) and transversal (ω_T) excitons are determined with bigger errors as compared to the reflectivity spectra. Therefore, we determined these parameters from the reflectivity spectra, not from the wavelength modulated ones. The structure of the wavelength modulated spectra was used for the determination of the Rydberg constant and the parameters of the energy bands. The structure of exciton lines is situated in the energy interval from 1.74 eV to 2.83 eV for all the compositions of the investigated solid solutions. The n = 1 and n = 2 lines of the A, B, and C excitons are observed in the wavelength modulated spectra of all the solid solutions. The main exciton parameters and the energy gaps have been determined from the energy position of the n = 1 and n = 2 exciton lines (Table 6 and Table 7).

The position of the n = 1 line demonstrates a non-liner dependence on X in the CuAl$_{1-x}$Ga$_x$Se$_2$ solutions (Figure 70). The exciton binding energy (Rydberg constant) decreases linearly with the transition from CuGaSe$_2$ to CuAlSe$_2$ (Table 1, Figure 5). A weak deviation from linearity is observed for the energy gap, i. e. the energy interval $\Gamma_7(V_1) - \Gamma_6(C_1)$ (Figure 70).

The dependence of the energy intervals for the A, B, and C excitons upon the composition is described by the polynomial

$$E_g(x) = E_g^1 X + E_g^2 (1 - X) - CX(X - 1) \qquad (32)$$

where E_g^1 and E_g^2 are the bandgaps of the CuGaSe$_2$ and CuAlSe$_2$ compounds, respectively, and C is the coefficient of non-linearity. The deviation from linearity is weak (C equals 0.07 – 0.1) for the investigated solid solutions. A similar deviation from linearity is inherent to most of the I-III-VI$_2$ solid solutions. The C parameter equals 0.39 and 0.31 for the CuGa(S$_x$Se$_{1-x}$)$_2$ and CuIn$_x$Ga$_{1-x}$S$_2$ compounds, while it is around 0.29 - 0.34 for CuAl$_x$Ga$_{1-x}$S$_2$ solutions. Therefore, the value of the C parameter is a little lower for the CuAl$_{1-x}$Ga$_x$Se$_2$ solid solutions.

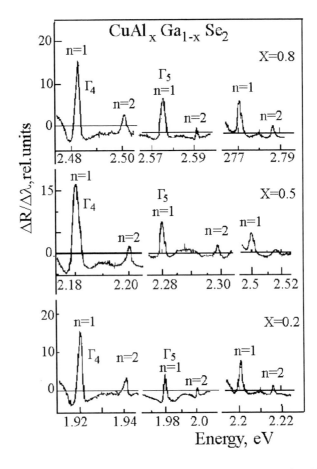

Figure 69. Wavelength modulated reflectivity spectra of *CuAl$_{1-x}$Ga$_x$Se$_2$* crystals with *X = 0.8*, X = *0.5*, and *X = 0.2* measured at T = 10K. Reprinted with permission from [171], N.N. Syrbu, A.V. Dorogan, V.V. Ursaki, A. Masnik (2010) Wavelength modulated optical reflectivity spectra of CuAl$_{1-x}$Ga$_x$Se$_2$ crystals. J. Opt. accepted. Copyright (2010), with permission from Institute of Physics.

Table 7. Dielectric constants, exciton parameters and bandgaps in $CuAl_{1-x}Ga_xSe_2$ crystals

X	1.0 [46]	0.8 [171]	0.5 [171]	0.2 [171]	0.0 [46]
$\varepsilon^{\parallel}/\varepsilon^{\perp}$	4.2/5.1	4.9/5.5	5.8/6.0	6.3/6.9	6.67/8.28
ε_b	7.0	7.2	7.3	7.4	7.4-8.2
μ^A	0.13 m_0	0.12	0.13	0.11	0.11 m_0
μ^B	0.13 m_0	0.12	0.11	0.11	0.099 m_0
μ^C		0.097	0.09	0.08	0.07 m_0
M_{exc}^A	1.4 m_0	1.3 m_0	1.2 m_0	1.1 m_0	(1.0-1.3) m_0
m_C	0.14 m_0	0.13 m_0	0.12 m_0	0.12 m_0	0.11 m_0
m_{V1}	1.26 m_0	1.24 m_0	1.23 m_0	1.20 m_0	1.20 m_0
m_{V2}	1.26 m_0				(0.4-0.7) m_0
m_{V3}					0.25 m_0

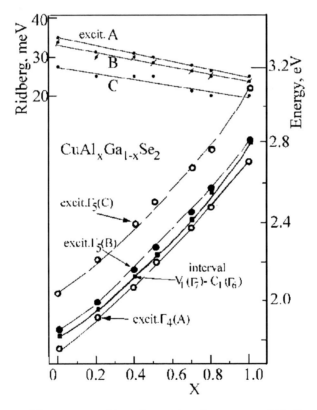

Figure 70. Dependence of the energy position of ground exciton states and binding energy of $\Gamma_4(A)$, $\Gamma_5(B)$ and $\Gamma_5(C)$ excitons upon the composition of solid solutions (X). Reprinted with permission from [171], N.N. Syrbu, A.V. Dorogan, V.V. Ursaki, A. Masnik (2010) Wavelength modulated optical reflectivity spectra of $CuAl_{1-x}Ga_xSe_2$ crystals. J. Opt. accepted. Copyright (2010), with permission from Institute of Physics.

8.1. Calculation of the Contour of Wavelength Modulated Reflectivity Spectra in the Case of Weak Polariton Effect

According to the theory of Hopfield and Toyozawa [3,156], the absorption coefficient in the n exciton state at ω frequency is proportional to the asymmetrical Lorentz coefficient

$$\frac{\Gamma_n + 2A_n[\hbar\omega - (E_{ex,n} + \Delta_n)]}{[\hbar\omega - (E_{ex,n} + \Delta_n)]^2 + \Gamma_m^2} \qquad (33)$$

where Γ_n is the half width at full maximum resulting from the broadening of the n level due to the scattering on phonons. Δ_n is the input to the exciton energy $E_{ex,n}$ from the lattice vibrations. A_n is asymmetry coefficient which differs from zero due to the weak dependence of the exciton density of states upon energy near the absorption peak.

The Γ_n and A_n parameters are considered to be independent on the $\hbar\omega$ photon energy in the interval $|\hbar\omega - E_{ex,n}| \leq \Gamma_n$ in the case of a weak exciton-phonon interaction. In the case of a strong exciton-phonon interaction, i. e. a high value of Γ_n, the contour of the line described by the expression (32) will change due to the dependence of Γ and A values upon ω in the interval $|\hbar\omega - E_{ex,n}| \approx \Gamma_n$, and will not demonstrate anymore an asymmetrical Lorentzian shape. The shape of the each line will be Gaussian in the limit of strong interaction.

Hopfield suggested that the most probable mechanism determining the line shape is the scattering on optical phonons [3]. Toyozawa performed a more thorough analysis taking into account the transitions between different exciton states due to the exciton-phonon interaction (the consideration of the inter-band scattering) [156]. It was shown that the exciton absorption spectrum can be considered as a superposition of individual asymmetrical Lorentzian (32) absorption bands (the additive rule). The asymmetry coefficient is basically determined by the inter-band scattering [156]. By taking the derivative from the real and imaginary components of the dielectric function ε one can obtain the following expressions for the ε_1 and ε_2 derivatives:

$$\frac{\partial \varepsilon_1}{\partial \omega} = \frac{(\hbar\omega - E_{ex})^2 - \Gamma^2 - 4A\Gamma(\hbar\omega - E_{ex})}{[(\hbar\omega - E_{ex})^2 + \Gamma^2]^2} \qquad (34)$$

$$\frac{\partial \varepsilon_2}{\partial \omega} = \frac{2A[\Gamma^2 - (\hbar\omega - E_{ex})^2] - 2\Gamma(\hbar\omega - E_{ex})}{[(\hbar\omega - E_{ex})^2 + \Gamma^2]^2} \qquad (35)$$

The main attention has been paid to the fitting of the central part of the excitonic band with the A and Γ parameters, taking into account the determining role of the LO phonons in the formation of this part of the excitonic band [156]. The reflectivity spectra change from minimum to the maximum in this region of the spectrum. The A and Γ parameters are determined from the position of zeros and minima in the spectrum. One should mention that the theoretical curves are strongly influenced even by small changes of these parameters, which is evidenced by a comparison with the experimental curves.

The interaction of excitons with lattice vibrations leads to the broadening and the shift of exciton lines. It was shown [156] that the shape of the exciton line is Lorentzian in the case of a weak exciton-phonon interaction and a low value of the effective exciton mass, while it is Gaussian in the case of a strong exciton-phonon interaction and a high value of the effective exciton mass. For a weak or intermediate exciton-phonon interaction the exciton dispersion is expressed as

$$\varepsilon(\omega) \cong \frac{2A-1}{(\hbar\omega - E_{ex}) - i\Gamma} \quad (36)$$

where Γ is the half width at full maximum resulting from the broadening of the exciton level due to the scattering on phonons, E_{exc} is the exciton transition energy, and A is asymmetry coefficient which differs from zero due to a weak dependence of the exciton density of states upon energy near the absorption peak.

A comparison of the experimental and the calculated spectra of wavelength modulated reflectivity for the n = 1 line of Γ_4(A) and Γ_5(B) excitons in CuGaSe$_2$ and CuAlSe$_2$ crystals is presented in Figure 71. The best fit is produced in the case of a classical Lorentz oscillator and the absence of spatial dispersion. The analysis of the reflectivity spectra in the considered crystals suggested that a weak polariton interaction is inherent to all the solid solutions, i. e. the damping parameter is bigger or equal to longitudinal-transversal exciton splitting. One comes to the same conclusion from the analysis of reflectivity spectra according to the dispersion relations taking into account the spatial dispersion [148].

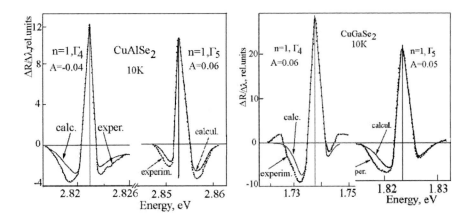

Figure 71. Experimental and calculated wavelength modulated reflectivity contours for the n = 1 state of Γ_4 (A) and Γ_5 (B) excitons in CuAlSe$_2$ and CuGaSe$_2$ crystals. Reprinted with permission from [171], N.N. Syrbu, A.V. Dorogan, V.V. Ursaki, A. Masnik (2010) Wavelength modulated optical reflectivity spectra of CuAl$_{1-x}$Ga$_x$Se$_2$ crystals. J. Opt. accepted. Copyright (2010), with permission from Institute of Physics.

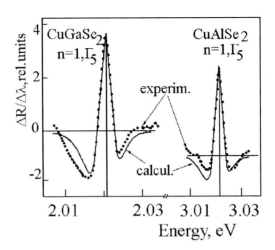

Figure 72. Experimental and calculated wavelength modulated reflectivity contours for the n = 1 state of $\Gamma_5(C)$ excitons in CuAlSe$_2$ and CuGaSe$_2$ crystals. Reprinted with permission from [171], N.N. Syrbu, A.V. Dorogan, V.V. Ursaki, A. Masnik (2010) Wavelength modulated optical reflectivity spectra of CuAl$_{1-x}$Ga$_x$Se$_2$ crystals. J. Opt. accepted. Copyright (2010), with permission from Institute of Physics.

As one can see, the calculated spectra fit very well the experimental ones in the central part of the curves, while there is some different at the wings. According to the Toyozawa theory, two types of oscillations contribute to the exciton band: one of which (LA or LO phonons) contributes to the central part and another one (TA phonons) contributes to the formation of wings [156]. Therefore, one can consider that the central part and the wings of the $\Gamma_4(A)$, $\Gamma_5(B)$, and $\Gamma_5(C)$ bands are produced by different phonons. The shape of the Γ_4 and Γ_5 exciton lines at T = 10 K in CuGaSe$_2$ and CuAlSe$_2$ crystals is asymmetrical. The asymmetry factor A is equal to 0.06 and 0.04 for the A excitons in CuGaSe$_2$ and CuAlSe$_2$ crystals, respectively. For the B excitotons, A equals 0.05 and 0.06 in CuGaSe$_2$ and CuAlSe$_2$ crystals, while it is around 0.01 for both compounds for the C excitons (Figure 72). This corresponds to the case of weak exciton-phonon interaction. One should mention that the damping parameter varies in the interval from 2.6 meV to 3.9 meV for the A, B, and C excitons, while the longitudinal-transversal splitting does not exceed 3.4 meV [148]. The shape of the exciton lines becomes less asymmetrical, and the absolute value of the asymmetry coefficient does not increase with increasing the temperature. This behavior of the shape of bands in the wavelength modulated reflectivity is due to the fact that the value of the damping parameter is higher than longitudinal-transversal exciton splitting. The damping parameter for the Γ_4 excitons in solid solutions with X = 0.5 equals 4.6 meV, while the longitudinal-transversal splitting does not exceed 3.4 meV.

Therefore, a weak polariton effect is observed for all the considered solid solutions. The value of the damping parameter for mixed compositions is higher as compared to its value in non-mixed compositions. The damping parameter does not exceed 3 meV for compounds with X = 1 and X = 0. The asymmetry parameter is nearly the same for all compositions, and it varies in the range of 0.02 – 0.06, i. e. the sign of the asymmetry coefficient for the n = 1 line of A, B, and C excitons is positive for all compositions of the solid solutions. The effective masses of electrons and holes in CuAl$_{1-x}$Ga$_x$Se$_2$ solid solutions change insignificantly with the variation of the X value from 1 to 0, their values being in the interval $m_C^* \sim (0.14\text{-}0.11)m_0$, $m_{V1}^* \sim (1{,}26\text{ -}1.20)m_0$.

9. RESONANCE RAMAN SCATTERING AND OPTICAL ORIENTATION OF EXCITONS IN $CuGa_xAl_{1-x}S_2$ CRYSTALS

Resonance Raman scattering (RRS) is a versatile tool for studying interaction of electrons and holes with elementary excitations in crystals and extracting information concerning the spin state of the excited electron-hole pairs. The spin orientation of carriers in semiconductors by means of optical pumping has been investigated experimentally and theoretically [157-159]. The excitation of excitons by polarized light in crystals leads to the change of the radiation polarization [159-164]. The optical orientation of free and bound excitons has been studied in CdSe [161], GaSe [160] and CdS [161-164] crystals. The degree of emission polarization observed during the optical pumping of excitons can be significantly higher as compared to that inherent to carrier recombination, provided that the polarized light produces direct excitation of excitons. The spin states of the excited carrier pairs remain always correlated due to their interaction in the exciton, while the energy dissipation by one or several longitudinal optical (LO) phonons occurs. The crystals investigated in Refs [135-137] have only one polar mode. The investigation of optical orientation of excitons in crystals with many polar modes has been reported in Ref. [165]

The efficiency of RRS depends on the energy of the excitation photons, and it is determined mainly by the denominator of the Raman tensor

$$I \propto \left| \frac{1}{(E_i - E_{ex})(E_s - E_{ex})} \right|^2 \qquad (37)$$

where E_{ex} is the exciton energy, E_i and E_s are the energies of the incident and scattered photons, respectively.

This expression exhibits maxima when $E_i = E_{ex}$ or $E_s = E_{ex}$.

Evidence for these multiple resonant scatterings can be obtained, in principle, by continuously varying the temperature to vary the intermediate-state energy and its damping constant. This variation of the resonance condition is expressed by the temperature dependence of the energy denominator in the Raman tensor. When interference effects are neglected, the resonant behavior for single Stokes scattering on either type of the intermediate states can simply be written as

$$I \propto \left| \frac{1}{(E_i - E_0 + i\Gamma_0)(E_s - E_0 + i\Gamma_0)} \right|^2 \qquad (38)$$

where Γ_0 is the damping constant for the intermediate state located at E_0, and E_i and E_s are, respectively, the excitation and scattered light energies.

Two emission bands at ω_L and ω_T as well as P_1-P_7 peaks are observed in the RRS spectra of CuGaS$_2$ crystals measured in the x(zz)x polarization at 10 K under the excitation by 4880 Å Ar laser line [99-101,125,126]. The excitation energy of 2.5403 eV (λ = 4880 Å) is lower than the energy of the continuum E_g = 2.5474 eV and it is a little higher than the energy of the n = 2 (2.5356 eV) excited state of the Γ_4 excitons in CuGaS$_2$. This photon energy results in the resonance excitation of exciton polaritons and the resonance one-phonon Raman scattering in the region of exciton polaritons (Figure 73).

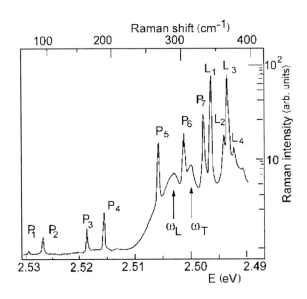

Figure 73. Resonance Raman scattering in CuGaS$_2$ crystals under the excitation by 4880 Å Ar$^+$ laser line (2.5403 eV) at 10 K. Reprinted with permission from [165], N.N. Syrbu, I.M. Tiginyanu, V.V. Ursaki, A.V. Dorogan (2008). *Resonance Raman scattering in CuGa$_x$Al$_{1-x}$S$_2$ crystals*. J. Opt. A: Pure Appl. Opt. 10 125002. Copyright (2008), with permission from Institute of Physics.

In contrast to the II-VI compounds, one A$_1$ mode, three B$_1$ modes, three polar B$_2$ modes and six polar modes of E symmetry are optically active in the investigated crystals. These modes are observed in our spectra and have also been reported previously [137-143]. According to theoretical considerations [1-5], the 1LO scattering lines are forbidden. The emergence of these lines is due to the presence of crystal defects, or to some other mechanisms resulting in the breakdown of the wave vector selection rules [3-5]. The intensity of one-phonon RRS lines (1LO) is several times lower than the intensity of two-phonon lines (2LO). The P_6 and P_7 lines are superimposed on the background of the thermalized luminescence ω_L and ω_T. The P_6 and P_7 lines are more intensive and are characterized by a small full width at half maximum (FWHM).

An effective relaxation of energy with the participation of LO-phonons occurs when the CuGaS$_2$ crystals are excited by the 4880 Å laser line which energy is higher than the energy of the ground exciton state. This excitation results in the formation of free excitons with different internal or kinetic energy. The energy of the scattering line E$_s$ differs from the energy of the excitation line E$_i$ by the LO-phonon energy and it equals to

$$E_s = E_{ex} - mE_{LOj}$$

where j is the type of the phonon and m is the number of LO phonon involved in the scattering process.

Therefore, the phonon lines are arranged according to their increasing energies (table 8). The bands P_1, P_2, and P_3 are due to E^1_{LO}, B^1_{2LO}, and E^3_{LO}, respectively (Figure 73, table 8). The symmetry and the energy of phonons were identified according to refs [136-137]. However, in this scattering mechanism, the character of the energy relaxation of excitons significantly changes near the conduction band bottom, as in the case of CdS and other II-VI compounds [1-6,144-146]. Different phonons (1LO, 2LO, etc) participate in the processes of relaxation of exciton energy in $CuGaS_2$ crystals.

The scheme of the cascade relaxation of the excitonic polaritons by the LO phonons of various symmetry at 9K for the 4880 Å excitation is shown in Figure 74. The central part shows the schematic of the upper and lower polariton branches n = 1 (UPB and LPB) and the position of n = 2 and E_g. The right-hand section illustrates the nonequilibrium population of the exciton band, dN/dE, and the relative intensity of the scattering lines. A schematic of the one-phonon scattering at the LO phonons of various symmetries is shown to the left. For convenience, the polariton dispersion law $E(\mathbf{k})$ is given in one direction of the wave vector $\mathbf{k}$.

"Photon" polaritons with a respective polarization are created when the crystal is excited by linearly polarized light with the energy E. These polaritons can be scattered from the "photon" branch to the exciton state with the emission of LO-phonons. According to theoretical considerations and experimental results [1-5], the scattering on LO-phonons with low values of the wave-vector (k=0) should not lead to the change of polarization. The observed emission results from the scattering of exciton polaritons on the "photon" branch and occurs with the participation of LO phonons. The observed change of the polarization (degree of polarization) of the RRS emission is explained by the following: (i) the interaction with fully symmetrical LO-phonons, and (ii) the small life-time of polaritons in excitonic states.

Table 8. Position (parameters) of one-phonon RRS lines in the region of exciton polaritons in $CuGaS_2$ crystals under the excitation by 4880 Å Ar$^+$ laser line (2.5403 eV) at 10 K

No	E, eV	$E_L - E_{Pj}$ cm^{-1}	Peak intensity	Phonons	$Q(\Gamma_5)$	$Q(\Gamma_4)$	$Z_\sigma(\Gamma_5)$	$Z_\sigma(\Gamma_4)$
P_1	2.5298	84.7	20	E^1_{LO}				
P_2	2.5285	95.2	50	B^1_{2LO}	0.76	0.89	0.58	0.68
P_3	2.5194	167.8	70	E^3_{LO}	0.79	0.85	0.54	0.67
P_4	2.5163	194	80	E^3_{LO}, B^2_1	0.78	0.84	0.59	0.69
P_5	2.5065	272.6	248	B^2_{2TO}, E^4_{LO}	0.76	0.80	0.53	0.63
P_6	2.5048	312	200	A_1	0.29	0.48	0.38	0.16
ω_L	2.5040		276					
ω_T	2.5011		274					
P_7	2.4982	339.6	350	E^5_{LO}	0.28	0.44	0.35	0.12

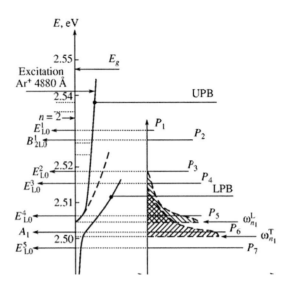

Figure 74. Schematic representation of the cascade relaxation of excitonic polaritons in CuGaS$_2$ at 1LO phonons of various symmetries at 9 K under the 4880 Å excitation.

Three valence bands with the symmetry Γ_7 (V$_1$), Γ_6 (V$_2$) and Γ_7(V$_3$) are inherent to CuGaS$_2$ crystals, from which the exciton series are formed [166-170]. In the long-wavelength region, the excitons of Γ_4 и Γ_5 symmetry (lettered as the A-series) are formed as a result of interaction of the electrons from the Γ_6(C$_1$) band with the holes from the Γ_7(V$_1$) band. The Γ_4 excitons are allowed in the E∥c polarization, while the Γ_5 excitons are allowed in the E⊥c polarization. The parameters of the Γ_4 excitons {2.5011 eV (ω_T) and 2.5045 eV (ω_L) for n = 1; 2.503 eV for n = 2; and 2.5357eV for n =3} practically coincide with the parameters of the Γ_5 excitons {2.5001 (ω_T) and 2.5017eV (ω_L) for n =1; 2.5252 eV for n = 2; and 2.5307 eV for n =3} [99-101,125-126]. The excitons of Γ_1, Γ_2 and Γ_5 symmetry are formed as a result of interaction of the electrons from the Γ_6(C$_1$) band with the holes from the Γ_6(V$_2$) band. The exciton series Γ_5 (lettered as the B-series) is allowed in the E⊥c polarization [125,126,148-151,165-171]. The Γ_1 and Γ_2 excitons are forbidden in both the polarizations. The Γ_4 exciton series allowed in the E∥c polarization and lettered as the C-series is formed as a result of interaction of the electrons from the Γ_6(C$_1$) band with the holes from the Γ_7(V$_3$) band. The ground and the excited states of the B (2.6217 eV) and C (2.6323 eV) excitons are situated in the high-energy region in respect to the A excitons. The excitons of the B and C series should not influence the polarization of the phonon emission of A excitons under the excitation with energies lower than the energy of the ground state of B and C excitons.

Taking into account the properties of CuGaS$_2$ crystals, for the determination of the polarization degree [165], the measurements have been performed in the following configuration: Q(Γ_5), Q(Γ_4), Z$^\sigma$(Γ_5) and Z$^\sigma$(Γ_4), where

$$Q(\Gamma_5) = \frac{J_{Z(XX)Z} - J_{Z(XY)Z}}{J_{Z(XX)Z} + J_{Z(XY)Z}}, \quad Q(\Gamma_4) = \frac{J_{Y(ZZ)Y} - J_{Y(ZX)Y}}{J_{Y(ZZ)Y} + J_{Y(ZX)Y}},$$

$$Z^\sigma(\Gamma_5) = \frac{J^{\sigma+} - J^{\sigma-}}{J^{\sigma+} + J^{\sigma-}}, \text{ and } Z^\sigma(\Gamma_4) = \frac{J^{\sigma+} - J^{\sigma-}}{J^{\sigma+} + J^{\sigma-}}.$$

$Z^\sigma(\Gamma_5)$ corresponds to the case when the wave vectors of the incident and scattering light are parallel to the C-axis of the CuGaS$_2$ crystals, while $Z^\sigma(\Gamma_4)$ corresponds to the case when the wave vectors of the incident and scattering light are parallel to the Y-axis of the CuGaS$_2$ crystals. The indexes of σ^+ and σ^- mean that the light waves are circularly polarized in the right and left directions, respectively.

As one can see from Figure 73 and Table 8 the intensity of the P$_5$, P$_6$, and P$_7$ lines is higher than the intensity of the P$_1$- P$_4$ lines. The intensity of the P$_5$, P$_6$, and P$_7$ lines is also higher that the intensity of the emission related to the upper ω_L and lower ω_T polariton branches. The increase of the intensity of RS lines with approaching the exciton resonance is observed practically in all the II-VI compounds [3-6]. Apart from that, the intensity of lines due to scattering of 2, 3, etc. phonons is higher than the intensity of one-phonon lines. LO-phonons with different symmetry participate in the RS processes in CuGaS$_2$ crystals. The Stocks shift, peak energy, and FWHM of P$_1$- P$_7$ RRS lines as well as ω_L and ω_T lines are presented in Table 8. Apart from that, the symmetry of the one-phonon vibration modes participating in the relaxation of exciton polaritons in CuGaS$_2$ crystals is shown in Table 8.

Figure 75 shows the RRS spectra under the excitation by 4765 Å laser line (2.602 eV) in the x(zz)x backscattering geometry. An intensive emission maximum is observed at 2.504 eV which is due to the upper ω_L and lower ω_T Γ_4 polariton branches. Apart from that, the e$_0$—e$_{11}$ lines due to the two-phonon RRS of exciton polaritons are observed in the emission spectra. The energy of scattering lines, their Stocks shift, peak intensity, as well as the combination of 2LO phonons which participate in the relaxation of exciton polariton energy are shown in Table 8. One can see from Figure 75 that the FWHM of the 2LO phonon lines is smaller than the FWHM of ω_L and ω_T lines. The same is true for the RRS with 1 LO phonons (Figure 74, Table 8). The intensity of 1 LO phonon lines is by a factor of 3 – 5 lower than the intensity of 2 LO phonon lines.

The scheme of the cascade relaxation of exciton-polaritons on 1 LO (the lines P$_1$-P$_7$) and 2 LO (the lines e$_1$-e$_{11}$) phonons with different symmetry under the excitation by the 4765 Å laser line (2.602 eV) at 10 K is presented in Figure 76. The scheme of upper and lower polariton branches for n = 1 excitons (UBP$_{n=1}$ and LBP$_{n=1}$), the position of n = 2 state, and the bandgap energy E_g is shown in the central part of the Figure. The non-equilibrium population of the exciton band dN/dE estimated from the FWHM of the scattering lines is shown on the right side of the Figure The one-phonon and two-phonon RRS are shown on the left side of the Figure The dispersion relation of polaritons E(k) is shown for one direction of the wave vector k for the convenience.

One can conclude on the basis of data presented in Table 9 and Figure 73 that the depolarization of P$_1$- P$_5$ scattering lines is around 65 – 68 %. For the P$_6$ and P$_7$ lines the depolarization is around 28 %. These lines are situated closely to the ω_L and ω_T lines of the Γ_5 and Γ_4 excitons. Such a strong change of the polarization of P$_6$ and P$_7$ scattering lines is due to their close vicinity to the thermalized luminescence, and possibly to the change of the intermediate state from Γ_4 to Γ_5 excitons, or vice-versa.

The change of polarization degree of the two-phonon e$_1$-e$_{11}$ RRS lines in CuGaS$_2$ crystals is by 15 % larger as compared to the change of polarization of the one-phonon P$_1$-P$_{11}$ RRS lines (Table 8,9). The change of linear (Q) and circular (Z) polarization is nearly the same. Predominantly LO phonons participate in two-phonon processes. The intensity of the e_{10} and e_{11} lines decreases due to their close position to the thermalized luminescence like the case of one-phonon region. The RRS lines in the long-wavelength region are not discussed due to the presence of intensive bound exciton L_1- L_4 lines (Figure 73).

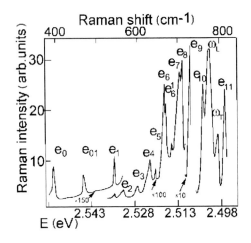

Figure 75. Two-phonon Resonance Raman scattering in CuGaS$_2$ crystals under the excitation by 4765 Å Ar$^+$ laser line (2.602 eV) at 10 K. Reprinted with permission from [165], N.N. Syrbu, I.M. Tiginyanu, V.V. Ursaki, A.V. Dorogan (2008). *Resonance Raman scattering in CuGa$_x$Al$_{1-x}$S$_2$ crystals*. J. Opt. A: Pure Appl. Opt. 10 125002. Copyright (2008), with permission from Institute of Physics.

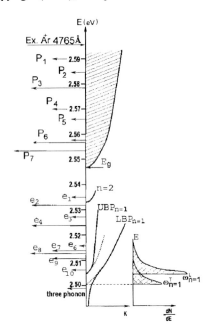

Figure 76. The energy position of the upper (UBP$_{n=1}$) and lower (LPB$_{n=1}$) polariton branches of the Γ_4 excitons and the position of RRS lines in CuGaS$_2$ crystals. Reprinted with permission from [165], N.N. Syrbu, I.M. Tiginyanu, V.V. Ursaki, A.V. Dorogan (2008). *Resonance Raman scattering*. J. Opt. A: Pure Appl. Opt. 10 125002. Copyright (2008), with permission from Institute of Physics.

It is known that the resonance conditions change with changing the excitation wavelength, since the ΔE value changes, i. e. the difference between the excitation energy (E_i) and the resonance exciton energy (E_{exc}). Usually, the change of the ΔE value is provided by the change of the excitation wavelength of a tunable laser. These experiments provide a most effective analysis of the arrangement of excitons, of the conversion of energy between the polariton modes, etc. In our case, the change of the ΔE value was performed by changing the composition of the solid solutions. The use of variable composition of solid solutions leads to the change of the energy position of exciton polariton levels with respect to the energy of laser line.

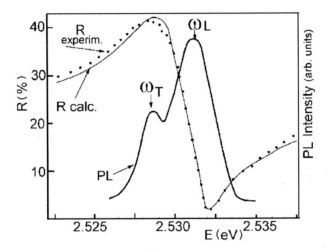

Figure 77. Reflection spectra of $CuGa_{0.95}Al_{0.05}S_2$ crystals measured in $E \parallel c$ polarization at 10 K, and the results of calculations. Reprinted with permission from [165], N.N. Syrbu, I.M. Tiginyanu, V.V. Ursaki, A.V. Dorogan (2008). *Resonance Raman scattering.* J. Opt. A: Pure Appl. Opt. 10 125002. Copyright (2008), with permission from Institute of Physics.

In $CuGa_{0.95}Al_{0.05}S_2$ crystals the resonance values of exciton transitions are situated at higher energies in comparison with $CuGaS_2$. Figure 77 presents the reflectivity spectra in $E \parallel c$ polarization and the luminescence spectra in these crystals.

The values of the energy of transversal exciton $\omega_0 = 2.528$ eV, the dielectric constant $\varepsilon = 7$, the exciton translation mass $M=1.0m_0$, the longitudinal-transversal splitting $\omega_{LT}=3.2$ meV, the damping parameter $\gamma=1.5$ meV and the thickness of the dead-layer 15 Å were obtained from the calculation of the contour of reflectivity spectra of the ground state of the Γ_4 exciton.

Intense luminescence is observed in the region of the maximum and minimum in the reflectivity spectra at 2.5312 eV and 2.528 eV which is due to the emission from the upper ω_L and lower ω_T polariton branches.

One-phonon $a_1 - a_{13}$ scattering lines are observed in the spectral region $E < \omega_L$ of the Γ_4 exciton in the RRS spectra of $CuGa_{0.95}Al_{0.05}S_2$ crystals under the excitation by the 4880 Å laser line (2.5403 eV) (Figure 78, Table 9). Bound excitons are not observed in the vicinity of Γ_4 excitons in these crystals. The depolarization of the a_2 and a_4 lines is quite low for the linear polarization (Q~87%), and it is significantly higher for the circular polarization (Z~30%). For the $a_5 - a_{13}$ scattering lines which are situated in the region $E < \omega_T$, the change of the sign of linear and circular polarization occurs. We suggest that this is due to the change of the intermediate branch responsible for the scattering of exciton polaritons. The depolarization for $a_5 - a_{13}$ lines is nearly the same for both the linear and circular polarization (Table 9).

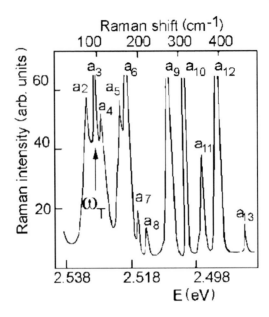

Figure 78. Resonance Raman scattering in $CuGa_{0.95}Al_{0.05}S_2$ crystals under the excitation by 4880 Å Ar^+ laser line. [Reprinted with permission from [165], N.N. Syrbu, I.M. Tiginyanu, V.V. Ursaki, A.V. Dorogan (2008). *Resonance Raman scattering in $CuGa_xAl_{1-x}S_2$ crystals*. J. Opt. A: Pure Appl. Opt. 10 125002. Copyright (2008), with permission from Institute of Physics.

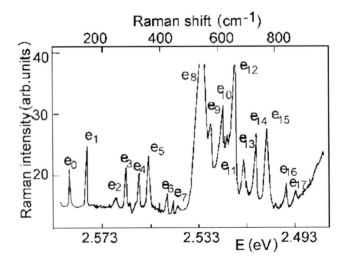

Figure 79. Resonance Raman scattering in $CuGa_{0.95}Al_{0.05}S_2$ crystals under the excitation by 4765 Å Ar^+ laser line at 10 K. Reprinted with permission from [165], N.N. Syrbu, I.M. Tiginyanu, V.V. Ursaki, A.V. Dorogan (2008). *Resonance Raman scattering in $CuGa_xAl_{1-x}S_2$ crystals*. J. Opt. A: Pure Appl. Opt. 10 125002. Copyright (2008), with permission from Institute of Physics..

Table 9. Position (parameters) of RRS lines in the region of exciton polaritons in $CuGaS_2$ crystals under the excitation by 4765 Å (2.602 eV) Ar$^+$ laser line (e_1-e_{11} lines), and in $CuGa_{0.95}Al_{0.05}S_2$ crystals under the excitation by 4880 Å (2.5403) Ar$^+$ laser line (a_1-a_{13} lines)

No	E, eV	ΔE, cm^{-1}	$E_{ph1}+E_{ph2}$	Modes	$Q(\Gamma_5)$	$Z_\sigma(\Gamma_5)$	No	E, eV	ΔE cm^{-1}	Modes	$Q(\Gamma_5)$	$Z_\sigma(\Gamma_5)$
e_1	2.5352	535	367+167	$E^6_{LO}+E^3_{LO}$	0.55	0.62	a_1	2.5327		ω_L		
e_2	2.5322	559	167+392	$E^3_{LO}+B^3_{2LO}$	0.56	0.60	a_2	2.5307	95	E^1_{TO}	0.86	0.29
e_3	2.5277	569	312+284	$A_1+E^4_{LO}$	0.59	0.63	a_3	2.5282		ω_T		
e_4	2.5238	627	345+284	$E^5_{LO}+E^4_{LO}$	0.60	0.65	a_4	2.5262	113	B^1_1	0.88	0.34
e_5	2.5188	667	392+273	$B^3_{2LO}+E^4_{LO}$	0.63	0.66	a_5	2.5213	153	E^2_{LO}	-0.10	-0.08
e_6	2.5163	688	286+401	$B^2_{2LO}+B^3_1$	0.62	0.56	a_6	2.5193	170	E^3_{LO}	-0.12	-0.08
e_7	2.5139	707	312+392	$A_1+B^3_{2LO}$	0.66	0.58	a_7	2.5158	197	B^2_1	-0.13	-0.09
e_8	2.5131	713	347+367	$E^5_{LO}+E^6_{LO}$	0.65	0.69	a_8	2.5133	217	$E^2_{LO}+E^1_{LO}$	-0.10	-0.07
e_9	2.5102	737	347+392	$E^5_{LO}+B^3_{2LO}$	0.59	0.60	a_9	2.5059	277	B^2_{2LO}	-0.13	-0.08
e_{10}	2.5060	771	367+401	$E^6_{LO}+B^3_1$	0.44	0.46	a_{10}	2.5016	312	A_1	-0.15	-0.10
ω_L	2.5040			ω_L			a_{11}	2.4962	355	E^5_{LO}	-0.18	-0.12
ω_T	2.5011			ω_T			a_{12}	2.4918	391	B^3_1	-0.20	-0.15
e_{11}	2.4987	830		3-phonon	0.33	0.24	a_{13}	2.4841	453			

Table 10. Position (parameters) of RRS lines in the region of exciton polaritons in $CuGa_{0.95}Al_{0.05}S_2$ crystals (e_0–e_{17} lines), and in $CuGa_{0.9}Al_{0.1}S_2$ crystals (x_1–x_{16} lines) under the excitation by 4765 Å Ar$^+$ laser line (2.602 eV)) at 10 K

No	E, eV	$\Delta\omega$, cm^{-1}	Modes	$Q(\parallel\perp)$	$Z(\sigma-\sigma)$	No	E, eV	$\Delta\omega$, cm^{-1}	Modes	$Q(\parallel\perp)$	$Z(\sigma-\sigma)$
e_0	2.5864	82	E^1_{LO}	0.45	0.06	x_1	2.5930	72	E^1_{LO}	0.53	0.07
e_1	2.5807	169	E^3_{LO}	0.50	0.07	x_2	2.5895	94	B^1_{2LO}	0.57	0.5
e_2	2.5683	269	E^4_{LO}	0.47	0.5	x_3	2.5822	141	E^2_{LO}	0.48	0.07
e_3	2.5636	306	A_1	0.50	0.07	x_4	2.5800	167	E^3_{TO}	0.50	0.03
e_4	2.5585	348	E^5_{LO}	0.50	0.03	x_5	2.5796	182	E^3_{LO}	0.49	0.04
e_5	2.5544	381	B^3_{2LO}	0.49	0.04	x_6	2.5770	196	B^2_1	0.43	0.04
e_6	2.5463	446		0.40	0.04	x_7	2.5690	259	E^4_{LO}	0.45	
e_7	2.5443	462				x_8	2.5672	277	B^2_{2LO}	0.50	0.04
e_8	2.5332		ω_L			x_9	2.5626	312	A_1	0.50	0.03
e_9	2.5292		ω_T			x_{10}	2.5584	348	E^5_{LO}	0.48	0.09
e_{10}	2.5242	624		-0.28	-0.09	x_{11}	2.5570		ω_L		
e_{11}	2.5223	640				x_{12}	2.5535		ω_T		
e_{12}	2.5198	660		-0.23	-0.12	x_{13}	2.5223	408	$A_1+B^1_{2LO}$	-0.10	
e_{13}	2.5153	696		-0.21	-0.14	x_{14}	2.5198	438	$B^2_{2LO}+E^2_{LO}$	-0.13	-0.12
e_{14}	2.5104	736		-0.24	-0.13	x_{15}	2.5153	444	$E^2_{LO}+B^1_1$	-0.11	-0.14
e_{15}	2.5060	771		-0.23	-0.12	x_{16}	2.5104	494	$E^2_{LO}+E^5_{LO}$	-0.14	-0.13
e_{16}	2.4977	838		-0.20	-0.13						
e_{17}	2.4938	870		-0.21	-0.11						

The lines $e_0 - e_{17}$ are observed in the spectra of $CuGa_{0.95}Al_{0.05}S_2$ crystals under the excitation by 4765 Å laser line (2.602 eV) (Figure 79, Table 10). The lines $e_0 - e_7$ have a depolarization of 45-50 % for the linearly polarized, and 5 % for circularly polarized light. The scattering lines with $E < \omega_T$, i. e. $e_{10} - e_{17}$ lines, change the polarization by 23% and 12% for Q and Z, respectively. The Q and Z values are positive for $a_1 - a_4$ lines, and negative for $a_5 - a_{13}$ lines.

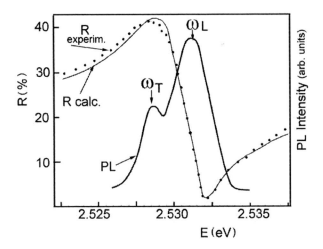

Figure 80. The measured and calculated reflection spectra of $n = 1$ state of Γ_4 excitons in $E \parallel c$ polarization, and the luminescence spectrum of $CuGa_{0.9}Al_{0.1}S_2$ crystals under the excitation by 4765 Å Ar^+ laser line. Reprinted with permission from [165], N.N. Syrbu, I.M. Tiginyanu, V.V. Ursaki, A.V. Dorogan (2008). *Resonance Raman scattering*. J. Opt. A: Pure Appl. Opt. **10** 125002. Copyright (2008), with permission from Institute of Physics.

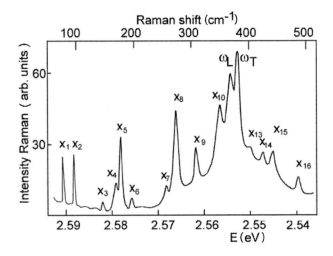

Figure 81. Resonance Raman scattering in $CuGa_{0.9}Al_{0.1}S_2$ crystals at 10 K. Reprinted with permission from [165], N.N. Syrbu, I.M. Tiginyanu, V.V. Ursaki, A.V. Dorogan (2008). *Resonance Raman scattering in $CuGa_xAl_{1-x}S_2$ crystals*. J. Opt. A: Pure Appl. Opt. **10** 125002. Copyright (2008), with permission from Institute of Physics.

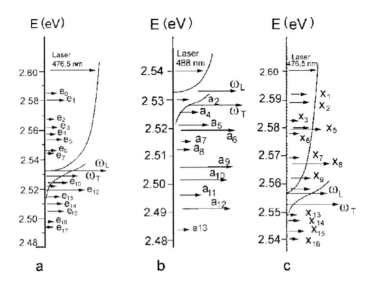

Figure 82. The position of RRS lines and the energy of exciton polaritons in CuGa$_{0.95}$Al$_{0.05}$S$_2$ (a.b) and CuGa$_{0.9}$Al$_{0.1}$S$_2$ (c) crystals under excitation with different wavelengths. Reprinted with permission from [165], N.N. Syrbu, I.M. Tiginyanu, V.V. Ursaki, A.V. Dorogan (2008). *Resonance Raman scattering.* J. Opt. A: Pure Appl. Opt. **10** 125002. Copyright (2008), with permission from Institute of Physics.

In CuGa$_{0.9}$Al$_{0.1}$S$_2$ crystals, the levels of the Γ_4 and Γ_5 excitons are shifted to high energy region. The reflectivity spectra for CuGa$_{0.9}$Al$_{0.1}$S$_2$ crystals in E∥c polarization and the luminescence spectra in the region of n = 1 exciton line are shown in Figure 80. The parameters of the Γ_4 exciton (ω_L = 2.5574 and ω_T = 2.5540 eV) are determined from the calculation of the contour of reflectivity spectra. The translation mass of the exciton M equals 1.1 m$_0$, while the damping parameter is 1.8 meV, and ε equals 7.1.

One-phonon x$_1$ – x$_{10}$ and two-phonon x$_{13}$ – x$_{16}$ RRS lines are observed in CuGa$_{0.9}$Al$_{0.1}$S$_2$ crystals under the excitation by 4765 Å laser line (2.602 eV) (Figure 81).

The change of polarization equals 50% (Q = 50%) for the linearly polarized emission, and 5% (Z = 5%) for the circularly polarized emission for the x$_1$ – x$_{10}$ lines. The values of Q and Z equal 12% and have an opposite sign for the x$_{13}$ – x$_{16}$ lines (Table 10). The position of scattering lines, exciton levels, and excitation lines for different composition of CuGa$_x$Al$_{1-x}$S$_2$ solid solutions and different excitation wavelengths are shown in Figure 82.

Therefore, the observation of linear and circular depolarization of the emission of free excitons suggests the existence of a predominant orientation of exciton spins in the direction of spins of the exciting photons. In CuGa$_x$Al$_{1-x}$S$_2$ crystals for x = 1.0; 0.95; and 0.9, the increase of the depolarization for the emission near the band gap is not connected with the splitting of the upper valence band by the crystal field into Γ_6 and Γ_7 subbands (the spin of holes ±3/2 and ±1/2, respectively). The long-wavelength exciton series A is energetically separated from the B and C exciton series. The depolarization of the emission is observed in the direction parallel to the C-axis of the crystal, as in the case of CdS compound [3-7]. Taking into account that the oscillator strength of the Γ_4 exciton is by a factor of 3000 higher than the oscillator strength of the Γ_5 exciton, the electron of the Γ_4 exciton should recombine with its own hole, and the scattering emission should not change the polarization. Since the depolarization occurs in the experiment with scattering along the Y-axis, it could be that the electron of the Γ_4 exciton recombines with the hole of the Γ_5 exciton. The Γ_4 and Γ_5 excitons are formed by the same couple of bands $\Gamma_7(V_1)$ and $\Gamma_6(C_1)$.

The depolarization of the LO phonon RRS lines occurs practically in the same way for phonons with different symmetries. This means that the observed peculiarities are intrinsic properties of crystals. The depolarization of the LO scattering is connected probably with the breakdown of the polarization selection rules, expected for k = 0, as a result of the participation of polariton states with higher wave vectors. The light excitons enter into all processes of RRS as initial and final states. Therefore, the depolarization connected with these states should influence in a similar way the polarization degree of all the LO lines with different symmetry in the sample. This is observed in the experiment. For all the investigated samples, there are insignificant differences in the depolarization of the LO- and 2LO scattering lines. We suggest that the observed different change of the polarization in different samples is due to the state of "photon" polaritons in crystals.

10. RESONANCE RAMAN SCATTERING IN $CUGA_xIN_{1-x}S_2$ SOLID SOLUTIONS

Resonance Raman scattering of exciton polaritons is usually investigate by means of tunable lasers. The resonance conditions are changed by changing the excitation wavelength. A similar effect can be attained by changing the composition of a solid solution [172]. Using variable compositions of the solid solution leads to the variation of the energy position of exciton polariton levels with respect to the energy of laser lines. The energy position of ground states of A, B, and C excitons in $CuGa_xIn_{1-x}S_2$ solid solutions and the position of Ar^+ laser lines are shown in Figure 83.

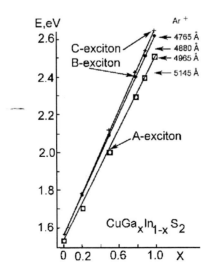

Figure 83. energy position of ground states of A, B. and C excitons in $CuGa_xIn_{1-x}S_2$ solid solutions and the position of Ar^+ laser lines. Reprinted with permission from [172], N.N. Syrbu, L.L. Nemerenco, A.V. Dorogan, V.E. Tezlevan, E. Arama (2009). *Multiple-phonon resonant Raman scattering*. Optical Materials 31, 970-975. Copyright (2009), with permission from Elsevier.

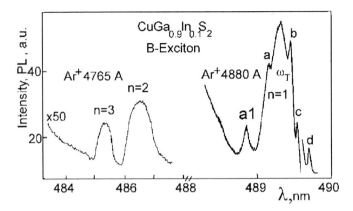

Figure 84. Luminescence spectra of $CuGa_{0.9}In_{0.1}S_2$ solid solutions. Reprinted with permission from [172], N.N. Syrbu, L.L. Nemerenco, A.V. Dorogan, V.E. Tezlevan, E. Arama (2009). *Multiple-phonon resonant Raman scattering*. Optical Materials 31, 970-975. Copyright (2009), with permission from Elsevier.

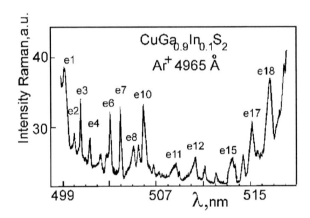

Figure 85. Resonance Raman scattering in $CuGa_{0.9}In_{0.1}S_2$ crystals. Reprinted with permission from [172], N.N. Syrbu, L.L. Nemerenco, A.V. Dorogan, V.E. Tezlevan, E. Arama (2009). *Multiple-phonon resonant Raman scattering in $CuGa_xIn_{1-x}S_2$ single crystals*. Optical Materials 31, 970-975. Copyright (2009), with permission from Elsevier.

The position of ground exciton states in solutions with x = 0.9 is shifted to the long-wavelength region. Emission lines of the n=2 (2.5413 eV) and n=3 (2.5458 eV) excited states of the B-exciton are observed (Figure 84) under the excitation by the 4765 Å laser line (the energy of this line corresponds to the region of the continuum). No structure characteristic for the resonant Raman scatteringis observed in the region of B-exciton excited states.

An intense band at 2.5267 eV associated with lower branch of the B-exciton polariton (ω_T) is observed under the excitation of these solid solutions by the 4880 Å laser line. This band contains a fine structure of a,b,c and d lines which are due to the resonance Raman scattering. One should note that a broader maximum (a_1) is observed in the short-wavelength region. It is due to the emission from the upper polariton branch. In this case, the longitudinal-transverse splitting is about 5 meV. A series of lines (e_1-e_{18}) associated with resonance Raman scattering of the Γ_4 exciton polaritons are observed in solid solutions with x=0.9 under the excitation by 4965 Å laser line (Figure 85, Table 11).

Table 11. Energies of the one-phonon and two-phonon resonance Raman scattering in CuGa$_{0.9}$In$_{0.1}$S$_2$ crystals under the excitation by 4965 Å Ar$^+$ laser line

Line numeration	Band energy, eV	ΔE, meV	
Laser line 4965 Å	2.4968	0	
e$_1$	2.4837	13.1	
e$_2$	2.4793	17.5	
e$_3$	2.4769	19.9	
e$_4$	2.4731	23.7	
e$_5$	2.4688	28.0	One-phonon region
e$_6$	2.4650	31.8	
e$_7$	2.4603	36.5	
e$_8$	2.4556	41.2	
e$_9$	2.4532	43.6	
e$_{10}$	2.4509	45.9	
e$_{11}$	2.4392	57.6	
e$_{12}$	2.4309	65.9	
e$_{13}$	2.4268	70.0	
e$_{14}$	2.4226	74.2	
e$_{15}$	2.4162	80.6	Two-phonon region
e$_{16}$	2.4117	85.1	
e$_{17}$	2.4085	88.3	
e$_{18}$	2.4026	94.2	
e$_{19}$	2.3986	98.2	
e$_{20}$	2.3968	100.0	
ω_L	2.3852		
ω_T	2.3817		

The excitation energy lies exactly in the region of the upper branch of the Γ$_4$ exciton polariton. The e$_1$-e$_{10}$ emission lines are due to the one-phonon Raman scattering, while the e$_{11}$-e$_{18}$ lines are due to the two-phonon Raman scattering. A broad band related to the luminescence from the ground state of the Γ$_4$ exciton is observed in the long-wavelength region of the spectrum. Broader emission bands as compared to the e$_1$-e$_{18}$ lines are observed at 2.3852 eV and 2.3817 eV. These bands are due to the luminescence from the upper and lower branches of the Γ$_4$ exciton polariton.

As mentioned above, the efficiency of resonance Raman scattering is described by the relation (37). The excitation of the investigate solid solution by the 5145 Å laser line allows one to detect narrow a$_1$-a$_{13}$ scattering lines and a broad emission band at 2.3786 eV (Figure 86). The broad line is due to the emission from the lower branch of the Γ$_4$ exciton polariton, while the narrow lines are due to the resonance Raman scattering (Table 12).

Resonance emission of bound excitons can be obtained in the configuration of Raman scattering under the excitation by 4965 Å and 5145 Å laser lines. The energy of 4965 Å and 5145 Å lines is higher than the resonance energy of the A-exciton for solutions with x = 0.9

(Figure 86 and Figure 83). The intensity of the E(LO) increases as compared to the intensity of the A1 mode with approaching the excitation energy to the exciton energy. The B- and C-excitons situated at 2.63 and 2.64 eV are in resonance with the 4965 Å excitation line for solutions with X = 0.8. The intensity of the E(L) mode and multi-phonon modes increases with approaching the excitation energy to the energy of B- and C-excitons. Luminescence related to the upper and lower branches of the Γ_4 polaritons (Figure 87) and luminescene due to the annihilation of excitons with emission of phonons is observed under the oblique excitation (with an angle of incidence of 45° with respect to the surface of the sample) by 5145 Å line. Figure 87 presents also exciton reflectivity spectra for the solutions woth x = 0.9 in the region of the ground (n = 1) state.

Table 12. Resonance Raman scattering line of CuGa$_{0.9}$In$_{0.1}$S$_2$ crystals under the excitation by 5145 Å Ar$^+$ laser line

	Band energy, eV	Raman shift, Δv, см$^{-1}$	
Laser line 5145 Å	2.4094	0	
a$_1$	2.4004	9.0	One-phonon region
a$_2$	2.3982	11.2	
a$_3$	2.3883	21.1	
a$_4$	2.3830	26.4	
a$_5$	2.3795	29.9	
	2.3786		ω_T Γ_{4exit}
a$_6$	2.3765	32.9	Multi-phonon region
a$_7$	2.3711	38.3	
a$_8$	2.3667	42.7	
a$_9$	2.3619	47.5	
a$_{10}$			
a$_{11}$			

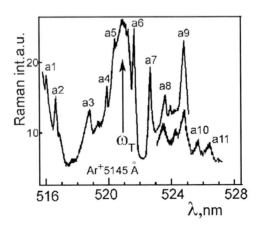

Figure 86. Resonance Raman scattering of CuGa$_{0.9}$In$_{0.1}$S$_2$ crystals. Reprinted with permission from [172], N.N. Syrbu, L.L. Nemerenco, A.V. Dorogan, V.E. Tezlevan, E. Arama (2009). *Multiple-phonon resonant Raman scattering in CuGa$_x$In$_{1-x}$S$_2$ single crystals*. Optical Materials 31, 970-975. Copyright (2009), with permission from Elsevier.

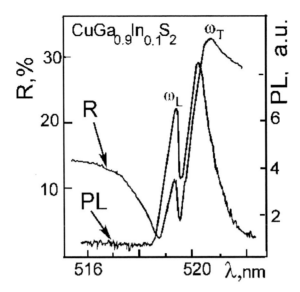

Figure 87. Reflectivity and luminescence spectra of $CuGa_{0.9}In_{0.1}S_2$ crystals in the region of the ground exciton state. Reprinted with permission from [172], N.N. Syrbu, L.L. Nemerenco, A.V. Dorogan, V.E. Tezlevan, E. Arama (2009). *Multiple-phonon resonant Raman scattering*. Optical Materials 31, 970-975. Copyright (2009), with permission from Elsevier.

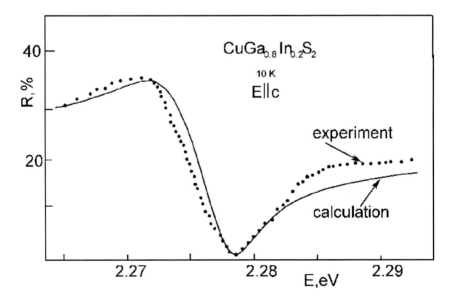

Figure 88. Calculated and experimental reflectivity spectra of $CuGa_{0.8}In_{0.2}S_2$ crystals in the region of the ground Γ_4 exciton state. Reprinted with permission from [172], N.N. Syrbu, L.L. Nemerenco, A.V. Dorogan, V.E. Tezlevan, E. Arama (2009). *Multiple-phonon resonant Raman scattering*. Optical Materials 31, 970-975. Copyright (2009), with permission from Elsevier.

The reflectivity spectra have a usual excitonic shape with a maximum in the region of 2.3780 eV (transverse exciton). A small maximum ("spike") is observed in the region of the reflectivity minimum. The energy position of this peak corresponds to the energy of the Γ_4 exciton. The energy position of A, B, and C excitons shift to lower energies with changing the composition of the solid solution to x = 0.8 (Figure 83). The reflectivity spectra of CuGa$_{0.8}$In$_{0.2}$S$_2$ solutions measured in the E∥c polarization at 10 K are shown in Figure 88.

The following parameters have been deduced from fitting of experimental reflectivity spectra: ε_b = 8, exciton mass M=0,7; energy of the transverse exciton ω_T=2.267 eV, damping parameter γ=3.5 meV.

Table 13. Resonance Raman scattering lines of CuGa$_{0.8}$In$_{0.2}$S$_2$ crystals under the excitation by 5145 Å laser line

Line numeration	Band energy, eV	ΔE, meV	ΔE, cm^{-1}			
Laser line 5145 Å	2.4094	0	0			
d$_1$	2.4008	8.6	69.4	E^1_{In}		One-phonon region
d$_2$	2.3982	11.2	90.3	$B^1_{2\,6a}$		
d$_3$	2.3892	20.2	163.0	E^3_{6a}		
d$_4$	2.3777	31.7	255.7	E^{4T}_{6a}		
d$_5$	2.3759	33.5	270.2	$B^2_{2\,6a}$		
d$_6$	2.3720	37.4	301.7	A_1		
d$_7$	2.3676	41.8	337.2	E^{5T}_{6a}		
d$_8$	2.3667	42.7	344.5	E^{5L}_{6a}		
d$_9$	2.3624	47.0	380.1	B^{3L}_{26a} or E^{6L}_{6a}		
d$_{10}$	2.3563	53.1	428.3	$B^1_{26a} + E^{5T}_{6a}$	90.3 + 337.2 = 427.5	Two-phonon region
d$_{11}$	2.3511	58.3	470.3	$B^2_1 + B^2_{2\,6a}$	200 + 270 = 470	
d$_{12}$	2.3425	66.9	539.7	$2B^2_{2\,6a}$	270.2 + 270.2 = 540.4	
d$_{13}$	2.3382	71.2	574.4	$B^2_{2\,6a} + A_1$	270.2 + 301.7 = 571.9	
d$_{14}$	2.3344	75.0	605.0	$2A_1$	302 + 302 = 604	
d$_{15}$	2.3284	80.5	649.4	$E^5_{6a} + A_1$	344.5 + 302 = 646.5	
d$_{16}$	2.3247	84.7	683.3	$E^{5T} + E^{5L}$	337.2 + 344.5 = 681.7	
d$_{17}$	2.3201	89.3	720.4	$B^{3L}_2 + E^{5L}$	379.1 + 344.5 = 723.6	
d$_{18}$	2.3150	94.4	761.5	$2B^{3L}_{26a}$	380 + 380 = 760	
d$_{19}$	2.3088	100.6	811.5			Three-phonon region
d$_{20}$	2.3030	106.4	858.3			
d$_{21}$	2.2939	115.5	931.7			
d$_{22}$	2.2849	124.5	1004.3			
ω^L_{n1A}	2.2800					
ω^T_{n1A}	2.2764					
d$_{23}$	2.2740					

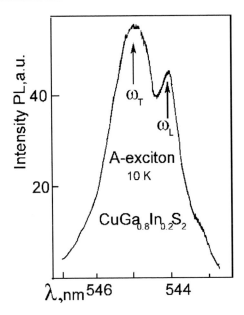

Figure 89. Luminescence spectra of CuGa$_{0.8}$In$_{0.2}$S$_2$ crystals in the region of the ground state of the Γ$_4$ exciton. Reprinted with permission from [172], N.N. Syrbu, L.L. Nemerenco, A.V. Dorogan, V.E. Tezlevan, E. Arama (2009). *Multiple-phonon resonant Raman scattering*. Optical Materials 31, 970-975. Copyright (2009), with permission from Elsevier.

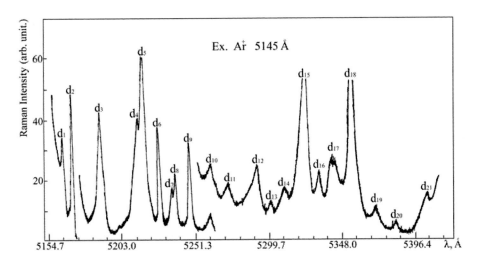

Figure 90. Resonance Raman scattering spectra of CuGa$_{0.8}$In$_{0.2}$S$_2$ crystals. Reprinted with permission from [172], N.N. Syrbu, L.L. Nemerenco, A.V. Dorogan, V.E. Tezlevan, E. Arama (2009). *Multiple-phonon resonant Raman scattering in CuGa$_x$In$_{1-x}$S$_2$ single crystals*. Optical Materials 31, 970-975. Copyright (2009), with permission from Elsevier.

Two broad luminescence bands due to the emission from the lower and upper exciton polariton branches are observed at 2.2738 eV и 2.2759 eV under the excitation of crystals with this composition by the 5145 Å laser line (Figure 89).

Lines d_1-d_{21} and bands ω_L (2.2759 eV) and ω_T (2.2738 eV) are observed in the Raman scattering spectra of CuGa$_{0.8}$In$_{0.2}$S$_2$ crystals measured in the x(zz)x geometry at 9K under the λ = 5145 Å excitation (Figure 90).

The excitation energy of the 2.4094 eV laser line is higher than the energy of the continuum of Γ_4 excitons for this composition of the solid solution. It is also higher than the energy of the n=1 state of the B-excitons, but is lower than the energy of the n=2 state. This leads to resonant excitation of B-exciton polaritons. Energy exchange occurs between the lower branch of the B-exciton and the upper branch of the Γ_4 exciton. This leads to the resonance one-phonon and two-phonon Raman scattering in the region of Γ_4 exciton polaritons.

Narrow Raman scattering lines a_1-a_{27} are observed for this solid solution under the excitation by 4965 Å laser line (Figure 91). The energy of the 4965 Å laser line excites the C-exciton polariton branch which leads to the one-phonon (a_1-a_7) and multi-phonon (a_8-a_{27}) scattering.

The lower (ω_T) branch of the C-exciton polariton and the upper (ω_L) branch of the B-exciton polariton are also observed in the interval of two-phonon a_{11}-a_{16} lines. These bands have the maximum intensity and coincide with the frequency values of longitudinal and transverse C- and B-excitons determined from the reflectivity spectra. This demonstrates the energy exchange between the lower and upper branches of different polaritons (C and B) in the investigated crystals.

Excitation close to the continuum of the B- and C excitons is produced by the 4880 Å laser line. Figure 92 shows the resonance Raman scattering and reflectivity spectra in the region of ground states of the B- and C-exciton series. The reflectivity spectra are typical for excitons, but with a small amplitude. The amplitude of reflectivity spectra from the minimum (ω_L) to the maximum (ω_T) is about 2% (Figure 92).

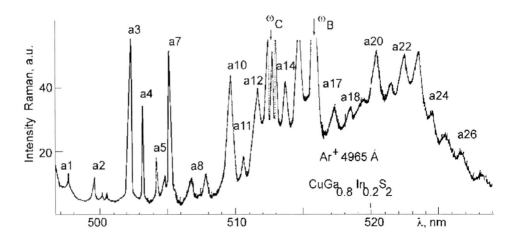

Figure 91. Resonance Raman scattering spectra under the excitation by 4965 Å laser line. Reprinted with permission from [172], N.N. Syrbu, L.L. Nemerenco, A.V. Dorogan, V.E. Tezlevan, E. Arama (2009). *Multiple-phonon resonant Raman scattering in CuGa$_x$In$_{1-x}$S$_2$ single crystals*. Optical Materials 31, 970-975. Copyright (2009), with permission from Elsevier.

Table 14. Multi-phonon resonance Raman scattering lines of CuGa$_{0.8}$In$_{0.2}$S$_2$ crystals under the excitation by 4965 Å laser line

Band numeration	Band energy, eV	ΔE, meV	ΔE, cm^{-1}			
Laser line 4965Å	2.4968	0	0			
a$_1$	2.4861	10.7	86.3	E_{In}^2 or $B_{2\,6a}^1$		One-phonon region
a$_2$	2.4769	19.9	160.5	E_{6a}^3		
a$_3$	2.4641	32.7	263.8	E_{6a}^{4L}		
a$_4$	2.4593	37.5	302.5	A_1		
a$_5$	2.4546	42.2	340.4	E_{6a}^{5T}		
a$_6$	2.4518	45.0	363.0	$B_{2\,6a}^{3T}$		
a$_7$	2.4499	46.9	378.3	$B_{2\,6a}^{3T}$		
a$_8$	2.4425	54.3	438.0	$E_{6a}^{5L} + B_{26a}^1$	90.3 + 344.5 = 434.8	Two-phonon region
a$_9$	2.4378	59.0	475.9	$B_2^1 + B_{26a}^{3L}$	378.3 + 90.3 = 468.6	
a$_{10}$	2.4291	67.7	546.1	$B_{26a}^2 + B_{26a}^2$	270.2 + 270.2 = 540.4	
a$_{11}$	2.4249	71.9		ω_{n1C}^L		
a$_{12}$	2.4203	76.5		ω_{n1C}^T		
a$_{13}$	2.4158	81.0	653.4			
a$_{14}$	2.4112	85.6	690.5		302.5 + 378.3 = 680.8	
a$_{15}$	2.4067	90.1	726.8	$B_2^{3L} + E^{5L}$	378.3 + 344.5 = 722.8	
a$_{16}$	2.4017	95.1		ω_{n1B}^L		
a$_{17}$	2.3950	101.8	821.2			Three-phonon region
a$_{18}$	2.3897	107.1	863.9			
a$_{19}$	2.3852	111.6	900.3			
a$_{20}$	2.3812	115.6	932.5			
a$_{21}$	2.3768	120.0	968.0			
a$_{22}$	2.3724	124.4	1003.5			
a$_{23}$	2.3676	129.2	1042.2			
a$_{24}$	2.3637	133.1	1073.7			
a$_{25}$	2.3593	137.5	1109.2			
a$_{26}$	2.3541	142.7	1151.3			
a$_{27}$	2.3472	149.6	1206.8			

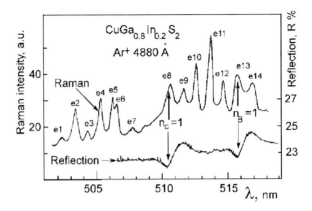

Figure 92. Resonance Raman scattering and reflectivity spectra of $CuGa_{0.8}In_{0.2}S_2$ crystals in the region of B- and C-excitons. Reprinted with permission from [172], N.N. Syrbu, L.L. Nemerenco, A.V. Dorogan, V.E. Tezlevan, E. Arama (2009). *Multiple-phonon resonant Raman scattering in $CuGa_xIn_{1-x}S_2$ single crystals*. Optical Materials 31, 970-975. Copyright (2009), with permission from Elsevier.

Narrow emission lines e_1-e_{13} (Table 15) due to the two-phonon Raman scattering (e_1-e_6) and three-phonon scattering (e_9-e_{11}) are observed in the emission spectrum. The e_7 and e_8 lines are probably due to the emission from the upper and lower branches of the C-exciton polariton. The e_{12} and e_{13} emission lines are supposed to be associated with the upper and lower branches of the B-exciton polariton. These data agree with the results obtained from reflectivity and wavelength-modulated reflectivity spectra. Narrow emission lines h_1-h_{12} are observed under the excitation of crystals with x=0.8 by the 4765 Å laser line. These lines are due to the multi-phonon Raman scattering of exciton polaritons. Scattering line practically disappear with increasing temperature to 40 K (Figure 93), while the emission from the upper and lower branches of the C-exciton polaritons is still observed (h_{11}, h_{12}).

Table 15. Raman scattering lines of $CuGa_{0.8}In_{0.2}S_2$ crystals under the excitation by 4880 Å laser line

Band numeration	Band energy, eV	Raman shift, Δv, cm^{-1}	
Laser line 4880 Å	2.5403	0	
e_1	2.4641	615	One-phonon region
e_2	2.4593	653	
e_3	2.4546	691	
e_4	2.4490	729	
e_5	2.4453	766	
e_6	2.4383	823	Two-phonon region
e_7	2.4254		
e_8	2.4208		
e_9	2.4162	1001	
e_{10}	2.4117	1037	
e_{11}	2.4072	1074	
e_{12}	2.4022		
e_{13}	2.3977		

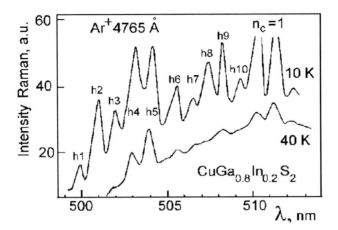

Figure 93. Resonance Raman scattering at 10 K and 40 K. Reprinted with permission from [172], N.N. Syrbu, L.L. Nemerenco, A.V. Dorogan, V.E. Tezlevan, E. Arama (2009). *Multiple-phonon resonant Raman scattering in CuGa$_x$In$_{1-x}$S$_2$ single crystals*. Optical Materials 31, 970-975. Copyright (2009), with permission from Elsevier.

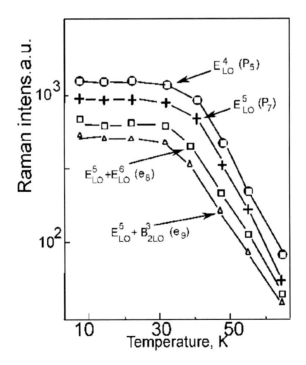

Figure 94. Dependence of the Resonance Raman scattering intensity on the temperature. Reprinted with permission from [172], N.N. Syrbu, L.L. Nemerenco, A.V. Dorogan, V.E. Tezlevan, E. Arama (2009). *Multiple-phonon resonant Raman scattering in CuGa$_x$In$_{1-x}$S$_2$ single crystals*. Optical Materials 31, 970-975. Copyright (2009), with permission from Elsevier.

The ground (n=1) and n=2 state of the C exciton are observed at 2.4565 eV and 2.4226 eV, respectively, for this solid solution. Therefore, the exciton binding energy is 45.2 meV and the energy of the continuum is 2.4678 eV for this composition.

These experiments demonstrate the dependence of the first-, second-, and third-order resonance Raman scattering upon the excitation energy. The resonance effect is significant at low temperatures (9K), but not at room temperature.

Table 16. Raman scattering lines of CuGa$_{0.8}$In$_{0.2}$S$_2$ crystals under the excitation by 4765 Å laser line

Band numeration	Band energy, eV	Raman shift, Δv, см$^{-1}$		
Laser line 4765Å	2.6016	0	0	
h$_1$	2.4755	0.1261	983.1	One-phonon region
h$_2$	2.4712	0.1304	1051.9	
h$_3$	2.4659	0.1363	1099.5	
h$_4$	2.4617	0.1399	1128.5	
h$_5$	2.4570	0.1446	1166.5	
h$_6$	2.4499	0.1517	1223.7	Four-phonon region
h$_7$				
h$_8$	2.4411	0.1605	1294.7	
h$_9$	2.4378	0.1638	1321.3	
h$_{10}$				
h$_{11}$	2.4281	ω_{n1C}^{L}		
h$_{12}$	2.4235	ω_{n1C}^{T}		
h$_{13}$				

In order to clarify the dependence of different mode intensity on temperature, the change of the intensity of LO(P$_5$, P$_7$) and 2LO(e$_8$, e$_9$) lines with increasing temperature from 8.6 to 60 K was analyzed (Figure 94). Main attention has been pad to modes excited by the 4880 Å laser line, as well as to two-phonon modes excited by the 4765 Å laser line. The energy of the A-exciton bandgap remains nearly constant up to 80 K, and shifts to lower energies with further temperature increase. At the same time, the Raman modes do not shift. The intensity of E$_{LO}$ modes (P$_3$, P$_4$, P$_5$, P$_7$) strongly depends on temperature. The intensity of the last modes decreases so fast that they are not observed at temperatures above 80 K. At the same time, the intensity of the A1 mode practically does not decrease above 120 K. If one compares the dependence of the exciton energy and the intensity of exciton emission from the lower and upper polariton branches with the intensity of Raman scattering, one can observe that the fast decrease of the intensity of LO and 2LO modes occurs in the same temperature interval where the decrease of exciton energy occurs. This behavior is characteristic for resonant scattering as a consequence of free excitons. Approaching of the resonance energy to the excitation energy and increasing of the intensity of LO and 2LO modes with changing the composition of the solid solution confirm the resonance of the photon scattering for the E(LO) modes. The insignificant variation of the A1 mode intensity demonstrates that the resonance effect is insignificant at temperatures above 120 K.

CONCLUDING REMARKS

Investigation of exciton states in mew materials is very important from the point of view of observing new effects, as well as for the determination of main material parameters, which are difficult to be determined by other methods. The experimental data presented in this chapter demonstrate that the exciton spectroscopy can be used both in materials with strong and weak exciton oscillator strength, or strong and weak polariton effect. Scattering of excitons by LO phonons influences many emission processes observed in the near band edge region. One should note that strong interaction with LO phonon is inherent to intrinsic electron excitations in polar semiconductors and multi-component materials. The investigation of multi-phonon exciton luminescence and resonant Raman scattering processes demonstrate that the main peculiarities of spectra are well explained in the framework of a general model assuming Froehlich scattering of intermediate states of the hydrogen-like exciton spectra. This simple model is suitable for a wide variety of A^2B^6, A^2B^5 and multi-component $A^1B^3C^6_2$ materials. Nevertheless, the quantitative characteristics of multi-phonon spectra are determined by the specific parameters of crystals. The effects of exciton arrangement are of especial interest in multi-component materials with a big number of active vibration modes. Multi-phonon Raman scattering on LO-phonons in the region of the fundamental absorption edge and the oscillating structure in the luminescence excitation spectra indicate on the existence of a mechanism of energy relaxation, according to which, the electron-hole pair created by the light loses the energy through the emission of many LO phonons as a whole without being divided into separate carriers. Such a relaxation mechanism which can be treated as relaxation of hot excitons is efficient in a wide energy range covering the energy of many LO-phonons above the fundamental absorption edge. This spectral range contains a significant part of the integral intensity of excitation and luminescence spectra. One can conclude that the energy relaxation of hot excitons with emission of LO-phonons plays an important role in the processes of conversion and scattering of the optical energy absorbed by a crystal.

REFERENCES

[1] Pekar, SI. *Zh. Eksp. Teor. Fiz..*,1957, 33, 1022.
[2] Pekar, SI. *Zh. Eksp. Teor. Fiz.*, 1958, 34, 1176.
[3] Hopfild, JJ; Thomas, DG. *Phys. Rev.*, 1963, 132, 563.
[4] Agranovich, VM; Ginzburg, VL. Spatial dispersion in crystal optics and the theory of excitons, *Interscience*, New York, Wiley, 1966.
[5] Born, M; Huang, K. *Dynamical theory of crystal lattices*, Clarendon Press, Oxford, 1954.
[6] Knox, RS. *Theory of Excitons*, Academ Press-New York and London, 1963.
[7] Agranovich, VM; Galanin, MD. Perenos energii elerctronnogo vozbujdenia v condensirovannih sredah, Наука: *Moskva*, 1978.
[8] Tait, WC; Weiher, RL. *Phys. Rev.*, 1969, 166, 769.
[9] Sumi, HJ. *Phys. Soc. Jap.*, 1976, vol. 41, 526.
[10] Agranovich, VM. *In Surface polaritons*, VM; Agranovich, DL. Mills, Ed; North-

Holand Publ.Comp.: Amsterdam, 1982.
[11] Agranovich, VM; Daramanyan, SA; Rupasov, VI. Zh. Eksp. *Teor. Fiz.*, 1980, 78, 656.
[12] Permogorov, S; Travnikov, V; Sel`kin, AV. Sov. *Phys. Solid State*, 1972, 14, 3642.
[13] Brodin, MS; Miasnikov, EN; Marisova, SV. *Polaritoni v cristalooptice*, Naukova dumka: Kiev, 1984.
[14] Askary, F; Yu, PY. *Sol. State Commun.*, 1983, 47, 241.
[15] Travnikov, VV; Krivolapciuk, VV., Zh. *Eksp. Teor. Fiz.*, 1983, 85, 2087.
[16] Pevtsov, AV; Sel`kin, AV. Sov. Phys. *Solid State*, 1981, vol. 23, 2814.
[17] Pevtsov, AB; Sel`kin, AV. Zh. Eksp. *Teor. Fiz.*, 1982, 83, 516 [Sov. Phys. JETF., 1982, 83, 516].
[18] Sel'kin, AV. Sov. Phys. *Solid State*, 1977, 19, 1433.
[19] Ivchenko, EL; Pevtsov, AB; Sel'kin, AV. *Sol. State Commun.*, 1981, 39, 453.
[20] Evangellsti, FR; Frova, A; Patella, F. *Phys. Rev. B*, 1974, 10, 4253.
[21] Permogorov, S; Sel`kin, AV. Sov. *Phys. Solid State*, 1973, 15, 3025.
[22] Ivchenko, EL. *In Exitons*, EA; Rasba, MD. Struge, Ed; North-Holand Publ.Comp.: Amsterdam, 1982.
[23] Pevtsov, AB; Permogorov, SA; Sel`kin, AV; Syrbu, NN; Umanets, AG. *Fiz. Tekh. Poluprovodn*, 1982, 16, 1399 [*Sov. Phys. Semicond.* 1982, 16, 897].
[24] Syrbu, NN; Mamaev, VM. Fiz. Tekh. *Poluprovodn.*, 1983, 17, 694 [*Sov. Phys. Semicond.,* 1983, 17, 433].
[25] Pevtsov, AB; Sel`kin, AV; Syrbu, NN; Umanets, AG. Zh. Eksp. *Teor. Fiz.*, 1985, 89, 1155 [*Sov. Phys. JETF*, 1985, 62, 665].
[26] Syrbu, NN; Khachaturova, SB; Radautsan, SI. *Dokl. Akad. Nauk SSSR*, 1986, 286, 345 [*Sov. Phys. Dokl.* 1986, 31, 53].
[27] Syrbu, NN; Stamov, IG; Radautsan, SI. *Dokl. Akad. Nauk SSSR*, 1982, 262, 1138.
[28] Syrbu, NN. Optoelectronnie svoistva soedinenii grupi A^2B^5, Stiintsa: *Kishinev*, 1983.
[29] Arimoto, O; Okamoto, S; Nakamura, K. *J. Phys. Soc. Jpn.*, 1990, 59, 3490.
[30] Arimoto, O; Tachiki, M; Nakamura, K. *J.Phys .Soc. Jpn.*, 1991, 60, 4351.
[31] Sugisaki, M; Arimoto, S; Nakamura, K. *J. Phys. Soc. Jpn.*, 1995, 65, 23.
[32] Arimoto, O; Takeuchi, H; Nakamura, K. *Phys. Rev. B.*, 1992, 46, 15512.
[33] Krochmali, AP; Gubanov, VA; Janchuk, ZZ. *Fiz. Tverd. Tela*, 2003, 45, 1177.
[34] Gorbani, IS; Krochmali, AP; Janchuk, ZZ. Fiz. *Tverd. Tela*, 1999, 41, 193.
[35] Gorbani, IS; Krochmali, AP; Janchuk, ZZ. Fiz. *Tverd. Tela*, 2000, 42, 1582.
[36] Berg, RS; Yu, PY; Mowles, T. Sol. St. *Commun.*, 1983, 46, 101.
[37] Radautsan, SI; Syrbu, NN; Khachaturova, SB; Stratan, GI; Peev, LG. Dokl. Akad. *Nauk SSSR*, 1990, 311, 866.
[38] Syrbu, NN; Stamov, IG; Radautsan, SI. Dokl. Akad. *Nauk SSSR*, 1982, 262, 1138.
[39] Syrbu, NN; Stamov, IG; Radautsan, SI. Dokl. Akad. Nauk Mold. SSR, *Ser. Fiz. Tekh. Mat. Nauk*, 1981, 3, 85.
[40] Stamov, IG; Syrbu, NN; Khachaturova, SB. Fiz. *Tverd. Tela*, 1984, 26, 2468.
[41] Syrbu, NN; Khachaturova, SB; Stamov, IG. Fiz. Tekh. *Poluprovodn.*, 1984, 18, 1498.
[42] Gorban, IS; Belyi, NM; Borbat, VA; Gubanov, VA; Dmitruk, IN; Yanchuk, ZZ. Dokl. Akad. *Nauk USSR. Ser. A*, 1988, 4, 48.
[43] Arimoto, O; Tachiki, M; Nakamura, K. *J. Phys. Soc. Jpn.*, 1991, 60, 4351.
[44] Frohlich, D; Schlierkamp, M; Schubert, J; Spitzer, S; Arimoto, O; Nakamura, K. *Phys. Rev. B*, 1994, 49, 10337.

[45] Gorban, IS; Krokhmal, AP; Yanchuk, ZZ. Fiz. *Tverd. Tela*, 1999, vol. 41, 193, Sov. *Phys. Solid State*, 1999, 41, 170.
[46] Taguchi, S; Goto, T; Takeda, M; Kido, G. *J. Phys. Soc. Jpn.*, 1988, 57, 3256.
[47] Goto, T; Taguchi, S; Cho, K; Nagamune, Y; Takeyama, S; Miura, N. *J. Phys. Soc. Jpn.*, 1990, 59, 773.
[48] Berg, RS; Yu, PY; Mowies, Th. *Solid Stale Commun.*, 46, 2, 101.
[49] Sobolev, VV; Kozlov, AI. *Phys. Status Solidi B*, 1984, 126, K59.
[50] Syrbu, NN. Fiz. Tekh. *Poiuprovodn.*, 1992, 26, 1069 [*Sov. Phys. Semicond.*, 1992, 26, 599]
[51] Sugisaki, M; Arimoto O; Nakamura, K. *J. Phys. Soc. Jpn.*, 1996, 65, 23.
[52] Bir, GL; Pikus, GE. *Symmetry and Deformation Effects in Semiconductors* [in Russian]; Nauka: Moscow, 1972.
[53] Sobotta, H; Neumann, H; Syrbu, NN; Riede, V. *Phys. Status Solidi B*, 1983, 115, K55.
[54] Permogorov, S; Travnikov, V. *Phys. Status Solidi B*, 1976, 78, 389.
[55] Hayashi, T; Watanabe, M. *Solid State Commun.*, 1989, 69, 959.
[56] Cho, K; Kawata, M. *J. Phys. Soc. Jpn.*, 1985, 54, 4431.
[57] Rudin, S; Reinecke, TL; Segall, B. *Phys. Rev. B*, 1990, 42, 11218.
[58] Hayashi, T; Watanabe, M. *Solid State Commun.*, 1989, 69, 959.
[59] Cho, K; Kawata, M. *J. Phys. Soc. Jpn.*, 1985, 54, 4431.
[60] Pevtsov, AB; Permogorov, SA; Sel'kin, AV. *JETP Lett.*, 1984, 39, 312.
[61] Arimoto, O; Okamoto, S; Nakamura, K. *J. Phys. Soc. Jpn.*, 1990, 59, 3490.
[62] Arimoto, O; Tachiki, M; Nakamura, K. *J. Phys. Soc. Jpn.*, 1991, 60, 4351.
[63] Taguchi, S; Goto, T; Takeyama, M; Kido, G. *J. Phys. Soc. Jpn.*, 1988, 57, 3256.
[64] Goto T; Goto, Y. *J. Lumin.*, 1991, 48/49, 103.
[65] Goto, T; Goto, Y. *J. Phys. Soc. Jpn.*, 1991, 60, 1763.
[66] Cotelynicov, VV; Tolpigo, KB. Sov. *Phys. Solid State*, 1980, 22, 2928.
[67] Bonnot, A. Benoit a la Guillaume. Proc. 1st *Taormina Res. Conf. Struct. Matter*, Pergamon: N.Y., 1974.
[68] Sermage, B; Voos, M; Schwab, C. *Phys. Rev. B*, 1979, 20, 3245.
[69] Yamawaki, M; Hamaguchi, C. *Phys. Stat. Sol. (b)*, 1982, 112, 201.
[70] Pevtsov, AB; Sel'kin, AV. *Sov. Phys. Solid State*, 1981, 23, 2814.
[71] Isimaru, A. *Rasprostranenie i rasseianie voln v sluciaino neodnorodnih sredah*, Mir: Moskva, 1981.
[72] Bisty, VE. *Short Note of the Physics FIAN SSSR*, 1977, 1, 34.
[73] Galanin, MD; Miasnikov, EN; Han-Magomedova, SD. Izv. AN SSSR, *seria fiz.*, 1980, 44, 730.
[74] Danilenko, DN; Karapetianc, MH; Danilenko, VE; Magomedjiev, GG. *Deponirovano*. №3919-77, VINITI, 1977.
[75] Agranovich, VM. *Teoria exsitonov*, Nauka: Moskva, 1968.
[76] Born, M; Volf, E. *Principles of optics*, Pergamon Press: N. Y., 1984.
[77] Azzam, R; Basara, N. *Elipsometria i polarizovannii svet*, Mir: Moskva, 1981.
[78] Schoper, H. *Optik*, 1953, 10, 426.
[79] Sell, DD; Stokowskl, SE; Dingle, R; DiLorenzo, JV. *Phys. Rev. B*, 1973, 7, 4568.
[80] Lagous, J; Hummer, K. *Phys. Stat. Sol. (b)*, 1975, 72, 393.
[81] Evangilisti, F; Fishbach, JU; Frova, A. *Phys. Bev. B*, 1974, 9, 1516.

[82] Broser, I; Rosenwtig, M; Broser, R; Richard, M; Birkicht, E. *Phys. Slat. Sol.* (b), 1978, 90, 77.
[83] Ivchenco, EL; Sel`kin, AV. Zh. Eksp. *Teor. Fiz*, 1979, 76, 1837.
[84] Pevtsov, AB; Permogorov, SA; Sayfulev, SR; Sel`kin, AV. *Sov. Phys. Solid State*, 1980, 22, 2400.
[85] Novicov, BV; Roppiser, F; Talalaev, VG. *Sov. Phys. Solid State*, 1979, 21, 823.
[86] Tell, B; Kasper, HM. *Phys. Rev. B*, 1973, 7, 740.
[87] Rite, JC; Center, RN; Bridenbough, PM; Veal, BW. *Phys. Rev. B*, 1977, 16, 4491.
[88] Bodnar, JV; Smirnova, GF; Karoza, AG; Chernyakova, AP. *Phys. Status Solid B*, 1990, 158, 469.
[89] Mrtzner, H; Eberhardt, J; Cieslak, J; Hahn, Th; Goldhahn, RU; Reislohner, U; Withuhn, W. *Thin Solid Films*, 2004, 451-452, 241.
[90] Honda, T; Hara, K; Yoshino, J; Kukimoto, H. *J. Phys. Chem. Solids*, 2003, 64, 2001.
[91] Botha, JR; Branch, MS; Chowles, AG; Leitch, AWR; Weber, J. *Physica B: Condensed Matter*, 2001, 308-310, 1065.
[92] Tsuboi, N; Uchiki, H; Ishikawa, H; Iida, S. *Jpn. J. Appl. Phys.*, 1993, Suppl. 32, 584.
[93] Jaffe, JE; Zunger, A. *Phys.Rev. B*, 1983, 28, 5822.
[94] Jaffe, JE; Zunger, A. *Phys.Rev. B*, 1984, 29,1882.
[95] Ahuya, R; Auluck, S; Eriksson, O; Wills, JM; Johansson, B. *Sol. Energy Mater. Sol. Cells*, 1998, 53, 357.
[96] Lazewski, J; Jochym, PT; Parlinski, K. *J. Chem. Phys.*, 2002, 117, 2726.
[97] Laksari, S; Chahed, A; Abbouni, N; Benhelal, O; Abbar, B. *Comput. Mater. Sci.*, 2006, 38, 223.
[98] Syrbu, NN; Ursaki, VV; Nyari, T. *In Proceedings of the International Semiconductor Conference CAS*, 2000, Sinaia, Romania.
[99] Syrbu, NN; Ursaki, VV; Tiginyanu, IM; Tezlevan, VE; Blaje, MA. *J. Phys. Chem. Solids*, 2003, 64, 2003, 1967.
[100] Syrbu, NN; Blaje, MA; Tezlevan, VE; Ursaki, VV. *Optics and Spectroscopy*, 2002, 92, 402.
[101] Syrbu, NN; Blaje, MA; Tiginyanu, IM; Tezlevan, VE. *Optics and Spectroscopy*, 2002, 92, 395.
[102] Tsuboi, N; Uchici, H; Sawada, M. *Physica B (Amsterdam)*, 1993, 185, 348.
[103] Susaki, M; Wakita, K; Yamomoto, N. *Jpn. J. Appl. Phys.*, 1999, 38, 2787.
[104] Susaki, M; Yamamoto, N; Prevot, B; Schwab, C. *Jpn. J. Appl. Phys.*, 1996, 35, 1652.
[105] Rocket A; Birkmire, RW. *J. Appl. Phys.*, 1991, 70, R80.
[106] Andriesch, AM; Syrbu, NN; Iovu, MS; Tezlevan, VE. *Phys. St. Sol.*(b), 1995, 187, 83.
[107] Syrbu, NN; Korzun, BV; Fadzeyeva, AA; Mianzelen, RR; Ursaki, VV; Galbic, I. *Physica B: Condensed Matter*, 2010, in Press.
[108] Syrbu, NN; Khachaturova, SB; Bodnar, IV; Rabotinskii, ND. *Optics and Spectroscopy*, 1989, 66, 693.
[109] Levcenko, S; Syrbu, NN; Tezlevan, VE; Arushanov, EK; Merino, JM; León, M. *J. Phys. D: Appl. Phys.*, 2008, 41, 0055403.
[110] Syrbu, NN; Bogdanas, M; Tezlevan, VE; Mushcutariu, I. *Physica B*, 1997, 229, 199.
[111] Mandel, L; Tomlinson, RD; Hampshire, MJ; Neuman, H. *Solid St. Commun.*, 1979, 32, 201.
[112] Betini, M. *Solid St. Commun.*, 1973, 13, 599.

[113] Shirakata, S; Ogawa, A; Isomura, S; Kariya, T. *Jpn. J. Appl. Phys.*, 1993, Suppl. 32-3, 94.
[114] Permogorov, SA; Selkin, AV; Travnicov, VV. Soviet. *Phys. Solid. State*, 1973, 14, 3051.
[115] Permogorov, AA; Selkin, AV; Travnicov, VV. *Soviet Phys. Solid. State*, 1973, 15, 1215.
[116] Tosatti, B; Harbece, G. *Nuovo Cimento*, 1974, 228, 87.
[117] Osaka, Y; Takeuti, Y. *J. Phys. Soc. Japan*, 1968, 24, 236.
[118] Agarwal, GS; Pattanayak, DN; Wolf. E. *Opt. Commun.*, 1971, 4, 255.
[119] Agarwal, GS; Pattanayak, DK; Wolf. E. *Opt. Commun*, 1971, 4, 260.
[120] Sketrup, T. *Phys. Stat. Sol.* (b), 1973, 60, 695.
[121] Kiselev, VA; Razbirin, BS; Uraltsev, IN. *Zh. Eksp. Teor. Fiz. Pisma*, 1973, 18, 504.
[122] Kiselev, VA; Razbirin, BS; Uraltsev, IN; Kochereshko, VP. *Fiz. Tverd. Tela*, 1975, 17, 640.
[123] Kiselev, VA; Razbirin, BS; Uraltsev, IN. *Phys. Status Solidi B*, 1975, 72, 161.
[124] Kiselev, VA; Makarenko, IV; Razbirin, BS; Uraltsev, IN. *Fiz. Tverd. Tela*, 1977, 19, 1348.
[125] Syrbu, NN; Tiginyanu, IM; Nemerenko, LL; Ursaki, VV; Tezlevan, VT; Zalamai, VV. *J. Phys. Chem. Sol.*, 2005, 66, 1974.
[126] Syrbu, NN; Nemerenco, LL; Stamov, IG; Bejan, VN; Tezlevan, VE. Opt. *Commun.*, 2007, 272, 124.
[127] Syrbu, NN; Tezlevan, VE; Galbich, I; Nemerenco, L; Ursaki, VV. *Opt. Commun.*, 2009, 282, 4562.
[128] Selkin, A. *Phys. Status Solidi* (b), 1977, 83, 47.
[129] Bonnot, A; Planel, R; Benoita la Guillaume, C. *Phys. Rev. B*, 1974, 9, 690.
[130] Permogorov, S. *Phys. Stat. Sol.*, (b), 1975, 68, 9.
[131] Permogorov, S; Morozenko, J; Kazencov, B. *Fizica Tverdogo Tela*, 1975, 17, 2970.
[132] Yu, PY; Shen, YR; Petroff, Y; Falicov LM. *Phys. Rev. Letts.*, 1973, 30, 283.
[133] Gross, E; Permogorov, S; Morozenko, Ya; Kharlamov, B. *Phys. Status Solidi* (b), 1973, 59, 551.
[134] Klein, MV. *Phys. Rev. B*, 1973, 8, 919.
[135] Carlone, G; Olego, D; Layaraman, A; Cardona, M. *Phys. Rev. B*, 1980, 22, 3877.
[136] Van der Ziel, JP; Meixner, AE; Kasper, HM; Ditzenberger, IA; *Phys. Rev. B*, 1974, 60, 4286.
[137] Koshel, W; Bettini, M. *Phys. Status Solidi*, (b), 1975, 72, 729.
[138] Yamamoto, N; Kitakuni, M; Susaki, M. *Jpn. J. Appl. Phys.*, 1995, 34, 3019.
[139] Yamamoto, N; Susaki, M; Huang, WZ; Wakita, K. *Crystal Res. Technol.*, 1996, 31, S369.
[140] Susaki, M; Yamamoto, N; Prevot, B; Schwab, C. *Jpn. J. Appl. Phys.*, 1996, 35, 1652.
[141] Terasako, T; Umiji, H; Shirakata, S; *Jpn. J. Appl. Phys.*, 1999, 38, L805.
[142] Matsumoto, T; Shimojo, T; Kumakyro, H; Uchiki, H; Iida, S. *Jpn. J. Appl. Phys.*, 2001, 40, 4077.
[143] Yamamoto, N; Susaki, M; Wakita, K. *Journal of Luminescence*, 2000, 87-89, 226.
[144] Permogorov, SA; Sel'kin, AV; Travnikov, VV. *Sov. Phys. Solid State*, 1973, 15, 1215 [Fiz. Tverd. Tela, 1973, 15, 1823].
[145] Permogorov, SA; Travnikov, VV; Sel'kin, AV. Sov. *Phys. Solid State*, 1972, 14, 3051

[Fiz. Tverd. Tela, 1972, 14, 3642].
[146] Ivchenko, EL; Sel'kin, AV. *Sov. Phys. JETP*, 1979, 49, 933 [Zh. Eksp. Teor. Fiz., 1979, 76,1837.
[147] Syrbu, NN; Nemerenco, L; Tezlevan, VE. *Optical Materials*, 2007, 30, 451.
[148] Levcenko, S; Syrbu, N; Tezlevan, V; Arushanov, E; Doka-Yamingo, S; Schedel-Niedrig, Th; Lux-Steiner, M. *Ch. J. Phys.: Condens. Matter,* 2007, 19 456222.
[149] Syrbu, NN; Nemerenco, L; Bejan, VN; Tezlevan, VE. *Opt. Commun*, 2007, 280, 387.
[150] Syrbu, NN; Tighinyanu, IM; Ursaki, VV; Tezlevan, VE; Zalamai, VE; Nemerenco, L. *Physica B*, 2005, 365, 43.
[151] Syrbu, NN. *Optics and Spectroscopy*, 1995, 79, 249.
[152] Bettini, M; Holzaptel, W. *Solid State Commun.*, 1975, 16, 17.
[153] Henry, CH; Nassau, K. *Phys. Rev.*, 1970, 2, 997.
[154] Weller, RG; Dimmock, JO. *Phys.Rev.*, 1965, 125, 1805.
[155] Lautenschlager, P; Garriga, M; Logothetidis, S; Cardona, M. *Phys. Rev. B*, 1987, 35, 9174, Cardona, M. *Modulation spectroscopy*, Academic: New York, 1969.
[156] Toyozawa, Y. *Progr. Theor. Phys*, 1958, 19, 214, Toyozawa, Y. *Progr. Theor. Phys.*, 1959, 20, 53, Toyozawa, Y. *Progr. Theor. Phys.*, 1962, 27, 89.
[157] Bir, GL; Picus, GE. *Pisma Zh. Eksp. Teor. Fiz.*, 1972, 15, 370.
[158] Bir, GL; Picus, GE. Zh. Eksp. *Teor. Fiz.*, 1974, 67, 788.
[159] Gross, EF; Erimov, AI; Razbirin, BS; Safarov, VI. *Pisma Zh. Eksp. Teor. Fiz.*, 1971, 14, 108.
[160] Vescunov, JP; Zacharcenia, VP; Leonov, EI. *Fiz. Tverd. Tela*, 1972, 14, 2678.
[161] Bonnot, A; Planel, R; Benoit a la Guillaume, C. *Phys, Rev. B*, 1974, 9, 690.
[162] Gross, E; Permogorov, S; Morozenko, Ya; Kharlamov, B. *Phys. Stat. Sol.*, (b), 1973, 59, 551.
[163] Permogorov, S. *Phys. Stat. Sol.* (b), 1976, 68, 9.
[164] Gross, E; Permogorov, S; Travnikov, V; Selkin, A. Proc. 2nd Int. Conf. *Light Scattering in Solids*, Paris, Flammarion, Ed. Balkanski, M. 1971.
[165] Syrbu, NN; Tiginyanu, IM; Ursaki, VV; Dorogan, AV. *J. Opt. A: Pure Appl.*, 2008, 10, 125002.
[166] Jaffe, IE; Zunger, A. *Phys. Rev. B*, 1983, 28, 5822.
[167] Rife, JC; Dexter, RN; Bridenbaugh, PM; Veal, BW. *Phys. Rev. B*, 1977, 16, 4491.
[168] Metzner, H; Eberhardt, J; Cieslak, J; Hahn, Th; Goldhahn, R; Reislohner, U; Withuhn, W. *Thin Solid Films*, 2004, 451-452, 241.
[169] Honda, T; Hara, K; Yoshino, J; Kukimoto, H. *J. Phys. Chem. Solids*, 2003, 64, 2001.
[170] Botha, JR; Branch, MS; Chowles, AG; Leitch, AWR; Weber, J. *Physica B*, 2001, 308-310, 1065.
[171] Syrbu, NN; Dorogan, AV; Ursaki, VV; Masnik, A. *J. Opt.*, 2010, 12, 075703.
[172] Syrbu, NN; Nemerenco, LL; Dorogan, AV; Tezlevan, VE; Arama, E. *Optical Materials*, 2009, 31, 970.

In: Exciton Quasiparticles
Editor: Randy M. Bergin

ISBN: 978-1-61122-318-7
© 2011 Nova Science Publishers, Inc.

Chapter 2

EXCITON RELAXATION DYNAMICS IN COLLOIDAL CORE AND CORE/SHELL CDSE NANORODS

A. Creti and M. Lomascolo

IMM-CNR, Institute for Microelectronics and Microsystems,
Lecce, Italy

ABSTRACT

In this chapter we focus on the exciton relaxation dynamics of colloidal CdSe core and core/shell nanorods. In particular we show how confinement effects and defect states affect linear and non linear optical properties. We present a brief introduction on the colloidal nanocrystals (NCs), namely on their electronic and optical properties, and on the recombination processes in these materials. Then we discuss the results of our systematic study of ultrafast exciton relaxation dynamics in CdSe core and CdSe/CdS/ZnS core/shell nanorods with a radius of few nm, different length (20-40 nm) and different shell thickness.

Femtosecond pump-probe spectroscopy, in the visible spectral range with non resonant / resonant pump energy has been performed to investigate the fast processes in NCs.

The effect of the shell thickness on Stimulated Emission (SE) and Photoinduced Absorption (PA) transitions in core and core/shell CdSe nanorods is exposed and the role of surface/interface defect states is pointed out. We show that the defect states distribution depends on the shell thickness and that the interface defects can be negligible for thin ones, resulting in a longer lifetime of SE. Furthermore we demonstrate that a resonant pumping increases the SE lifetime and enhances Auger scattering, clarifying that PA processes, involving defect states, are the main obstacle to sustain the SE. In the case of resonant pumping measurements the role of defect states on Auger processes and the presence of coherent confined acoustic phonons in CdSe core nanorods are also discussed. In particular we find that the modulation frequency observed in photobleaching and photoabsorption dynamics of core NCs, corresponds to the coherent radial breathing modes of the nanorods. Finally we quantitatively investigate, by time-resolved photoluminescence (PL) spectroscopy, the shell thickness dependence of exciton trapping and its effects on the PL quantum yield.

1. INTRODUCTION

Colloidal CdSe semiconductor NCs display interesting electronic and optical properties that depend strongly on size and shape [1-6]. Of particular interest for scientific and tecnological applications are spherical- and rod-shaped NCs, tetrapod and core/shell nanostructures. They result very promising as novel materials for biological labels [6-10], for optoelectronic as photovoltaic devices [11-21] and lasers [20, 22-29] and for gas-sensing technologies [30-33]. Under this perspective they have been intensively investigated over recent years in order to clarify their photophysics [34-43] and to study the exciton and/or multi-exciton relaxation mechanisms [20, 44, 45] eventually related to surface and interface defect states [35-40, 46, 47].

The aim of this chapter is to report on ultrafast exciton dynamics in core and core/shell CdSe Quantum Rods (QRs) and it is organized as follow: in Sect.1 some very general issues related to CdSe core and core/shell NCs are shortly discussed and the exciton relaxation processes are introduced. In Sect.2 we present the general feature of the three sets of QRs samples used in this study. Additionally, a brief introduction to the linear and non-linear optical techniques employed in their investigation, are given. In Sect.3, we describe the recombination mechanisms in these nanostructures, highlighting the role of surface and interface defect states. The shell-thickness effects on Stimulated Emission (SE) and Photo-Absorption (PA) processes competition are clarified in Sect.4. In the Sect.5 the role of defect states on radiative recombination process is studied and the shell-thickness dependence of emission quantum yield (QY) is reported. Finally in the Sect.6 the recombination processes are studied in the case of energy excitation resonant to the band-edge. In this frame we clarify the role of defect states on Auger processes and we demonstrate that the main obstacle to sustain SE can be attributed to PA processes rather than Auger ones.

1.a. Colloidal Nanocrystals

Strong interest in colloidal semiconductor NCs arises from their unique electronic and optical properties, which can be controlled by shape and size, allowed by the advances in the synthesis techniques achieved in the last years [10, 48-65]. A variety of interesting effects have been predicted and observed in nanoscale systems in which the electrons are confined in 3 - 2 and 1 dimension. Nanometer-size inorganic NCs in wide variety of shapes including spheres, rods, wires and tetrapods are intensively studied.

Colloidal chemical procedure ensures NCs with nearly atomic precision, with a size of few nanometers and a narrow size distribution (tipically ~ 5%). It is very important to obtain very small particles because size, comparable or lower than the material's Bohr-exciton radius a_B, induces strong confinement of the electron and hole pair, non correlated by Coulomb interaction. In particular, in the strong-confinement limit ($R/a_B << 1$), linear and non linear optical properties exhibit a greatest enhancement [3, 4, 66]. In this case a large level spacing and a strong size dependence on the band gap are observed. The scaling of band gap versus size induces many effects, intensively investigated in CdSe quantum dots (QDs), including variations of Coulomb and exchange interaction, band mixing, tunneling of electrons and holes through finite confinement barriers and changes in the dielectric constant.

In this confinement regime the different shape also, leads to different properties affecting the electronic state symmetries [67]. Rod-like semiconductor NCs, as example, induces a polarized emission along their long axis [53, 68, 69], resulting very valuable in biological tagging experiments where the orientation of the tag is required for the identification. Furthermore, in CdSe nanorods the gap between emission and absorption energies (Stokes shift) is larger than in NCs QDs [68], reducing the re-absorption process in optoelectronic applications such as LEDs.

In order to go more insight in the comprehension of the photo-physical processes occurring in these systems, several theoretical models have been developed to analyze energy levels and the optical and electronic properties as a function of NCs size and shape.

Quantum rods are an intermediate case between QDs and quantum wires structures, providing the opportunity to study the transition from zero-dimensional (0D) to one-dimensional (1D) quantum-confined system. In CdSe nanorods this transition occurs on a scale fixed by the Bohr radius. In particular, the ratio between the Bohr radius and nanoparticles length rules all the electronic and optical properties of the nanorods [54, 70-72]. In the case of strong confinement regime the band-gap of the QRs is only width dependent and then it is possibile to tune their emission wavelength in the visible spectral range only through the control of the rod diameter.

The electronic structure of nanorods is well described in literature by several models [73-76], including effective mass approximation and semiempirical pseudo-potential calculations, which provide a description of:

(i) band-edge variation vs diameter and length;
(ii) non monotonic change of global Stokes shift [68];
(iii) highly linearly polarized emission [69,77], in contrast to the plane-polarized emission from spherical quantum dots [78].

The exciton fine structure has been also investigated. Theoretical calculation have been used to describe the experimental results obtained in single nanorod optical spectroscopy [79, 80]. Effects of the electric field and shape on excitonic properties are also studied by effective-mass envelope function theory taking into account the spin-orbital coupling [81]. The exciton binding energy decreases with the increase of both internal electric field and aspect ratio. In the QRs with larger aspect ratio (> 2) the increased average distance between electron and hole, results in a less confined quantum structure.

1.b. Core/Shell Nanostructures

The quantum confinement effects induced in nanostructures by size and shape confer them the well known appealing features. On the other side, due to the high surface-to-volume ratio, the surface atoms play a very important role expecially in very small nanostructures. In particular the electronic states related to surface defects, strongly affect the physical and chemical properties of the semiconductor NCs, inducing, as example, reduced emission efficiency and/or photodegradation. Surface states strongly affect also the photoinduced processes of NCs [28, 38].

In their respect, surface structure are intensively investigated by both experimental [36-40, 82, 83] and theoretical approches [84]. As a matter of fact, in many semiconductor NCs applications, such as biological marker or optoelectronic devices, high emission efficiency from band-edge states and stability against photodegradation are strongly required.

Additionally also the interesting properties of QRs as laser active materials have driven the research for improvements in their emission QY and photostability.

Overgrowing the NCs with a few monolayer thick shell of a suitable higher band-gap semiconductor, improves the photoluminescence (PL) QY and the stability [85] mainly due to surface defect states passivation. This improvement depend on the shell thickness [86], and in addition the shell type and the shell thickness [85, 87-88] provide further control for tailoring the optical, electronic and chemical properties. In fact, to fully exploit NCs potentialities one should know in detail the electronic structure and its relationship to their composition.

PL QY in NCs tends to increase by increasing the shell thickness up to a critical thickness, typically in the range of 2-4 ML, which depends on both the core size and the lattice mismatch between core and shell material [88, 89]. A further increase in the shell thickness causes a noteworthy decrease in the QY. This behavior has been ascribed, in both QDs and QRs, to the generation of defects at the core/shell interface due to the lattice strain relaxation [83, 85, 90]. In fact, the strain relaxation at the interface introduces new extended defects that change NCs optical properties [36-40, 85, 86, 91]. Transmission Electron Microscopy (TEM) observations give evidence of misfit dislocations in such core/shell structures [88, 92].

To optimize the advantages due to shell overgrown around nanostructured materials, several methods have been proposed [82, 85-87]. Concernig the QRs, in particular, CdSe/CdS/ZnS and CdSe/ZnSe/ZnS core-shell-shell has been obtained [87, 89]. The growth of both middle and outer shell occurs in the regime of coherent epitaxy and with controllable thickness of both shells. Such double shell structures allow a stepwise change of lattice spacing from the emitting CdSe core and the protecting ZnS shell. This design considerable reduces the strain inside NCs, allowing the PL QY to increase. Furthermore highly emissive colloidal CdSe/CdS heterostructures of mixed dimensionality has been obtained in the last years, due to the strongly efficient shape control. In fact, different shape may be achieved in the shell of colloidal grown semiconductor NCs (independent on the core) allowing the combination of a 0-D spherical CdSe core with a 1-D rodlike CdS shell [10, 58, 63, 64, 93]. The formation of an elongated CdS epitaxial shell on spherical CdSe core results in a considerable increase of PL QY (70%-40% for nanorods with an aspect ratio of 2 and 4 respectively) as well as photostability [10, 58].

As the interface can drastically affects the optical properties, optical spectroscopy results useful in the investigation of these coherency-strain effects. They can be verified by measurements of PL QY changes, PL lifetime and studies of carrier/exciton dynamics, although the exact nature of the defects is not possible to derive only by the photophysical parameters.

1.c. Exciton Relaxation Processes

The quantization of electronic and phonon energies and the large surface to volume ratio, in these nanostructures strongly modify the energy relaxation mechanisms with respect to the bulk material.

Studies of intraband carrier dynamics by means of ultrafast optical spectroscopy, in CdSe nanostructures in regime of strong confinement, show extremely fast electron relaxation, despite the "phonon bottleneck effect" previsions [94-97]. The energy relaxation occurs, in fact, not only by phonon energy-loss mechanism, but also by intrinsic Auger type interactions, which mediate the e-h energy transfer. Additionally at low pump intensities, namely at low photocarrier excitation density, carrier trapping at surface/interface states competes with intrinsic decay of the interband relaxation processes. At high pump intensities, in the regime of two or multiple e-h pairs excitation, non radiative Auger recombination dominates the interband relaxation processes involving anyway the defect states [39].

The enhancement of Auger process in NCs, with respect to the bulk material, originates from the confinement effect, which enhances the Coulomb interactions, and from the relaxation of the translation momentum conservation, due to the reduced translation symmetry [66].

The shape of NCs also affects the relaxation processes. In nanorods the electron-hole dynamics is quite different respect to the QDs. The larger number of exciton states, arising from the splitting of the energy states, due to the decrease in symmetry, the restoration of momentum conservation along the c-axis and the increasing ratio between size and volume, leads to difference in intraband and interband exciton decay [98, 99]. The difference in the band-gap relaxation of QRs is ascribed to a less effective surface trapping with respect to the QDs one. In fact, the geometry plays an important role in the formation of topological surface states, due to the different surface curvature [100]. As a consequence, different surface states are expected in cylindrical and spherical nanoparticles. The faster intraband relaxation in QRs is also modified, due to the increase in the electronic states density and to the smaller energy separation, which increase electron-phonon and/or electron coupling processes. Finally, in the QRs the role of Auger recombination processes is also strongly modified. Despite the quasi-continuum density of energy states which simplify the energy-conservation, the efficiency of Auger process results strongly reduced, due to the restoration of momentum conservation along one direction, and in particular to the change from strong confinement regime to Coulomb-correlated particles (1D exciton), with increasing QRs lenght.

In the case of 1D exciton regime the recombination-energy transfer occurs by an exciton-exciton process resulting less probable. The reduced Auger recombination efficency has important consequence in the optical gain mechanism. Multiparticle Auger recombination and PA processes are in fact the limiting factors to sustain optical gain and SE in 0D semiconductor.

Concerning PA processes, they result strong sensitive to NCs size [25, 28]. In particular, the size dependence of gain threshold observed for CdSe QDs in solutions, results as a direct consequence of the size-dependent interplay between gain and PA. The PA contribution, arising from re-excitation of photo-carrier trapped at NCs interfaces, completely suppresses optical gain in NCs of small sizes. On the other hand it is reduced in QRs, due to the decreasing of NCs size/volume ratio.

Finally its magnitude is also sensitive to the type and to the quality of surface passivation, as well as, to the solvent/matrix material [28].

CdSe QRs show beneficial features for lasing applications compared to spherical nanoparticles, such as a reduced lasing threshold due to enhanced absorption cross section, and an increased lifetimes of optical gain, due to suppressed Auger decay and PA losses. They are largely investigated due to the ability to obtain optical gain and tunable amplified spontaneous emission (ASE) [12, 26, 101, 102].

2. Experimental

In this section we describe the colloidal core and core/shell CdSe QRs samples investigated in this chapter and, in particular we introduce their optical properties, their energy level-structure, and we discuss the role of the shell on the surface/interface quality.

Additionally a briefly presentation of the experimental techniques employed in the relaxation dynamics measurements is reported.

2.a. Samples: Core and Core/Shell Nanorods

CdSe core QRs samples of different radius and length and the corresponding core/shell CdSe/CdS/ZnS nanorods, here investigated, were prepared following the procedure described in ref [89]. In this chapter, in particular three sets of samples have been considered and their average sizes, as resulting from transmission electron microscopy (TEM) measurements, are reported in table 1.

All samples have been dispersed in chloroform solution and characterized, at room temperature, by conventional continuous wave optical spectroscopy. Absorption (abs), photoluminescence (PL) and photoluminescence excitation (PLE) spectroscopy measurements were recorded in order to have preliminary information about the electronic structures and the surface/interface quality.

The discrete structure of quantized states of semiconductor nanorods, resulting from abs measurements, in particular, has been compared to theoretical calculations (see in the next). PL QY, also reported in table 1 for each sample, were determined by comparing the integrated emission with that of rodamine 6G dye in ethanol solution with equal optical density at the excitation energy (2.54 eV). The QY values were corrected for the differences in refractive index between chloroform and ethanol.

The optical spectra of QRs NCs are dominated by quantum confinement effects: a series of discrete features appear in the abs spectra and optical excitation leads to band edge luminescence, as reported in the Figure 1, where abs, PL and PLE spectra of set_1 and set_2 samples are shown.

The broadening of optical transitions is due to the inhomogeneity arising from size dispersion and as a consequence from the contribution of different energy levels (see in the next). At highest spectral energy the broadening is also due to the increasing spectral density of electron –hole states. No evidence of impurity is observed in both abs and PL spectra. Nevertheless the observed small QY values confirm that in these samples the non radiative mechanism has a predominant role in the relaxation processes, which are the subject of this investigation.

Table 1. Sample sizes estimated by TEM images. The luminescence QY is also reported. More details are reported in the text. Concerning the shell thickness we assume that 1ML = d_{111} lattice spacing of the ZnS material of about 3.1 Å

	Sample	Diameter (nm)	Length (nm)	Aspect ratio L/R	Shell Thickness (ML)	QY (%)
Set_1	Core_D	3.5 ± 0.5	22 ± 3	6	-	0.4
	Core/Shell Ds	3.8 ± 0.6	24 ± 3	"	0.50 ± 0.15	4.0
Set_2	Core_L	3.6 ± 0.1	29 ± 5	8	-	0.3
	Core/Shell Ls	4.1 ± 0.7	30 ± 4	"	0.80 ± 0.25	3.0
Set_3	Core G	6.0 ± 1.0	37 ± 3	6	-	1.5
	Core/Shell_A	6.4 ± 1.2	"	"	0.60 ± 0.10	6.5
	Core/Shell_B	6.8 ± 1.6	"	"	1.30 ± 0.20	0.8

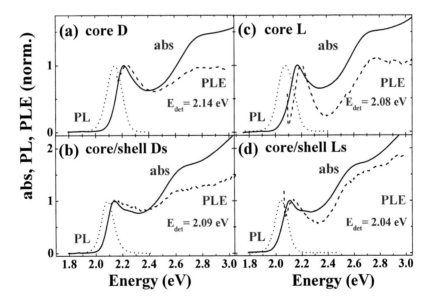

Figure 1. Absorption (solid-line), PL (dotted-line) and PLE (dashed-line) spectra of CdSe core (a)/(c) and core/shell (b)/(d) QRs of set1/set2 respectively. All spectra are normalized for comparison.

In order to clarify the nature of these processes PLE spectroscopy detected at the PL maximum has been performed at room temperature [more details in ref.37], suggesting trapping from the high energy states. In fact they follow the abs profile only in the energy range around the first abs peak in all samples. In particular a departure of the PLE spectrum signal from the abs lineshape of the core and core/shell samples of set_1, is observed for energy larger than 2.4 eV. This reduction results to be comparable for both samples and it becomes more and more evident as the energy increases. At highest energy (3.0 eV) the reduction is of about 30% and 25% in core and core/shell samples respectively.

The PLE lineshape for the core and core/shell samples of set_2, also follows the absorption lineshape only around the first absorption band. At energy larger than > 2.3 eV the signals are always below the abs one and the reduction of the signal increases as the energy increases.

In the PLE spectrum of core (core/shell) sample the signal reduction is of about 20% (10%) at 2.4 eV and of about 40% (20%) at 3.0 eV. Therefore in core/shell sample the presence of the shell reduces the difference between absorption and PLE signal of about 50%.

The strong deviation of PLE spectrum from the optical absorption profile is a strong argument in favor of trapping from the high energy states in these samples. These results in fact, clearly demonstrate that the carriers populating the high excited states can be efficiently trapped before relaxing to the first excited state.

Furthermore the reduced deviation in the core/shell samples of set_2, suggest that trapping of highly excited carriers is reduced by the thick-shell overgrowth, pointing out that trapping processes take place from surface states, passivated in core/shell sample [see in ref. 37].

The role of surface and interface defect states in the exciton dynamics is the subject of the next sections, where time resolved techniques are employed to study how confinement effects and shell-thickness affect the relaxation processes in QRs samples.

The different inter and intraband relaxation mechanism in core and core/shell samples, are studied and a model of defect states distribution for surface and/or interface defects is proposed.

2.b. Theoretical Model

As previoulsy exposed, QRs nanostructures are an intermediate case between 0D and 1D quantum-confined system. In particular strong confinement is induced in these nanostructures by a short radius value (R < a_B = 60 Å in the bulk CdSe) and the band-gap is tuned by the lateral confinement which play an important role even when the rods are very long. Nevertheless the elongation modifies the band-mixing [70] and as a consequence the band-gap also depends on the length.

As reported in Table 1, the core D and the core L, for example, have the same diameter but different length (20 nm vs 30 nm), and as a consequence the confinement energy in core L results reduced by the larger length (see abs spectra in the Figure 1 (a) and (c)). On the contrary, the abs spectra of these samples result strongly blu-shifted with respect to the one of core G (see Figure 21 in Sec.5), due to the different radius, as expected.

The transition between 0D to 1D regime occurs in the case of longer samples and as a general trend, the suitable parameter for estimate the confinement regime is the aspect ratio L/R.

Theoretical model calculate that in QRs with L/R >2, the electron-hole pair result less confined [79, 81] and 1D exciton appears. Furthermore it has been observed that nanorods with short radius (few nm) and L/R > 8 clearly show 1D excitons behaviour in the relaxation dynamics [99].

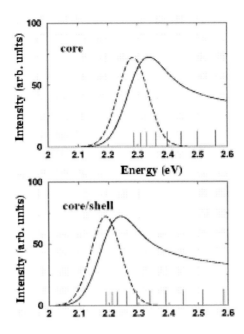

Figure 2. Calculated absorption (solid-line) and PL (dashed-line) spectra of CdSe core (top panel) and core/shell (bottom panel) QRs of set2. Reprinted with permission from [37], copyright 2005, American Physical Society.

In order to estimate the Coulomb interaction in our samples, we have determined by numerically solving the three-dimensional single-particle effective-mass Schrödinger equation [103] the hole-electron transition giving rise to the abs resonances, in core and core/shell samples of set_2 (L/R = 8) [for more details see ref. [37]].

The excitation energies have been calculated by considering single-particle electron-hole transitions with an exciton correction due to the electron-hole Coulomb attraction. For the rods here investigated, the calculated exciton correction was around 30-40 meV.

The presence of the shell reduces the lowest excitation energy by about 100 meV (comparable to the experimental value of = 80 meV). The shape of the absorption spectra is the result of the convolution of several energy transitions (shown as narrow lines in Figure 2) and as a consequence its peak does not correspond to the lowest energy transition, from which the PL originates. All the reported transitions have S-orbital character, due to the length of the rod. In particular, the calculated optical spectra accurately reproduce the experimental first main peak and the higher-energy side decay as well as the difference between the core and the core-shell sample.

The core-shell absorption and PL spectra are red-shifted with respect the core ones because in the former the electron wave function can tunnel into the shell, increasing the delocalization of the electrons that lowers the confinement energy and consequently the energy of the excited states [85, 88, 90]. Moreover, the red-shift of the absorption peak definitely rules out the formation of an alloy of the core and shell materials. In this case a blue-shift is expected due to larger energy gap of the alloy compared with the band gap of the CdSe. The Stokes shift between the PL maximum and the first absorption peak is assigned not only to the dark-exciton effect [104] but also to the shift between the lowest excited state (from which the PL originates) and the absorption peak which is the result of the convolution of the S-type states close to the QRs band-gap [37].

2.c. Spectroscopic Methods

Optical spectroscopy combined with ultrashort laser pulses is a powerful tool for the investigation of a wide range of phenomena related to non-equilibrium and non-linear properties of semiconductor. Fundamental informations about exciton and phonon relaxation dynamics can be obtained on femtosecond and picosecond time scales.

In our studies, for the measurements of exciton dynamics in nanoparticles, time-resolved laser techniques, including fs-transient differential transmission (pump and probe technique) and ps-time-resolved fluorescence have been employed.

The basic approach in the pump and probe experiment is to first excite the nanoparticles with a short femtosecond / picosecond light pulse generated from an ultrafast laser system. Then in the transient transmission, the excited-exciton is monitored with a second laser pulse (probe) that is delayed with respect to the excitation (pump) pulse. The change in transmission of second pulse indicates change in excited-state exciton population.

The action of the pump pulse on the sample may be analyzed in two different mode: i) by observing new effects created by the probe itself before and after the action of the pump pulse (Raman scattering, laser-induced fluorescence); ii) by comparing the modification of the probe pulse characteristics after crossing the sample, before and after the action of the pump pulse (time-resolved absorption technique).

The simplest pump and probe spectroscopy technique consists in the measurement of the pump induced change in the transmitted probe as a function of the time delay between the pump and the probe beams.

Another method to monitor the dynamics of excited state population is to time-resolve the luminescence from the excited states. This technique provides a very efficient investigation tool of the exciton dynamics in radiative and non radiative relaxation mechanisms. In this study, two dimensional streak camera in combination with a spectroscope has been used in the picosecond range.

In particular time-resolved PL was excited by 2-ps pulse delivered at an 82 MHz repetition rate by the second harmonic 360 nm line of a Ti: sapphire laser, in conjunction with a streak camera (time resolution ~20 ps).

Concerning the time resolved pump and probe experimental setup, it was based on an amplified Ti-sapphire laser which provides 150 fs, 750 µJ pulses at 780 nm, at 1 KHz repetition rate. The pump beam, at high energy measurements, was obtained by second harmonic generation (390 nm, 3.18 eV) in 1-mm thick lithium triborate crystal and it was linearly polarized at the magic angle (54,7°) with respect to the probe. In the case of measurements with energy resonant to the first abs peak of the sample, the output beam was used as the pump exploiting parametric amplification in a noncollinear optical parametric amplifier. The large spectrum probe beam was obtained by white supercontinuum generation in a 2-mm thick plate of sapphire and it is collected and focused onto the sample only by mirrors to minimize frequency chirp effect. Differential transmission spectra at a given pump and probe delay were recorded by an optical multi-channel analyzer averaging on a large number of laser pulses. Differential transmission dynamics were obtained using a standard lock-in technique and 10 nm band width interference filters to select the probe wavelength.

All the pump and probe experiments discussed in this paper were performed with nanorods dispersed in chloroform solution (room temperature) at the National Laboratory for Ultrafast and Ultraintense Optical Science (ULTRAS) facilities of CNR-INFM, Politecnico di Milano (Italy).

3. EXCITON RELAXATION PROCESSES: ROLE OF SURFACE AND INTERFACE DEFECT STATES

In this section we will focus on the study of optical nonlinearities induced in the QRs by photoexcitation with pump energy higher than the first absorption peak one (in non resonant excitation regime). Ultrafast pump and probe spectroscopy has been performed on CdSe core and core/shell nanorods of set_2 in the visible spectral range, in order to determine the energy relaxation processes and to investigate the role of surface and interface defect states[1].

3.a. Differential Transmission Spectra

In the pump and probe experiment, as previously introduced, the pump pulse excites the sample inducing population redistribution among the electronic levels and the probe pulse (time delayed) is used to measure the induced transmission changes, assuming that the perturbation due to the probe itself is negligible.

The data are typically presented in the form of normalized differential transmission given by: $\frac{\Delta T}{T_0} = \frac{T - T_0}{T_0}$, namely the change in transmission ΔT induced by the pump pulse is divided by the transmission of the probe in absence of the pump T_0. Changes in the transmission (absorption) spectrum are the signature of the sample. In particular, in the case of small signal changes (< 10%), the experimental observable, is the change of the optical density or absorbance:

$$\frac{\Delta T}{T} = \exp(-\Delta\alpha d) - 1 \approx -\Delta\alpha d \qquad (1)$$

Where d is the sample thickness and $\Delta\alpha$ the absorption change.

If the pump pulse duration is much shorter than the kinetic process under study (impulsive approximation), the change of the optical density can be described as:

$$\Delta\alpha d = \sum \sigma_{KJ}(\omega)\Delta N_K(t)d \qquad (2)$$

---

[1] The contents of Section 3 are related to the ref.[37].

Where σ_{KJ} is the absorption cross section and $\Delta N_K(t) = N_K(t) - N_K$ is the pump-induced change of the k-state population. Then the experimental signal $\frac{\Delta T}{T}$ accounts for the change in state population. We observe that $\frac{\Delta T}{T}$ signal depends also on the frequency ω of the pump pulse, which can excite different electron states [105].

In our discussion we will assume that the laser pulses duration (100 fs) are longer compared to the dephasing time of the sample. Then the differential transmission signal resulting from the measurements can contains three different types of signal:

(i) Photobleaching (PB)
(ii) Photoinduced Absorption (PA)
(iii) Stimulated Emission (SE)

PB is the reduction of the optical density in the absorption spectral range of the sample. In this picture it results to be a positive $\frac{\Delta T}{T_0}$ signal, due to two contributions both related to the absorption of pump beam: depletion of the ground absorption state and/or filling of the final state.

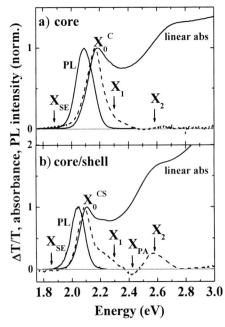

Figure 3. Linear abs and $\Delta T/T$ differential transmission spectra (dotted lines) recorded 4 ps after excitation at 3.18 eV with 200 fs – 160 µJ/cm^2, of (a) CdSe core and (b) CdSe /CdS/ZnS core/shell QRs of set_2. PL spectra obtained after excitation at 3.18 eV are also shown for both samples. All spectra are normalized for comparison. Reprinted with permission from [37], copyright 2005, American Physical Society.

PA is associated to a negative $\frac{\Delta T}{T_0}$ signal and it is due to the absorption of probe beam, involving an occupied photo-excited state (by pump absorption) and the higher lying empty levels.

Finally a positive signal in the spectral range of photoluminescence is attributed to SE.

For a more detailed discussion of pump and probe experiment and modeling see ref.105.

The normalized chirp-free differential transmission spectra of core and core/shell QRs of set_2, for an excitation energy of 3.18 eV and a pump and probe delay of 4 ps are reported in Figure 3 (dotted line). In these measurements an excitation fluence of 160 µJ/cm^2 has been used, corresponding to initial QRs population N_0 of few e-h pair per QRs on average (~3.5). This value is obtained as $N_0 = j_p\sigma_0$ where j_p is the excitation fluence and σ_0 is the absorption cross section of the nanorod at a pump spectral energy [98].

The linear absorption and PL spectra are also depicted for comparison in Figures 3a) and 3b), for CdSe core and CdSe/CdS/ZnS core/shell QRs, respectively.

We observe that in both samples a structured linear absorption spectrum with defined features is present. The absorption spectra show a first absorption maximum at 2.19 eV for CdSe core and at 2.11 eV for the core/shell sample. The redshift of the first absorption peak in the core/shell sample is an indication that a shell has been grown on the cores. Additionally the PL maximum is Stokes shifted from the first absorption peak by 100 and 70 meV, respectively. The PL signal provides an additional confirmation of the presence of the shell. In fact in the core/shell sample the luminescence quantum yield increases tenfold, going from 0.3%, in the core sample, to 3% [36, 85, 87-90].

In the differential transmission spectra presented in the Figure 3, we observe that three absorption bleaching peaks can be identified for both core and core/shell samples: $X_0^{C(CS)}$, X_1, X_2, centered at 2.17, (2.09) eV, 2.29 eV, and 2.58 eV, respectively. The first peak is redshifted of ~20 meV for both samples, compared to the linear absorption maximum. X_1, X_2 bands are less intense in core QRs, suggesting that higher photoinduced absorption is competing with bleaching effect.

The observed well-defined bleaching signals $X_0^{C(CS)}$ are attributed in both samples to the state filling mechanism [34]. In particular, the peaks observed are assigned to the convolution of the absorption bleaching of the different transitions present in the lowest energy side of the absorption spectra, as previously observed (see in the Figure 2.). Similarly, also the further bands observed at the higher energy side of the spectra X_1, X_2 are attributed to the convolution of the bleaching of higher lying states.

In NCs, band-edge optical nonlinearities arise primarily from state filling, leading to bleaching of optical transition and Coulomb multiparticle interactions, namely, the Stark effect, which results in transition energy red-shift. However, despite Stark effects should dominate at shorter delay time (<1 ps) [as reported in Refs. 34 and 106], we observe an energy red-shift also for pump and probe delay higher than 4 ps. We are currently unable to explain this effect in our experiment.

The bleaching band at the lower energy side is more broadened in the core sample spectrum with respect to the core/shell sample. In core QRs a region of positive $\frac{\Delta T}{T}$ is visible at energy lower than 2.0 eV while no clear evidence of positive signal is present in core/shell QRs below the absorption onset.

The positive $\frac{\Delta T}{T}$ signal in the emission energy range of both samples is attributed to SE as it is much larger that the expected absorption bleaching. The latter can be easily estimated by remembering that, in the limit of small signals, the absorption bleaching is proportional to the sample absorption (equation 1). For this reason, in the absence of PA and SE, the absorption bleaching and the linear absorption line shapes must be identical. Then the absorption bleaching contribution at the emission peak energy $\Delta T/T_{em}^{bl}$ can be estimated from the absorption bleaching at the X$_0$ energy $\Delta T/T_{X_0}^{bl}$, and from the linear abs at the corresponding energies Abs_{em} and Abs_{X_0}, by the relation:

$$\frac{\Delta T/T_{em}^{bl}}{\Delta T/T_{X_0}^{bl}} = \frac{Abs_{em}}{Abs_{X_0}} \qquad (3)$$

In our case the obtained value $\frac{Abs_{em}}{Abs_{X_0}}$ for the core (core/shell) sample is about 0.6% (0.2%), which is smaller that the $\frac{\Delta T/T_{em}^{bl}}{\Delta T/T_{X_0}^{bl}}$ observed signal, of about 2% (0.5%).[2]

Finally, concerning the PA signal we observe that the bleaching bands, at the higher energy, have a very small signal in the core sample with respect to the core/shell sample, suggesting a less important PA effect in core/shell sample in this energy range. On the contrary a well defined PA band centered at 2.42 eV is present only in the differential transmission spectrum of the core/shell sample.

As previously discussed, the PA processes attribution is an important step for the understanding of the light amplification processes in colloidal NCs. It is known that PA from defect states competes with the optical gain [25, 28, 107]. In our samples, in particular, we attribute the observed PA signal, modified by the presence of the shell, to processes involving superficial defect states. We observe in particular that the shell overgrown strongly reduces the PA processes at high energy, and it enhances them at low energy (see in the next).

The difference observed between the differential transmission spectra of core and core/shell sample indicates that SE and PA processes at higher energy are strongly modified by the presence of the shell.

The details of the defect states filling and depletion dynamics and of their energy position are discussed in the next sections, where on time-resolved differential transmission measurements at probe energy corresponding to the bleaching, photo-absorption and SE signals are reported.

---

[2] We observe that in our measurements the $\Delta T/T_{X_0}^{bl}$ is 15% (43%) for the core (core/shell) sample, then the approximation of small signal, is not rigorously true, then the use of equation 1 leads to overestimate $\Delta T/T_{em}^{bl}$ of about 8% (20%) of its value for the core (core/shell). As a consequence the value $\Delta T/T_{em}^{bl}$ should be even smaller than the estimated one, resulting anyway smaller than the signal observed in the emission region of both samples, which can be attributed to SE process [37].

3.b. Bleaching Dynamics

Time-resolved differential transmission measurements at high excitation energy (3.18 eV) and at several probe energies were performed on core and core/shell samples. The relaxation mechanisms inter and intraband and their dependence on the surface (interface) of core (core/shell) sample are investigated.

We first study the excited states associated with the bleaching signal, pumping with 160 µJ/cm^2 and probing the $X_0^{C(CS)}$, X_1, X_2 peaks. All the traces, shown in the Figure 4 for core (top panel) and for core/shell (bottom panel) samples, present a multi-exponential decay.

A quantitative analysis of the bleaching decay curves is performed using a four-exponential fit, which indicates that different relaxation processes are present. As a matter of fact, the experiments were performed on an ensemble of size-dispersed QR, hence all the features are inhomogeneously broadened. The observed relaxation dynamics arise from the contribution of all the single rod dynamics, which might depend on the rod dimensions [108] and the above mentioned process may contribute to the overall multi-exponential behaviour.

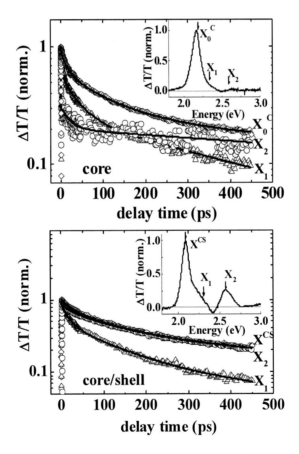

Figure 4. Normalized bleaching kinetics recorded at X_0^C = 2.17eV (X_0^{CS} = 2.10eV), X_1 = 2.29 eV and X_2 = 2.58 eV for the core (top panel) and core/shell (bottom panel) QRs samples of set_2, at an excitation fluence of 160 µJ/cm^2, together with multi-exponential best fits (continuous lines). Inset: normalized ΔT/T spectrum of core (top) and core/shell (bottom) sample respectively.

Table.2. Time constants for core and core/shell samples of set_2 [37]. All the constants are expressed in picoseconds

Core L	τ_{rise}	τ_{d1}	τ_{d2}	τ_{d3}	τ_{d4}
X_0 (N_0 =0.3)	0.280±0.025	0.80±0.010	15±1	220±45	1250±100
X_0 (N_0 =3.5)	0.480±0.020	1.5±0.1	15±5	128±2	1300±100
X_1 (N_0 =3.5)	0.345±0.025	0.7±0.1	20±5	130±10	1250±100
X_2 (N_0 =3.5)	0.210±0.030	0.220±0.010	22±5	140±10	1300±100
Core/shell Ls	τ_{rise}	τ_{d1}	τ_{d2}	τ_{d3}	τ_{d4}
X_0 (N_0 =0.8)	0.280±0.020	0.60±0.03	28±10	200±50	660±100
X_0 (N_0 =3.5)	0.990±0.030	9±1	27±10	122±10	565±20
X_1 (N_0 =3.5)	0.530±0.030	0.820±0.05	17±5	100±10	550±20
X_2 (N_0 =3.5)	0.280±0.015	0.487±0.014	22±5	120±10	550±20

The best fit values of the decay and the rise times, defined as time interval from the 10% to the 90% of the rising signal, are reported in Table 2 for core and core/shell samples respectively [37].

The main results can be summarized as follows:

(i) In the core and core/shell samples the rise time and the fastest decay time τ_{d1} continuously increase as the energy level decreases.

(ii) In the core/shell sample all rise times and all fastest decay times τ_{d1} are longer than the corresponding ones of the core QRs.

(iii) In the core/shell sample the fastest decay time τ_{d1} of the X_2 and X_1 feature coincides, within the fitting error, with the rise time of the nearest lower level, X_1 and X_0^{CS}, respectively.

(iv) Finally the three longer decay times τ_{d2}, τ_{d3} and τ_{d4} have the same value (within the statistical error) for all states in each sample.

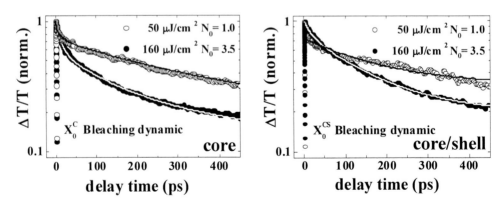

Figure 5. Normalized bleaching kinetics recorded at $X_0^{C(CS)}$ band in the core (left panel) and core/shell (right panel) samples of set_2 for different pump fluences together with multi-exponential best fits (continuous lines).

In order to investigate the role of the nonlinear relaxation processes, such as Auger recombination and/or exciton-exciton scattering on the system relaxation, we also performed differential transmission measurements at low excitation fluences on the $X_0^{C(CS)}$ and X_2 bands of the samples. Figure 5 shows the bleaching dynamics of the lowest exciton transition $X_0^{C(CS)}$ for the core (left panel) and core/shell (right panel) QRs, at high (160 µJ/ cm^2) and low (50 µJ/ cm^2) excitation fluences, corresponding to different values of the average numbers N_0 of e-h pairs excited.

The signal rise time and the different decay times show different dependence on the excitation fluence. As a general trend we observe that:

(i) As the excitation fluence increases from 16 µJ/cm^2 (N_0=0.3) to 160µJ/cm^2 (N_0=3.5) the signal rise time and the fastest decay time τ_{d1} increase from 0.280 to 0.480 ps and from 0.80 to 1.5 ps, respectively, in core sample. Also in the core/shell sample as the excitation fluence increases (from 50 to 160 µJ/cm^2) both the rise time and the first decay time constant increase from 0.280 to 0.99 ps and from 0.60 to 9ps, respectively.

(ii) The decay time τ_{d2} and the longest decay time τ_{d4} are instead independent of the pump fluence and they have the same value within the statistical error for all excitation fluences in each sample.

(iii) The decay time τ_{d3} decreases with the pump fluence and the contribution to the decay increases in both samples.

3.b.1. Intraband relaxation

From the analysis of the bleaching dynamics at high pump fluence, and in particular, from the analysis of the rise times (see in Table 2), we can deduce information about the intraband relaxation processes.

The normalized bleaching kinetics in the first 4 ps recorded at $X_0^{C(CS)}$, X_1, and X_2 band, for core (left panel) and core/shell (right panel) QRs samples, at high excitation fluence, with multi-exponential best fit are shown in the Figure 6 for more clarity.

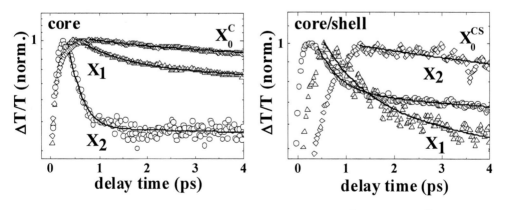

Figure 6. Normalized bleaching kinetics in the first 4 ps, recorded at X_0^C = 2.17eV (X_0^{CS} = 2.10eV), X_1 = 2.29 eV and X_2 = 2.58 eV for the core (left panel) and core/shell (right panel) QRs samples of set_2, at an excitation fluence of 160 µJ/cm^2, together with multi-exponential best fits (continuous lines).

We observe that in core sample, the initially pump populated states relax to the highest X_2 band in about 210 fs. It is interesting to observe that the fastest relaxation time of X_2 (τ_{d1}=220 fs) is faster than the X_1 rise time (~345fs), suggesting that the thermalization process to X_1 band, taking place in about 345 fs (rise time of X_1), is in competition with another depletion channel of X_2, which we attribute to carrier surface trapping. In particular we estimate the trapped carrier fraction by using the simple rule: $\tau_{d1}^{-1}/[\tau_{d1}^{-1}+(\tau_{rise})^{-1}]$ [37]. It results that only the 40% of photo-excited carriers reach the X_1 band, and subsequently they reach by thermalization the X_0^C band.

On the other hand from the analysis of the core/shell data it results that in the early 280 fs the carriers distribute themselves among few energy states at high energy (around X_2 feature). In the following 500 ps the population of the higher energy states relaxes rapidly to the lower states (X_1), and finally in the first ps in X_0^{CS} state. The fastes decay time τ_{d1} of X_2 and X_1 in fact coincides with the rise time of the following band, as previously observed. Then after the first ps the system relaxes by interband processes.

3.b.2. Trapping mechanism

Trapping in defect states is responsible for the fastest bleaching decay time (1.5 ps) of X_0 feature, in core sample. This is deduced by the comparison of the bleaching dynamics recorded in the first 4 ps and performed at low and high excitation fluence, as reported in the Figure 7 (left panel). In particular the fastest bleaching decay time τ_{d1} increases from 0.80 to 1.5 ps as the fluence increases from 16 to 160 µJ/cm^2.

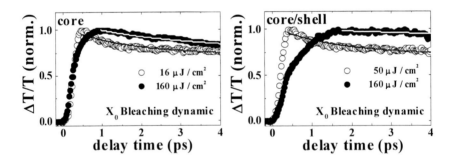

Figure 7. Normalized bleaching kinetics in the first 4 ps, recorded at $X_0^{C(CS)}$ band in the core (left panel) and core/shell (right panel) QRs samples of set_2 for different pump fluences together with multi-exponential best fits (continuous lines).

Note that the trapping mechanism becomes less important as the state energy decreases (τ_{d1} bleaching decay time becomes longer). We thus conclude that the defects trapping mechanism is more probable when the photo-excited carriers populate the highest energy states, suggesting an higher density of defects states close in energy to X_2.

The nature of the fastest relaxation mechanism from X_2 is confirmed by differential transmission measurements performed at different excitation fluences.

In Figure 8 (left panel) we report the relaxation dynamics of the X_2 bleaching band of core sample recorded in the first 4ps, at low (50 µJ/cm^2) and high (160 µJ/cm^2) excitation fluence.

At low excitation fluence the bleaching signal presents a very fast decay in the first 0.5 ps and then a clear increase occurs. This behaviour indicates that the absorption bleaching of the X_2 feature is spectrally overlapped to some photoinduced absorption signal that initially overcomes. As the excitation density increases the fast relaxation becomes slower and the photoinduced absorption is less evident. This process, which becomes slower by increasing the pump intensity, cannot be attributed to nonlinear recombination processes such as, for instance, Auger recombination, because this mechanism would lead to the opposite behavior.

This result, together with the observed reduced importance of the photoinduced absorption at the highest excitation density, strongly suggests that the rapid decay process comes from excited states depletion due to trapping, which is partially saturated at the highest excitation density.

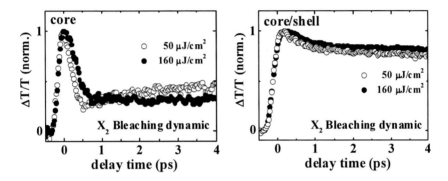

Figure 8. Normalized bleaching kinetics recorded at X_2 band in the core (left panel) and core/shell (right panel) QRs samples of set_2 for different pump fluencies.

In the core/shell sample the faster decay times τ_{d1} are longer than the corresponding ones of core sample in all the bleaching bands at high pump fluence.

This means that carriers depletion channel has been suppressed/reduced as we can see in the Figure 7. (right panel), where the X_0 bleaching dynamics of core/shell sample in the first 4 ps, for different pump fluence, are shown.

This conclusion is supported by the comparison of the feature relaxation dynamics in X_2 band for both samples (see in the Figure 8.)

In particular, this comparison clearly shows that the fraction of the bleaching signal recovered in the first 500 fs is larger in the core than in the core/shell sample, at the same laser excitation intensity. At 50 μJ / cm^2, 80% of the core signal recovers in 500 fs while only 15% of the bleaching signal recovers in the core/shell in the same time; at 160 μJ/cm^2, the 55% of the excitation recovers in the core sample, while only 10% of the excitation recovers in the core/shell one. We can thus conclude that different relaxation paths exist in the two samples.

The effect of fast carrier trapping by surface defects, quite evident in core QRs, is not present in the core/shell sample. The absence of further relaxation pathways in addition to intrinsic intraband relaxation processes in X_2 core/shell dynamics, clearly demonstrates the passivation of these surface defects by the shell overgrowth.

In summary we can deduce that in the core sample the bare surface defects introduce energy states which result resonant to the high energy intrinsic levels. These energy states are passivated by the shell and then the large increase of QY in the core/shell sample is due to an increase in the population of the emitting state, by eliminating electron trapping from higher excited states.

The reduction of the trapping from highly excited states due to the shell growth has been previously observed in the comparison between absorption and PLE spectra (reported in the Figure 1.c and 1.d in the Sec.2). The figure shows that also for the core/shell sample the PLE spectrum follows the absorption spectrum only for the X_0^{CS} band, but at higher energy the difference between the absorption and PLE spectra is clearly reduced with respect to the core sample, indicating that the trapping of highly excited carriers is reduced by the shell overgrowth.

Finally, at low excitation density, we observe that the fastest relaxation process (τ_{d1}) at X_0^{CS} dynamic is faster in core-shell sample, than the corresponding one at X_0^C in core sample. This suggests that in the core/ shell sample the carriers in X_0^{CS} can be trapped in new defect states, close in energy to this band, introduced by the shell. This conclusion is confirmed by the increase of τ_{d1} with the excitation fluence, due to defect saturation. Further evidence of the presence of new defect states in the core/shell sample will be discussed in the next.

3.b.3. Auger processes

The analysis of the bleaching decay dynamics of the lowest exciton transition $X_0^{C\,(CS)}$ as a function of the excitation fluence, shows also that in core and core/shell samples the Auger recombination process is responsible for the long decay time τ_{d3}.

In the core sample the relaxation from defects states takes place in about 130 ps (τ_{d3}) due to nonradiative processes. The nature of these nonradiative relaxation processes is clarified by the intensity dependence observed in the bleaching decay dynamics of the lowest exciton transition.

The decrease of τ_{d3}, as the excitation fluence increases, together with the increasing importance of this process (extimated by the amplitude of this component as resulting from the fitting procedure) suggest that the relaxation is due to Auger recombination. In the Figure 9 the best fit values (open squares) and the corresponding amplitude (solid squares) of the τ_{d3} component, as a function of the excitation fluence, are reported for clarity.

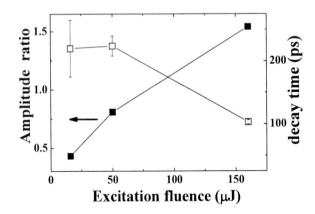

Figure 9. Best fit τ_{d3} value (open squares) and the amplitude (solid squares) of τ_{d3} component, relative to the slow component of bleaching kinetic recorded at X_0^C in core sample of set_2, as a function of the excitation fluence.

As the trapped carriers populate localized states, an Auger recombination between trapped carriers results to be very unlikely. The observed results can be instead explained by Auger relaxation of trapped carriers, involving scattering with free carriers [109,110]. This Auger process offers then an alternative carrier recombination mechanism after the initial surface trapping and, as the pump intensity increases, it becomes faster in time and more and more important. As this relaxation component is still present at the lowest pump fluence, we conclude that even under this excitation condition carrier-carrier scattering cannot be neglected.

The τ_{d3} value increases as the excitation fluence increases, also in the core/shell sample, but we can suppose that the Auger process involves different energy levels in this sample, where defect levels are passivated (see in the next).

3.c. Stimulated Emission and Photo-Absorption Processes

Finally, we studied the relaxation dynamics in the emission energy range of the samples where SE is expected to occurs. As observed in the differential transmission spectra, at low energy, positive signal is present, larger than the expected one only due to the bleaching, as previously discussed. However the differential transmission signal at the emission energy range of the sample ($\Delta T / T_{em}$) is, in general, given by the superposition of absorption bleaching ($\Delta T / T^{bl}$), stimulated emission ($\Delta T / T_{em}^{SE}$), and photo-induced absorption ($\Delta T / T_{em}^{PA}$). In the limit of small signals, which is valid in the emission range of both samples, these signals are additive.

In order to determine how long after the pump pulse the SE signal is stronger than the PA one, namely as long as SE is larger than PA, in our analysis we determine the time interval in which:

$$\frac{\Delta T / T_{em}^{bl} + \Delta T / T_{em}^{SE} + \Delta T / T_{em}^{PA}}{\Delta T / T_{X_0}^{bl}} \geq \frac{Abs_{em}}{Abs_{X_0}} = \frac{\Delta T / T_{em}^{bl}}{\Delta T / T_{X_0}^{bl}} \quad (4)$$

(the validity limit of the last identity has been already discussed), equivalent to the condition :

$$\Delta T / T_{em}^{SE} + \Delta T / T_{em}^{PA} \geq 0 \Rightarrow \left| \Delta T / T_{em}^{SE} \right| \geq \left| \Delta T / T_{em}^{PA} \right| \quad (5)$$

In conclusion, we consider the bleaching signal ratio as a net SE threshold, which allow us to estimate the upper limit for the optical gain lifetime (see in the next) [37].

3.c.1. Core sample

The kinetic recorded at probe energy of 1.88 eV (X_{SE} in Figure 3) features at an excitation fluence of 160 µJ/cm^2 shows, as evident from the Figure 10. (left panel), that a photoinduced absorption signal dominates in the first few hundreds of femtoseconds, while positive signal, possibly due to SE, is visibile for larger delays. The positive signal rise time is about 600 fs, while the decay is bi-exponential, with a fast decay time τ_{S1} of about 14 ps and a longer decay time τ_{S2} of about 1250 ps. The positive signal formation time is longer than the rise time of all the bleaching bands, while the two decay times are very close to the bleaching decay time constants τ_{d2}, τ_{d4}. The same trend has been observed for probe energy of 2.00 eV not shown.

The matching of the decay times for the bleaching and the positive signal at 1.88 eV suggests that the same relaxation process is involved in both signal relaxations. We then attribute τ_{d2} and τ_{d4} to the system relaxation from the emitting excited states.

Finally we analyze the ratio between the transient dynamics recorded at 1.88 eV and at X_0^C respectively in order to determine the interval of net SE, by means the procedure previously exposed. As reported in Figure 10 (right panel), the ratio results to be above the net SE threshold of $\frac{Abs_{em}}{Abs_{X_0}} \cong 0.6\%$ only in the first 50 ps after the pump pulse. As a stimulated emission stronger than the PA is a necessary condition to have gain in the sample, 50 ps can be considered as an upper limit for the gain lifetime in the core QR. In fact as the optical gain is essentially a biexciton phenomenon, its lifetime is limited by the Auger relaxation, whose rate increases as the excitation density is increased to reach the population inversion.

In our case net SE is observed in a time interval, shorter than the observed Auger decay time (130 ps), and comparable to the gain lifetime observed for this system in literature.

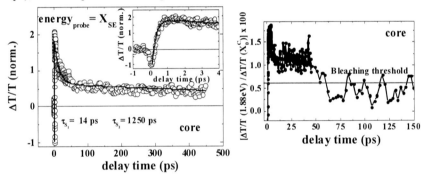

Figure 10. Left panel: ΔT/T dynamic recorded at 1.88 eV for the core QRs sample of set_2, at an excitation fluence of 160 μJ/cm² and best fit (continuous lines). Inset: zoom in the first 4 ps. Reprinted with permission from [37], copyright 2005, American Physical Society. Right panel: ratio between ΔT/T dynamic (in percentage) of core sample recorded at 1.88 eV and X_0^C. The continuous line represents the SE threshold resulting from linear abs spectrum [37].

3.c.2. Core/shell sample

We also recorded the $\frac{\Delta T}{T}$ dynamics of core/shell sample in the low energy side spectrum, where SE is expected. This is shown in Figure 11 (left panel) where we report the trace recorded at 1.88 eV, at an excitation fluence of 160 μJ/cm².

This $\frac{\Delta T}{T}$ dynamic shows a photoinduced absorption in the first picosecond, and then a positive signal as previously observed in core sample. The positive signal relaxation is biexponential, with a fast decay times τ_{S1} of about 19 ps and a longer one τ_{S2} of about 150 ps which are similar to the bleaching decay times τ_{d2} and τ_{d3}. This suggests that the time constant τ_{d2} is related to relaxation from the emitting excited states, as already observed in the core sample and τ_{d3} is related to the Auger recombination process. At delay time higher than about 70 ps the signal becomes negative, indicating that the PA process is stronger than both the absorption bleaching and the SE.

The SE maximum lifetime in the core/shell QRs (150 ps) is reduced with respect to the core sample (1250 ps) and this result clearly indicates that in the core/shell sample nonradiative relaxation channels from X_0^{CS} are present, that reduce its lifetime.

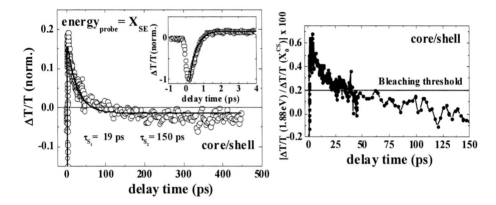

Figure 11. Left panel: ΔT/T dynamic recorded at 1.88 eV for the core/shell QRs sample of set_2, at an excitation fluence of 160 μJ/cm² and best fit (continuous lines). Inset: zoom in the first 4 ps. Reprinted with permission from [37], copyright 2005, American Physical Society. Right panel: ratio between ΔT/T dynamic (in percentage) of core/shell sample recorded at 1.88 eV and X_0^{CS}. The continuous line represents the SE threshold resulting from linear abs spectrum [37].

In order to estimate how long SE is stronger than PA, we analyze the ratio between the transient dynamics recorded at 1.88 eV and at X_0^{CS} respectively. The ratio, reported in Figure 11. (right panel), results to be above the net SE threshold of 0.2%, only in the first 25 ps. Then, in the core/shell sample the SE signal is smaller in amplitude, with a faster decay relaxation, and it is overcome by PA processes for time delay longer than 25 ps. Also in the core/shell sample 25 ps can be considered an upper limit for the optical gain lifetime.

Finally photoinduced absorption processes are observed in core/shell sample in the traces recorded at 1.77 and 2.43 eV (X_{PA} feature in Figure 3), at an excitation fluence of 160 μJ/cm² (Figure 12).

Concerning the PA attribution it is important to observe that at both the probed energies (1.77 and 2.43 eV) the PA relaxation dynamics is different from the SE one. The PA dynamic at 1.77 eV shows a very fast rise time and a nonexponential decay with a fastest decay time of about 300 fs and a longest one of about 464 ps; the signal at 2.43 eV is overcome by the bleaching in the first 1 ps, then PA dominates and the decay is biexponential with decay times τ_{PA1} of about 9 ps and τ_{PA2} of about 455 ps. We then conclude that the PA signals do not come from the intrinsic emitting states of the QRs and we attribute them to PA from defect states.

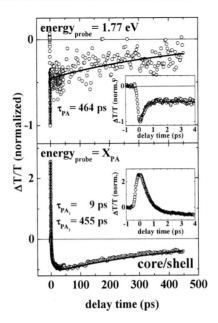

Figure 12. Photoinduced absorption at 1.77 eV (top panel) and at 2.43 eV (bottom panel) of core/shell QRs sample of set_2, at an excitation fluence of 160 Jµ/cm² and best fit (continuous lines). Inset: zoom in the first 4 ps. Reprinted with permission from [37], copyright 2005, American Physical Society.

In more details, we observe that the 2.43 eV PA shows a rise time (1 ps) similar to the X_0^{CS} bleaching and the SE rise time, and both the decay times (9 and 455 ps) comparable to the bleaching decay times τ_{d1} and τ_{d4} of the X_0^{CS} dynamics. This suggests that the 2.43 eV PA can be ascribed to PA from defect states close in energy to X_0^{CS}.

On the contrary the 1.77 eV PA shows a very fast rise time (about of 300 fs) and a non-exponential decay with a fast relaxation (300 fs) followed by a slow decay similar to the 2.43 eV PA. Moreover the signal intensity is much lower than the 2.43 eV one. These results suggest that 1.77 eV PA is partially due to PA from defects close to X_0^{CS} (as in the case of 2.43 eV feature) and partially due to PA from defect states close to X_2, as observed in the core, which are not completely passivated by the shell and with low density (as suggested by the low signal). These states are quickly populated from X_2 after excitation and then rapidly relax to lower lying defect states in about 300 fs.

Finally τ_{d1} and τ_{d4} bleaching time constants are assigned to relaxation from defect states to ground state.

The defect states close to X_0^{CS} involved in the PA processes, in the core/shell sample, can be ascribed to lattice strain relaxation, through the formation of extended defects at the core/shell interface, that cannot be ruled out in our samples, where a 0.80 ML thick shell has been overgrown, even if the graded CdS/ZnS shell is expected to reduce the lattice misfit between core and shell, compared to the usual ZnS shell. These defects can bring additional states in the midgap that act as traps and/or nonradiative recombination sites competing with the development of gain. So the shell decreases the high density of surface states and avoids carriers trapping which occurs mainly from the higher excited state, as found in the core sample. However the shell introduces additional states in the midgap, which are involved in the PA processes [37].

4. SHELL THICKNESS EFFECTS ON DEFECT STATES DISTRIBUTION

In order to study how length and shell thickness of the QRs affect the intra-inter band relaxation processes, we compare the ultrafast dynamics of QRs of set_2 with those of core and core/shell samples of set_1, which have the same diameter but different length (22 nm vs 29 nm) and shell thickness (0.5 ML vs 0.8 ML)[3].

The different length does not strongly modify the confinement energy and, as a consequence, the life time of the energy levels results to be only slightly affected. Nevertheless Auger relaxation mechanisms and PA processes strongly depend on the QRs aspect ratio [28, 99]. Then we expect that a length decreasing can enhance the probability of Auger and PA processes, and as a consequence SE results to be reduced or suppressed.

Finally SE and PA competition should be also modify in core/shell sample of set_1 (core/thin-shell), by means the reduced stress in the structure, resulting by a reduced shell-thickness and length, with respect to the core/shell of set_2 (core/thick-shell).

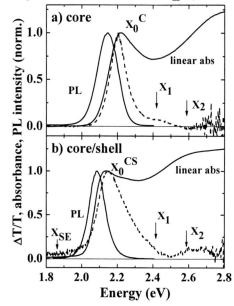

Figure 13. Linear abs and $\Delta T/T$ differential transmission spectra (dotted lines) recorded 4 ps after excitation at 3.18 eV with 200 fs – 160 μJ/cm^2, of (a) CdSe core and (b) CdSe /CdS/ZnS core/shell QRs of set_1. PL spectra obtained after excitation at 3.18 eV are also shown for both samples. All spectra are normalized for comparison.

4.a. Differential Transmission Spectra

The normalized chirp-free differential transmission spectra (dotted line) for a pump and probe delay of 4 ps and an excitation fluence of 160 μJ/cm^2, are shown in the Figure 13. for a) core and b) core/shell QRs sample of set_1 respectively. The absorption and PL spectra

---

[3] The contents of Section 4 are related to the ref.[38].

(continuous line) are also shown for comparison. We observe that the absorption spectra show a first absorption maximum at 2.21 eV for CdSe core and at 2.14 eV for the core/shell sample. The confinement energy in short core results lightly increased with respect to the long core, due to the smaller length, as expected [70]. The redshift of the first absorption peak, in the core/thin-shell spectrum, also results smaller than the one of core/thick-shell, due to the reduced shell thickness [85, 88, 90].

The presence of the shell, also in these samples, increases tenfold the PL QY, which goes from 0.4% in the core sample to 4% in the core/shell sample.

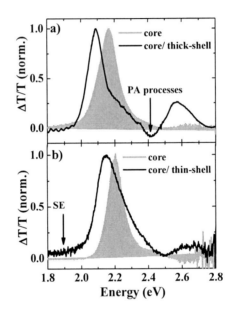

Figure 14. ΔT/T differential transmission spectra recorded 4 ps after excitation at 3.18 eV with 200 fs – 160 µJ/cm², of (a) CdSe core and core/shell of set_2 (b) core and core/shell of set_1. All spectra are normalized for comparison.

Concerning the differential transmission spectra, we observe that the bleaching band of the first exciton transition dominates the spectrum in both samples. At the energies of the allowed optical transition the nonlinear optical response is dominated by state filling leading to pronunciated bleaching band [34]. Absorption bleaching and photoinduced processes competition is present at higher energy. The overlap of PA signal decreases the positive bleaching signal with respect to the expected one (by looking at the absorption spectra). This is more evident in the spectrum of the core sample which shows a negative band at about 2.6 eV. In the spectrum of the core/thin-shell sample the differential trasmission signal at 2.6 eV become positive, due to PA reduction.

Finally the thin shell overgrowth leads to a positive signal in the emission region of the sample, which can be attributed to SE. Moreover the SE is not evident in the low energy region of core sample.

In the Figure 14 we highlight the difference between the differential transmission spectra of set_2 (Figure 14 a)) and set_1 (Figure 14b)) samples. The main results can be summarized as follow:

(i) No evidence of SE is present in short core spectrum, in contrast with the case of long core.
(ii) The bleaching band is much broadened in the emission range of core/thin-shell sample, indicating a remarkable increase of SE, with respect to the corresponding core sample and with respect to the previously investigated samples.
(iii) PA processes in the spectral range from about 2.2 eV to about 2.4 eV in core/thin shell result strongly modified with respect to the core/thick-shell. In this last sample the bleaching signal is strongly reduced and PA is more evident respect to the spectrum of the core/thin-shell, presenting a PA band centered at 2.43 eV.

It is important to note that in these samples the exciton levels are basically the same because the QRs have the same diameter and, as previously observed, the electronic structure is not affected by the different length [54], so the differences observed in the differential transmission spectra have to be ascribed to the differences in the defects distribution or density, which gives rise to different PA transitions. In particular the reduced SE in short core suggests higher Auger relaxation processes, due to the higher confinement [99]. On the contrary the increased SE in core/thin-shell suggests that the low energy defect states competing with SE in thick shell samples are not present in this sample. Finally the reduced PA at high energy, can be ascribed to a partial saturation of surface defect states, due to the thin shell. Then we can conclude that the differences observed between the two set of the samples are mainly related to the shell-thickness effects.

4.b. Bleaching Dynamics

Time-resolved differential transmission measurements were performed at several probe energies with a pump fluence of 160 µJ/cm^2, corresponding to an average number of electron-hole pairs excited N_0=2.5, in order to determine the role of defect states and their energy distribution.

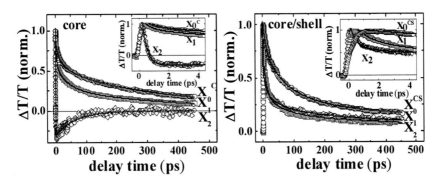

Figure 15. Normalized bleaching kinetics recorded at X_0^C = 2.21 eV, X_1 = 2.43 eV and X_2 = 2.58 eV of core QRs sample of set_1 (left panel). Normalized bleaching kinetics recorded at X_0^{CS} = 2.14 eV, X_1 = 2.38 eV and X_2 = 2.63 eV of core/shell QRs sample of set_1 (rigth panel) at an excitation fluence of 160 µJ/cm^2, together with multi-exponential best fits (continuous lines). Inset: zoom in the first 4 ps.

Concerning the bleaching dynamics recorded at $X_0^{C(CS)}$, X_1 and X_2 peaks and shown in the Figure 15 (left and right panel), we observe as a general trend, that all the dynamics follow a multiexponential decay, as previously observed in the core and core/shell samples of set_2.

The best fit values of the decay and the rise times are reported in table 3, for both samples.

The analysis of the bleaching dynamics in core and core/shell samples is in agreement with the results obtained in the previous section. In particular we observe that in both samples, X_2 band is immediately populated (within the time resolution of the experiment), then rapid intraband relaxation occurs by intrinsic thermalization processes from X_2 to X_1 and from X_1 to $X_0^{C(CS)}$ band, as we can deduce by the comparison between the fastest decay time and the rise time of the corresponding lower energy band (see in table 3).

Table 3. Time constants for core and core/shell samples of set_1 [38]. All the constants are expressed in picoseconds

Core D	τ_{rise}	τ_{d1}	τ_{d2}	τ_{d3}	τ_{d4}
X_0 (N_0 =0.3)	0.300±0.040	0.9±0.03	9.5±1.0	-	390±5
X_0 (N_0 =2.5)	0.460±0.030	1.2±0.05	13±1.0	79±5	448±24
X_1 (N_0 =2.5)	0.300±0.030	0.5±0.2	7.5±0.1	74±2	500±15
X_2 (N_0 =2.5)	0.190±0.030	0.360±0.001	17±10	42±10	500±100
Core/shell Ds	τ_{rise}	τ_{d1}	τ_{d2}	τ_{d3}	τ_{d4}
X_0 (N_0 =0.3)	0.350±0.008	-	12.0±2.0	-	687±9
X_0 (N_0 =2.5)	0.950±0.030	-	10.0±1.0	74±3	810±70
X_1 (N_0 =2.5)	0.360±0.030	1.00±0.05	10.0±0.5	63±9	613±38
X_2 (N_0 =2.5)	0.240±0.030	0.40±0.01	8.5±0.5	63±12	640±40

On the contrary, the lower exciton level $X_0^{C(CS)}$ shows a different behavior in core and core/shell samples. In fact a rapid decay time, of about 1 ps, is observed only in core sample signal, attributed to trapping in defect states, located at the core surface and passivated in the core/shell sample. In particular, we observe that the shell does not introduce new defect states in the midgap, in contrast to the thick shell, where these new defects are responsible for the fastest relaxation process also observed in X_0^{CS} (see for details in the table 2). The presence of trapping mechanism is also confirmed by measurements performed at lower excitation fluence as we will see in the next.

Concerning the X_2 bleaching band dynamics we observe that in core sample PA processes overcome and the bleaching signal is observed only in the first 500 fs, then a very fast decay occurs, comparable to the rise time of the bleaching bands, and the signal become negative. See in the Figure 15 (left panel). The reduced (increased) bleaching (PA) signal in short core sample, with respect to the long one, can be ascribed to an increased role of defect states, due to the reduced dimension of the sample. In fact, also in the X_2 bleaching band dynamics of long core PA processes and bleaching signal compete, but in that sample, as reported in the previous section, the PA overcomes only in the first 500 fs.

In order to investigate the role of the non-linear relaxation processes, such as Augur recombination and/or exciton-exciton scattering, on the energy relaxation, we also performed time-resolved differential transmission measurements at low excitation fluences on the $X_0^{C(CS)}$

of core and core/shell samples of set_1. Figure 16 (a) shows the bleaching dynamics of the lowest exciton transition of core (circles) and core/thin shell (squares) at 160 µJ/cm² (solid symbols) and 16 µJ/cm² (open symbols) excitation fluences. In the Figure 14 (b) and (c) the zoom in the first 4 ps are also reported for both samples. As the excitation fluence increases from 16 µJ / cm² to 160 µJ / cm² we observe that:

(i) The rise time and the fastes decay time τ_{d1} (in core sample) increase
(ii) The decay time τ_{d2} and the longest one τ_{d4} do not depend on the pump fluence
(iii) The decay time τ_{d3} is present only at high excitation fluence

The increase of the fast decay τ_{d1} in core sample and the absence in core/thin shell also at low excitation fluence confirm that the decay process are related to defect states at the core surface. It increases, as the pump fluence increases, due to the saturation defect states.

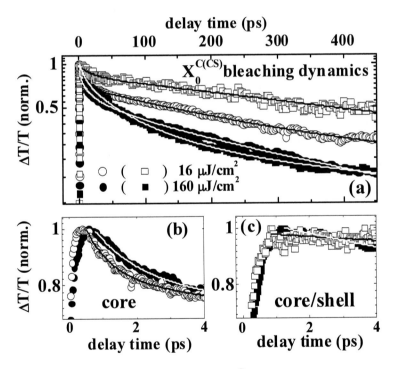

Figure 16. (a) Normalized bleaching kinetics recorded at X_0^C band of core (circles symbols) and at X_0^{CS} band of core/shell (square symbols) QRs samples of set_1, at low (empty symbols) and high (filled symbols) pump fluence, together with multiexponential best fits (continuous lines). Zoom in the first 4 ps for (b) core and (c) core/shell QRs samples.

The decay time τ_{d3}, observed at high excitation fluence, in both samples, is attributed to Auger recombination. In particular it is important to note that this process is faster in shorter rods with respect to the longer ones, even though the population of excited carriers is reduced (2.5 vs 3.5), as expected from the higher confinement [99].

4.c. Stimulated Emission and Photo-Absorption Processes

Finally we study the relaxation dynamics at low energy, in the emission range of the samples.

SE was measured at 2.03 eV for the core and at 1.98 eV for the core/thin-shell sample. The traces are shown in Figure 17. in the left and right panel, respectively. These dynamics show PA signal in the first 700 fs, involving defect states, and then a positive signal. The PA signal in thin-shell sample is smaller than the one observed in core QRs. This confirms the reduction of the trapping from highly excited states due to the shell growth. The positive signal can be fitted with a bi-exponential decay. The decay time constants are similar to the bleaching ones τ_{d2} and τ_{d3} in both samples. This matching suggests that the relaxation processes involved in both signals are the relaxation from the emitting exciton state (τ_{d2}) and Auger recombination process (τ_{d3}). We observe that in both samples the strong Auger recombination prevents to observe in the SE signal the typical long radiative lifetime of about 1 ns in these structures [111].

Finally, following the procedure explained in the previous section we determine the upper limit for gain lifetime, by determining how long the SE signal is stronger than the PA one, after the pump pulse. SE is achieved in core and in core/thin-shell samples, but it is sustained for longer time in the core/thin-shell sample (150 ps vs 40 ps of the core sample) as we can see in the Figure 18. (a) and (b). Moreover in core/thin-shell the SE lifetime is strongly increased with respect to thick-shell QRs, characterized by a very short SE lifetime (25 ps) due to defects which trap the carriers below the absorption onset, resonant to the lowest energy emitting states. So it appears that thin shell enhances SE processes, with respect to the bare core and the thick-shell QRs, because it reduces defect states that trap the excited carriers, causing PA which may compete with SE.

Finally, we observe that the τ_{d4} bleaching lifetime is instead attributed to relaxation from the defect states. This decay time is in fact not observed in the SE dynamics (thus ruling out a relaxation process of the emitting states) and it is in agreement with the lifetime of the PA at low probe energy (1.90 eV and 1.98 eV) in the core sample shown in the Figure 19.

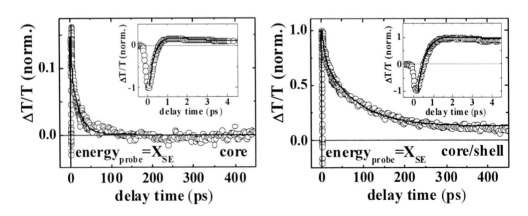

Figure 17. Time evolution of the SE signal for the core (left panel) and core/shell (right panel) QRs samples of set_1, recorded at 2.03 eV and 1.98eV respectively, together with multiexponential best fits (continuous lines). Inset: zoom in the first 4 ps.

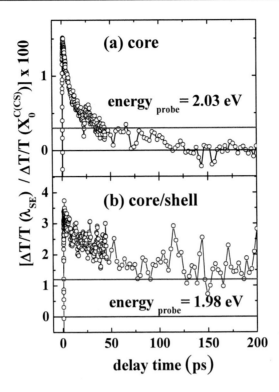

Figure 18. Time evolution of SE signal (more details in the text) for the (a) core and (b) core/shell QRs samples of set_1. The continuous lines represent the SE threshold resulting from linear absorption spectrum for each sample.

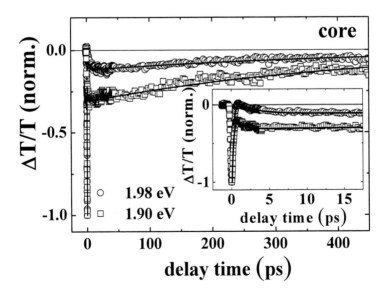

Figure 19. Photoinduced absorption at 1.98 eV (circles symbols) and at 1.90 eV (squares symbols) of core QRs sample of set_1, at an excitation fluence of 160 J/cm^2 together with multiexponential best fits (continuous lines). Inset: zoom in the first 4 ps.

In summary we can deduce the surface and interface defect states distribution in core and core/shell nanorods, schematized in the Figure 20.

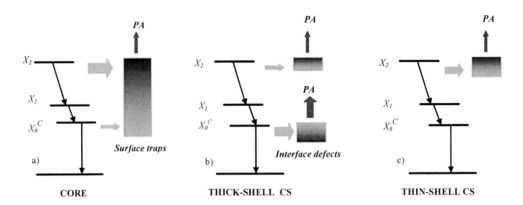

Figure 20. Energy level diagram for (a) the core sample, (b) the core/ thick-shell sample and (c) the core/thin-shell sample. Reprinted with permission from [38], copyright 2006, American Physical Society.

In core QRs the bare surface defect states are resonant to the high energy intrinsic levels and they can be passivated by the shell. However, the shell can introduce new defects which are resonant to the midgap. As a result the shell reduces carrier trapping from high energy defects and increases the PL QY, but causes new PA processes from midgap defect states that compete with SE, as observed in core and core/thick-shell sample. When QRs are coated with a thin-shell, the reduction of high energy defects density is less efficient, nevertheless the QY is enhanced and the thin-shell does not introduce new defect states at the midgap, because the strain relaxation at the core-shell interface results reduced with respect to the one in the core/thick-shell.

As a consequence PA processes are not present at low energy and SE lifetime is longer than the one observed in core/thick-shell. These results demonstrate that the improvement of both the PL QY and the optical gain requires an effective passivation of the high energy surface states without the introduction of low energy defect states. As the last states are due to surface strain relaxation, a thinner shell can ensure a good compromise between the two processes.

5. ROLE OF THE SHELL THICKNESS ON EXCITON EMISSION PROCESS

The role of surface and interface defect states in core and core/shell QRs has been demonstrated in the previous sections, by pump and probe experiments. However, the overlap of interband and intraband signals in pump and probe signal makes extremely difficult to quantitatively investigate the defect states effects on the interband relaxation processes, like light emission.

In this section we study the different contribution of surface and interface defect states on exciton emission decay in colloidal CdSe core and CdSe/CdS/ZnS core/shell QRs by time-resolved photoluminescence (PL) spectroscopy, at room temperature.

The role in PL QY of surface trapping mechanism and the contribution of defect states introduced by strain release at the core/shell interface are evaluate quantitatively.

Additionally the PL decay traces investigated in the PL broadening of the sample give information about the confinement regime in QRs samples[4].

5.a. Experimental Results

We analyzed the PL relaxation dynamics in the samples of set_3, namely three selected samples: a bare CdSe core QR and two core/shell A and B, with shell thickness of 0.6 monolayer (ML) and 1.3 ML respectively (for more details see in Table 1, section 2.a.).

The abs and PL spectra of the samples of set_3 are shown in the Figure 21. The red shift between the first abs peak of core and core/shell sample increases as the shell thickness increases (60 meV vs 90 meV), as expected [85, 88, 90]. The luminescence QY, as reported in table 1, increases from 1.5% in the core sample to 6.5% in the core/shell A sample (core / thin-shell sample), but it results reduced in the core/shell B sample (core / thick-shell sample) down to 0.8%.

It is known that in QDs and QRs core/shell samples, the PL QY tends to increase by increasing the shell thickness up to a critical thickness, which depends on both the core size and the lattice mismatch between core and shell materials [85, 87, 89]. In spherical quantum dots with a diameter of few nm the typical critical ZnS or CdS shell thickness results of about 2ML [85, 86, 93]. In the case of elongated nanoparticles, the critical shell thickness depends on radius, length and their aspect ratio [87], due to an increased interfacial strain respect to spherical dots, induced by the increased surface. Also a partial surface coverage can affect the properties in these samples. In fact as previously observed in section 3, a different percentage of coverage (0.5 ML vs 0.8 ML) gives different effects on SE lifetime and PA processes.

Concerning the origin of the emitting states we observe that in general radiative relaxation can be observed from intrinsic and defect states. However, defects-related emission, for example from dangling bonds, is often evidenced by clear emission bands red shifted with respect to the intrinsic emission, which is not observed in any of the investigated samples. Furthermore all samples exhibit a luminescence band with a FWHM of about 100 meV (150 meV for the core/shell B) which are comparable to the FWHM of the first abs peak band. For this reason we attribute the observed PL to radiative relaxation from band-edge recombination process and its broadening is due to the size dispersion. The increased broadening in core/shell B arises from the increased size distribution (see in table 1).

In order to determine the shell effects on the PL relaxation processes we carried out time-resolved PL measurements with an average excitation power of about 35 mW, corresponding to an average number of excited electrons in each QR $N_0 \ll 1$. A reduced excitation power reduces the biexciton formation and consequently Auger processes.

---

[4] The contents of Section 5 are related to the ref. [40], copyright 2008, with permission from Elsevier.

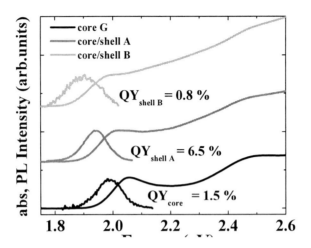

Figure 21. Linear absorption and PL spectra obtained after excitation at 2.54 eV of CdSe core and CdSe/CdS/ZnS core/shell QRs samples of set_3. All spectra are normalized and translated for comparison. The PL QY is also reported for each sample.

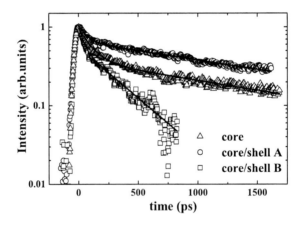

Figure 22. Time-resolved PL-normalized traces recorded at the peak of the PL emission, at an excitation of 35 mW, for CdSe core (triangles symbols), core/shell A (circles symbols) and core/shell B (squares symbols) QRs samples of set_3. The best-fit (continuous lines) are also shown. Reprinted from [40]copyright 2008,with permission from Elsevier.

The PL dynamics of the three samples, registered at the PL peak of each samples are shown in the Figure 22. The traces show a resolution-limited rise time, and a non-exponential relaxation, with a faster decay in the first 100 ps after the laser pulse, and a slower decay for larger times.

The experimental data are well reproduced by a bi-exponential fit function:

$$I_{PL}(t) = A_1 e^{-(t/\tau_1)} + A_2 e^{-(t/\tau_2)} \qquad (6)$$

It results that in core and in core/thin-shell sample very similar decay times are present. We observe a fast decay time $\tau_{1C}=115\pm 18$ ps in core sample and $\tau_{1CSA} = 85 \pm 20$ ps in core/shell A sample and the same slow decay time constant $\tau_2 = 1800 \pm 200$ ps in both samples.

On the contrary faster relaxation is observed in the core/thick-shell sample, with a fast and slow decay time reduced to $\tau_{1CSB} = 30\pm18$ ps (comparable to the time resolution of the system) and $\tau_{2CSB}= 320 \pm 27$ ps, respectively.

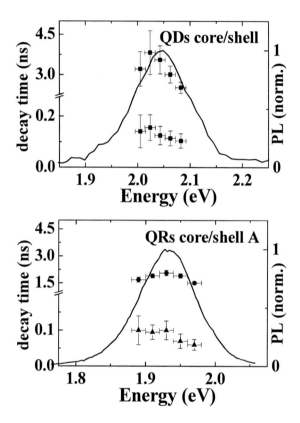

Figure 23. Time constants as a function of emission energy for CdSe core/shell QDs (top panel) and core/shell QRs of set_3 (bottom panel). The time –integrated PL spectra (continuous lines) are also shown.

Finally the dependence of the time constants on the emission energy within the inhomogeneous PL broadening has been detected in all samples and the resulting trend for the core/shell A are reported in Figure 23 (bottom panel) for example. The confinement energy in the nanostructures in fact results size-dependent and these samples have large size dispersion, then we can study the recombination dynamics in particles that are progressively smaller, by monitoring decay times at increasing emission energies.

5.b. Discussion

The very similar values of τ_1 and τ_2 in the core and in the core/shell A suggest that the same relaxation processes are present in both samples. However, the relative contribution to the relaxation of the fastest process is clearly reduced in the core/shell A with respect to the core (see Figure 22). This result together with the PL QY increase in the core/shell A suggest that the faster process is of non-radiative origin, which is reduced by the shell overgrowth, while the slowest process is related to samples emission. We suggest that τ_1 is due to electron trapping in defect states localized at the nanocrystal surface and/or at the core/shell interface [35].

A quantitative estimate[5] of the trapping in the relaxation mechanism can be performed by observing that the collected PL is proportional to the population of the emitting state, through the radiative rate $1/\tau_{rad}$ and a scale factor B, depending on the collection setup geometry and efficiency. Assuming that the radiative rate is the same for all the NCs, we can rewrite the fitting function in the form:

$$I_{PL}(t) = B\left(\frac{n_1}{\tau_{rad}}e^{-(t/\tau_1)} + \frac{n_2}{\tau_{rad}}e^{-(t/\tau_2)}\right) \qquad (7)$$

where n_1 and n_2 are the number of excited QRs that relax due to electron trapping and light emission, respectively. The fraction $f_i = \dfrac{n_i}{n_1+n_2}$ of QRs decaying, due to the two processes, can be then estimated from the best-fit values of A_i in Eq. (6) as $\dfrac{A_i}{A_1+A_2}$.

In order to estimate the QY variations induced by both the variations of the trapped excitons fraction and the decay times, we observe that the PL QY is given by:

$$QY = f_2 \frac{\tau_2}{\tau_{rad}} \qquad (8)$$

In this formula we considered that only the fraction of non-trapped excitons f_2 contributes to the emission. After some algebra, and assuming that the shell modifies only the non-radiative decay rate, while leaving the radiative rate unaffected (the radiative rate only depends on intrinsic properties of the samples, which are almost identical in the three investigated samples) it is possible to rewrite the QY of core/shell samples as:

$$QY_{CS} = \frac{f_{2CS}}{f_{2C}} \times \frac{\tau_{2CS}}{\tau_{2C}} \times QY_C \qquad (9)$$

---

[5] The following discussion is taken from ref. [40], copyright 2008, with permission from Elsevier.

We then estimate that the percentage of excitons that decay non-radiatively due to electron trapping is about 60% in the core sample, and it is reduced down to about 30% in the core/shell A. We also observe that, even if the percentage of trapped excitons is reduced by a factor of 2 in the core/shell A, the PL QY increases by a factor of about 4.3 (from 1.5% to 6.5%). This indicates that the trapping reduction from the emitting states, namely from defect states in the band gap, is not enough to fully explain the observed PL QY increase.

We suggest then that the overgrowth of a thin shell results not only in a two times reduction of the exciton trapping from the emitting states, as demonstrated by the f_1 decrease, but also in a reduction of about two times of the exciton trapping from higher energy defects states during the exciton relaxation from the pumped state to the emitting one. These two effects then lead to the observed 4.3 times increase of the PL QY.

These conclusions are in agreement with pump probe results in similar samples, previously discussed in the Sec.3 and Sec.4 [37, 38].

Concerning the sample B, we observe that the overgrowth of a thicker shell results in a reduction of both the trapping time and the emission one. Moreover, a thicker shell also increases the fraction of trapped excitons f_1 up to about 62%, which is comparable with the value of the core sample. These results confirm that the main effect of the thick shell is the introduction of new defects, in the midgap, as previously observed in sec.3 [37], thus allowing a stronger trapping of the increased population of band gap excited states and further non-radiative decay processes for the emitting excitons. The generation of defects in the core/shell B sample is attributed to the lattice strain released [85, 87, 89, 90] at the core/shell interface.

Additionally substituting in Eq. (9) the experimental values, we find $QY_{CSB} \approx 0.6\%$, which is in reasonable agreement with the experimental value of 0.8%.

Finally, we observe that radiative recombination from band-edge states is independent from the nanorods size in all investigated samples as we can see in the Figure 23 (bottom panel) where the time constants resulting from the different emission energy decay, registered within the inhomogeneous PL broadening, for the core/shell samples A are reported. The time constants are almost the same within the inhomogeneous PL broadening. The same results have been obtained in core and core/shell B (not shown) [40].

This result suggests that in the nanorods samples, which have an aspect ratio of 6, the small radius does not assure the strong confinement regime, but the levels structure evolves to levels structure of one-dimensional system [54, 70].

In the case of 0D nanostructure in fact the lifetime increases linearly with the radius, as we can see in the Figure 23 (top panel) where the dependence of the time constants on the emission energy within the inhomogeneous PL broadening for a core/shell QDs sample, with the same diameter of QRs here investigated, is reported. The linear dependence on R is related to the confinement effect: in smaller dots the e/h pairs are closer to each other, inducing to an enhanced overlap integral of their wave function, proportional to the optical trend rate. Furthermore the oscillator strength in CdSe QDs for the first transition also results proportional to the radius [36].

6. EXCITON RELAXATION PROCESSES AT RESONANT PUMP ENERGY

In order to study the exciton relaxation processes in core and core/shell nanorods at the band-edge, we have performed femtosecond pump and probe experiments with a pumping energy resonant to the first abs peak of each sample.

It deserves to be stressed that the resonant pumping is a non trivial experiment by itself and that the initial electronic/hole distribution among the available energy states opens a completely different scenery concerning the relaxation paths, if compared to the non resonant pumping condition.

In this way we are able to clarify the competition between Auger and PA processes. In particular we demonstrate that Auger processes involve defect states at high energy and that contra-intuitively they are not the main obstacle to sustain SE (namely optical gain). In fact even though it results increased in the case of resonant pumping the SE lifetime increases, due to the reduction of PA processes.

6.a Differential Transmission Spectra

The normalized chirp-free differential transmission spectra, for a pump and probe delay of 4 ps, an excitation fluence of 160 µJ/cm^2 and an excitation pulse of 90 fs, resonant (R) to the first abs peaks of each sample, are reported in Figure 24 (filled-spectra) for core and core/shell QRs of set_1(a)/(b) and set_2(c)/(d). The differential transmission spectra registered at non resonant (NR) excitation energy (3.18 eV) (gray continuous lines) and previously discussed in Sec.3 and Sec.4 for set_2 and set_1 respectively, are also reported for comparison.

The linear absorption and PL spectra are also depicted in the figure for CdSe cores and CdSe/CdS/ZnS core/shell QRs samples.

In the differential transmission spectra with R pumping excitation of all samples we observe one bleaching band, due to state filling effects and characterized by a full width at half-maximum, comparable to those of the absorption and PL spectra, as expected. In the high energy range, the PA processes dominate while the PB signal is strongly reduced, with respect to the one observed in the NR measurements.

In the short core and core/thin shell samples (set_1 samples, left panels in Figure 24) the PA prevails at high energy. In particular the signal, at energy larger than 2.3 eV, is much smaller than the expected absorption bleaching and the strongest reduction is present in core sample.

PA prevails and in particular a negative signal at an energy of about 2.45 eV (2.40 eV) is observed in the long core (core/thick-shell) spectrum (set_2 samples, right panels in Figure 24).

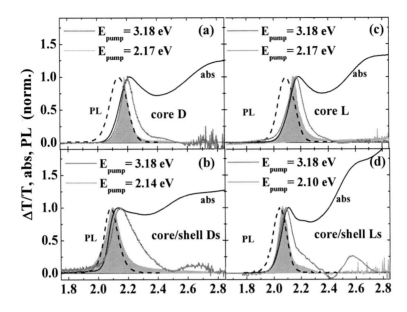

Figure 24. Linear abs (dark line) and ΔT/T differential transmission spectra recorded 4 ps after excitation at 3.18 eV (grey line) and at pump energy resonant to the first abs peak (gray area) of (a) / (c) CdSe core and (b) / (d) CdSe/CdS/ZnS core/shell QRs of set_1 / set_2. PL spectra obtained after excitation at 3.18 eV are also shown for all samples. All spectra are normalized for comparison.

The PB signal at high energy is more reduced in the core/thick-shell spectrum with respect to that of the long core one (set_2). In the core/thick-shell spectrum in fact we observe an almost zero signal due to the competition between PA and PB processes, thus confirming that PA processes are enhanced in this sample, by means the presence of mid-gap defect states localized near the interface and which are filled during the first ps after the pump pulses [101].

6.b. Bleaching Dynamics

In order to clarify the effects of R pumping in the exciton relaxation mechanisms, we perform time-resolved differential transmission measurements in the region of bleaching signal (X_0 band). The pump fluence dependence of the bleaching dynamic of the samples, are also investigated, comparing the X_0 band signal with a R pumping energy at 160 μJ / cm^2 to the one at 16 μJ / cm^2. The quantitative analysis of the bleaching dynamics at low and high pump fluence shows that all the kinetics of core and core/shell samples follow a multi-exponential decay. The traces are reported in Figure 25.

The comparison with the results obtained for the same samples in the previously reported pump and probe experiment with NR pumping energy and at similar pump fluences (Sec.3-Sec.4), allow us to identify the different relaxation processes by means the related time constant resulting from the signal decays.

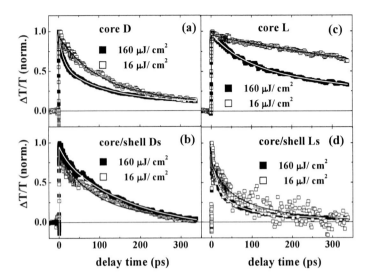

Figure 25. Normalized bleaching kinetics recorded at $X_0^C = 2.29$ eV in the core (a) / (c) and at $X_0^{CS} = 2.21 / 2.17$ eV in the core/shell (b) / (d) QRs samples of set_1 / set_2, for different pump fluences. The best fits (continuous lines) are also shown.

We observe that as general trend, at both high and low pump fluences, the same relaxation processes are present in the samples at R and NR pumping energy. In fact, strong similarity of the best fit decay times is observed, as we can see in table 4, where the time constants and the rise-time are reported for the core and core/shell samples of set_1 and set_2.

In more details we observe that:

(i) Trapping in the mid-gap states is evident in core samples and in the core/thick-shell one ($\tau_1 \sim 1$ ps). The shorter time costant and the related relaxation process, is not present in the core/thin shell sample dynamics.

(ii) Relaxation from emitting states occurs in all samples ($\tau_2 \sim 15$ ps), independent of pump fluence, as previously observed in NR experiments.

(iii) Auger processes are observed in all samples ($\tau_3 \sim 70 - 100$ ps), but different trend results in core and core/shell sample behaviour (see in the next).

(iv) Finally, at high pump fluence, a long time is observed in the core samples and in the core/thick-shell one, due to relaxation from high energy defect states and from near-gap states respectively ($\tau_4 \sim 500 - 1000$ ps). The relaxation from defect states is not present in core/thin-shell sample.

This is well explained by the distribution in energy of defect states in this sample. In fact defect states are not present in the mid-gap, due to the reduced strain at the core/thin-shell interface, the shell passivates the states at high energy and as a consequence trapping is not observable.

When the excitation density is decreased the short and the long decay time, due to trapping and relaxation from defect states, are not present in core samples, showing that these defect states are empty. We then conclude that in these samples, R pumping allows to avoid the direct trapping of carriers at high energy defect states, while the presence of trapped carriers, even under R pumping at high excitation density, suggests the presence of a defect states population mediated by some excitation density dependent process (see in the next).

Table 4. Time constants for core and core/shell samples of set_1 (core D and core/shell Ds) and of set_2 (core L and core/shell Ls). All the constants are expressed in picoseconds

Core L	τ_{rise}	τ_{d1}	τ_{d2}	τ_{d3}	τ_{d4}
X_0 (N_0 =0.2)	0.100±0.005	-	12±2	129±23	-
X_0 (N_0 =2.0)	0.090±0.007	1.0±0.5	14±3	70±5	1000±220
Core/shell Ls	τ_{rise}	τ_{d1}	τ_{d2}	τ_{d3}	τ_{d4}
X_0 (N_0 =0.2)	0.156±0.004	0.9±0.05	12±2	170±15	600±70
X_0 (N_0 =2.0)	0.125±0.003	1.0±0.08	15±3	130±20	530±100
Core D	τ_{rise}	τ_{d1}	τ_{d2}	τ_{d3}	τ_{d4}
X_0 (N_0 =0.2)	0.104±0.003	-	15±6	150±2	-
X_0 (N_0 =2.0)	0.085±0.002	0.60±0.08	10.0±0.5	75±10	460±120
Core/shell Ds	τ_{rise}	τ_{d1}	τ_{d2}	τ_{d3}	τ_{d4}
X_0 (N_0 =0.2)	0.075±0.005	-	9±2	168±5	-
X_0 (N_0 =2.0)	0.063±0.003	-	11±6	123±3	-

Trapping in core/thick-shell is observed also at low pump fluence, as expected, due to the presence of defect states in the mid-gap.

6.b.1. Role of defect states on Auger processes and competition with SE

We have observed that at high pump fluence a costant decay time of about 70 ps (>100ps) is present in core (core/shell) samples. This decay constant and then the related relaxation process, results dependent on pump fluence. In fact it increases as the pump fluence decreases, as previously observed in NR pumping measurements and as a consequence it can be attributed to Auger processes. In particular we observe that in R pumping dynamics the Auger process results enhanced in core samples with respect to the core/shell sample ones[6]. Furthermore, it is important to observe that the average number of photo-excited excitons N_0 is, at any pump fluence, about two times smaller under R rather than NR pumping (1.5 vs 3.5 and 1.0 vs 2.5 in set_2 and set_1 respectively). As a consequence, since the Auger lifetime under R pumping and at high pump fluence is similar (reduced) to the one obtained under NR pumping in short (long) core, despite the smaller value of N_0, we conclude that R pumping also results in an increase of the Auger efficiency in the core samples. This Auger enhancement, together with the observed strong excitation density dependence of the defects population, clearly suggests that the surface defect states can be involved in the Auger process. Actually the main difference between the R and NR cases is that, in the first ps after the pump pulse, the surface defect states at high energy result, respectively, in empty and filled states (see Figure 26, left panel). As a consequence, when

---

[6] The content of sec 6.b.1 is taken from ref. [39], copyright 2007,with permission from American Institute of Physics.

the excitation density is increased leading to a non-negligible Auger rate, the carriers can be scattered in high energy defect states only under R pumping.

Under NR pumping, on the contrary, these defect states are filled, thus reducing the number of available states for the scattered carriers and then decreasing the Auger rate for a given N_0 (see Fig. 26 right panel).

This effect is strongly reduced in core/shell sample where the defect states at high energy are passivated by the shell.

Finally, as nonradiative Auger recombination is generally considered the dominant decay channel of optical gain in NCs [12], we also investigate in the short-core (set_1) the effects of R pumping also in the range of the spectrum where SE is expected. We observe that in core sample the SE signal at the probe energy of 2.03 eV shows the same relaxation processes observed in the case of NR pumping, namely, relaxation from the emitting exciton state with time constant τ_2^R and Auger recombination with time constant τ_3^R. Furthermore it results that in short core the SE signal is stronger than the PA one, as resulting from the comparison between the ratio of the signal at 2.03 eV to the bleaching signal at X_0 with the ratio of linear abs signal at the same energies (see in the previous sections) for a longer time than the one achieved in the case of NR pumping (~ 70 vs ~ 40 ps) despite the observed increase of the Auger efficiency.

In the Figure 27 the comparison of SE lifetime at R (a) and NR (b) pumping experiment in short core sample is shown.

This is a very important result, which suggests that, in core QRs, the main obstacle to sustain SE signal is not the Auger recombination, as generally believed, but defect states related PA processes in the emission energy range.

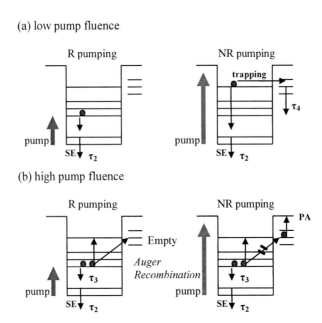

Figure 26. Schematic representation of relaxation processes at (a) low and (b) high pump fluences in the case of R and NR pumping. Reprinted from [39], copyright 2007, with permission from American Institute of Physics.

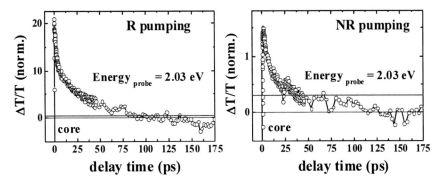

Figure 27. Time evolution of SE (more details in the text) of core QRs sample of set_1, recorded at 2.03 eV, with R (left panel) and NR (right panel) pumping energies at high pump fluence. The continuous lines represent the SE thresholds resulting from linear abs spectra.

6.b.2. Confined acoustic phonons

As previously observed, the analysis of ultrafast dynamics in the 400 ps range with a R pumping energy confirms all the results observed in the case on NR pumping and in particular, it allows us to evaluate the role of defect states in Auger processes and to clarify the competition between Auger recombination and SE emission process. Additionally, R energy pumping induces in the first ps after excitation, a new interesting feature in core samples dynamics, not present in the NR pumping measurements, which is the subject of this sub-section[7].

The time traces of PA and PB signals of both the core samples, at high pump fluence, exhibit in fact a superimposed modulation, which disappeared after 4ps. This is clearly visible in the Figure 28 (a) and (b) where X_2 bleaching dynamics of core samples are reported.

Oscillations in the time-resolved differential transmission spectrum are usually attributed in nanoparticles to the modulation of the bandgap and absorption cross section by the coherent acoustic phonons, which are excited mainly by coupling of excitons to acoustic phonons by a deformation–potential interaction [113].

In a bulk semiconductor, coupling of excitons to acoustic phonons occurs via the deformation–potential interaction. In nanoparticles, the confinement effects influence not only the emission properties but also the coupling to optical modes (which is reduced) and to acoustic phonons (which is actually increased) [114]. In QDs on the other hand the acoustic vibrational modes are localized [113] and the deformation potential induces new discrete vibrational modes that depend on the size [113] but not on the shape [115]. Spherical and rod-shaped semiconductor NCs with comparable diameters show similar oscillation periods, indicating that in the nanorods transverse (radial) modes predominate, due to the strong dependence of the bandgap on the width rather than on the length.

The periodic oscillation of the absorption spectrum due to deformation potential coupling leads in the modulation of the transmitted signal observed in our dynamics.

The modulation is present in almost all probe energies investigated in long core. We don't observe any dependence of the oscillation frequency on probe energy. The modulation in long core is well visible in the dynamics recorded at 2.58 and 2.53 eV, while it became less

---

[7] The content of sec 6.b.2. is taken from ref. [112], copyright 2008, with permission from IOP Publishing.

distinct at 2.29 and 1.88eV and it disappeared at 1.70 eV. The relative amplitude of the modulation became progressively weaker, as for instance at 1.88 eV it was reduced to about 30% of the corresponding dynamics at 2.58 eV. The amplitude dependence on the energy probe can be explained by the trend in the absorption spectrum. At higher energy the cross section is larger and therefore the oscillations are stronger.

In the short core the modulation is observed only at high energy probe of 2.58 eV (Figure 28 (a)).

The Fourier transform of the oscillatory signal indicated a frequency of modulation of 0.73 THz (24 cm^{-1}) which is in good agreement with that reported in the literature for both spherical and rod-shaped CdSe NCs [113,115] with diameters comparable to those of our samples. This confirms that the observed oscillations are due to the radial mode of coherent phonons, which are independent from the NCs shape [115]. It is nevertheless important to observe that the excitation density in our experiments (about 200 nJ/pulse) is about five times smaller than the one reported in the literature and which was regarded as the threshold density for triggering the oscillations via nonresonant pumping (3.18eV) [113,115]. This indicates that the mechanism of damping is reduced by resonant pumping and suggests that the carrier relaxation processes induced by the pumping at high energy influence the coherent oscillation in the lattice.

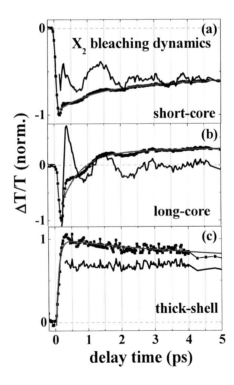

Figure 28. ΔT/T kinetics of the (a) short core (set_1), (b) long core and (c) core/thick-shell QRs samples of set_2, recorded at 2.58 eV in the first 4 ps, with an excitation fluence of 160 μJ/cm^2, together with multi-exponential best fits compared with their isolated oscillatory feature.

Furthermore we observe that the damping time of oscillation results dependent on the type of nanorod surface, as the oscillations could not be observed in the QRs samples coated with a CdS/ZnS shell, as we can see in the Figure 28, where the dynamic registered at 2.58 eV of core/thick-shell is also shown. In the figure the residual oscillation is also reported for each sample. Usually, a matrix that mediates the coupling of NCs vibrations with the environment increases the damping time with respect to the case of NCs in solution [113]. However, in our work, the surface coverage of CdSe QRs with a different inorganic material induces a loss of coherence even in the case of core/thin-shell sample (not shown). In particular, the different elastic constant should influence the frequency and the phase of the oscillation, and the different lattice constant (4.3 Å versus 3.8 Å) should influence the amplitude, therefore leading to a drastic damping of the oscillations that originates from the CdSe core.

7. Conclusion

In this chapter we have clarified the role of the surface and interface defect states in the exciton relaxation dynamics and the competition between stimulated emission and photoabsorption processes in CdSe core and core/shell nanorods. Additionally the controversial attribution of PA processes has been discussed.

In particular, in the Section 3, 4 and 5 we have presented the result of a systematic study of ultrafast carrier relaxation in CdSe core and core/shell nanorods, with different length and shell thickness. Analysis of inter and intraband carrier dynamics has evidenced the presence of different carriers relaxation processes and it has provided a strong evidence that the relaxation processes involve different defect states in core and core/shell samples. At higher energy states, resonant unpassivated surface states are responsible to the faster decay time in core than in the core/shell samples. Relaxation data of the lowest electron levels, on the contrary, indicates that the depopulation dynamic is strongly affected by trapping at the defect states in the mid-gap in core/shell sample, particularly in core/shell sample with a thick shell. This suggests a relaxation mechanism that is due to trapping at localized interface defects, rather at surface defect states. These states, not present in the core and core/thin shell, which are attributed to strain relaxation at the interface of the core/shell sample, engages an important role in the achievement of optical gain. In core/thick shell, in fact, the SE is reduced, despite the increased in spontaneous emission QY, with respect to the corresponding core and core/thin shell ones, due to PA processes involving those interfacial defect states in the mid-gap.

In the Section 6 we have clarified the competition between Auger and PA processes, showing that, contra-intuitively, Auger processes are not the main obstacle to sustain SE in CdSe QRs samples.

Acknowledgments

The pump and probe experiments presented in this chapter have been performed at ULTRAS-CNR-INFM (Milan), under the supervision of Prof. G. Lanzani and Dr. Ing. M.

Zavelani-Rossi. We thank them for the hospitality and for the highlightening discussions. Many thank are due to Dr. L. Manna and Dr. F. Della Sala (NNL-CNR, Lecce) for samples delivering and theoretical supporting. Special thanks are also due to Dr. G. Leo (ISMN-CNR, Rome). Finally we are very grateful to Dr. Marco Anni for many critical discussions.

REFERENCES

[1] Efros, AlL; Efros, A. *Sov. Phys. Sem.*, 1982, 16, 772.
[2] Brus, L. *J. Chem. Phys.*, 1983, 79, 5566.
[3] Brus, L. *Appl. Phys. A*, 1991, 53, 465.
[4] Nirmal, M; Norris, DJ; Kuno, M; Bawendi, MG; Efros, AL; Rosen, M. *Phys. Rev. Lett.*, 1995, 75, 3728.
[5] Alivisatos, AP. *Science*, 1996, 271, 933.
[6] Aslam, F; von Ferber, C. *Chem Phys.*, 2009, 362, 114.
[7] Bruchez, M; Moronne, M; Gin, P; Weiss, Sh; Alivisatos, AP. *Science*, 1998, *281*, 2013.
[8] Chan, WCW; Nie, SM. *Science*, 1998, *281*, 2016.
[9] Taton, TA; Mirkin, CA; Letsinger, RL. *Science*, 2000, 289, 1757.
[10] Deka, S; Quarta, A; Lupo, MG; Falqui, A; Boninelli, S; Giannini, C; Morello, G; De Giorgi, M; Lanzani, G; Spinella, C; Cingolani, R; Pellegrino, T; Manna, L. *J. Am. Chem. Soc.*, 2009, *131*, 2948.
[11] Colvin, VL; Schlamp, MC; Alivisatos, AP. *Nature*, 1994, *370*, 354.
[12] Klimov, VI; Mikhailovsky, AA; Xu, S; Malko, A; Hollingsworth, JA; Leatherdale, CA; Eisler, HJ; Bawendi, MG. *Science*, 2000, *290*, 314.
[13] Huynh, WU; Dittmer, JJ; Alivisatos, AP. *Science*, 2002, *295*, 2425.
[14] Banin, U. Nat. *Photonics*, 2008, *2*, 209.
[15] Alivisatos, AP. *Science*, 1996, *271*, 933.
[16] Sun, B; Marx, E; Greenham, C. *Nano Lett*, 2003, *3*, 961.
[17] Landi, BJ; Castro, SL; Ruf, HJ; Evans, CM; Bailey, SG; Raffaelle, RP. *Solar Energy Materials and Solar Cells*, 2005, *87*, 746.
[18] Klimov, VI. *J Phys Chem B*, 2006, *110*, 16827.
[19] Nair, G; Bawendi, MG. *Phys Rev B*, 2007, *76*, 081304 (R).
[20] Klimov, VI. *Annu. Rev. Phys. Chem.*, 2007, 635 and reference therein
[21] Hillhouse, HW; Beard, MC. *Current opinion in colloid & interface Science*, 2009, *14*, 245.
[22] Arakawa, Y; Sakaki, H. *Appl. Phys. Lett*, 1982, *40*, 939.
[23] Asada, M; Miyamoto,Y; Suematsu, Y. *IEEE J. Quantum Electron*, 1986, *22*, 1915.
[24] Klimov, VI; Mikhailovsky, AA; Xu, S; Malko, A; Hollingsworth, JA; Leatherdale, CA; Eisler, HJ; Bawendi, MG. *Science*, 2000, *290*, 314.
[25] Klimov, VI; Bawendi, MG. *MRS Bull*, 2001, *26*, 998.
[26] Kazes, M; Lewis, DY; Ebenstein, Y; Mokari, T; Banin, U. *Adv. Mat*, 2002, *14*, 317.
[27] Klimov, VI. In Semiconductor and Metal Nanocrystals: *Synthesis and Electronic and Optical Properties*, VI. Klimov, Ed; Marcel Dekker: New York, NY, 2003, *159*.
[28] Malko, AV; Mikhailovsky, AA; Petruska, MA; Hollingsworth, JA; Klimov, VI. *J. Phys. Chem. B*, 2004, *108*, 5250.

[29] Kazes, M; Saraidarov, T; Reisfeld, R; Banin, U. *Adv. Mater*, 2009, *21*, 1716.
[30] Kong, J; Franklin, NR; Zhou, C; Chaplin, MG; Peng, S; Cho, V; Dai, H. *Science*, 2000, *287*, 622.
[31] Aslam, F; von Ferber, C. *Chem Phys.*, 2009, *362*, 114.
[32] Vassiltsova, OV; Zhao, Z; Petrukhina, MA; Carpenter, MA. *Sens. Act. B*, 2007, *123*, 522.
[33] Somers, RC; Bawendi, MG; Nocera, DG. Chem. *Soc. Rev.*, 2007, *36*, 579.
[34] Klimov, VI. *J. Phys. Chem. B*, 2000, *104*, 6112.
[35] Zhang. J. *J. Phys. Chem. B*, 2000, *104*, 7239.
[36] Lomascolo, M; Cretì, A; Leo, G; Manna, L; Vasanelli, L. *Appl. Phys. Lett*, 2003, *82*, 418.
[37] Creti, A; Anni, M; Zavelani-Rossi, M; Lanzani, G; Leo, G; Della Sala, F., Manna, L; Lomascolo, M. *Phys. Rev. B*, 2005, *72*, 125346.
[38] Creti, A; Zavelani-Rossi, M; Anni, M; Lanzani, G; Manna, L; Lomascolo, M. *Phys. Rev. B*, 2006, *73*, 165410.
[39] Creti, A; Anni, M; Zavelani-Rossi, M; Lanzani, G; Manna, L; Lomascolo, M. *Appl. Phys. Lett*, 2007, *91*, 093106.
[40] Creti, A; Lomascolo, M; Leo, G; Manna, L; Anni, M. *J of Lum.*, 2008, *128*, 361.
[41] Malkmus, S; Kudera, S; Manna, L; Parak, WJ; Braun, M. *J. Phys. Chem. B*, 2006, *110*, 17334.
[42] Lupo, MG; Della Sala, F; Carbone, L; Zavelani-Rossi, M; Fiore, A; Luer, L; Polli, D; Cingolani, R; Manna, L; Lanzani, G. *Nano Lett.*, 2008, *8*, 4582.
[43] Sitt, A; Della Sala, F; Menagen, G; Banin, U. *Nano Lett.*, 2009, *9*, 3470.
[44] Rajadell, F; Climente, JI; Planelles, J; Bretoni, A. *J. Phys. Chem. C*, 2009, *113*, 11268.
[45] Hilczer, M; Tachiya, M. *J. Phys. Chem. C*, 2009, *113*, 18451.
[46] Carey, CR; Yu, Y; Kuno, M; Hartland, GV. *J. Phys. Chem. C*, 2009, *113*, 19077.
[47] Jones, M; Lo, SS; Scholes, GD. *J. Phys. Chem. C*, 2009, *113*, 18632.
[48] Murray, CB; Norris, DJ; Bawendi, MG. *J. Am. Chem. Soc.*, 1993, *115*, 8706.
[49] Micic, OI; Curtis, CJ; Jones, KM; Sprague, JR; Nozik, AJ. *J. Phys. Chem.*, 1994, *98*, 4966.
[50] Guzelian, AA; Katari, JEB; Kadavanich, AV; Banin, U; Hamad, K; Juban, E; Alivisatos, AP; Wolters, RH; Arnold, C; Heath, JR. *J. Phys. Chem.*, 1996, *100*, 7212.
[51] Hines, MA; Guyot-Sionnest, P. *J. Phys. Chem. B*, 1998, *102*, 3655.
[52] Manna, L; Scher, EC; Alivisatos, AP. *J. Am. Chem. Soc.*, 2000, *122*, 12700.
[53] Peng, X; Manna, L; Yang, W; Wickham, J; Scher, E; Kadavanich, A; Alivisatos, AP. *Nature*, 2000, *404*, 59.
[54] Mokari, KT; Rothenberg E; Banin, U. *Nature*, 2003, *2*, 155.
[55] Manna, L; Milliron, DJ; Meisel, A; Scher, EC; Alivisatos, AP. *Nature Mater*, 2003, 2, 382.
[56] Dai, Y; Zhang, Y; Li, QK; Nan, CW. *Chem. Phys. Lett*, 2002, *358*, 83.
[57] Fiore, A; Mastria, R; Lupo, MG; Lanzani, G; Giannini, C; Carlino, E; Morello, G; De Giorgi, M; Yanqin, L; Cingolani, R; Manna, L. *J. Am. Chem. Soc.*, 2009, *131*, 2274.
[58] Talapin, DV; Koeppe, R; Gtzinger, S; Kornowski, A; Lupton, JM; Rogach, AL; Benson, O; Feldmann, J; Weller, H. *Nano Lett*, 2003, *3*, 1677.
[59] Milliron, DJ; Hughes, SM; Cui, Y; Manna, L; Li, J; Wang, LW; Alivisatos, AP. *Nature*, 2004, *430*, 190.

[60] Talapin, DV; Mekis, I; Gotzinger, S; Kornowski, A; Benson, O; Weller, H. *J. Phys. Chem. B*, 2004, *108*, 18826.
[61] Mokari, T; Rhotenberg, E; Popov, I; Costi, R; Banin U. *Science*, 2004, *304*, 1787.
[62] Shieh, F; Saunders, AE; Korgel, BA. *J.Phys.Chem B Lett*, 2005, *109*, 8538.
[63] Kudera, S; Carbone, L; Casula, MF; Cingolani, R; Falqui, A; Snoeck, E; Parak, J; Manna, L. *Nano Lett*, 2005, *5*, 445.
[64] Carbone, L; Nobile, C; De Giorgi, M; Della Sala, F; Morello, G; Pompa, P; Hytch, M; Snoeck, E; Fiore, A; Franchini, IR; Nadasan, M; Silvestre, AF; Chiodo, L; Kudera, S; Cingolani, R; Krahne, R; Manna, L. *Nano Lett*, 2007, *7*, 2942.
[65] Choi, CL; Alivisatos, AP. *Annu Rev Phys Chem.*, 2010, *61*, 369
[66] Klimov, VI; Mikhailovsky, AA; Mc Branch, DW; Leatherdale CA; Bawendi, MG. *Science*, 2000, *287*, 1011.
[67] Li, J; Wang, LW. *Nano Lett*, 2003, *3*, 1357.
[68] Hu, J; Li, L; Yang, W; Manna, L; Wang, L; Alivisatos, AP. *Science*, 2001, *292*, 2060.
[69] Chen, X; Nazzal, A; Goorskey, D; Xiao, M. *Phys. Rev. B*, 2001, 64, 245304.
[70] Li, L; Hu, J; Yang, W; Alivisatos, AP. *Nano Lett*, 2001, 1, 349.
[71] Steiner, D; Katz, D; Millo, O; Aharoni, A; Kan, SH; Mokari, T; Banin, U. *Nano Lett*, 2004, 4, 1073.
[72] Millo, O; Steiner, D; Katz, D; Aharoni, A; Kan, SH; Mokari, T; Banin, U. *Phys. E*, 2005, 26, 1.
[73] Hu, J; Wang, LW; Li, L; Yang, W; Alivisatos, AP. *J. Phys. Chem. B*, 2002, 106, 2447.
[74] Katz, D; Wizansky, T; Millo, O; Rothenberg, E; Mokari, T; Banin, U; *Phys. Rev. Lett*, 2002, 086801.
[75] Li, XZ; Xia, JB. *Phys. Rev. B*, 2002, 66, 115316.
[76] Li, L; Yang, W. *Nano Lett*, 2003, 3, 1357.
[77] Chen, X; Nazzal, AY; Xiao, M; Peng, ZA; Peng, X. *J. of Lum.*, 2002, 97, 205.
[78] Empedocles, SA; Norris, DJ; Bawendi, MG. *Phys. Rev. Lett*, 1996, 77, 3873.
[79] Shabaev, A; Efros, AlL. *Nano Lett*, 2004, 4, 1821.
[80] Le Thomas, N; Herz, E; Schops, O; Woggon, U; Artemyev, MV. *Phys. Rev. Lett*, 2005, 94, 016803.
[81] Li, XZ; Xia, JB. *Phys. Rew. B*, 2003, 68, 165316.
[82] Borchert, H; Talapin, DV; McGinley, C; Adam, S; Lobo, A; Moller, T; Weller, H. *J. Chem. Phys.*, 2003, 119, 1800.
[83] Choi, CL; Koski, KJ; Sivasankar, S; Alivisatos, AP. *Nano Lett*, 2009, *9*, 3544.
[84] Aruguete, DM; Marcus, MA; Li, L; Williamson, A; Fakra, S; Gygi, F; Galli, GA; Alivisatos, AP. *J. Phys. Chem C*, 2007, *75*.
[85] Dabbousi, BO; Rodriguez-Vejo, J; Mikulec, FV; Heine, JR; Mattousi, H; Ober, R; Jensen, KF; Bawendi, MG. *J. Phys. Chem. B*, 1997, 101, 9463 and reference therein.
[86] Chen, X; Lou, Y; Samia, AC; Burda, C. *Nano Lett*, 2003, 3, 799.
[87] Mokari, T; Banin, U. *Chem Mater*, 2003, 15, 3955.
[88] Cao, YW; Banin, U. *J. Am. Chem. Soc.*, 2000, 122, 9692.
[89] Manna, L; Scher, EC; Li, L.-Shi & Alivisatos, A. P. *J. Am. Chem. Soc.*, 2002, 124, 7137.
[90] Peng, X; Schlamp, MC; Kadavanich, AV; Alivisatos, AP. *J. Am. Chem., Soc.*, 1997, 119, 7019.
[91] Jiang, Z; Jie; Leppert, V; Kelley, DF. *J. Phys. Chem. C*, 2009, 113, 19161.

[92] Ricolleau, C; Audinet, L; Gandais, M; Gacoin, T. *Thin Solid Films*, 1998, 336, 213.
[93] Talapin, V; Mekis, I; Gotzinger, S; Kornowski, A; Benson, O; Weller, H. *J. Phys. Chem. B*, 2004, 108, 18826.
[94] Klimov, VI; McBranch, DW. *Phys. Rev. Lett*, 1998, 80, 4028.
[95] Klimov, VI; Schwarz, Ch.J; McBranch, DW; Leatherdale, CA; Bawendi, MG. *Phys. Rev. B*, 1999, 60, R2177.
[96] Guyot-Sionnest, P; Shim, M; Matranga, C; Hines, M. *Phys. Rev. B*, 1999, 60, R2181.
[97] Klimov, VI; Mikhailovsky, AA; McBranch, DW; Leatherdale, CA; Bawendi, MG. *Phys. Rev. B*, 2000, 61, R13349.
[98] Htoon, H; Hollingworth, JA; Malko, AV; Dickerson R; Klimov, VI. *Appl. Phys. Lett*, 2003, 82, 4776.
[99] Htoon, H; Hollingsworth, JA; Dickerson, R; Klimov, VI. *Phys. Rev. Lett*, 2003, 91, 227401.
[100] Mohamed, MB; Burda, C; El-Sayed, MA. *Nano Lett*, 2001, 11, 589.
[101] Malko, AV; Mikhailovsky, AA; Petruska, MA; Hollingsworth, JA; Htoon, H; Bawendi, MG; Klimov, VI. *Appl. Phys. Lett*, 2002, 81,1303.
[102] Kazes, M; Oron, D; Shweky, I; Banin*http://pubs.acs.org/doi/abs/10.1021/jp070075q?journalCode=jpccck&quickLinkVolume=111&quickLinkPage=7898&volume=111 -jp070075qAF1#jp070075qAF1*, U. *J. Phys. Chem. C*, 2007, 111, 7898.
[103] Li, TS; Kuhn, KJ. *J. Comput. Phys.*, 1994, 110, 292.
[104] Efros, AlL; Rosen, M; Kuno, M; Nirmal, M; Norris, DJ; Bawendi, M. *Phys. Rev. B*, 1996, 54, 4843.
[105] Gadermaier, C; Lanzani, G. *J. Phys. Condens, Matter*, 2002, 14, 9785.
[106] Klimov, VI; Hunsche, S; Kurz, H; *Phys. Rev. B*, 1994, 50, 8110.
[107] Link, S; El-Sayed, M. *J. of Appl. Phys.*, 2002, 92, 6799.
[108] Klimov, VI; Mc Branch, DW; Leatherdale, CA; Bawendi, MG. *Phys. Rev. B*, 1999, 60, 13740.
[109] Ai, X; Jin, R; Ge, Ch; Wang J; Zou, Y; Zhou, X; Xiao, X. *J. Chem. Phys.*, 1997, 106, 3387.
[110] Landsberg, PT. *Recombination in semiconductors*, Cambridge Univ. Press: 2003, Chap 5.3.
[111] Wang, XY; Zhang, JY; Nazzal, A; Darragh, M; Xiao, M. *Appl. Phys. Lett*, 2002, 81, 4829.
[112] Creti, A; Anni, M; Zavelani-Rossi, M; Lanzani, G; Manna, L; Lomascolo, M. *J. Opt. A: Pure Appl. Opt*, 2008, 10, 064004.
[113] Cerullo, G; De Silvestri, S; Banin, U. *Phys. Rev. B*, 1999, 60, 19284 and reference therein
[114] Valerini, D; Creti, A; Lomascolo, M; Cingolani, R; Manna, L; Anni, M. *Phys. Rev. B*, 2005, 71, 235409.
[115] Son, DH; Wittenberg, JS; Banin, U; Alivisatos, AP. *J. Phys. Chem. B*, 2006, 110, 19884.

In: Exciton Quasiparticles
Editor: Randy M. Bergin

ISBN: 978-1-61122-318-7
© 2011 Nova Science Publishers, Inc.

Chapter 3

RADIATION-ASSISTED PREPARATION OF POWDER MATERIALS AND THEIR EXCITON LUMINESCENCE

Václav Čuba[1] and Martin Nikl[2]

[1]CTU in Prague, Faculty of Nuclear Sciences and Physical Engineering, Břehová 7, Prague 1, 115 19, Czech Republic
[2]Institute of Physics AS CR, v.v.i., Cukrovarnická 10, Prague 6, Czech Republic

ABSTRACT

This chapter provides overview of preparation of nanopowder materials using ionizing and UV irradiation. Basic principles and current status of research in this field are briefly discussed. The processes related to the radiation reduction of metal ions to lower valences or to metallic particles and radiation induced oxidation are described. Stabilization of nanoparticles in solutions and preparation of materials from aqueous or micellar solutions is overviewed. Exciton luminescence in nanopowders is discussed, including their advantages/disadvantages and comparison with the bulk materials. Preparation of selected nanomaterials with excitonic luminescence via irradiation route is described in detail, followed by discussion of their characterization and luminescence properties.

1. INTRODUCTION - EXCITON LUMINESCENCE IN NANOPOWDERS

Exciton luminescence has been systematically studied in many semiconductor and dielectric material, both organic and inorganic ones [1,2], and became of particular interest after the boomed research in the quantum confinement phenomena field [3]. There are two basic kinds of excitons: (i) Wannier exciton of larger radius which exceeds even several times the lattice constant and is typical for the direct-gap semiconductors as GaAs, ZnO, GaN,

CuCl, CdS and CdSe and (ii) self-trapped exciton of smaller radius which typically occurs in wide band-gap dielectrics as alkali halide and some binary and complex oxide compounds. Wannier exciton is rapidly moving through the lattice, shows very narrow absorption and emission bands with small Stokes shift (up to tens of meV) and very short subnanosecond decay times due to superradiance effect. Its binding energy is in the range of units – several tens meV and when thermal energy kT becomes comparable to it, the exciton becomes thermally disintegrated and the equilibrium between the exciton and free electron and hole states is established. Self-trapped exciton is immobile and shows luminescence characteristics more similar to luminescent ions based on 4f-5d, d-d and ns2-nsnp transitions, i.e. broader emission bands with large Stokes shift up to a few eV, the decay times in ms- tens of ns time scale and in most cases the bigger binding energies and related thermal stability with respect to Wannier exciton systems. Quantum confinement of Wannier exciton emission occurs in material systems where one or more dimensions of the host become comparable with the exciton radius [3]. In such geometries the motion of exciton becomes limited and it results in change of its characteristics. Namely, the energy of exciton transition increases and its binding energy as well which results in higher thermal stability of confined exciton state. Moreover, better coherence of exciton state often occurs within smaller volumes of such nanostructures and exciton-polariton coupling is inhibited which results in shorter decay times of confined excitons in comparison with their counterparts in the bulk single crystal systems [4,5].

2. PREPARATION OF NANOPOWDER MATERIALS USING IONIZING AND UV RADIATION

Both ionizing and non-ionizing radiations present powerful tools for synthesis or processing of various materials. For preparation of materials from aqueous solution using ionizing radiation, the radiolysis of water plays important role. It results in formation of reactive intermediates, summarized as [6,7]:

$$H_2O + h\nu \rightarrow e^-_{aq}, H, OH, H_2, H_2O_2, H^+.$$

Preparation using non-ionizing UV radiation or visible light encompasses usage of sensitizers, which produce radicals upon irradiation, reacting with target compound [8,9]; direct photolysis of studied compounds, which results in formation of desired material [10,11]; and reactions with products of water photolysis [12,13]:

$$H2O + h\nu \rightarrow H_2O^* \rightarrow \cdot OH + \cdot H.$$

In the last decades, radiation or photochemical reduction was successfully utilized for preparation of small-sized particles or composite clusters consisting of one or more metals. Both noble and non-noble metals were successfully reduced to form nanocolloids, predominantly silver, copper, platinum, or nickel [14-18]. Similarly, various oxides were prepared from aqueous solutions, using UV [19,20] or ionizing [21] radiation. Radiation processes have some advantages over common chemical methods; they are mostly independent of temperature and yield material of high purity with narrow size distribution of particles. The mechanism of Cu_2O formation from microemulsions was studied by Chen et al. [22]. Shen et al. also prepared [23] octahedron cuprous oxide of size less the 100 nm using irradiation of microemulsions containing copper nitrate and Triton X-100. Nanocrystalline cuprous oxide preparation from buffered solutions containing cupric sulfate and acetate buffer using gamma irradiation was reported [24]. The authors observed formation of single phase Cu_2O with 14 nm particles.

There are basically few possible mechanisms of oxide particles formation; one of them is radiation induced precipitation of solid phase in the radiation field. Another possible mechanism is partial or total radiation reduction of dissolved metal ions to metallic particles via reactions with hydrated electrons

$$Me^{2+} + 2e^-_{aq} \rightarrow Me^0,$$

followed by their oxidation via reaction with oxygen under suitable conditions:

$$2Me + O_2 \rightarrow 2MeO.$$

The formation of metallic oxides in the field of radiation may also occur due to partial reduction of metal ions

$$Me^{2+} + e^-_{aq} \rightarrow Me^+,$$

followed by reaction with hydroxyl ions

$$Me^+ + OH^- \rightarrow Me(OH) \rightarrow Me_2O + H_2O,$$

or by disproportionation

$$2Me^+ \rightarrow Me^0 + Me^{2+}.$$

3. PREPARATION OF ZNO BASED MATERIALS USING RADIATION METHODS

Crystalline ZnO occurs mostly as hexagonal wurtzite. ZnO in the form of powder, thin film, quantum wells, wires or dots and single crystal bulk systems became intensively studied due to the number of emerging applications, namely opto-electronic ones, in which it could be used [25-33]. At room temperature in the undoped ZnO its Wannier exciton luminescence is observed as the 1-LO phonon replica of free exciton situated at 3.26 eV (380 nm) and its photoluminescence decay time was measured as 400 ps [34]. In small enough (few nm) nanocrystals of ZnO clear high energy shift of exciton emission and shortening of the decay time has been reported [35]. It is worth mentioning that intrinsic donor character of ZnO and various n-doping attempts give rise to the donor-bound exciton emission lines at the low energy side of free exciton emission [36]. Such donor-bound excitons are observed only at low enough temperatures (typically below 100 K) and show giant oscillator strength resulting in even faster emission decay with respect to free exciton [34]. Such a fast luminescence at the edge of UV-visible spectral range became of interest for superfast scintillators [37] and the Ga-doped ZnO appeared as the most promising candidate [38]. ZnO based scintillators are best applicable in the powder or thin layer forms [39], because small Stokes shift of the Wannier exciton emission results in significant reabsorption in the bulk material of the thickness of 1mm or higher [40].

3.1. Preparation Methods Using Visible Light, Ultraviolet and Ionizing Radiation

Various authors studied utilization of UV or visible light in formation of nanostructures containing zinc oxide. Crystallization and reduction of sol-gel prepared zinc oxide films was studied by [41]. Irradiated solutions were prepared by hydrolysis after dissolution of zinc acetate hydrate in methanol. Gel films were produced by spinning of the precursor solutions on glass slides, silica glass or silicon wafers, and then were heated in air at 60, 100 or 150°C for 1 hour. Irradiation took 25-400 hours; low pressure mercury lamp emitting light at predominant wavelengths 185 and 254 nm was used in the experiments. It was found that UV irradiation induced formation of hexagonal ZnO crystals from amorphous ZnO films pre-heated at 100°C, while irradiation of porous ZnO films pre-heated at 60°C led also to formation of metallic zinc, i.e. these results indicate that the UV-induced reactions depend on the original structure of the sol-gel ZnO films. Amorphous ZnO thin films were also prepared by direct UV (254 nm) irradiation of b-diketonate Zn^{2+} precursor complexes spin-coated on Si(100) and fused silica substrates [19]. The as-deposited films were amorphous and isothermal heat treatment at 800 °C resulted in ZnO crystallization. It was observed that thermal treatment of the photodeposited films favors the stoichiometric formation of ZnO; the resulting polycrystalline film has excellent optical properties (Figure 1).

Composite ZnO/Cu and ZnO/Ag/Cu nanostructures were prepared via the photocatalytic reduction (wavelength 310-390 nm) of cuprous chloride and silver nitrate over the chemically prepared ZnO nanoparticles in aqueous solution [17]. It has been also found that silver nitrate can be photochemically deposited onto the surface of ZnO nanoparticles under the illumination with the visible light in the presence of the sensitizer – methylene blue [18]. Two-step solution-based method was successfully used for the fabrication of the Co_3O_4/ZnO nanowire heterostructures [20]. First, ZnO nanowires were grown by ammonia solution hydrothermal method. Afterward, Co_3O_4 was coated on the ZnO nanowires using a photochemical reaction.

Recently, direct photochemical preparation of zinc peroxide was also reported [42,43]. This compound was prepared by irradiation of aqueous solution containing zinc acetate and hydrogen peroxide using 35 or 75W Xe lamp:

$$Zn(CH_3COO_2)_2 + H_2O_2 + h\nu \rightarrow ZnO_2 + 2CH_3COOH.$$

Consequently, nanocrystalline zinc oxide was prepared by thermal decomposition of zinc peroxide [42].

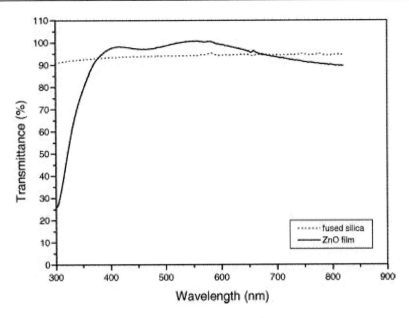

Figure 1. Optical transmittance spectra of a photodeposited ZnO film grown on a fused silica substrate [19].

In the last ten years, high energy radiation techniques were employed to prepare amorphous or crystalline zinc oxide. Ren et al. [44] used consequential Zn/F ion implantation in amorphous silica matrix to form Zn/ZnO core–shell nanoclusters. They irradiated the matrix with 160 keV zinc ions and 40 keV fluorine ions. The doses for both elements were 2×10^{17} ions cm^{-2} and the current densities less than 2μA cm^{-2} to reduce heating effect. The authors observed formation of crystalline Zn core with amorphous ZnO shell and concluded that zinc oxide forms mainly via reaction of zinc with oxygen released from silica due to F ions bombardment. Single-crystalline ZnO hexangular prism was successfully prepared by irradiation of alkaline micellar solution, containing zinc sulphate, sodium hydroxide, cetyltrimethylammonium bromide (CTAB) and 2-propanol (OH radical scavenger) at room temperature [45]. As a source of gamma radiation, ^{60}Co radioisotope was used and the solutions were irradiated with the dose 100 kGy. XRPD and SEM/TEM analyses revealed, that solid phase formed during irradiation consists of single crystalline wurtzite nanorods with a rim 230 nm and a length up to 8.5μm. It was found that presence of CTAB is crucial for the shaping of nanorods. Without CTAB, only a ZnO conglomeration was formed. Room temperature photoluminescence spectrum showed broad green emission with maximum at 577 nm, indicting a significant amount of oxygen vacancies in prepared zinc oxide (Figure 2).

Other authors [21] studied formation of zinc oxide nanostructures under similar conditions. They irradiated aqueous solution containing zinc sulphate, tert-Butanol and sodium hydroxide with dose 130 kGy, using ^{60}Co gamma source. It was found that the product is nanocrystalline wurtzite zinc oxide with particle size 7-9 nm. The emission and excitation spectra of irradiated samples were also studied. Contrary to [45], the main emission peak of irradiated samples was found in UV area, at 395 nm (Figure 3). This could be attributed to the differences in the shape and size of prepared zinc oxides.

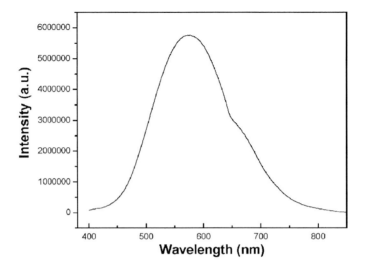

Figure 2. Room-temperature PL spectrum of ZnO hexangular prism [45].

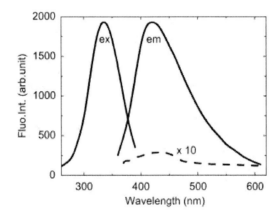

Figure 3. Steady-state fluorescence and excitation spectra of unirradiated and irradiated aqueous t-BuOH solutions, containing 5 10^{-3} M ZnSO4 and 6M NaOH. Solid lines represent the irradiated and dashed lines represent the unirradiated samples. Gamma radiolysis, dose 130 kGy [21].

Aside from gamma irradiation, the authors used pulse radiolysis for the study of mechanism of zinc oxide formation. They concluded that the reduction of zinc (II) ions is followed by the oxidation of metallic zinc with oxygen:

$$[Zn(OH)_4]^{2-} \text{ or } Zn^{2+} \xrightarrow{e^-_{aq}} Zn \xrightarrow{O} ZnO \xrightarrow{ZnO} (ZnO)_{np}$$

$$O_2 \longrightarrow O_2^{\cdot -}$$

3.2. Preparation and Excitonic Luminescence of Powder ZnO Prepared under UV-VIS and Ionizing Radiation

Based on recent research [46,47], a comparative study dealing with characterization and radioluminescence properties of zinc oxide prepared from various precursors under ionizing or non-ionizing radiation could be performed. It reveals many similarities between materials prepared under both types of irradiation. The aims of performed comparison is to evaluate the possibilities of radiation methods for preparation of powder crystalline zinc oxide (both pure and doped with lanthanum) from aqueous solutions, to determine basic characteristics of formed solid phase, to evaluate the effect of heat treatment on the morphology of solid phase and to compare the scintillation properties of prepared materials.

All chemicals used for the study were of analytical grade (Sigma Aldrich, Fluka); ultra pure deionized water (conductance $\leq 1\mu S$) was used for irradiation experiments. Hydrogen peroxide, strong oxidizing agent and source of hydroxyl and hydroperoxyl radicals due to photolysis or radiolysis, was added to some solutions. Following aqueous systems with the concentration of zinc 0.046 mol dm^{-3} were prepared:

- Zinc nitrate dissolved in aqueous solution of 1.3 mol dm^{-3} propan-2-ol. (**I**)
- Zinc formate dissolved in aqueous solution of 1 mol dm^{-3} hydrogen peroxide. (**II**)
- Zinc formate dissolved in aqueous solution of 1 mol dm^{-3} hydrogen peroxide and 10^{-5} mol dm^{-3} polyvinyl alcohol (PVA). (**II**)

Irradiation of solutions and characterization of solid phase

The high energy irradiation was performed using linear electron accelerator with electron energy 4.5 MeV, pulse width 3 µs and repeating frequency 500 Hz. Samples were irradiated in polypropylene vials, absorbed dose was 70 kGy. The dose determination was performed using alanine dosimeters. For irradiation in UV-VIS region, the medium pressure mercury lamp emitting photons at wavelengths 200-580 nm (70% of the intensity in UV region 200-400 nm), with variable power output 140-400 W, was used. The lamp in quartz tube was immersed in 2 dm^3 of solutions in glass reactor with thermometer. During irradiation, the solutions were continually stirred and the reactor was cooled with water, so that the temperature of irradiated solution did not exceed 45°C. Potassium ferrioxalate actinometry [48] was performed under the same conditions. Irradiation took 1 hour at photon flow 2 10^{19} hv s^{-1}.

To better understand processes occurring in the irradiated systems, chemical speciation of prepared solutions was calculated, using Visual MINTEQ software v. 2.60. Speciation calculations showed predominant presence of zinc ions, nitrates and non-dissociated propan-2-ol in solution (**I**). From the intermediates of water radiolysis, propan-2-ol reacts predominantly with OH radicals [6]. The final products of its degradation are hydrogen, acetone, hydrogen peroxide, acetaldehyde, and methane [49,50]. Besides scavenging OH radicals, propan-2-ol probably also stabilizes small particles of ZnO in the solution and prevents their agglomeration. Hydrated electrons react with both nitrates and zinc(II) ions [6,51], producing a metal zinc. According to supposed mechanism discussed in section 3.1, zinc is then oxidized to zinc oxide via reaction with oxygen.

Speciation diagrams of solutions (II) and (III) are similar; the main forms are zinc ions, formate and zinc-formate ions and hydrogen peroxide, according to speciaton calculation. Both formic acid and especially formates scavenge OH radicals [6] and thus provide optimum condition for the reduction of zinc(II) ions. During the radiolysis, hydroperoxyl radical may be formed. The products of the formic acid/formate degradation are water, oxygen, hydrogen, hydrogen peroxide and carbon dioxide [52-55]. Presence of additive strong oxidizing agents, namely hydrogen peroxide and hydroperoxyl radical probably contribute to high overall yield of solid phase from solutions (II) and (III), which applies to both types of studied radiation. The only difference between solutions (II) and (III) is the presence of PVA in the latter. According to Zhang and Yu [56], PVA undergoes radiolysis mainly via reaction with OH radicals, which can either abstract H atoms from the alpha position to the OH group, or from the neighboring methylene group. Formed reactive radical intermediates undergo chain scission, disproportionation or crosslinking. PVA thus probably plays dual role in studied solutions: it acts as scavenger for OH radicals [57], and it also stabilizes small particles of metal/oxide and prevents their further growth. Due to the length of molecular chain, it may retain its properties even at high doses.

While probable mechanism of zinc oxide formation in radiolysed solutions was established by some authors, as discussed in sec. 3.1, the same is not true to the same extent for photolysed solutions, as the mechanism of zinc oxide (or peroxide) formation under non ionizing radiation was not studied in detail, yet. However, we may suppose that partial or total photochemial reduction of metal ions under convenient conditions [10,11], followed by oxidation of metallic particles could be proposed for UV-irradiated system, too. Reduced zinc may be oxidized by oxygen dissolved in water; hydrogen peroxide, present in solutions **b** and **c**, is another source of oxidative species, as its photolysis results in formation of ·OH radicals [9]. Both oxidative (·OH) and reductive (·H) species are also formed via direct photolysis of water [12,13].

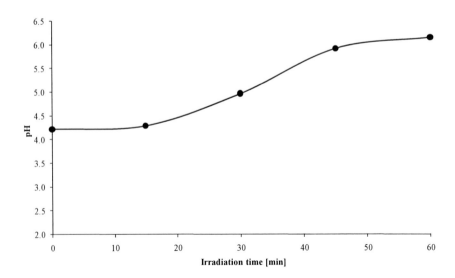

Figure 4. Typical development of pH with irradiation time (i.e. absorbed energy).

Evaluation of pH and absorption spectra of irradiated solutions reveals no noticeable difference between ionizing and non-ionizing radiation. Typically, pH increases during irradiation up to value ~ 6, as is documented in Figure 4. At this pH, the dissolution of formed zinc oxide does not occur and the finely dispersed solid phase is stable.

Figure 5 shows the changes in absorption spectra of the solutions during irradiation. Again, the spectra are similar. The absorption of irradiated solutions monotonously increases with irradiation time due to increasing amount of formed solid phase. The spectra of solution (**I**) show two peaks at 241 and 350 nm (irradiation with high energy electrons) or 246 and 355 nm (UV-VIS irradiaton) that increase with irradiation time, probably indicating the formation of colloidal and bulk ZnO [58,59]. Li et al. [60], found absorption maximum at 346 nm for ZnO prepared by hydrothermal method and attributed it to surface plasmon resonance of ZnO nanocrystals.

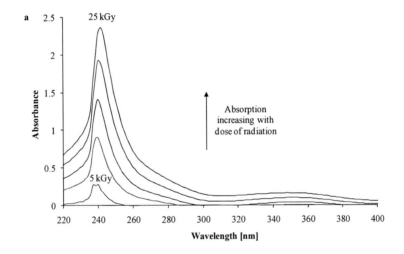

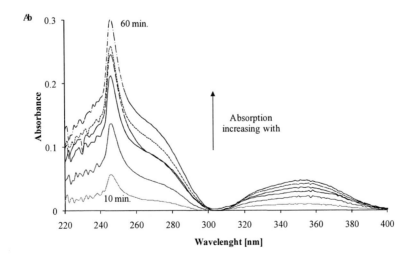

Figure 5. Absorption spectra of solution (**I**), comparison of irradiation with accelerated electrons (**a**) and UV light (**b**) [46,47].

After irradiation, the formed finely dispersed solid phase was separated via microfiltration and dried at 40°C to the constant weight. Comparison of XRPD spectra of all discussed materials is shown in Figure 6. Regardless of the type of irradiation the solid phase obtained from solution (**I**) consists of hexagonal zinc oxide and possibly also of rare tetragonal volume-centered zinc oxide modification studied by various authors [61-63].

Similarly, the diffractogram of material (II) shows diffraction lines of both hexagonal and tetragonal zinc oxide and few lines probably corresponding to cubic zinc peroxide. Diffraction lines of sample (III) dried at 40°C correspond to pure cubic zinc peroxide. The presence of zinc peroxide is not very surprising, as its formation was observed previously under UV-VIS irradiation of similar solutions [42,43]. At higher calcination temperatures, the decomposition of remaining organic impurities, stoichiometrization and/or recrystallization of zinc oxide occur. Zinc peroxide decomposes to oxide at ~ 200°C [42]. It was confirmed that at higher temperatures, well developed zinc oxide (hexagonal) crystalline nanoparticles are formed, regardless of the irradiation type or initial precursors.

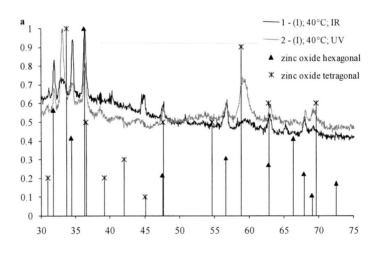

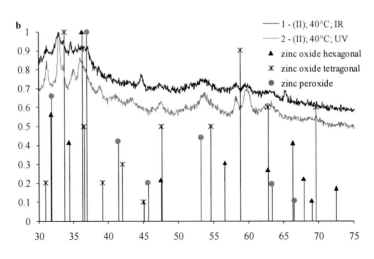

Figure 6. Continued

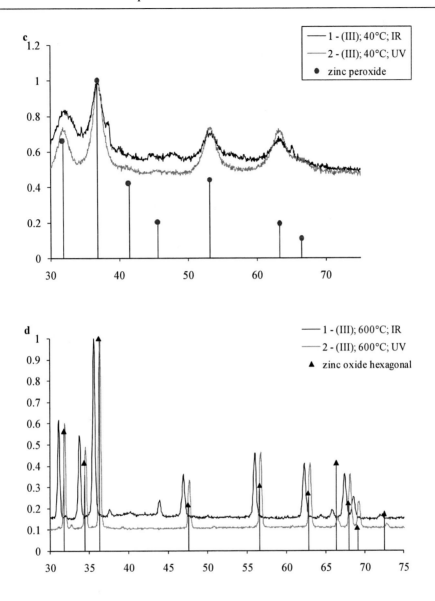

Figure 6. comparison of solid materials prepared from various precursors under ionizing or UV irradiation, dried at 40 °C or calcinated at 600 °C [46,47].

High resolution transmission electron microscopy (HRTEM), selected area electron diffraction (SAED) and scanning electron microscopy (SEM) were used as additional methods for solid phase characterization. TEM/SAED images show that prepared material is polycrystalline (Figure 7). This was observed in all studied systems; other authors [45] irradiated micellar solutions and obtained crystalline material consisting of well developed single crystals. SEM images of ZnO formed under either of both types of radiation (Figure 7) show regularly shaped structures with size ranging from 100 to 400 nm. The size of the particles and specific surface area (SSA) of solid phase depend strongly on calcination temperature and varies somewhat between materials, but it does not seem to depend too much on the type of used radiation (Table 1).

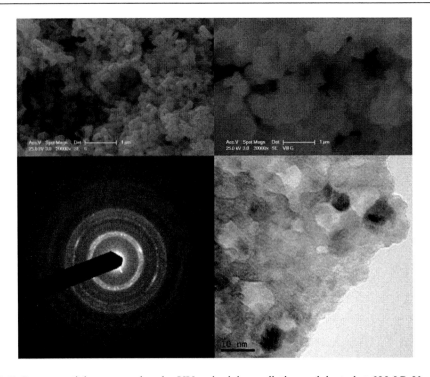

Figure 7. ZnO nanoparticles prepared under UV or ionizing radiation, calcinated at 600 °C. Upper left: SEM image of ZnO from solution (I) prepared under UV irradiation. Upper right: SEM image of ZnO from solution (I) prepared under ionizing radiation. Lower left: SAED image of ZnO from solution (I) prepared under UV irradiation. Lower right: HRTEM image of ZnO from solution (I) prepared under UV irradiation.

Table 1. Specific surface area and particle size of ZnO prepared under ionizing and UV radiation from various precursors [46,47]

Solid phase from solution (I)			
Temperature [°C]	SSA [$m^2 g^{-1}$][*]	Diameter [nm][*]	Diameter [nm][**]
40	75	7.1	28
650	11	48.6	168
1000	2.5	214	224
Solid phase from solution (II)			
Temperature [°C]	SSA [$m^2 g^{-1}$][*]	Diameter [nm][*]	Diameter [nm][**]
40	5.4	99	N/A
650	3	178	232
1000	2.4	223	250
Solid phase from solution (III)			
Temperature [°C]	SSA [$m^2 g^{-1}$][*]	Diameter [nm][*]	Diameter [nm][**]
40	136	3.9	N/A
650	3.2	167	112
1000	27.5	20	25

[*] - values obtained for materials prepared under UV irradiation
[**] - values obtained for materials prepared under ionizing irradiation

Luminescence properties

To evaluate scintillation characteristics of prepared materials, radioluminescence spectra were measured under excitation by X-ray tube. Spectra were corrected for the spectral dependence of sensitivity of the detection part. From the luminescence intensity point of view, the calcination conditions, namely temperature and atmosphere, are more important than the precursors for material preparation. Generally, the higher temperature, the higher luminescence intensity was observed, most probably due to thermal decomposition of impurity phases, changes in stoichiometry and/or point defect characteristics of zinc oxide particles. Samples calcinated in air generally showed better results than those treated under vacuum, probably due to lower oxygen vacancy concentration in the former. All samples

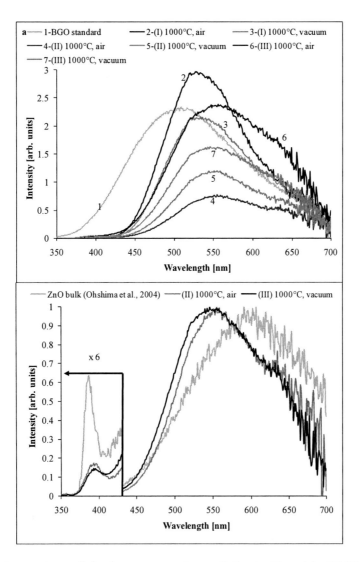

Figure 8. Room-temperature radioluminescence spectra, excitation by X-ray tube (40 kV, 15 mA); **a** - ZnO prepared from solutions (**I**), (**II**), (**III**), treated at 1000 °C under vacuum or on air, see the legend; **b** – comparison of selected ZnO nanopowders and ZnO bulk single crystal [46].

calcinated at temperatures higher than 650°C showed under X-ray excitation the intensive visible luminescence, when compared with the standard $Bi_4Ge_3O_{12}$ (BGO) scintillator sample as for the absolute intensity (Figure 8). Well shaped visible luminescence spectra are typical for the defect-related emission centers in zinc oxide [64]. Moreover, zinc oxide from solutions (II) and (III) showed well-shaped exciton luminescence at 390-400 nm, belonging to emission of (phonon replica of) free exciton.

The good performance of zinc oxide from solutions (II) and (III) can be attributed to the fact, that heat treatment of almost amorphous material formed during irradiation yields small crystals with high purity, and, in the case of solution (III) also to the fact, that PVA envelope stabilizes very small particles in solution [65,66]. Calcination of solid phase then decomposes PVA and enables formation of very small ZnO nanocrystals. Radioluminescence spectra of these samples are qualitatively similar to the spectrum of single crystal of ZnO - hydrothermally grown sample in Tokyo Denpa [67], which confirms the purity and quality of prepared material (Figure 8). Small spectral shifts observed both in exciton and defect emission bands in ZnO nanopowders can be attributed to the influence of (perturbation by) surface states which become of enhanced importance due to high surface/volume ratio in nanocrystalline material. Moreover, variable positioning of the defect luminescence is known due to lattice defects which can participate in these emission centers [68]; then the difference in the position of emission maxima of visible luminescence when comparing single crystal and powder samples in Figure 8 can be explained by the existence of different defects and their concentrations in the bulk and powder samples prepared by very different technologies. Figure 9 shows the radioluminescence spectra of identical samples, but prepared under UV irradiation and calcinated at 1000°C in air. The results are practically identical with those obtained for samples prepared under ionizing radiation (compare Figures 8 and 9).

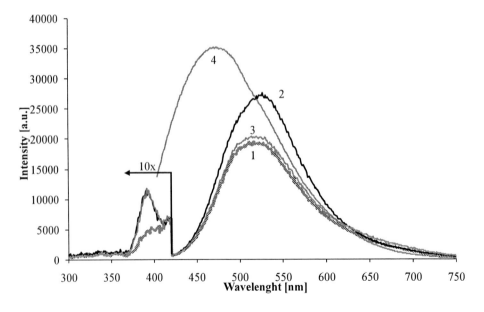

Figure 9. Room-temperature radioluminescence spectra, samples treated 2h at 1000°C in air. Sample (**I**) (1), sample (**II**) (2), sample (**III**) (3), BGO/10 (4) [47].

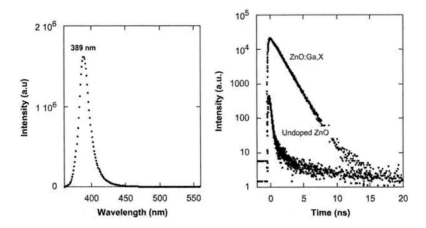

Figure 10. Room-temperature photoluminescence spectrum of a ZnO:G a sample made by reaction of ZnO with Ga_2O_3 (left), and decay time of the room-temperature X-ray excited luminescence of the same sample compared to an undoped sample (right) [38].

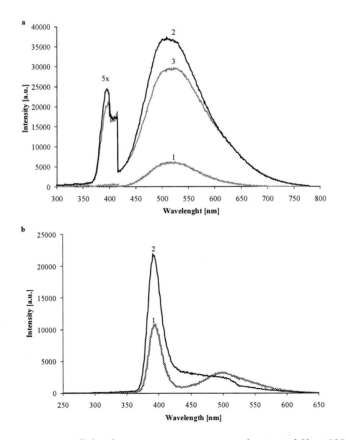

Figure 11. Room-temperature radioluminescence spectra; **a** - samples treated 2h at 1000°C in air; sample (**I**) doped with La (1), sample (**II**) doped with La (2), sample (**III**) doped with La (3); initial lanthanum acetate concentration 2.3×10^{-4} mol dm^{-3}; **b** - samples treated 2h at 250°C in air and 0.5h at 800°C in H_2/Ar atmosphere; sample (**II**) (1) and sample (**II**) doped with La (2); initial lanthanum acetate concentration 2.3×10^{-3} mol dm^{-3} [47].

Recently, the doping of zinc oxide with Ga has been used to enhance scintillation efficiency of ZnO [37,38]. The method for further increasing the excitonic luminescence of zinc oxide doped with gallium, based on treating the samples for 0.5 h at 800°C in the stream of H_2/Ar mixture (1:20), was used with good results [38]. The authors tested various methods for ZnO:Ga preparation and various annealing conditions. They found that the intensity of ultra fast UV luminescence was highest (up to 54% of standard scintillator $YAlO_3$:Ce, product QM58/ N-S1, made by Phosphor Technology, Inc.) when the stoichiometry of prepared material was balanced and luminous emissions from either excess oxygen or oxygen vacancies were minimized. The effect of annealing in reduction atmosphere on radioluminescence spectra for ZnO:Ga is shown in Figure 10. The luminescence spectrum shows a narrow emission at 389 nm. There is no 'green' or 'yellow' luminescence usually observed in undoped ZnO samples and attributed to native defects - an ionized oxygen vacancy and an oxygen interstitial, respectively. The decay of the X-ray excited luminescence is fast (0.8 ns), monoexponential in time over four decades which indicates that the luminescence arises from a homogeneous excited state.

Based on these results, the lanthanum doped samples were prepared by UV irradiation of aqueous solutions (I-III). Lanthanum acetate was added to the solutions prior irradiation, so that its concentration was 2.3×10^{-4} or 2.3×10^{-3} mol dm^{-3}. When compared to Ga^{3+}, significantly higher local lattice distortion can be expected in the case of La^{3+} due to its much bigger radius. In the lanthanum doped samples the XRPD spectra showed the presence of La_2O_3 and $La(OH)_3$, and relative ratio of ZnO:La_2O_3:$La(OH)_3$ was 77:1:22 or 95:1:4 in the samples initially containing 2.3×10^{-3} or 2.3×10^{-4} mol dm^{-3} of lanthanum acetate, respectively. Consistently with XRPD results, ICP/MS analysis of the former samples provided the content of La in ZnO of about 20%. This result indicates that only a fraction of percent of La could enter the ZnO structure and that the formation of lanthanum compounds under UV irradiation is much more efficient than the formation of La-doped zinc oxide. Effect of lanthanum doping on luminescence intensity of zinc oxide does nevertheless exists and is shown in Figure 11a. Both the visible and UV luminescence of ZnO is significantly increased. Furthermore, the effect of a post-growth annealing treatment in reductive H_2-containing atmosphere was studied following the description given by Bourret-Courchesne et al. [38], see Figure 11b. Consistently with the results reported in this reference, the significant increase of the excitonic UV emission in both undoped and La-doped samples was observed.

4. PREPARATION OF YAG BASED MATERIALS WITH EXCITON LUMINESCENCE

Yttrium aluminium garnet $Y_3Al_5O_{12}$ (YAG) is synthetic crystalline material with cubic structure [69,70]. Powder YAG may be used for preparation of transparent optical ceramics. When doped with foreign ions (e.g. lanthanides), such ceramics may serve as cheaper alternative to single crystalline YAG for applications in solid-state lasers [71-74], as phosphor in cathode ray tubes [75,76], in radiation dosimetry [77-79] or as a scintillating material also in an optical ceramic form [80,81]. Recently, also the application of YAG:Ce phosphor in white LED light source R&D was reported [82].

Intrinsic luminescence of YAG was studied mostly in single crystals grown from high temperature melt (about 2000°C) using Czochralski or Bridgman methods. In such crystals the presence of two isovalent cations and high temperature growth result in the appearance of atomic disorder and so called "anti-site defects" where especially the larger yttrium cation substitutes also at the octahedral aluminum site [83-85]. Such defects can efficiently anchor the excitons which gives rise to emission band around 4.8 eV observable below 150 K and which is gradually transformed to 3.95 eV band above this temperature [86]. Self-trapped exciton emission at 4.77 eV in YAG can be observed in high quality single crystalline films grown by the Liquid Phase Epitaxy at much lower temperatures which are free from antisite defects [87]. Similar effect as for the exciton trapping in YAG can be achieved by the isovalent dopants the radius of which is sufficiently different from that of Y^{3+}: the doping by La^{3+} and Sc^{3+} ions was studied in this respect [88] resulting in the appearance of emission bands at 4.1 eV and 4.5 eV, respectively. As the aluminum garnets are suitable host for fast scintillators when doped by Ce^{3+} or most recently by Pr^{3+} ions [89-91] the presence of antisite defects and shallow electron traps related to them is an unwanted issue as it slows down the energy transfer to fast Ce^{3+} and Pr^{3+} emission centers and scintillation response of the material is significantly deteriorated [92,93]. The gallium admixture was successfully applied in $Lu_3Al_5O_{12}$ (LuAG) host to suppress this unwanted trapping phenomenon [94]. Nanocrystalline YAG and LuAG powders are most frequently employed to prepare optical ceramics of these materials which were originally intended for solid state lasers using the Nd doping [95]. The Ce-doping was also tested in YAG [80,96] and LuAG [97] host to prepare fast scintillation ceramics. Despite of absence of the antisite defects the scintillation response and light yield of such ceramic materials are deteriorated most probably due to trapping states at the grain surfaces and interfaces [98].

Pure or doped powder YAG has been prepared by various methods involving preparation of solid precursors, usually mixture of oxides, and subsequent calcination. Calcination temperatures usually exceed 1000°C for preparation of pure YAG phase, but under strict control of reaction conditions, the lowering of calcination temperature below 1000°C is possible, especially for sol gel processes. For example, the powder YAG was synthesized via sol-gel method using ethanol as a solvent [99]. Precursors for the reaction were aluminium and yttrium nitrates and citric acid. Obtained solid phase was precalcinated at 500°C, YAG was first observed after calcination at temperature above 800°C. Neodymium doped YAG was prepared via sol-gel method from yttrium and neodymium oxides dissolved in acetic acid, and aqueous solution of aluminium nitrate with added 1,2-ethandiol. Prepared powder was calcinated at 900°C [100]. Modified sol-gel process – sol-gel combustion in solution - was used for powder YAG synthesis, due to reaction of yttrium and aluminium nitrates and glycine [101]. YAG garnet was subsequently obtained by calcination of formed solid phase – highly porous mixture of oxides – at 1100°C. Similarly, YAG doped with neodymium was prepared. Sol-gel combustion in solution of aluminium and yttrium nitrates in the presence of citric acid with subsequent calcination at 1000°C was used [102].

Powder YAG was also successfully prepared by coprecipitation methods. According to [103], the YAG precursors were prepared by coprecipitation in the presence of alcohol and water. Under these conditions, yttrium and aluminium are distributed homogenously in solid phase, and calcination at 900°C leads to YAG formation. YAG nanopowder was also synthesized by coprecipitation of nitrates of both compounds in the presence of ammonium hydrogen carbonate and surfactants [104]. Pure YAG was formed after calcination at 900°C.

Solid state reactions are the last significant group of methods for powder YAG preparation. YAG doped with Eu^{3+} ions was prepared from aluminium, yttrium and europium oxides, which were mixed in high energy ball-mill [105]. YAG was prepared after sintration at 1400°C. Composite ceramics YAG/ZrB$_2$ was prepared in the process consisting of three-steps: preparation of yttrium and aluminium oxides mixture by their coprecipitation from aqueous solution containing nitrates of both elements; milling of dried Y$_2$O$_3$-Al$_2$O$_3$/ ZrB$_2$ mixture; sintration of materials at 1500 – 1800 °C [106].

4.1. Preparation and Excitonic Luminescence of Powder YAG Prepared under UV-VIS and Ionizing Radiation

Recently, novel simple method for YAG synthesis via irradiation of precursors in aqueous solutions was investigated [107]. The aim of the work was to prepare solid phase from aqueous solution containing YAG precursors using ionizing or UV radiation, to prepare and characterize YAG after calcination and to perform preliminary study of luminescence properties of prepared material.

Irradiation of two aqueous systems was studied with components mixed at molar ratios corresponding to YAG stoichiometry:

- Solution (**IV**)- aqueous solution containing 7.5 10^{-3} mol dm^{-3} aluminium choride, 1.25x10^{-2} mol dm^{-3} yttrium nitrate and 0.18 mol dm^{-3} potassium formate.
- Solution (**V**)- aqueous solution containing 7.5 10^{-3} mol dm^{-3} aluminium nitrate, 1.25x10^{-2} mol dm^{-3} yttrium nitrate and 0.1 mol dm^{-3} potassium formate.

The irradiation experiments were carried out with equipment discussed in section 3.2. Applied dose was 75 kGy for solution (**IV**), and 90 kGy for solution (**V**); alanine dosimetry was used for the dose determination. UV-VIS irradiation took 60 minutes for solution (**IV**), and 120 minutes for solution (**V**) at photon flow 5x10^{19} hv s^{-1}.

It was found, that under either of both types of used irradiation, solid phase is quantitatively formed; After reaching dose 75 - 90 kGy under accelerated electron irradiation or after 60 - 120 minutes under UV irradiation, the maximum amount of solid phase was formed from both solutions (**IV**) and (**V**). In this respect, no differences were observed between ionizing and UV radiation. Solution (**IV**) yields finely dispersed precipitate; solution (**V**) yields a semi transparent gel-like material. No differences were observed in the color of filtered and dried solid phase; both solutions (**IV**) or (**V**) yield fine white powder, regardless of the type of irradiation.

It was observed that pH of both irradiated solutions increases with absorbed energy similarly to previously discussed systems, regardless of the type of radiation. Total absorption of solutions increases with irradiation time due to formation of solid phase, but the shape of the spectra does not change, except for broad band with maximum at 350 nm, indicating the formation of yttrium based colloids (Figure 12).

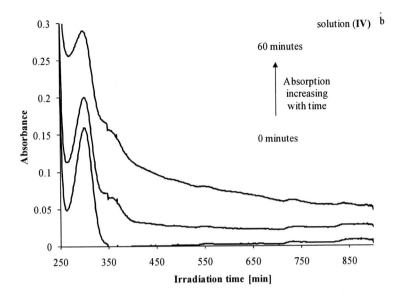

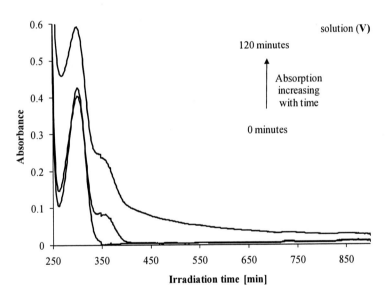

Figure 12. Changes in the absorption spectra of solutions (**IV**) and (**V**) under UV irradiation.

After irradiation, the finely dispersed solid phase was separated, dried and thermally treated at various temperatures. TA-MS analysis of materials formed under both types of irradiation suggests that yttrium is presented in different form, probably as a carbonate in sample (**IV**), while one may expect the presence of non-stoichiometric oxo-hydroxides in sample (**V**).

After calcination for 1h at 1000°C or more, the XRPD spectra of materials formed under UV irradiation show pure YAG phase with well developed crystals (Figure 13).

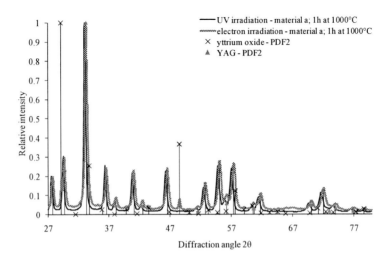

Figure 13. XRPD analysis of material formed from solution (**IV**), calcinated 1h at 1000°C.

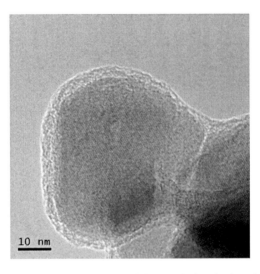

Figure 14. HRTEM image of crystalline YAG formed from solution (**IV**) under UV irradiation, calcinated 1h at 1000°C.

On the other hand, XRPD spectra of materials prepared under accelerated electrons irradiation show some low intensity lines corresponding to yttrium oxide or, at higher calcination temperatures, also lines of alpha – aluminium oxide. XRPD results indicate that yttrium and aluminium based components of formed solid phase blend much better under UV irradiation than under electron irradiation. When compared to the UV irradiation, the deposition of energy under accelerated electrons irradiation is very fast and highly non-uniform. Used linear electron accelerator implants the energy to the matter in short pulses (~ 3 μs), followed by comparatively long relaxation time (2 ms). Thus, the formed solid phase may feature non-uniformities with regards to both major components. TEM images show that prepared YAG consists of aggregates formed by smaller (probably single crystalline) particles with ~50 nm size; (see example in Figure 14).

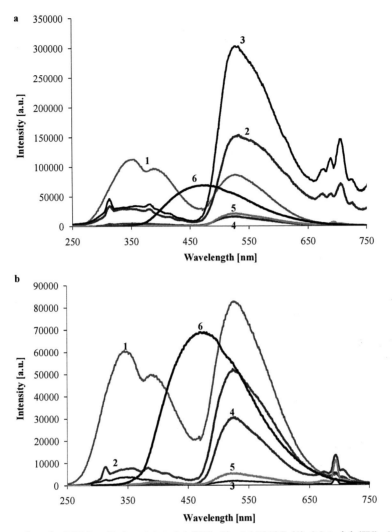

a - YAG prepared under UV irradiation. Material (**IV**), 1h at 1000°C (1). Material (**IV**), 1h at 1300°C (2). Material (**V**), 1h at 1300°C (3). Material (**IV**), 6h at 1300°C (4). Material (**V**), 6h at 1300°C (5). BGO standard (6).

b - YAG prepared under electron irradiation. Material (**IV**), 1h at 1000°C (1). Material (**IV**), 1h at 1300°C (2). Material (**V**), 1h at 1300°C (3). Material (**IV**), 6h at 1300°C (4). Material (**V**), 6h at 1300°C (5). BGO standard (6).

Figure 15. Room-temperature radioluminescence spectra of solid phase from solutions (**IV**) and (**V**) calcinated at 1000 or 1300 °C [107].

Radioluminescence (RL) spectra of prepared materials are shown in Figure 15. The dominating broad band at about 530 nm can be ascribed to Ce^{3+} center [108,109] which is due to Ce^{3+} impurity in raw materials (used precursor yttrium nitrate contains 100 ppm of CeO_2 and 300 ppm of La_2O_3). The band in UV area can be most probably ascribed to a defect-trapped exciton; e.g. the La and Sc related emissions in YAG lattice occur in 4-5 eV region [88] and emission due to antisite defects or vacancies are also reported in UV region [110]. The line emissions around 310 nm and 670-720 nm can be ascribed to Gd^{3+} and unidentified rare earth impurity, respectively. Apparently, the samples prepared under UV irradiation, with the calcination temperature of 1300 ^{0}C and calcination time of 1 hour, show the highest RL intensity. It can be taken as a measure of scintillation efficiency exceeding several times that of $Bi_4Ge_3O_{12}$ (BGO) powder reference scintillator sample. This indicates very good quality of YAG material containing low concentration of quenching sites of any kind. The samples prepared under electron irradiation and/or samples after 6h calcination show considerably decreased RL intensity which can be due to the previously mentioned structural non-uniformities arising in this kind of manufacturing process and resulting in high concentration of defects at atomistic scale, or due to the sintration of nanocrystalline material at longer calcination times.

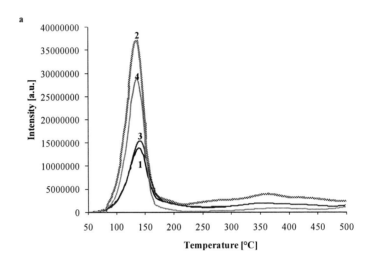

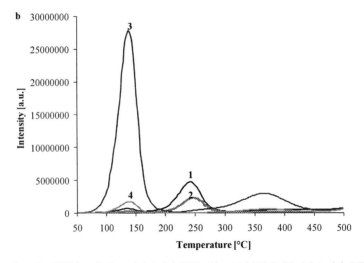

a - YAG prepared under UV irradiation. Material (**IV**), 1h at 1300°C (1). Material (**V**), 1h at 1300°C (2). Material (**IV**), 6h at 1300°C (3). Material (**V**), 6h at 1300°C (4).

b - YAG prepared under electron irradiation. Material (**IV**), 1h at 1300°C (1). Material (**V**), 1h at 1300°C (2). Material (**IV**), 6h at 1300°C (3). Material (**V**), 6h at 1300°C (4).

Figure 16. Thermoluminescence spectra of solid phase from solutions (**IV**) and (**V**) calcinated at 1300 °C. Irradiation by ^{60}Co gamma source at room temperature, dose 5Gy; heating rate 1°C min^{-1} [107].

Thermoluminescence spectra of prepared materials are shown in Figure 16. The peaks with maxima around 135-140°C correspond to those found in undoped powder YAG prepared by solution combustion method, after irradiation under the similar conditions [77]. Though the integral intensity depends on the composition of irradiated solution, temperature and time of calcination, and on the applied dose, the maximum does not shift significantly with any of the above mentioned parameters. Moreover, electron irradiated samples exhibit peak at 240 – 250°C. Similar peak was found, too [78] in YAG nanopowder prepared by sol-gel and precipitation and irradiated with beta radiation (10 – 600 Gy). TSL glow curve peaks above room temperature in aluminum garnets are usually ascribed to electron trap based on oxygen vacancies [111].

5. CONCLUSION

Convenient simple method for preparation of crystalline nanoparticles with interesting luminescence properties was discussed.

ZnO nanocrystals were prepared by various authors via irradiation of aqueous or (reverse) micellar solutions. The properties of prepared materials strongly depend on precursors used; amorphous, single crystalline and polycrystalline materials were prepared by radiation method under various conditions. It seems that both ionizing and non ionizing radiations are comparable when used for ZnO preparation, as only minor differences in the morphology and the shape of radioluminescence spectra were observed. Most intensive excitonic luminescence at 395 nm showed materials prepared from aqueous solution containing zinc formate, lanthanum acetate and hydrogen peroxide, treated at 1000°C. Radiation method may be used for preparation of doped materials, too; zinc oxide was successfully doped with lanthanum by adding lanthanum acetate to UV irradiated solutions. Further enhancement of excitonic emission was achieved by the subsequent annealing in reduction H_2-containing atmosphere at elevated temperature.

YAG is more complex material than the zinc oxide. It was successfully prepared from aqueous solutions containing yttrium nitrate, aluminium chloride or nitrate and potassium formate, using UV or accelerated electrons irradiation. The former is more convenient for YAG synthesis; calcination of the material formed under UV irradiation of solution containing yttrium nitrate, aluminium chloride and potassium formate for 1h at 1000°C yields powder YAG with well developed single crystals. After accelerated electrons irradiation, YAG phase is still the major component of prepared material, but some amounts of yttrium oxide or alpha-aluminium oxide were also detected after calcination at 1000°C or 1300°C. All prepared materials show leading radioluminescence peak at 530 nm typical for the Ce^{3+} doped YAG structure and appears due to CeO_2 impurity in the starting raw materials. YAG's formed under UV irradiation have more intensive radioluminescence in both visible and UV area than those formed under accelerated electrons irradiation. Thermoluminescence glow curves show distinctive peaks at 135 - 140°C °C and 240 – 250°C which are most probably due to oxygen vacancy-based electron traps. Again, the most intensive thermoluminescence is shown by materials prepared under UV irradiation.

Presence of organic compounds, scavenging OH radicals and/or protecting small particles of metal/oxide from agglomeration is needed to achieve high quality ZnO nanocrystals when using this technology. Prepared materials with homogenous distribution of particles size and good luminescence properties could serve in future as a basis for preparation of ZnO of YAG powder scintillators doped with various donor ions, either using radiation method alone, as was demonstrated in this chapter, or in combination with other preparative methods. Further optimization of donor ion concentration and a post-growth annealing treatment in various atmospheres (including reductive H_2-containing atmosphere) thus indicates the way of material optimization as far as the scintillation application is considered. Doped scintillators with sufficiently fast and intensive luminescence in UV area could be further press compacted to prepare transparent ceramics, or embedded in a host matrix (e.g. SiO_2) to prepare technologically applicable scintillating materials.

ACKNOWLEDGMENT

Authors gratefully acknowledge support of Grant Agency of AS CR, project KAN300100802.

6. REFERENCES

[1] Song, A. K. S. & Williams, R. T. (1996). *Self-trapped excitons*. Springer, Berlin.
[2] Agranovich, V. M. (2008). *Excitations in organic solids*. Oxford university press, Oxford.
[3] Henneberger, F., Schmitt-Rink, S. & Göbel, E. O. (1993). *Optics of semiconductor nanostructures.* Akademie Verlag, Berlin.
[4] Nikl, M., Nitsch, K., Polak, K., Pazzi, G. P., Fabeni, P., Citrin, D. S. & Gurioli, M. (1995). Optical Properties of Pb^{2+}-aggregated phase in CsCl host crystal. Quantum confinement effect. *Phys. Rev. B, 51*, 5192-5199.
[5] Nikl, M., Mihokova, E., Nitsch, K., Polak, K., Rodova, M., Dusek, M., Pazzi, G. P., Fabeni, P. & Gurioli, M. (1994). Photoluminescence and decay kinetics of $CsPbCl_3$ single crystals. *Chem. Phys. Lett, 220*, 14-18.
[6] Buxton, G. V., Greenstock, C. L., Helman, W. P. & Ross, A. B. (1988). Critical review of rate constants for reactions of hydrated electrons, hydrogen atoms and hydroxyl radicals in aqueous solution. *J. Phys. Chem. Ref. Data, 17*, 513-886.
[7] Ferradini, C. & Jay-Gerin, J. P. (2000). The effect of pH on water radiolysis: a still open question - a minireview. *Res. Chem. Intermed, 26*, 549-565.
[8] Kapoor, S., Palit, D. K. & Mukherjee, T. (2002). Preparation, characterization and surface modification of Cu metal nanoparticles. *Chem. Phys. Lett, 355*, 383-387.
[9] An, Y. J., Jeong, S. W. & Carraway, E. R. (2001) Micellar effect on the photolysis of hydrogen peroxide. *Water Res., 13*, 3276-3279.
[10] Loginov, A. V., Gorbunova, V. V. & Boitsova, T. B. (2002). Photochemical synthesis and properties of colloidal copper, silver and gold adsorbed on quartz. *J. Nanopart. Res., 4*, 193-205.
[11] Giuffrida, S., Costanzo, L. L., Ventimiglia, G. & Bongiorno, C. (2008). Photochemical synthesis of copper nanoparticles incorporated in poly(vinyl pyrrolidone). *J. Nanopart*

Res., *10*, 1183-1192.

[12] Mallick, K., Witcomb, M. J. & Scurrell, M. S. (2004). Polymer stabilized silver nanoparticles, a photochemical synthesis route. *J. Mater. Sci.*, *39*, 4459-4463.

[13] Azrague, K., Bonnefille, E., Pradines, V., Pimienta, V., Oliveros, E., Maurette, M. T. & Benoit-Marquié, F. (2005). Hydrogen peroxide evolution during V-UV photolysis of water. *Photochem. Photobiol*, *4*, 406-408.

[14] Belloni, J., Mostafavi, M., Remita, H., Marignier, J. L. & Delcourt, M. O. (1998). Radiation-induced synthesis of mono- and multi-metallic clusters and nanocolloids. *New J. Chem.*, 1239-1255.

[15] Zhou, R., Wu, X., Hao, X., Zhou, F., Li, H. & Rao, W. (2008). Influences of surfactants on the preparation of copper nanoparticles by electron beam irradiation. *Nucl. Instrum. Meth, B*, *266*, 599-603.

[16] Kumar, M., Kapoor, S. & Gopinathan, C. (1999). Study on the radiolytic reduction of mixed Cd^{2+}/Cu^{2+} ions in aqueous medium. *Radiat. Phys. Chem.*, *54*, 39-44.

[17] Shvalagin, V. V., Stroyuk, A. L. & Kuchmii, S. Y. (2007). Photochemical synthesis of ZnO/Ag nanocomposites. *J. Nanopart Res.*, *9*, 427-440.

[18] Shvalagin, V. V., Stroyuk, A. L. & Kuchmii, S. Y. (2004). Photochemical synthesis and spectral-optical characteristics of ZnO/Cu and ZnO/Ag/Cu nanoheterostructures. *Theor. Exp. Chem.*, *40*, 378- 382.

[19] Buono-Core, G. E., Cabello, G., Klahn, A. H., Del Río, R. & Hill, R. H. (2006). Characterization of pure ZnO thin films prepared by a direct photochemical method. *J. Non-Cryst. Solids*, *352*, 4088-4092.

[20] Tak, Y. & Yong, K. (2008). A novel heterostructure of Co_3O_4/ZnO nanowire array fabricated by photochemical coating method. *J. Phys. Chem. C*, *112*, 74-79.

[21] Rath, M. C., Sunitha, Y., Ghosh, H. N., Sarkar, S. K. & Mukherjee, T. (2009). Investigation of the dynamics of radiolytic formation of ZnO nanostructures materials by pulse radiolysis. *Radiat. Phys. Chem.*, *78*, 77-80.

[22] Chen, Q. D., Shen, X. H. & Gao, H. C. (2007). Formation of solid and hollow cuprous oxide nanocubes in water-in-oil microemulsions controlled by the yield of hydrated electrons *J. Colloid Interface Sci.*, *312*, 272-278.

[23] He, P., Shen, X. H. & Gao, H. C. (2005). Size-controlled preparation of Cu_2O octahedron nanocrystals and studies on their optical absorption. *J. Colloid. Interf. Sci.*, *284*, 510-515.

[24] Zhu, Y., Qian, Y., Zhang, M., Chen, Z. & Xu, D. (1994). Preparation and characterization of nanocrystalline powders of cuprous oxide by using γ-radiation. *Mater. Res. Bull*, *29*, 377-383.

[25] Klingshirn, C. (2007). ZnO, From basics towards applications. *Phys. Status Solidi B*, *244*, 3027-3073.

[26] Pearton, S. J., Norton, D. P., Ip, K., Heo, Y. W. & Steiner, T. (2003). Recent progress in processing and properties of ZnO. *Superlattice. Microst*, *34*, 3-32.

[27] Ehrentraut, D., Sato, H., Kagamitani, Y., Sato, H., Yoshikawa, A. & Fukuda, T. (2006). Solvothermal growth of ZnO. *Prog. Cryst. Growth Ch.*, *52*, 280-335.

[28] Xu, C. X., Sun, X. W., Chen, B. J., Shum, P., Li, S. & Hu, X. (2004). Nanostructural zinc oxide and its electrical and optical properties. *J. Appl. Phys.*, *95*, 661-666.

[29] Klingshirn, C. (2007). ZnO: Material, physics and applications. *Chem. Phys. Lett*, *8*, 782-803.

[30] Wang, Z. L. (2007). Novel nanostructures of ZnO for nanoscale photonic, optoelectronics, piezoelectricity, and sensing. *Appl. Phys. A*, *88*, 7-15.
[31] Izumi, T., Izumi, K., Kuroiwa, N., Senjuh, A., Fujimoto, A., Adachi, M. & Yamamoto, T. (2010). Preparation of electrically conductive nano-powder of zinc oxide and application to transparent film coating. *J. Alloy. Compd*, *480*, 123-125.
[32] Vayssieres, L. (2003). Growth of arrayed nanorods and nanowires of ZnO from aqueous solutions. *Adv. Mater*, *15*, 464-466.
[33] Yuan, K., Yin, X., Li, J., Wu, J., Wang, Y. & Huang, F. (2010). Preparation and DSC application of the size-tuned ZnO nanoarrays. *J. Alloy. Compd*, *489*, 694-699.
[34] Wilkinson, J., Uce,r, K. B. & Williams, R. T. (2004). Picosecond excitonic luminescence in ZnO and other wide-gap semiconductors. *Radiat. Meas*, *38*, 501-505.
[35] Yamamoto, S., Yano, H., Mishina, T. & Nakahara, J. (2007). Decay dynamics of ultraviolet photoluminescence in ZnO nanocrystals. *J. Lumin*, *126*, 257-262.
[36] Meyer, B. K., Alves, H., Hofmann, D. M., Kriegseis, W., Forster, D., Bertram, F., Christen, J., Hoffmann, A., Straßburg, M., Dworzak, M., Haboeck, U. & Rodina, A. V. (2004). Bound exciton and donor-acceptor pair recombinations in ZnO. *Phys. Status Solidi B*, *241*, 231-260.
[37] Derenzo, S. E., Weber, M. J. & Klintenberg, M. K. (2002). Temperature dependence of the fast, near-band-edge scintillation from CuI, HgI$_2$, PbI$_2$, ZnO:Ga and CdS: In. *Nucl. Instrum. Meth. A*, *486*, 214-219.
[38] Bourret-Courchesne, E. D., Derenzo, S. E. & Weber, M. J. (2009). Development of ZnO:Ga as an ultra-fast scintillator. *Nucl. Instrum. Meth. A*, *601*, 358-363.
[39] Lorenz, M., Hochmuth, H., Lenzner, J., Nobis, T., Zimmermann, G., Diaconu, M., Schmidt, H., von Wenckstern, H. & Grundmann, M. (2005). Room-temperature cathodoluminescence of n-type ZnO thin films grown by pulsed laser deposition in N$_2$, N$_2$O, and O$_2$ background gas. *Thin Solid Films*, *486*, 205-209.
[40] Ehrentraut, D., Sato, H., Kagamitani, Y., Yoshikawa, A., Fukuda, T., Pejchal, J., Polak, K., Nikl, M., Odaka, H., Hatanaka, K. & Fukumura, H. (2006). Fabrication and luminescence properties of single-crystalline, homoepitaxial zinc oxide films doped with tri- and tetravalent cations prepared by liquid phase epitaxy. *J. Mater. Chem.*, *16*, 3369-3374.
[41] Asakuma, N., Fukui, T., Toki, M., Awazu, K. & Imai, H. (2003). Photoinduced hydroxylation at ZnO surface. *Thin Solid Films*, *2*, 284-287.
[42] Sun, M., Hao, W., Wang, C. & Wang, T. (2007). A simple and green approach for preparation of ZnO$_2$ and ZnO under sunlight irradiation. *Chem. Phys. Lett*, *443*, 342-346.
[43] Sebok, D., Szabo, T. & Dekany, I. (2009). Optical properties of zinc peroxide and zinc oxide multilayer nanohybrid films. *Appl. Surf. Sci.*, *255*, 6953-6962.
[44] Ren, F., Guo, L. P., Shi, Y., Chen, D. L., Wu, Z. Y. & Jiang, C. Z. (2006). Formation of Zn–ZnO core–shell nanoclusters by Zn/F sequential ion implantation. *J. Phys. D, Appl. Phys.*, *39*, 488-491.
[45] Hu, Y., Chen, J., Xue, X., Li, T. & Xie, Y. (2005). Room-temperature irradiation route to synthesize a large-scale single-crystalline ZnO hexangular prism. *Inorg. Chem.*, *44*, 7280-7282.
[46] Cuba, V., Gbur, T., Mucka, V., Nikl, M., Kucerkova, R., Pospisil, M. & Jakubec, I. (2010). Properties of ZnO nanocrystals prepared by radiation method. *Radiat. Phys.*

Chem., 79, 27-32.

[47] Gbur, T., Cuba, V., Mucka, V., Nikl, M., Knizek, K., Pospisil, M. & Jakubec, I. Photochemical preparation of ZnO nanoparticles. *J. Nanopart. Res.*, in press, DOI: 10.1007/s11051-011-0407-y.

[48] Lee, J. & Seliger, H. H. (1964). Quantum yield of the ferrioxalate actinometer. *J. Phys. Chem.*, 40, 519-523.

[49] Allan, J. T. & Beck, C. M. (1964). The radiolysis of deaerated aqueous aolutions of 2-propanol containing nitrous oxide. *J. Am. Chem. Soc.*, 86, 1483-1488.

[50] Van der Linde, H. J. & Freeman, G. R. (1971). The influence of temperature on the γ-radiolysis of isopropyl alcohol vapor. Effect of molecular structure on the nonchain and chain decompositions of alcohol vapors. *J. Phys. Chem.*, 75, 20-24.

[51] Meyerstein, D. & Mulac, W. A. (1969). Effects of ligand on reactivity of metal cations towards the hydrated electron. Part 2: effect of glycine, ethylenediamine and nitrilotriacetic acid. *Trans. Faraday Soc.*, 65, 1812-1817.

[52] Hart, E. J. (1951). Mechanism of the γ-ray induced oxidation of formic acid in aqueous solution. *J. Am. Chem. Soc.*, 73, 68-73.

[53] Hart, E. J. (1954). Gamma-ray induced oxidation of aqueous formic acid – oxygen solutions. Effect of oxygen and formic acid concentrations. *J. Am. Chem. Soc.*, 76, 4312-4315.

[54] Hart, E. J. (1954). γ-Ray induced oxidation of formic acid – oxygen solutions. Effect of pH. *J. Am. Chem. Soc.*, 76, 4198-4201.

[55] Smithies, D. & Hart, E. J. (1960). Radiation chemistry of aqueous formic acid solutions. Effect of concentration. *J. Am. Chem. Soc.*, 82, 4775-4775.

[56] Zhang, S. J. & Yu, H. Q. (2004). Mechanistic study on the radiolysis of dilute PVA aqueous solutions. *Chem. Lett*, 33, 562-563.

[57] Ulanski, P., Bothe, E., Rosiak, J. M. & von Sonntag, C. (1994). OH-radical-induced crosslinking and strand breakage of poly(vinyl alcohol) and its monomeric model pentane-2,4-diol. A pulse radiolysis and product study. *Macromol. Chem. Phys.*, 195, 1443-1461.

[58] Bahnemann, D. W., Kormann, C. & Hoffmann, M. R. (1987). Preparation and Characterization of Quantum Size Zinc Oxide, A Detailed Spectroscopic Study. *J. Phys. Chem.*, 91, 3789-3798.

[59] Baruah, S., Rafique, R. F. & Dutta, J. (2008). Visible light photocatalysis by tailoring crystal defects in zinc oxide nanostructures. *NANO - Brief reports and reviews*, 3, 399-407.

[60] Li, F., Liu, X., Qin, Q., Wu, J., Li, Z. & Huang, X. (2009). Sonochemical synthesis and characterization of ZnO nanorod/Ag nanoparticle composites. *Cryst. Res. Technol.*, 44, 1249-1254.

[61] Rykl, D. & Bauer, J. (1968). Hydrothermal synthesis of zincite. *Krist. Tech.*, 3, 375-384. in German.

[62] Tsvigunov, A. N. (2001). A new modification of zinc oxide synthesized by the hydrothermal method. *Glass Ceram.*, 58, 17-19.

[63] Guo, L., Cheng, J. X., Li, X. Y., Yan, Y. J., Yang, S. H., Yang, C. L., Wang, J. N. & Ge, W. K. (2001). Synthesis and optical properties of crystalline polymer-capped ZnO nanorods. *Mat. Sci Eng. C*, 16, 123-127.

[64] Borseth, T. M., Svensson, B. G., Kuznetsov, A. Y., Klason, P., Zhao, Q. X. & Willander, M. (2006). Identification of oxygen and zinc vacancy optical signals in ZnO. *Appl. Phys. Letters*, *89*, 262112.

[65] Pomogailo, A. D. & Kestelman, V. N. (2005). *Metallopolymer Nanocomposites*. Springer, Berlin.

[66] Belloni, J. (2006). Nucleation, growth and properties of nanoclusters studied by radiation chemistry: *Application to catalysis. Catal.* Today, *113*, 141-156.

[67] Ohshima, E., Ogino, H., Niikura, I., Maeda, K., Sato, M., Ito, M. & Fukuda, T. (2004). Growth of the 2-in-size bulk ZnO single crystals by the hydrothermal method. *J. Cryst. Growth*, *260*, 166-170.

[68] Reshchikov, M. A., Morkoc, H., Nemeth, B., Nause, J., Xie, J., Hertog, B. & Osinsky, A. (2007). Luminescence properties of defects in ZnO. *Physica B*, *401-402*, 358-361.

[69] Warshaw, I. & Roy, R. (1959). Stable and metastable equilibria in the systems Y_2O_2-Al_2O_3, and Gd_2O_3-Fe_2O_3. *J. Am. Ceram. Soc.*, *42*, 434-438.

[70] Abell, J. S., Harris, I. R., Cockayne, B. & Lent, B. (1974). An investigation of phase stability in the Y_2O_3-Al_2O_3 system. *J. Mater. Sci.*, *9*, 527-537.

[71] De With, G. & Van Dijk, H. J. A. (1984). Translucent $Y_3Al_5O_{12}$ ceramics. *Mater. Res. Bull*, *19*, 1669-1674.

[72] Ikesue, A., Kinoshita, T. & Kamata, K. (1995). Synthesis of Nd^{3+},Cr^{3+}-codoped YAG ceramics for high-efficiency solid-state lasers. *J. Am. Ceram. Soc.*, *78*, 2545-2547.

[73] Ikesue, A., Kinoshita, T., Kamata, K. & Yoshida, K. (1995). Fabrication and optical properties of high-performance polycrystalline Nd,YAG ceramics for solid-state lasers *J. Am. Ceram. Soc.*, *78*, 1033-1040.

[74] Lu, J. R., Prabhu, M., Song, J., Li, C., Xu, J., Ueda, K., Kaminskii, A. A., Yagi, H. & Yanagitani, T. (2000). Optical properties and highly efficient laser oscillation of Nd,YAG ceramic. *Appl. Phys. B*, *71*, 469-473.

[75] Kang, Y. C., Lenggoro, I. W., Park, S. B. & Okuyama, K. (1999). Photoluminescence characteristics of YAG:Tb phosphor particles with spherical morphology and non-aggregation. *J. Phys. Chem. Solids*, *60*, 1855-1858.

[76] Zhou, Y., Lin, J., Yu, M., Wang, S. & Zhang, H. (2002). Synthesis-dependent luminescence properties of $Y_3Al_5O_{12}$:Re^{3+} (Re=Ce, Sm, Tb) phosphors. *Mater. Lett*, *56*, 628-636.

[77] Kulkarni, M. S., Muthe, K. P., Rawat, N. S., Mishra, D. R., Kakade, M. B., Ramanathan, S., Gupta, S. K., Bhatt, B. C., Yakhmi, J. V. & Sharma, D. N. (2008). Carbon doped yttrium aluminum garnet (YAG:C)-a new phosphor for radiation dosimetry. *Radiat. Meas*, *43*, 492-496.

[78] Rodriguez, R. A., De la Rosa, E., Salas, P., Melendrez, R. & Barboza-Flores, M. (2005). Thermoluminescence and optically stimulated luminescence properties of nanocrystalline Er^{3+} and Yb^{3+} doped $Y_3Al_5O_{12}$ exposed to β-rays. *J. Phys. D: Appl. Phys.*, *38*, 3854-3859.

[79] Rodriguez, R. A., De la Rosa, E., Melendrez, R., Salas, P., Castaneda, J., Felix, M. V. & Barboza-Flores, M. (2005). Thermoluminescence characterization of nanocrystalline and single $Y_3Al_5O_{12}$ crystal exposed to β-irradiation for dosimetric applications. *Opt. Mater*, *27*, 1240-1244.

[80] Zych, E., Brecher, C., Wojtowitcz, A. J. & Lingertat, H. (1997). Luminescence properties of Ce-activated YAG optical ceramic scintillator materials. *J. Lumin.*, *75*,

193-203.

[81] Zych, E., Brecher, C., Wojtowitcz, A. J. & Lingertat, H. (1998). Host-associated luminescence from YAG optical ceramics under gamma and optical excitation. *J. Lumin*, *78*, 121-134.

[82] Mueller-Mach, R., Mueller, G., Krames, M. R., Hoppe, H. A., Stadler, F., Schnick, W., Juestel, T. & Schmidt, P. (2005). Highly efficient all-nitride phosphor-converted white light emitting diode. *Phys. Status Solidi A*, *202*, 1727-1732.

[83] Ashurov, M. Kh., Voronko, Yu. K., Osiko, V. V., Sobol, A. A. & Timoshechkin, M. I. (1977). Spectroscopic study of stoichiometric deviation in crystals with garnet structure. *Phys. Status Solidi A*, *42*, 101-110.

[84] Lupei, V., Lupei, A., Tiseanu, C., Georgescu, S., Stoicescu, C. & Nanau, P. M. (1995). High resolution optical spectroscopy of YAG:Nd: A test for structural and distribution models. *Phys. Rev. B*, *51*, 8-17.

[85] Kuklja, M. M. (2000). Defects in yttrium aluminium perovskite and garnet crystals: atomistic study. *J. Phys.-Condens. Mat*, *12*, 2953-2967.

[86] Babin, V., Blazek, K., Krasnikov, A., Nejezchleb, K., Nikl, M., Savikhina, T. & Zazubovich, S. (2005) Luminescence of undoped LuAG and YAG crystals. *Phys. Status Solidi C*, *2*, 97-100.

[87] Zorenko, Yu., Voloshinovskii, A., Savchyn, V., Voznyak, T., Nikl, M., Nejezchleb, K., Mikhailin, V., Kolobanov, V. & Spassky, D. (2007). Exciton and antisite defect-related luminescence in $Lu_3Al_5O_{12}$ and $Y_3Al_5O_{12}$ garnets. *Phys. Status Solidi B*, *244*, 2180-2189.

[88] Murk, V. & Yaroshevich, N. (1995). Exciton and recombination processes in YAG crystals. *J. Phys.-Condens. Mat*, *7*, 5857-5864.

[89] Moszynski, M., Ludziewski, T., Wolski, D., Klamra, W. & Norlin, L. O. (1994). Properties of the YAG:Ce scintillator. *Nucl. Instrum. Meth. A*, *345*, 461-467.

[90] Nikl, M., Mihokova, E., Mares, J. A., Vedda, A., Martini, M., Nejezchleb, K. & Blazek, K. (2000). Traps and timing characteristics of LuAG:Ce^{3+} scintillator. *Phys. Status Solidi B*, *181*, R10-R12.

[91] Pejchal, J., Nikl, M., Mihóková, E., Mareš, J. A., Yoshikawa, A., Ogino, H., Schillemat, K. M., Krasnikov, A., Vedda, A., Nejezchleb, K. & Múčka, V. (2009). Pr^{3+}-doped complex oxide single crystal scintillators. *J. Phys. D. Appl. Phys.*, *42*, 055117.

[92] Nikl, M. (2005). Energy transfer phenomena in the luminescence of wide band-gap scintillators. *Phys. Status Solidi A*, *202*, 201-206.

[93] Nikl, M., Mihokova, E., Pejchal, J., Vedda, A., Zorenko, Yu. & Nejezchleb, K. (2005). The antisite Lu_{Al} defect-related trap in $Lu_3Al_5O_{12}$:Ce single crystal. *Phys. Status Solidi B*, *242*, R119-R121.

[94] Nikl, M., Pejchal, J., Mihokova, E., Mares, J. A., Ogino, H., Yoshikawa, A., Fukuda, T., Vedda, A. & D'Ambrosio, C. (2006). Antisite defect-free $Lu_3(Ga_xAl_{1-x})_5O_{12}$:Pr scintillator. *Appl. Phys. Lett*, *88*, 141916.

[95] Lu, J., Ueda, K., Yagi, H., Yanagitani, T., Akiyama, Y. & Kaminskii, A. A. (2002). Neodymium doped yttrium aluminum garnet $Y_3Al_5O_{12}$ nanocrystalline ceramics - a new generation of solid state laser and optical materials. *J. Alloy. Compd*, *341*, 220-225.

[96] Mihokova, E., Nikl, M., Mares, J. A., Beitlerova, A., Vedda, A., Nejezchleb, K., Blazek, K. & D'Ambrosio, C. (2007). Luminescence and scintillation properties of YAG:Ce single crystal and optical ceramics. *J. Lumin*, *126*, 77-80.

[97] Nikl, M., Mares, J. A., Solovieva, N., Li, H., Liu, X., Huang, L., Fontana, I., Fasoli, M., Vedda, A. & D'Ambrosio, C. (2007). Scintillation characteristics of $Lu_3Al_5O_{12}$:Ce optical ceramics. *J. Appl. Phys.*, *101*, 033515.

[98] Nikl, M., Mihokova, E., Pejchal, J., Vedda, A., Fasoli, M., Fontana, I., Laguta, V. V., Babin, V., Nejezchleb, K., Yoshikawa, A., Ogino, H. & Ren, G. (2008). Scintillator materials -achievements, opportunities, and puzzles. *IEEE T. Nucl. Sci.*, *55*, 1035-1041.

[99] Yang, L., Lu, T., Xu, H. & Wei, N. (2009). Synthesis of YAG powder by the modified sol–gel combustion method. *J. Alloy. Compd*, *484*, 449-451.

[100] Barbaran, J. H., Farahani, M. F. & Hajiesmaeilbaigi, F. (2005). Synthesis of highly doped Nd:YAG powder by sol-gel method. Semiconductor Physics, *Quantum Electronics & Optoelectronics*, *8*, 87-89.

[101] Kakade, M. B., Ramanathan, S. & Roy, S. K. (2002). Synthesis of YAG powder by aluminum nitrate–yttrium nitrate–glycine reaction. *J. Mater. Sci. Lett*, *21*, 927-929.

[102] Costa, A. L., Esposito, L., Medri, V. & Bellosi, A. (2007). Synthesis of Nd-YAG material by citrate-nitrate sol-gel combustion route. *Adv. Eng. Mater*, *9*, 307-312.

[103] Tong, S., Lu, T. & Guo, W. (2007). Synthesis of YAG powder by alcohol–water co-precipitation method. *Mater. Lett*, *61*, 4287-4289.

[104] Xu, G., Zhang, X., He, W., Liu, H., Li, H. & Boughton, R. I. (2006). Preparation of highly dispersed YAG nano-sized powder by co-precipitation method. *Mater. Lett*, *60*, 962-965.

[105] Yang, H. K. & Jeong, J. H. (2010). Synthesis, crystal growth, and photoluminescence properties of YAG:Eu^{3+} phosphors by high-energy ball milling and solid-state reaction. *J. Phys. Chem. C*, *114*, 226-230.

[106] Song, J. G., Li, J. G., Song, J. R. & Zhang, L. M. (2007). Preparation of high-density YAG/ZrB2 multi-phase ceramics by spark plasma sintering. *J. Ceram. Process. Res.*, *8*, 356-358.

[107] Cuba, V., Indrei, J., Mucka, V., Nikl, M., Beitlerova, A., Pospisil, M. & Jakubec, I. Radiation induced synthesis of powder yttrium aluminium garnet. *Radiat. Phys. Chem.*, in press, DOI: 10.1016/j.radphyschem.2011.04.009.

[108] Weber, M. J. (1973). Nonradiative decay from 5d states of rare earths in crystals. *Solid State Commun*, *12*, 741-744.

[109] Bachmann, V., Ronda, C. & Meijerink, A. (2009). Temperature quenching of yellow Ce^{3+} luminescence in YAG:Ce. *Chem. Mater*, *21*, 2077-2084.

[110] Zorenko, Y., Gorbenko, V., Konstankevych, I., Voloshinovskii, A., Stryganyuk, G., Mikhailin, V., Kolobanov, V. & Spassky, D. (2005). Single-crystalline films of Ce-doped YAG and LuAG phosphors, advantages over bulk crystals analogues. *J. Lumin*, *114*, 85-94.

[111] Xing, L., Xu, X. B., Gu, M., Tang, T. B., Mihokova, E., Nikl, M. & Vedda, A. (2009). Dielectric relaxations in undoped, Ce-doped and Ce,Zr-codoped $Lu_3Al_5O_{12}$ single crystals. *J. Phys. Chem. Sol.*, *70*, 595-599.

In: Exciton Quasiparticles
Editor: Randy M. Bergin

ISBN: 978-1-61122-318-7
© 2011 Nova Science Publishers, Inc.

Chapter 4

THE SEMICLASSICAL MOLECULAR EXCITON: PATH/TIME ORDERED PROPAGATORS VS. STOCHASTIC DYNAMICS

William R. Kirk
Mayo Clinic College of Medicine, Rochester, Minnesota, USA

The semi-classical model of the molecular exciton as the short-time resonance ("evanescence") between electromagnetic and matter-field is reviewed. From a simple 'bubble diagram' quasi-physical model a richer and more evocative physical picture emerges of an electron/hole loop around a set of current sources (nuclear normal modes) which effectively trap the excitation into an electronic excited state. In reverse, the exciton is the source for the emergent electromagnetic field during the emission process. This picture leads to an account of 'mapping' vibrational modes vs. 'bath' modes, and an account of gaussian broadening at a fundamental level is obtained. The recent exposition by Adler of stochastic dynamics in the Schroedinger picture leading instead to effective Lorentzian profiles (arXiv:quant-ph/0208123v4) is contrasted. The 'loop'='bubble' picture makes other contacts with nonequilibrium quantum dynamics, gauge theory and time dependent density functional theory as well.

The word 'exciton' often means different things to different kinds of investigators. Herein we will not be concerned with the localized storage of electromagnetic field excitation in a medium[1], e.g. a crystal lattice, as is very commonly meant, still less shall we be concerned with the localized cross-sectional excitation of a (super)string![2]; instead, we will mean the excitation of an isolated molecule, a *fluorophore*, which is expected to be able to re-emit this energy back to the e-m field as light. That is, both 'up' and 'dowm' processes are of interest to us. The symmetry between these two processes is revealed in a deep and cogent manner in the molecular exciton model.

Keywords: exciton, path dependent gauge, correlation functions

INTRODUCTION – THE TRANSITION DENSITY AND SECOND QUANTIZATION

The molecular exciton model begins with the introduction of an object called the *transition density* – this is a scalar density, or 3-dimensional distribution function representing a single electron making the transition from the ground to the excited state. .One takes the ground state wavefunction ψ_{gr} and multiplies it into the excited state wavefunction ψ_{ex}, and then integrates over all electronic coordinates but one, or:

$\rho_{tr} = (\psi^*_{gr} \psi_{ex})$. Integrating instead over *all* electron coordinates one would have:

$\langle \psi_{gr} | \psi_{ex} \rangle = \langle \rho_{tr} \rangle = 1$ This object ρ_{tr} is then employed to derive the expression for the absorption (or emission) spectrum:

For, the intensity of (number of) absorbed photons per unit time is given by[3]:

$$\int dt \iint d^3r\, d^3r'\, r'(t-t_0) \cdot A(r', t-t_0)\, r(t_0) \cdot A(r,t)\, \rho_{tr}(t-t_0)\rho_{tr}(t_0) e^{-iEt/\hbar} = I(t) \quad (1)$$

where A(r,t) is the external vector potential of the e-m field evaluated at the coordinates r and t, E is the energy of the molecular system. This expression is a convolution of functions of r and r' at two different times, and allows for the vector potentials and the transition densities to be spread out, not point-like.

If one performs the indicated Fourier Transform (F.T.) embedded in the expression, one has the *spectrum* – the absorbed intensity as a function of frequency, or $I(\omega)$. To elucidate this idea, we will take things in a few separate steps.

First make a F.T. with respect to space coordinates into momentum or wavevector coordinates:

$$\int dt \iint d^3k\, d^3k'\, k'(t-t_0)k(t_0)\hat{A}^2(k)\, e^{-iEt/\hbar}\hat{\rho}_{tr}(t-t_0)\hat{\rho}_{tr}(t_0) \quad (2a)$$

the spatial wave character of the applied e-m field in the vector potentials is now 'embedded'. The differentials $d^3k\, d^3k' = dE$[4], for the energy of the molecular system. Since absorption occurs at resonance between the field and the molecule-system, and since $\hat{A}^2(k) = E_{e-m\,field}$, we can write:

$$\int dt \iint dE\, g(k,t) k^2_{eff}\, E_{e-m\,field}\, e^{-iEt/\hbar} \hat{\rho}_{tr,eff}^{\;2}(t). \quad (2b)$$

In momentum space the wavefunctions are not necessarily localized. The convolution product between the 'r' and 'r'' operators and between the ρ's, which are nonetheless not independent of each other, but correlated, are turned into a pure product by the F.T with respect to time, and their mutual correlations are accounted for by a correlation function g(k,t). Then altogether one obtains (converting the momenta back into dipole moments – since the transition densities integrate to an effective charge, their correlation, and even more the "momentum squared into energy renormalization" [1]prevents us from insisting that the integrated densities = one electron charge e_0):

$$<\mu_{tr,eff}^2> \iint E_{e-m\,field} dE\, g(k,\omega)\, \hat{\rho}_{tr,eff}^{\,2}(\omega) = I(\omega) \quad (2c)$$

The part of the spectrum independent of the effective transition dipole strength is then the Franck-Condon weighted density of states[2]. The further elucidation of the $g(k,\omega)$ will be a major objective of this report. Indeed the whole question of how to evaluate correlation functions (C.F.'s) arises again and again herein.

In second-quantization language, our transition density can be written as $ab^\dagger$ where $b^\dagger$ creates an electron in the 'b' (say excited) state and annihilates an electron in the 'a' (say, ground) state. Actually as we have written it, we should employ $\rho_{tr} = \left(ab^\dagger a^\dagger c^\dagger d^\dagger ...z^\dagger |vac\rangle\langle vac|c^\dagger d^\dagger ...z^\dagger|\right)$ where $c^\dagger$, $d^\dagger$ etc. are creation operators for all the other electrons in the system onto the 'vacuum', and the parentheses indicate integrations with respect to all those core electrons. In the second-quantization language, the spectrum seems very much to be a single *bosonic* matrix element (an energy expectation value), pairing an electron with a hole. This is just another statement of the exciton picture itself. The transition density acts like a single boson wavefunction – it is not to be confused with the wavefunction of a *state* however: it is that of a *resonance* between a molecular system and the e-m field, or perhaps we could call it an *evanescence*, instead of a state.

The correlation between electronic densities invoked above in $g(k,\omega)$ has a counterpart in the more ordinary employment of state-specific electron densities. – the *exchange/correlation hole*. It is worthwhile spending an aside to develop this concept.

First, one introduces the idea of the electron-pair distribution function,

$$\rho^{(2)}(r_a, r_b)$$

---

[1] because we did not integrate over all momenta or all values of A (or the so-called 'covariant momentum') $i\left(\hbar\frac{\partial}{\partial x} - e_0 A\right)^{(5)}$ - the interaction of an electron with light precludes conducting the real part of a F.T. into space-like separations of electron /electron -density or self-energy terms from the electron-photon 'collision' event, i.e. $ds^2 > 0$ implies ds is real and the real part of a F.T. goes over to the Imag part of the transformed function, i.e. the pole structure and not the real-valued residue - namely $<p^2>$. These points are covered in greater detail in the course of this paper. Note also that, as in (**2b**): $\int dE \int dt - \int dt \int dE = 2\pi i\hbar$ which is just a restatement of the Born-Heisenberg principle. In this form the expression practically begs for consideration as an analytic continuation, the *residue* being the difference between the two integrals – and is equal to the phase 'enclosed' in a contour along t, then E, and back again (and equal to the intrinsic angular momentum of one photon).

[2] Another point of interest is that this methodology allows one to write the transition matrix element – *squared* – as if it were just *one* single operation. Thus, for example in *Circular Dichroism*, the effective operator is a **m**∘**p**, or magnetic dipole moment dotted into an electric dipole moment, which can also be represented by a helicity operator or **l**∘**p** with **l** = **r**x**p**. The expression vanishes in the molecular exciton approximation unless one now refers to three different one-electron operators (that is: $\mathbf{r}_1 \wedge \mathbf{p}_2 \wedge \mathbf{p}_3$) – indeed, such considerations lead to the suggestion of a novel CD mechanism employing an intermediate electronic transition : $\mathbf{p}_{12} \times \mathbf{p}_{1'2'} \bullet \mathbf{p}_{1''2''}$. Rosenfeld[6] in fact originally discussed CD in terms of an expansion of the vector potential over an extended molecule.

which is not unlike those used in statistical physics.[7,8] It represents the probability of finding an electron at r_b given that there is one at r_a. This is then seen to be related to the single-electron densities at centers 'a' and 'b' by:

$\rho^{(2)}(r_a, r_b) = \frac{1}{2}\rho^{(1)}(r_a)\rho^{(1)}(r_b)(g(r_a, r_b) - 1)$, wherein $g(r_a, r_b)$ is again a correlation function.

The so-called total correlation $h(r_a, r_b)$ (an object also commonly used in statistical mechanics [77]), is just this $h(r_a, r_b) \equiv g(r_a, r_b) - 1.0$, that is to say the radial correlation 'g' minus the value it would have (i.e. "1.00") under the assumption of complete statistical independence of the two electrons. Pursuing the analogy with statistical mechanics[8], we can write this correlation also in terms of the commutator of 'a' and 'b' creation operators:

$$h(r_a, r_b) = [a^\dagger, b^\dagger] - 1 \qquad (3)$$

Let us look at the following object[9]:

$$\int \rho^{(1)}(r_b) h(r_a, r_b) = -1 \Rightarrow \langle bb^\dagger ([a^\dagger, b^\dagger] - 1) vac \rangle$$

but $[a^\dagger, b^\dagger]^+ = 0$; or: $a^\dagger b^\dagger = -b^\dagger a^\dagger$, thus (4a,b)

$\langle bb^\dagger ([a^\dagger, b^\dagger] - 1) vac \rangle = \langle -2bb^\dagger b^\dagger a^\dagger - 1 | vac \rangle = \langle -2ba^\dagger b^\dagger b^\dagger - 1 | vac \rangle$; while:

$\langle b^\dagger b^\dagger | vac \rangle = 0$; so that $\langle bb^\dagger ([a^\dagger, b^\dagger] - 1) vac \rangle = -1$, as asserted

(4c,d)

This is a concise derivation of the exchange-correlation hole[10], i.e. when account is taken of the anti-commutativity of electron creation operators (i.e. the Pauli principle), then the 'h' integrated over a single electron density is equivalent to an electron hole, or:

$$\int \rho^{(1)}(r_b) h(r_a, r_b) = -1. \qquad (5)$$

Similarly, in an exciton we have a hole and an electron, but though they propagate to some degree independently at first, in response to the impinging field, they later become bound together again -- namely in a new electronic state, with energy raised by the amount of bound, or absorbed, energy from the e-m field. If we consider the 'exciton' then as $ab^\dagger$, it is also important to take cognizance of the anticommutator relation[9]

$$[a, b^\dagger]^+ = \langle a | b \rangle \qquad (6)$$

one finds the spatial overlap of two wavefunctions for an *atom*, or $\langle a | b \rangle = 0$, but for two atoms in a molecule $\langle a | b \rangle \neq 0$, i.e. it is not necessarily zero, hence

$[a, b^\dagger]^+ = \langle a | b \rangle$, not δ_{ab} as in the atomic case.

At the beginning of the exciton's evanescent existence, <a|b> might be considerable, if we include the whole wavefunction, that is, including also the product of vibrational wavefunctions <a;$\phi_1\phi_2...\phi_\varphi$| b; $\chi_1\chi_2...\chi_j$> since it is not located at just one atom. Yet we also expect this overlap to *evolve in time*, and eventually, at the end of the process, to vanish (the wavefunctions becoming orthogonal). Thus the anticommutator really is expected to diminish in time along with the overlap.

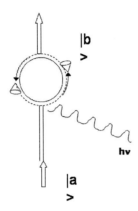

Figure 1. 'Feynman'-like diagram of an exciton. Double line is the evolution of a state, time increasing towards the top of the figure; beginning in the ground state |a>, at certain time a photon impinges on the system, an electron and a hole propagate from there (the hole is 'backwards in time'). The respective light cones of each are shown. At the initial time the electron and hole are dressed by the normal modes of the ground state. At a later time the electron and hole become dressed with the upper state's normal modes, effectively the new state |b> is established and the exciton 'bubble' is extinguished.

THEORY

A. The Exciton Propagator

Consider the Feynnan diagram in Figure 1, depicting the absorption process. When light interacts with the state |a> it opens a 'bubble' graph consisting of an electron and hole pair (unlike usual diagrams, we designate with double lines *a bound stationary state*, the bubble is at first the propagation of a 'freed up' electron/ 'freed up' hole pair due to the e-m field, but they become 'bound' into the excited state) which propagate until they 'recombine' at some future point, now in the new electronic state |b>. This binding event is closely correlated with the pair becoming 'dressed' by molecular vibrations.

We beg the reader's indulgence to pursue a rather lengthy descriptive analogy here, one which would probably havebeen incomprehensible in any period before the advent of video games. Imagine a game in which the operator controls an 'aircraft' say. This aircraft has a number of roughly outlined routes to a destination to choose from, the object of the game being to pilot one's craft successfully to the destination. The vehicle must, however, negotiate deep canyons, riverbeds, etc. in a quite challenging landscape whichever route may be chosen. Now the 'canyon' walls are not just stable structures, but are actually vibrating. This vibration can become so violent and chaotic in places that that region or passage cannot be mapped out accurately, which is why the routes can only be roughly sketched. Moreover the medium in which this 'aircraft' travels consists of elastic spheres, smaller than the canyon walls, but larger than the craft, whose motion is imparted to the canyon walls and is in fact mostly responsible for their moment-to-moment dynamics. The motion of these spheres gives the 'bath temperature'. At low vehicle-speeds, the craft can maneuver easily enough through these spheres, but at high velocities they provide quite a hindrance in themselves. At very low wall+solvent bath temperatures, many possible routes can be observed/attempted, as the features of the landscape are more well-defined, and the characteristic vibrations are of very low amplitude. But as temperatures increase, they become less so, and fewer and fewer major routes are 'allowed" (i.e. relatively narrow canyon channels can become effectively closed off by their vibrational amplitude). Each 'run', each trajectory of the vehicle would be unique, so that choosing a route and a temperature might only roughly characterize each 'run'. Thus it can be expected that the major routes are analogous to the major Franck-Condon (F.-C.) progressions in the mapping to the destination, that is, the excited state, and each route has for any given controller a more or less peaked value for expected arrival time at the destination, corresponding to the peaks of the F-C progression -- while each route is subject to a 'dispersion' in arrival times (with 'fuel expenditure' for each run analogous to 'e-m field work accomplished' – the total action being the fuel consumed times the 'time' of each run) which depends on the wall+bath spheres' motion. This arrival-time dispersion is thus the 'width' of each peak in the F-C progression. Here we have a classical model for optical transitions in a many-nuclei system. Quantum mechanics only enters in our description in the sense that each 'event' is really a sum over many trajectories run through the same route, i.e. a 'path integral', or a 'sum over histories'.

We will switch from a wavefunction based viewpoint to a path integral viewpoint, but we note that the exciton, though described by a path over the molecular system, is, in many ways, in analogy with the depiction in the bubble diagram, a sum of the electron's path *plus* the electron-hole's path: they both describe a contour around the molecule beginning and ending at the same place, but the electron traverses in one direction, while the hole traverses the other way. Moreover the path of the hole is 'about' the ground state, and the electron path is a contour about the excited state. For aromatic molecules, the *characteristic path* looks like a loop about the molecular structure and is similar in some ways to Platt's[11] ring system quantum mechanics. The ground state has, then, a certain number n of nodes in the wavefunction, corresponding to a certain *homotopy class* [12] of the paths (loops); while the excited state has $n+2$ nodes, corresponding to the next available higher homotopy class. The 'obstacles' in these loops we associate with local features of the electron's phase conditioning – *we expressly take the atoms composing the molecular structure to be sources of (intrinsic) vector potential, like wound solenoids* – the loops will avoid one or more of these sources; since electron density is largest near the atoms, the outer lying electrons will tend to be excluded from their vicinity. In fact things are a bit more complicated, in that we parameterize the underlying space to the space of normal coordinates for the nuclei motion (again, the nuclear centered electron densities are the sources of the intrinsic, molecular, vector potential[13]), since these diagonalize the kinetic energy of the nuclei, and are often simply related to the irreducible representations of the point group of the molecule[14].

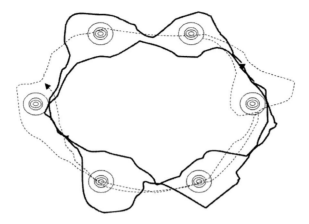

Figure 2. Characteristic paths about a molecule for an exciton– 'benzene-like' molecule is depicted, The concentric circles represent core electron density about nuclear positions. The upper state (electron) moves by convention counterclockwise. The electron-hole moves clockwise. Note that a characteristic path *avoids* two or more nuclei (no single *classical path* can represent the superposition involved in avoiding a nucleaus) – these are the equivalents of *nodes* of a wavefunction, but in path language they correspond to different *homotopy classes,* i.e. paths in two such classes cannot be deformed into each other. The first homotopy class of paths can be deformed to a single point, the second class has one 'lacuna' etc. Note also that the two contours share the same group irrep, and therefore represent a forbidden transition.

The electron (path) propagator takes the wavefunction from the ground to the excited state [15]:

$$G_{1\to 2}(t)\Psi_1(0) \underset{t\to\infty}{\to} \Psi_2 \quad \int dx G_{1\to 2}(x,t)\Psi_1(x,0) \underset{t\to\infty}{\to} \Psi_2 \qquad (7a)$$

The 0'th order propagator is defined in terms of an expansion over the eigenfunctions of the zero'th order Hamiltonian H_0:

$$G^0{}_{1\to 2}(t-t') = \sum_n |n\rangle e^{iH_0 t'} \langle n| e^{-iH_0 t} = \sum_n e^{-iE_n(t-t')} \qquad (7b)$$

Formula (**7c**) is a translation into path-integral language, that is, the propagator of **7b** changes the wavefunction with time, and the propagator of **7c** changes the particle path (from x_a to x_b) with time, the integrand $D(x(t))$ standing for 'integration over paths'[16]:

$$G^0{}_{a\to b}(x_a,t,x_b,t') = \int_{x_a}^{x_b} D(x(t)) \exp\left(-i\int L dt\right) \qquad (7c)$$

The *complete* propagator is found from the Dyson series expansion:

$$G_{1\to 2}(t) = G^0{}_{1\to 2} + G^0{}_{1\to 2} V_{12}(t) G_{1\to 2} \quad \text{or:} \qquad (8a)$$

$$G_{1\to 2}(t) = G^0{}_{1\to 2} + G^0{}_{1\to 2} T\{V(t-t_1)V'(t-t_2)...V'(t-t_n)\} G_{1\to 2} \qquad (8b)$$

where $T\{...\}$ means a time ordered product of the enclosed perturbations, "Normal ordering" is that the largest term is first to act.

The complete exciton then could be written as

$$\oint dx \int \Psi^*_1(x_a,t) G_{1\to 2}(t) \Psi_1(x_a,0) dt \qquad (9)$$

and what we shall argue here is that if we complete now a path integral over the whole system, the incoming path integral (or new state wavefunction) is propagated from time zero from the old state set-of-paths, and then is multiplied by the 'old' state to yield our exciton as a function of time.

The F.T. of expression (8) yields the typical form of the lorentizian spectrum:

$$G_{1\to 2}(\omega) = \sum_n \frac{iV_{1,2}(\omega)}{E_n + \varepsilon_m - E - V_{1,2}(\omega)} \qquad (10)$$

The additional term ε_m represents all the minor eigenvalues of the other perturbation terms, e.g. vibrational energies.

The F.T allows us to write the convolution in (8b) above as a simple product of $V(\omega)$'s If the V's which *map* the old ground state onto the excited state and those which *dress* the two states are considered separately, one option is to suppose that the dressing is done in a noisy, stochastic manner, such that the correlation between the mapping and dressing or *bath* modes decays before the action of the mapping modes is complete, i.e. before we have completed the path integral 'about' the molecular structure, or completed one 'loop' in a given mapping mode. In that case, as in the Adler – Dousi[17] formalism, we still recover the above formula, but now 'renormalized' by the dressing indicated by the curled braces $\{...\}$:

$$e^{-i\{H'_0 t\}} = e^{-i(H_0 t)} e^{-i(V_1 t')} e^{-i(V_2 t'')} e^{-i(V_3 t''')} .... \qquad (11)$$

Where we have assumed a causal order (or 'path ordered') product and subsumed the bath modes into an effective, 'renormalized' H_0 and V_{12}[18]. This implicitly assumes that the bath modes' dressing does follow upon influence of the causal 'G' on the one hand, and that thus there is no forcing, or feedback of the bath onto the mapping. Generally speaking, mapping interactions would be 'spacelike' separated from each other, but the relevant bath modes are timelike relative to the mapping modes which they dress. The complete separation into the two sorts for each possible choice of V mode and for all times (frequencies) seems *a priori* unlikely, however.

$$G_{1\to 2}(\omega) = \frac{i\{V_{1,2}\}(\omega)}{\{H_0\} - E - i\{V_{1,2}\}(\omega)} \qquad (12)$$

When we include some of these considerations, we realize that the 'path ordering' or time ordering instruction must be modified to take account of such correlations. Accordingly if we return to (11) we must now consider that we do not *know* in general whether the effect of some V_i (a mapping mode) on V_j (possibly a bath mode) really 'leads' V_j. Thus, if we consider them effectively correlated in the same normal mode contour of the molecular system, then the F.T. of the two must be considered together, and not as the separated product of Fourier-transformed terms; that is, instead of a V_{ij} term which is time independent, or which can be written just as another term in the product, such as: $V_{ij}(t) = V_{ij} \exp(-\gamma(t-i\beta\Box))$. Instead, however, one has recourse to the Campbell-Baker Hausdorff (CBH) expression

$$e^{-i\{H_0 t\}} = e^{-i(H_0 t)} e^{-i(V_i t')} e^{-i(V_i t'')} e^{-i(V_j t''')} =$$

$$e^{-i(H_0 t)} e^{-i(V_i t')-i(V_i t'')-i(V_j t''')-\frac{1}{2}[V_i t'', V_j t''']} \quad (13)$$

The commutator in "t^2" leads, upon Fourier transform, to a *gaussian* spectrum, not the Lorentzian as usually derived in texts. The coupling in the commutator can take the form, e.g[19].:

$$\left[\frac{\partial^2}{\partial Q_i^2}, H_{elec-vib,j} \right] = \frac{\partial^2 U_{el-j}}{\partial Q_i^2} + 2 \frac{\partial U_{el-j}}{\partial Q_i} \frac{\partial}{\partial Q_i} \quad (14)$$

where Q_i is a given normal mode (this time a bath mode, instead of a mapping mode) and U is a contribution to electronic potential energy within a vibrational well labeled by 'j'. The distinction made by Lin[20] between promoting and accepting modes is similar to the distinction between bath and mapping modes. Certainly the term: 'mapping modes' sounds more 'active' than the more passive-sounding term 'accepting modes'. Nonetheless, both mapping and bath modes belong to the set of Lin's 'accepting' modes.

It is a curious fact that many authors routinely employ gaussian broadened peaks, especially for the representation of Franck-Condon progressions, and yet the justification for their use is usually not terribly convincing[21], and indeed the usual spectral-lineshape treatments that do occur in textbooks[21,22] only suffice to predict *lorentizian* shapes.

Thus, the ubiquitous employment of a gaussian distribution seems generally less well motivated – and the generality of our treatment would seem to be a worthwhile correction to this situation. The distribution associated with this C.F. can be, as above, quadratic in time, or it can become linear or even more-than-quadratic in some circumstances. These results are expected to follow upon invoking the *Argument Principle.* which is considered in further detail below.

B. The Gauge-Theoretic Picture

There is an alternate way of looking at the loop' exciton which illuminates certain important features of the discussion so far, though unfortunately, it also obscures certain other features already elucidated. This is the gauge-theory/path-phase viewpoint. Here the important object is the phase change of a particle traversing a loop, corresponding to the phase of an electronic wavefunction, and measured by the path ordered integral of the vector potential, around the whole loop[23], or $\frac{ie_0}{\hbar}\oint dx_\mu A_\mu$. This object is an added phase on to the wavefunction, i.e. $e^{-i\frac{e_0}{\hbar}\int A dx}$.

For a small change in loop path, there is induced a change in the wavefunction, due to the electromagnetic field (the derivative of the vector potential) $F_{\mu\nu}$ thus[24]:

$$F_{\mu\nu} = \partial_\mu A_\nu^\alpha - \partial_\nu A_\mu^\alpha + \frac{1}{2}\left[A_\nu^\alpha, A_\mu^\alpha\right] \qquad (15a)$$

applying Stokes' theorem to the phase factor one obtans:

$$\Psi_{\delta Path} = \Psi_0 e^{\frac{ie_0}{\hbar} F_{\mu\nu}\sigma_{\mu\nu}} \qquad (15b)$$

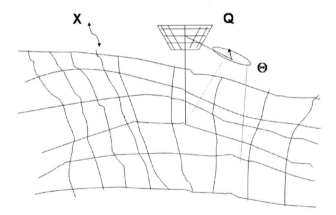

Figure 3. Bundle-space for a molecule: the underlying (Cartesian) coordinates for a molecule is the base space or **X**. A 'projection' is given by the fine line – so attached to any point in **X** is another space of normal modes **Q** for each vibrational state. A number of points in **X** can share the same vibrational state **Q**, they differ only in terms of what *phase* each normal mode possesses. These phases are given by the circle (one such circle for each normal mode at each point in **Q**) of Θ. It is seen that a particular value of Θ (from a particular point of **Q**) corresponds to a point of **X**. So a particular circle projects to a number of points in **X**. The **Q**'s are *fibers* above **X** and the Θ's (one for each normal mode) are *fibers* above **Q** (so **Q** is the base space for Θ). The fibers and base space are together the *bundle* space.

In Figure 3) is depicted the rather complicated series of mappings and relations by which the molecular system is parameterized here. The ordinary space in which we recognize a molecular structure, or consider the motion of a single electron is *X*, and commonly called the *base space*.[25, 26] 'Above' this space is a 3N-6 space of normal coordinates *Q*, depicted as a flag flying above any particle point in X. This will be called the '"Q-fiber above X". Points close in the configuration space are generally not really close in Q. In turn, 'above' Q for each of the normal modes is a little circle, representing the U(1) group *phase* of each normal mode – this is the "θ-fiber above Q" (one value of phase angle $θ_□$ for each normal mode j). Thus, many different instantaneous configurations in X map to the same point in Q and merely represent different phases in θ. Indeed all the points mapped to one such θ-fiber are *one vibrational state* – a particular value of $|n_1 n_2 ... n_j\rangle$ for each $j \in Q$. Each electronic state has a unique set vibrational modes, and in particular the zero point energies $|0_1 0_2 0_3 ... 0_j\rangle$ are different for each electronic state.

The nuclei, and ultimately the linear composition of their motions into normal coordinates, are taken to be sources of *internal* vector potential, or *j(r)* with each source given by:

$$j_i(r) = \frac{i\hbar}{2\sqrt{m}} \left(\psi^*_{core} \nabla \psi_{core} - \psi_{core} \nabla \psi^*_{core} \right) \quad (16)$$

here, the electronic momentum is taken to be that of *core* electrons, associated with each nucleus, and not the outer electrons which compose the molecular orbitals. Each such current source produces a vector potential A_i given by[27]:

$$A_i(r) = \int \frac{j_i(r)}{|r - r_i|} d^3 r \quad (17)$$

and the linear transform which takes the set of 'j' currents into normal coordinates **Q** for the j's is formally: $\mathbf{Lj} = \mathbf{Q}(j_1, j_2, ... j_N)$. In the excited state, a new set of normal coordinates is typical. These upper state coordinates **Q'** are rotated, and they are displaced from the original coordinates:

$$\mathbf{Q'} = \mathbf{L'j'}; \quad \mathbf{j'} = \mathbf{j} + \Delta; \quad \mathbf{L'} = \mathbf{D^{-1}LD}; \quad (18,a,b,c)$$

D is the so-called Duschinsky rotation matrix[28], which rotates ground state normal modes into new excited state normal modes. The Duschinsky effect, though usually small, thus also changes the normal mode frequencies to some degree. This effect changes the measure (the integrand) of the path integral of the two eigenstates. Δ is a (diagonal) matrix of displacements for each 'j$_i$' written in the underlying Cartesian coordinates.

The vector potential is then written in two parts, the external (due to applied e-m field) and the intrinsic: $A = A^{rad} + A^{int}$.

In fact since we are dealing with two components, the lower and upper state, or $\begin{bmatrix} \phi_1 \\ \phi_2 \end{bmatrix}$ to our exciton, we will be using the SU(2) group for the gauge A_μ although that itself is a simplification. The ordinary 2x2 matrix of SU(2) (actually SO(2))[29,30]: namely that occurring in $\begin{bmatrix} 0 & -A_\mu \\ A_\mu & 0 \end{bmatrix} \begin{bmatrix} \phi_1 \\ \phi_2 \end{bmatrix}$ stands in place of a much larger object, such that each matrix element in $\begin{bmatrix} \phi_1 \\ \phi_2 \end{bmatrix}$ represents an electronic part implicitly multiplied by an array (or multivector) of vibrational parts with different possible quantum numbers in each vibrational mode. The external vector potential (from the e-m field) is however, still in the U(1) group.

Written out, there may be a phase difference between the ground '1' state and the excited state '2' contours. This phase difference can be expressed as $[A^+{}_\mu, A^-{}_\nu] \approx \Delta x'y' - x'\Delta y'$., where we take only the 'x' and 'y' coordinates as an example. The new coordinate system (in the '2' state for example) is *primed* and $\Delta x'$ for instance represents the phase difference between the two paths in the new coordinate "x'". If, for example, one atom is excluded in the new coordinates (i.e. a the path takes a detour around the nucleus, as would be expected for the next higher homotopy class of paths), the external vector potential-phase factor is changed by $2\pi(R_i^2 - R_i^{0,2})A^{ext}$ [31] The mapping of one intrinsic vector potential to the other is represented by $A_2 = A_1 + ie_0/\hbar\ g^{-1}{}_{1,2}\partial g_{1,2}$ [32,33,] which also an expresses canonical gauge invariance. Here, however we will see there is a *path dependent phase factor* in the phase difference between the two paths, i.e. the two components to our exciton.

The *measure* of the path integral is related to the determinant of the differential operator which generates the phase function. Here the measure of the path integral for the double-contour of the exciton is the square root of the determinant of the matrix [33]

$$|\partial_t - V(t)| = \prod_i m_i\omega_i^2, \qquad (19)$$

is the general formula for harmonic oscillators with V(t) being $\sim \omega^2$.

In turn, the difference between the new and old path-measures is the expression

$$\|\delta X\|_{\Delta X} = \int e^{-\int_0^T [Xb]iT} \mathfrak{d} = \|\overline{\delta X}\| \|\overline{\delta X}\| = e^{Tr\int dX \int g_{12}^{-1}Ag_{12} + g^{-1}{}_{12}dg_{12}} = \|g_{12}\|e^{\Delta X^2/2}e^{\Delta A'} \qquad (20)$$

since $\dfrac{d\|g_{12}\|}{\|g_{12}\|} = d\ln g_{12} = Tr g^{-1}{}_{12}dg_{12}$., where, as before $\mathbf{g_{12}}$ is the coordinate transform between the two 'states'. This gives us again the prefactor $\dfrac{\prod_i\sqrt{m_i}\omega_i}{\prod_i\sqrt{m_i'}\omega_i'}$

The commutator $[A^+{}_\mu, A^-{}_\nu] \approx \Delta t'\tau' - t'\Delta\tau'$ among other terms, but the 'time' part is proportional to the E-field, i.e.

$$F_{\mu\nu}\Delta\sigma \cong \left(E_{Q_i} + \frac{1}{2}[\hat{A}_\mu^{\dagger(2)}, \hat{A}_\nu^{(1)}]\right)T_i\Delta Q \qquad (21)$$

where we now write the "A" commutator as a commutator between ladder operators for the internal vector potentials associated with the states '1' and '2' $\hat{A}^{(1)}{}_\nu$, $\hat{A}^{\dagger(2)}{}_\mu$, namely: $[\hat{A}^{\dagger(2)}{}_\mu, \hat{A}^{(1)}{}_\nu]$. Following Dirac[34] we evaluate this commutator as a 3-d delta function:

$$\frac{1}{2}[\hat{A}^{(2)}{}_\mu^\dagger, \hat{A}^{(1)}{}_\nu] = \Delta(\mathbf{x_1} - \mathbf{x_2}) \qquad (22)$$

Whilst:

$$\Delta(\mathbf{x}_1 - \mathbf{x}_2) \underset{Q \text{ space}}{\Rightarrow} |0_1, 0_2, 0_3, \ldots 0_N\rangle_{Q1} |n_1, n_2, \ldots n_N\rangle_{Q2} = \delta(n_{k2} - n_{k1})\delta(\Theta_{k2} - \Theta_{k1}) \quad (23)$$

expressing the Δ function in the normal coordinates Q, we thus require agreement between vibrational quanta and between phases.

$$\frac{e_0}{2}[\hat{A}^{(2)\dagger}{}_\mu, \hat{A}^{(1)}{}_\nu]|0_1, 0_2, 0_3, \ldots 0_N\rangle_{Q1} \langle n_1, n_2, \ldots n_N|_{Q_2} = (1)\hbar\omega\delta(n_{k1} - n_{k2})\delta(\Theta_1 - \Theta_2) \quad (24)$$

In addition the period T is given by $1/\omega$. The normal mode displacement ΔQ_i is written out in operators as $-i\dfrac{(a^\dagger_{lower} - a_{upper})}{2} S_i$ for the ladder operators "a" for the given normal mode, and S_i being the displacement in units of the normal mode displacement $\sqrt{\dfrac{\hbar}{m\omega}}$. Which vibrational part of the exciton does not vanish under the action of which "a" operator is indicated in the subscripts.

With harmonic oscillator wavefunctions for ϕ_I we have an expression like

$$e^{aS} = 1 + (aS) + \frac{1}{2}(aS)^2 + \ldots \frac{1}{n!}(aS)^n \ldots \text{ or i.e. more precisely:}$$

$$\prod_{i, \text{mod es}} \sum_{k, \text{quanta}} a^{\dagger k}{}_i S_i^k \sqrt{\frac{m_i \omega_i}{\hbar}} \bigg/ 2n! + \prod_{i \text{ mod es}} \sum_{k, \text{quanta}} a^k{}_i S_i \sqrt{\frac{m_i \omega_i}{\hbar}} \bigg/ 2n! \quad (25)$$

Also for the harmonic oscillator, the change in vector potential with the changed loop length per each "j" source, or $2\pi(R_j^{\prime 2} - R_j^{0,2})$ yields an effective phase factor per normal mode "i" of $\exp(-\Delta\Box_i^2/2)$. Thus, between these two factors, the ordinary Franck-Condon factor for a ground state vibrational array of |0,0,0,0,0...0⟩ can be recovered. The specific formula for Morse oscillators F-C. factors[35] ought to be derivable from the more general formula when expressed in terms of ladder operators for the Morse case.

There are two further interesting results worth mentioning. First, the gauge invariance equation by which the phase in one state can be transformed to the other state by means of a local-to-global coordinate mapping implies *current conservation*([29]. As long as the gauge invariance is true, then the transition current is the same. This means that, even when there is no mirror image symmetry in the excitation mapping vs the emission mapping, still the derived dipole strength (oscillator strength) is still correct, i.e. the mapping is effectively invertible even when it lack obvious inversion symmetry. Therefore also, the derived value of k_{rad} or the Einstein 'A' coefficient should still be correct. The mirror image symmetry often observed, because the vibrational state of the ground state before excitation, or of the excited state before emission are both likely to be 0,0,0,0,0...0⟩, is seen to be a special case when $\mathbf{g}_{12}^{-1} = \mathbf{g}_{12}^\dagger$. When, however charges or currents are not conserved (as in electron transfer or especially proton transfer reactions in the excited state as might be more likely in *polar protic solvent systems*) then the absorption spectrum-derived value of the Einstein 'A' coefficient, based on the measured oscillator strength, is not accurate for the emission – a rather well-known effect. In fact whenever the coordinate transform g_{12} is not invertible, the transition current will not be conserved, and the Einstein coefficient will not be accurately obtained from emission. This would also be true if there were carbon-skeletal rearrangements, group migrations etc.

Secondly, it is pointed out in Flanders[36] that the transition function g_{12} though defined *locally* must nonetheless be true globally, which suggests in turn that 1) the coordinate mapping is itself non-local and 2) immediately suggests further that something very like normal coordinates must be employed to describe the internal phase factors in terms of the local sources of vector potential, and 3) as long as a given external vector potential (i.e. radiation field) can excite currents in the system, then any exciton, of any energy, up to the point that the local current sources (the'j's') themselves are undisturbed, can be found from the same basic set of normal coordinates, subject to the rotation and displacement matrices already invoked. Since there are only 3N-6 normal modes (if the number of atoms is 'N') and there are considerably more electrons than this number, the Kuhn-Thomas rule at the least would suggest that not all transitions could indeed be so mapped. but perhaps only the lowest few.

DISCUSSION

A. Loop Theory vs. Platt's Ring Theory, M.O. Theory

The Platt[11] theory as well as the usual Molecular Orbital theory for ring systems, both suggest that the lowest lying transitions of the 'ring electrons' or ring M.O.s will occur in 'pairs' of similar energy. In Platt's theory the number of nodes in the wavefunction determined the overall energy (the total 'angular' momentum), and the degeneracy is thought to be lifted by the nodes occurring either at atomic positions or at bonds, In the development here, the overall energy is determined again by the number of nodes (the homotopy class). But the nodal pattern may in fact be separable in most ring systems into two irreducible representations of the symmetry group, and the normal modes may also sort themselves out into the two irreps, recruited, so to speak, into two different transition densities. For example, in benzene the H.O.M.O. contains two nodes, these can be placed at opposite ends ('para') to each other in one irrep, and vicinal ('ortho') in another irrep.. The LUMOs would then be ones with four nodes at 1,2,4,5, and another at 1,2,3,4. The 'allowed' transitions would be between the 'para' irrep and the 1,2,3,4 irrep ($B_{2g} \rightarrow B_{1u}$), or between the 'ortho' and the 1,2,4,5, irrep ($B_{2u} \rightarrow B_{1g}$), utilizing the available set of normal modes. Indeed a linear combination of these two is possibly responsible for the vibronically allowed' transition observed, and forbidden transitions would then be linear combinations of the remaining two possible transitions, with a slightly different set of normal modes.

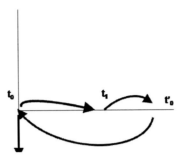

Figure 4. a-Typical Keldysh contour (see Jauho, ref. 37). A propagator is taken from t_0 to two later times, and then is dressed by the bath $i\beta\hbar$ after all other interactions. b- modified version of Keldysh contour. Note first that *this* contour is a simple closed curve. The contour is for calculating correlation functions of the various mode-specific propagators with each other and/or bath modes. As explained in the text, the hatched region is *inside* the light-cones of Figure 1. At 'infinite' time the electron propagator is 'connected' (the dashed lines) to the outgoing, initial, electron-hole. This can be considered a momentum jump (the instantaneous force-displacement of nuclei in the mapping of upper to lower state), at zero time, a similar jump is made from the electron-hole to the outgoing electron (this corresponds to a 'cooling off' at zero time, i.e. in the presence of an external vector potential). At longer times or lower temperatures, more of the momentum/interaction space is *excluded* from overlap with the light cones, thus more interactions effectively commute with a particular mapping, and the spectrum narrows/becomes more lorentzian.

B. The Keldysh Contour and the Argument Principle

We have associated the normal mode-normal mode C.F. with the spectral linewidth. As a distribution, over all the possible bath and/or mapping modes, it is best to consider an analytic continuation of the two-time commutator, to include the virtual bath 'time' $\pm i\beta\hbar$. Indeed, we expect mapping modes not to be timelike related to each other, instead we expect bath modes to be causally related to mapping modes. Nonetheless saying that the interaction of one mode with a bath mode is complete (with respect to the spectral density function of the path-propagator) at one time, without feedback from the bath (which, as a reservoir, is much larger in terms of quantum-number occupancy than any mapping mode is likely to be), and can be unequivocally time-ordered with respect to any other mode-coupling, seems quite a large assumption. A useful heuristic device for path-ordering or time-ordering a set of interactions is the *Keldysh contour*[37, 38]. In Keldysh formalism, events in the imaginary (bath) time axis are considered to happen after every event along the 'real time' axis. We have taken some liberties with the usual presentation: significantly, we want the contour to enclose an actual region, to take full advantage of the *residue theorem*, and the *Riemann mapping theorem*.[39] Note that the contour runs in the opposite orientation to the usual direction: thus the shaded region is technically *inside* the contour. We can think of the contour as graphing the mode-mode correlation in complex time, as if we were looking (and then graphing inside-out!) through the light cone-of the particular path of an exciton as shown in Figure 1. We have already obtained the gaussian function for the spectral density by taking the F.T. of

$\exp^{-[H_{map}t, H_{bath}t']\frac{1}{\hbar^2}}$, where it was clear that the exponent of the phase correlation function (~ t^2) is 2.0. By looking at the analytically continued C.F. and mapping it (via the Riemann mapping theorem) to the complex unit circle (or if we wish to continue to infinite time, the Riemann sphere), we must conclude that, in case the commutator does not vanish (i.e. within the light cone of the 'interaction' of interest, in our case the perturbation of the exciton propagator by the mapping normal mode coupling), this function wraps *twice* around the origin. The *Argument Principle* [39] informs us that the exponent of this C.F. takes the value of 2, that is, the number of zeros minus the number of poles = Z − P = 2 − 0 = 2. An ordinary C.F. of this sort in real time would begin at zero and at long times return to zero, with no poles between. But, are there perhaps zero's or poles we haven't counted? Indeed, in some circumstances there are. One has to think of this Keldysh contour as embedding just the C.F.(s) of one normal mode with all other nuclear terms that affect the electron's path, or i.e. the electronic potential energy surface and/or kinetic energy. There are many such modes involved in the mapping, each of them, by assumption here, *not* causally interconnected. But as long as an interaction is *inside* this contour, it *will* affect the C.F. In addition, each sample will actually have a distribution of such contours, since the detailed interactions possibly vary due to complicated environmental rearrangements. Sometimes such a mode will be anharmonic, and dissociate. If the strength of the coupling to the bath doesn't fall off quicker than the normal mode's changes in energy during dissociation events, a *phase pole* will result. This may or may not occur within the contour, depending on how much time we spend in developing the spectrum (depending on the lowest frequency mapping oscillator) and the temperature. Effectively the mode's period increases without bound while the correlation stays finite, and thus the phase function has a singularity somewhere in the interior. Then for that mode, N-P < 2.0 An *atom,* of course, has no nuclear mapping modes to speak of, and the bath-*electron* C.F. (commutator) vanishes, and the spectrum of an atomic transition is *essentially lorentzian*, as has long been known.

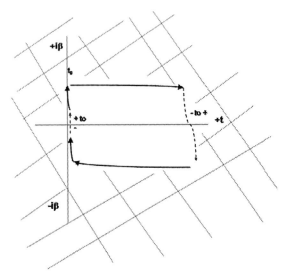

Figure 5.

Similarly, if the mapping mode is *forced* by bath modes, then extra *phase zero's* can occur in the C.F., and the spectrum can be closer to lorentzian than a pure gaussian. And this would tend to happen on the high frequency side of emission and low-frequency side of excitation, where there is exceptional work done by the environment + molecular system onto the exciton. Say there is a coupling constant γ between a harmonic oscillator nuclear mapping mode, and another oscillator, possibly a Drude oscillator within the molecule which responds to the varying e-m field with an induced dipole moment $\Delta\mu = \alpha \bullet E$ for the polarizability α. This Drude oscillator has a potential energy $U = \frac{1}{2}\Delta\mu^2/\alpha$. The coupling is written as $V_{coupl} = \gamma(t)\Delta\mu Q_j$, so that $\gamma(t)$ has the units of a charge density, for the 'j''th mode.

Then the probability of our target oscillator starting in quantum state 'n' and[30] ending in quantum state 'm', P(n—>m) is proportional to[40]:

$$\langle m|Q_j|n\rangle^2 \left[\delta(\hbar\omega + E_m - E_n) + \delta(\hbar\omega - E_m + E_n)\right] \times \iint dt ds \gamma(t)\gamma(s)\exp(-\omega(t-s)) \quad (26a)$$

with proportionality q_{exc}^2/M; or, the rate per unit time being:

$$\langle m|Q_j|n\rangle^2 \frac{q_{exc}^2}{M}\left[\delta(\hbar\omega + E_m - E_n) + \delta(\hbar\omega - E_m + E_n)\right] \times \int ds \gamma(t)\gamma(s)\exp(-\omega(t-s)) \quad (26b)$$

where $\omega = \frac{e_0}{\sqrt{M|\alpha|}}$ with M the mass of the Drude oscillator, the $\delta(\omega - E_m + E_n)$ being Dirac delta-functions about the difference and sum frequencies of the two oscillators, and $|\alpha|$ being1/3 the trace of the polarizability matrix.. The term 'q_{exc}' is the effective charge of the exciton (i.e the charge of the transition dipole moment). So each mode that is forced is split into two parts at these frequencies, for each such coupling. This effect, we argue, introduces more *zeroes* into the C.F., each mode has twice as many opportunities for the C.F. to go to zero. Moreover, the correlation is continued beyond the intersection of light cones of the 'exciton' and the nuclear vibration – such forcing is 'acausal', in other words, the Hamiltonians may already at zero time not commute. Thus there is *another* crossing in the 'future', and, as we say, an additional zero in the correlation function. So the exponent is 'stretched' beyond 2, as before it was 'shrunk' to a value below 2. Other authors[41] have used 'Brownian'-type expressions for nuclear mode coupling, and still others[42] have investigated damping effects with similar attributes. Nonetheless the 'acausal' nature of damping phenomena seems to preclude the consideration of it as Markov-like processes. In addition, most of these authors are concerned to leave the H_{el} and H_{Nuc} terms as a *time-ordered* product. One recognizes that the electromagnetic-electronic interaction is *causal*, that is, we restrict attention to the overlap of light cones for the exciton with the nuclear modes. But we treat the whole interaction as one operator, which is why we invoke the CBH expression. That treatment in fact corresponds with *one* concept of 'dressed states– there is but one exciton' with one kind of evolution operator, not a series of states, one following another. We interrogate the molecular system by sweeping an e-m field through a large frequency range: and we then analyze the output (disturbed) signal. The molecular system responds to an external vector potential by forming an exciton, at the end of which process a photon has been absorbed and the molecule is now in the excited state.

Modes, which, at thermal equilibrium, exchange energy between themselves and the environment, probably do not generally exchange during the exciton's duration. However modes whose rearrangement is largely responsible for the exciton (mapping modes) can certainly radiate *into* these bath modes, e.g. as part of the process they would normally undergo (or, in the forcing, anti--causal sense instead, like the Drude oscillator we discussed above). In another sense of 'dressing'[18], one considers the coupling of modes as effectively a constant 'background' throughout the process, this is the 'stochastic dynamics' approach[17r]. In that approach, these couplings are averaged over (using the "Trace formula") before the F.T. is taken. In the Keldysh formalism, in distinction, one specifically assumes a particular ordering of perturbations/operators, and the bath modes are always applied after all other operators.. To the extent that the modulation of the e-m field is much slower than the application or 'switching-on' of the various perturbations, the later (usually weaker) ones can be considered effectively folded-in to the earlier, and usually (in so-called 'normal ordering') stronger ones. The dressing then is taken to mean all the bath modes just add their imaginary time term to the stronger operators application times, the order of these and the distinctions invoked between advanced, retarded etc. propagators remain intact. As long as there are some modes (bath modes e.g.) that are indeed timelike separated from the stronger ones, then the CBH algorithm can be employed. The stochastic dynamics, in this manner of speaking, ignores time-like vs space-like separations, that is, they could all just as well be space-like -- to the extent that *no* particular time-ordering among them is prescribed. That is why perhaps it is best suited to an assumption of 'slow sweep' in the F.T. In the full time-ordered Keldysh formalism, a fast sweep F.T. would have only the bath temperature folded in to effective perturbation operators (and only to the latest one applied from the *negative* imaginary side – i.e. later than infinity, whereas in our Figure we assume that the bath modes can be radiated into/out of from before a particular mode-perturbation is applied, i.e. from the infinite past). As we see in the appendix, this gives the time dependent Green's Function (strict Keldysh) formalism and the stochastic dynamics formalism the same form of individual spectral density functions: i.e. they both present *lorentzian profiles*, though of course, with very different individual widths and total envelopes.

Our modified Keldysh contour thus invokes a different kind of 'dressing' of interactions. It even (see Figure) invokes the observed displacement of normal modes by which Franck Condon factors arise, to explain the closure of the contour as physically relevant between the incoming and outgoing propagators. This closure then allows meaningful treatments via the various theorems of complex analysis we have further invoked. A careful comparison of retarded and advanced Green's functions is necessary to resolve the order of nonequilibrium processes, as was found[43]. It was further found that such comparisons can reveal the presence of singular terms, which are expected in nonequilibrium situations[43]. And thus, analysis of *simple closed* contours is often advisable, not, that is, contours violating the provisions of the Riemann mapping theorem (as does the usual Keldysh contour), in order to take full advantage of this far-reaching complex-analytic apparatus. In particular, the idea that the last-acting operator can influence the 'bath' in what is effectively the 'indefinite future' can be matched with the influence of the bath on the first acting operator from the indefinite past', i.e. because it might radiate into said bath. And said bath may in turn have supplied radiation in the indefinite past.

CONCLUSION

The focus of this paper has been on redescribing the exciton as a gauge-field object (indeed as a pair of loops), employing some of the apparatus of modern quantum field theory, and is rather removed from the usual descriptions in terms of Molecular Orbitals. This focus might itself be equally-well, or perhaps better, adapted to the time-dependent density functional formalism[38], with the density being reinterpreted as 'density of contours/paths'.

In the Appendix we present the Adler-Diosi's[17] version of stochastic dynamics and the spectral linewidth problem, wherein the lorentzian spectral density function reappears in a stronger form. There being few strongly convincing arguments for the almost ubiquitously observed gaussian, we have in this paper provided a rather long exposition to justify it. It should be noted that this dressing for each single molecule corresponds nicely to the results that would obtain, assuming that the entire transition matrix for the exciton were partitioned into diagonal blocks, within each of which the instantaneous eigenvalues were chosen from a gaussian ensemble. In other words, exactly the results that would be obtained after assuming the validity, in the exciton case, of *random matrix theory*[44] in a *gaussian orthogonal ensemble*, which specifically requires that the system have integer spin (a boson) and that it possess time reversal invariance. One of the more intriguing predictions of this theory is that the *interval frequency distribution*, i.e. the distribution of intervals between eigenvalues (spectral peaks), would not be smoothly declining (Poisson statistics) from zero, but would be peaked (roughly a beta distribution).

We can redescribe the excitation process now with a view to these above considerations. The initial ground state is 'dressed' by the nuclear modes (including bath modes) to form a smeared –out gaussian orthogonal ensemble of mode-phases encompassing one true electronic state. Upon interaction with the e-m field the ground state has a distribution of possible maps to the excited state, with weight factors depending on the size of the distortions induced along normal modes. At first, the phases of the nuclear vibrations will coincide with the ones they possessed in the ground state – while the Duschinsky rotation of the mapping produces correlations in phase between some of the originally independent vibrations (to the extent that it mixes the originally orthogonal nuclear modes). In the unlikely cases where the most significant vibrations are all extended to their maximum (minimum) the energy of the excitation is minimal (maximal) since the field-work is partitioned into changing the 'electronic phase' at each nucleus and changing the nuclear kinetic energy. One can also now define the duration of existence of the exciton – it lasts as long as the 'memory' of vibrational phase correlations between normal modes, due to the Duschinsky effect, persists. As soon as this correlation dissipates, the new state is effectively dressed in its new normal modes. This should be the identical amount of time the bath modes need to 'decorrelate' the normal modes from the electronic Hamiltonian.

An exciton is fundamentally not in equilibrium with anything, *a fortiori* it isn't in thermal equilibrium with a bath. Bath modes dress the initial state, or the final state, but in some sense they actually *fail* to dress the intermediate exciton (that is, when once the bath modes have 'caught up' to the final state, the exciton process is 'over'); they are 'folded in' to the spectrum instead by means of their nonvanishing commutators with the mappings via the CBH expression, or by looking for the singular parts of the stochastic coupling (see Appendix). The displacements of the 'intrinsic vector-potential sources' in mapping from the ground to the excited state are, indeed, singular jumps in nuclear momenta (positiob within potential wells). To the extent that these jumps carry the whole density of paths with them, and not just the characteristic path itself, they are 'dressed via dispersion' by the bath.

APPENDIX

The stochastic quantum dynamics formalism begins with a short derivation of the Ito calculus[17, 45].

One begins with the simple integral, an expectation value of the operator $A(X_s,s)$ which depends on a path X_s, the path itself dependent on 'a time' parameter 's'. Thus the expectation of an observable: $E(A(X_s,s))$ depends on the path measure dX_s as well as the total time averaged.

$$E\{A\} \equiv \int_0^t A\, dX_s = \lim_{\Delta s \to 0} \sum A\Delta X_s \tag{A1}$$

$$s_m = mt/2^n \tag{A2}$$

$$\Delta_m = X_{m+1}(s_{m+1}) - X_m(s_m) \tag{A3}$$

$$\int_0^t dX_s A(X_s,s) = \lim_{n \to \infty} \sum_{m=0}^{2^n-1} \Delta_m A_m \tag{A4}$$

$$E\left\{\int_0^t dX_s A(X_s,s)\right\} = 0 \tag{A5}$$

which is meant to motivate the major wrinkle of Ito calculus, that is, that the since X_m and A_m are statistically independent, and so the expectation value of the integral vanishes since $E(X_s A_{s,m}) = E(X_s)E(A_{s,m}) = 0$ since $E(X_s) = 0$ for a stochastic one-dimensional Wiener process.

One further asserts for the differential of the quadratic process

$$d(X_{s,i} X_{s,j}) = X_{s,i} dX_{s,j} + X_{s,j} dX_{s,i} + ds\, \delta_{ij} \tag{A6}$$

in distinction with the ordinary Leibniz rule. This hypothesis is meant to be reminiscent of diffusion dynamics, namely that $dX_s \approx \sqrt{ds}$

A significant formula for the differential of an exponential process is the following:

$$d(\exp(aX)) = d\left(1 + aX + \frac{1}{2}a^2 X^2 + \frac{1}{6}a^3 X^3 ...\right) =$$

$$dX\left(a + a^2 X + \frac{1}{2}a^3 X^2 + ...\right) + \frac{1}{2}a^2 ds + \frac{1}{6}3a^3 Xds + \frac{1}{24}\binom{4}{2}a^4 X^2 ds ... =$$

$$adX\left(1 + aX + \frac{1}{2}a^2 X^2 + ...\right) + \frac{1}{2}a^2 ds\left(1 + aX + \frac{1}{2}a^2 X^2 + ...\right) =$$

$$\left(adX + \frac{1}{2}a^2 ds\right)\exp(aX) \quad (A7)$$

The 'stochastic' equation for the time dependence of a given state $|\varphi\rangle$ is:

$$d|\varphi\rangle = -iH|\varphi\rangle dt - \frac{1}{8}\sigma^2 H^2 |\varphi\rangle dt + \frac{1}{2}i\sigma H|\varphi\rangle dX_s \quad (A8)$$

If we let the "a" of (**A7**) stand for: $a = \frac{1}{2}i\sigma H$, then we can integrate (A8)

$$|\varphi(t)\rangle = \exp\left(-iH\left(t - \frac{1}{2}\sigma X_s\right)\right)|\varphi(0)\rangle \text{ or, since:} \quad (A9a)$$

$$d\rho = d|\varphi\rangle\langle\varphi| + |\varphi\rangle d\langle\varphi| + d|\varphi\rangle d\langle\varphi|, \text{ then:} \quad (A9b)$$

$$\frac{d}{dt}|m(t)\rangle\langle m(0)| \equiv \acute{\rho}_m(t) = \exp\left(-\Gamma\left(1 - \frac{1}{8}\sigma^2\Gamma\right)t\right)\rho_m(0) \text{ where } (A9c)$$

$$\Gamma \approx iV^2{}_{mm}\,\delta(E - E_m)/\hbar \quad (A9d)$$

when "V" is put into "H = H$_0$ + V"., while H$_0$ doesn't propagate the state $|m\rangle$ (i.e. the state stays in $|m(0)\rangle$ upon the action of H$_0$. The Γ is, of course, the decay rate out of "m".

Now it should be noted that there is an implicit $\frac{1}{\hbar}$ multiplying every occurrence of 'H' in the above. Therefore, both σ and dX$_s$ are in units of $\sqrt{t}$. Presumably the stochastic coupling parameter σ is a pure number times $\sqrt{\beta\hbar}$. But, one can, just as easily 'analytically continue' a thermal time $\beta\hbar$ into "it" as one can continue a real time into $i\beta\hbar$. And in this way a gaussian formula (upon F.T. the above expression) could arise within the stochastic dynamics as well. This is an important point: our exciton corresponds to 'very short times' compared with atomic transitions, because the *exciton is not a state itself but a process*, that is, at best it is a *resonance* and so it is not dressed by effective interactions with bath phonons etc. like a stationary state, with discrete energies for each bath mode, but rather as a resonance, with imaginary components, i.e. energy dispersions for each bath mode.

ACKNOWLEDGMENTS

Financial support was from an NIH grant GM34847 awarded to Franklyn Prendergast.

REFERENCES

[1] Frenkel, I. (1931). *Physical Review*, *37*, 17-30.
[2] Penrose, R. (2005). *The Road to Reality*. A. Knopf, N.Y. *904*.
[3] Kirk, W. R. (2007). *Biophysical Chemistry*, *125*, 13-23.
[4] Bethe, H. & Salpeter, E. (1957). *Quantum Mechanics of One- and Two- Electron Atoms* Plenum Pr. N.Y.(1977 edition –160-165
[5] Felsager, B. (1998). *Geometry, Particles*, and Fields Springer N.Y. *128*.
[6] Rosenfeld, L. (1928) *Z. Physik*, *52*, 161-174.
[7] Hansen, J. & McDonald, I. (1986). *Theory of Simple Liquids* 2nd ed. Acad. Pr. N.Y. 106-110.
[8] Foerster, D. (1975). Hydrodynamic Fluctuations, *Broken Symmetry and Correlation Functions*. 1990 Perseus Book ed. 44-109.
[9] Jørgenson, P. & Simons, J. (1981). *Second Quantization-Based Methods in Quantum Chemistry Acad*. Pr. N.Y., 2-5.
[10] Parr, R. & Yang, W. (1989). *Density Functional Theory of Atoms and Molecules*, Oxford Pr. N.Y. 33-35.
[11] Platt, J. (1949). *J. Chem. Phys.*, *17*, 484-496.
[12] Naber, G. (1997). *Topology, Geometry, and Gauge Fields Springer*, N.Y. 101-154.
[13] Greiner, W. & Reinhardt, J. (1994). *Quantum Electrodynamics* 2nd ed. Corrected Springer, *Heidelberg*, 98 –footnote 7.
[14] Wilson, E; Decius, J. & Cross, P. (1955). *Molecular Vibrations Dover* ed. 1980, *106*.
[15] Felsager, B. (1998). *op. cit.* 43-44.
[16] Roepstorff, G. (1993). *Path Integral Approach to Quantum Physics Springer*, Heidelberg. 102-108.
[17] Adler, S. (2002). "*Weisskopf-Wigner Decay Theory for Energy-Driven Stochastic Schroedinger Equation*" arXiv:quant-ph/0208123v4 7 Nov.
[18] McComb, W. (2004). *renormalization methods: a guide for beginners* Oxford. 134-142.
[19] Fischer, G. (1984). *Vibronic Coupling Acad*. Pr. N.Y. 210-211.
[20] Lin, S. (1966). *J. Chem. Phys.*, 44, 3759-3767.
[21] McQuarrie, D. (2000). *Statistical Mechanics Univ. Sci*. Sausolito. 501-507.
[22] Schatz, G. & Ratner, M. (2002). *Quantum Mechanics in Chemistry*, Dover. Mineola N.Y. *220*.
[23] Mandelstam, S. (1962) *Ann. Physics.*, *19*, 1-24.
[24] Barrett, T. (2008). *Topological Foundations of Electromagnetism World Sci*. Singapore. 106-133.
[25] Naber, G. *op. cit.* 16-66.
[26] Penrose, R. *op. cit.* 325-349.
[27] Jackson, J. (1975). *Classical Electrodynamics Wiley*, N.Y. 215.
[28] Fischer, G. (1984). *op. cit.* 120-125.

[29] Felsager, B. op. cit. 124-129.
[30] Naber, G. *op cit*. 261-263.
[31] Jackson, J. (1975). *op. cit.* 259.
[32] Creutz, M. (1983). *Quarks, Gluons, and Lattices*, 29-38.
[33] Felsager, B. *op. cit* 204-206.
[34] Dirac, P. (1958). *The Principles of Quantum Mechanics pprback reprint 1991*, Oxford Pr. 239-243.
[35] Kirk, W. (2008) *J. Phys. Chem.*, 112, 13009-13021
[36] Flanders, H. (1963). *Differential Forms with Application to the Physical Sciences*, (1989) Dover ed. 143-148.
[37] Jauho, A. *Introduction to the Keldysh nonequilibrium Green function technique.* www.nanohub.org/resources
[38] van Leeuwen, R. & Dahlen, N., et al. (2006). *in: Time Dependent Density Functional Theory*, M. Marques *et al.* eds. Springer Berlin. 32-57.
[39] Sidorov, Y., Fedoryuk, M. & Shabunin, M. (1985). *Lectures on the Theory of Functions of a Complex Variable Mir*, Moscow 220-290.
[40] Feynman, R. & Hibbs, H. (1965). *Quantum Mechanics and Path Integrals*, McGraw-Hill N.J. 232-236.
[41] Yan, Y. & Mukamel, S. (1988). *J. Chem. Phys.*, 89, 5160-5170.
[42] Gu, Y., Widom, A. & Champion, P. (1994). *J. Chem. Phys.* 100, 2547-2560 These authors also find an expression for spectral lines with damped oscillators, however, they use a time-ordered formalism which, it seems to me, is not perfectly suited to the problem. That is, the driving force is encapsulated in a 'correlator' function, appropriately acting before the radiation, but its Fourier transform is taken to be complete before that comprising the spectral density function is performed. So low frequency modes, possibly representing local disorder, would not seem to be accurately treated. While this nonetheless may be an excellent approximation for many cases it is at variance with the techniques used herein. Then again, an acausal coupling would seem to be at variance with our own presumed 'dressed-states'/noncommuting variables approach itself, as well, hence the introduction of those arguments surrounding the use of the modified Keldysh contour and the Argument Principle.
[43] Kirk, W. (2005). *J. Theoretical & Computational Chemistry*, 4, 475-492.
[44] Mehta, M. (2004). *Random Matrices* 3rd ed. Elsevier, Amsterdam, 36-37, 146-181.
[45] Roepstorff, G. *op cit*. 197-201.

In: Exciton Quasiparticles
Editor: Randy M. Bergin

ISBN: 978-1-61122-318-7
© 2011 Nova Science Publishers, Inc.

Chapter 5

SOLITONS AND EXITONS IN SEISMOTECTONIC PROBLEMS

A.V. Vikulin[*]

Institute of Volcanology and Seismology, Far East Branch of the Russian Academy of Sciences, Petropavlovsk-Kamchatsky, Russia

ABSTRACT

The problem of an elastic stress field in a rotating medium is formulated and solved analytically within the limits of the classical theory of elasticity with a symmetrical stress tensor. This is a rotation elastic field of action at a distance. There are two specific types of elastic waves with a moment in rotating media: solitons and excitons, or rotation waves. The soliton solutions to the wave equation represent waves of global earthquake migration (slow tectonic waves) which are no faster than (1-10) cm/s, i.e., approach the migration velocity of large and great earthquakes ($M \approx 8$ and more). The exciton solutions correspond to waves of local migration of foreshocks and aftershocks in earthquake sources (fast tectonic waves) and have their maximum velocity comparable to *S*-*P*-wave velocities.

Keywords: Solid earth; seismic focal blocks; Earth's rotation; rotation elastic field; waves of earthquake migration

INTRODUCTION

The available geological, geophysical, and astronomic data prompt a new geophysical model of the solid earth as an hierarchic system of blocks with their sizes spanning a range of twelve to fourteen orders of magnitude. This global system eludes description in terms of the

---

[*] Corresponding author: Email: vik@kscnet.ru

classical linear elastic continuum model, and new approaches are thus required (Problem, 2002; Sadovsky, 2004; Teisseyre, Takeo, Majewski, 2006).

The solid earth is intrinsically nonlinear. All its constituent blocks and their hierarchy are in perpetual relative motion and are able either to uptake energy from outside or to exchange energy with one another. Some blocks in systems of any scale can exists in a state close to instability. Unstable blocks and systems that receive additional energy can loose stability and either form new smaller systems or consolidate as separate elements into larger systems (Nikolaev, 2003; Ponomarev, 2008).

Therefore, the lithosphere can be treated as an ever changing alive medium (Goldin, 2003; Sadovsky, 2004). Modeling principles and approaches for this kind of actually nonclassical nonlinear media were suggested in (Ostrovsky, 2005; Teisseere, Nagahama, Majewski, 2008; Vikulin, 2004).

There is ample evidence that the motions of crustal blocks, lithospheric plates, and other geological structures are rotary, twisting, and vortical (Lee, 1928; Milanovsky, 2007; Tveritinova and Vikulin, 2005; Vikulin, 2004; Vikulin and Tveritinova, 2007, 2008; Uyeda, 2003). Many geophysicists note that these motions are "independent" (Sleznak, 1972), "own" (Peive, 1961), "with nonzero divergence and vortices" (Luk'yanov, 1999; Teisseyre, Takeo, Majewski, 2006), and elastic (Teisseere, Nagahama, Majewski, 2008; Ustinova et al., 2005) directly related to the planet rotation (Chao and Gross, 1995; Milanovsky, 2007; Vikulin and Tveritinova, 2007). Furthermore, it is the presence of strong nonlinearity that allows viewing geological structures as a medium with its own sources of elastic energy (Ponomarev, 2008). Rotation of macrostructures through large angles (to 1–10°) is known also in polycrystalline materials (Panin, 2002; Vladimirov and Romanov, 1986).

Interest to vortex structures and their interaction with the Earth's rotation has rekindled recently (Huang, 2003; Vikulin, 2003, 2004). For instance, every seventh presentation at the 35th Tectonic Workshop of 2002 (Koryakin, 2002) concerned with theoretical, global or regional problems of the Earth's rotation. Several contributions dealt with the effect of this rotation on modern geodynamics. A number of recent publications (Bulletin, 2009; Lee, Igel, Trifunac, 2009; Teisseere, Nagahama, Majewski, 2008; Teisseyre, Takeo, Majewski, 2006; Vikulin and Tveritinova, 2007, 2008) support the efficiency of these studies, which have essential tectonic and geodynamic implications and often drive at unexpected results. The vast collection of data on the problem available for the time being (Bulletin, 2009; Lee, Igel, Trifunac, 2009; Milanovsky, 2007; Uyeda, 2003; Vikulin, 2004; etc.) requires a special line of rotation tectonics to study structures produced by the rotation effect (Teisseyre, Nagahama, Majewski, 2008; Tveritinova and Vikulin, 2005; Vikulin, 2006; Vikulin and Tveritinova, 2007).

Thus, we arrive at a specific problem of an elastic stress field in a rotating block medium (Teisseyre, Takeo, Majewski, 2006; Vikulin, 2003, 2009; Vikulin and Ivanchin, 1998; Vikulin and Tveritinova, 2008).

PROBLEM FORMULATION

The problem of a stress field arising around a small macroscopic volume in a solid rotating at the angular velocity Ω, for the block V being elastically coupled with the solid and subject to rotation driven by its internal energy sources, was formulated and solved analytically (Vikulin, 2003, 2006, 2009; Vikulin and Ivanchin, 1998) in the case of nonlinearblock media (Goldin, 2003; Nikolaev, 2003; Ostrovsky, 2005; Sadovsky, 2004; Teisseyre, Takeo, Majewski, 2006) within the limits of the classical theory of elasticity with a symmetric stress tensor (Feynman, Leighton, Sands, 1964; Landau and Lifshits, 2003). The basic idea of the solution is that elastic stress arises around a macroscopic block (volume V) elastically coupled with the ambient medium (matrix) once the block angular momentum changes under an internal effect, and this stress has a moment of force by the conservation law. The idea is consistent with the postulate in vortex dynamics (Saffman, 1992) that vorticity is proportional to the angular momentum of particles.

Therefore, the "own block moment", or the spin, is meant in this study as the most proximal to the meaning of an "own angular momentum of a finite volume of a continuum" according to Sedov (1973) and Peive (1961). This approach to seismotectonic problems, when the elastic field around a macroscopic volume (block) in a rotating medium inherits the block spin (circulation) differs basically from other approaches which either neglect the Earth's rotation (Goldin, 2003; Kurlenya and Oparin, 2000; Sadovsky, 2004) or include it nominally, within the limits of the moment elasticity theory (Bulletin, 2009; Nikolaevsky, 1996; Teisseyre, Nagahama, Majewski, 2008; Xie Xin-Sheng, 2004). Unlike the Cosserat continuum commonly used to explain the Earth's rotation effect on geophysical processes, the suggested rotation model implies that blocks have their moment (spin), rather than merely rotational degrees of freedom, and the existence of this moment drives at a number of specific consequences associated with the Earth's rotation.

PROBLEM SOLUTION

The problem is solved in three steps (Vikulin, 2003, 2006; Vikulin and Ivanchin, 1998).

1. Consider two systems of coordinates rotated through the angle β relative each other about the common axis Y. The Z axis of one (initial) system is parallel to the rotation axis of the body (Ω) and directed from the southern to northern pole of the latter, and the $\overline{Z}$ axis of the other system is parallel to the angular momentum of the block V after its rotation through β. Both systems originate at the gravity center of V (Figure 1).

   The elastic stress that arises around the rotating block V can be estimated using a mental experiment. First one stops the rotation of V by applying the elastic stress σ_1 with the moment of force $\mathbf{K}_1$ along the negative direction of $\overline{Z}$, assuming that the kinetic energy of the V rotation fully converts to the potential energy of the elastic stress σ_1. Then one applies the elastic stress σ_2 with the moment of force $\mathbf{K}_2$ along Z in order to again set the block V into rotation at the body rotation velocity (Figure 1).

   In other words, when one stops the rotation of V, its kinetic energy

$$W = 1/2 I\Omega^2 \tag{1}$$

converts into the elastic energy defined by the stress tensor σ_1, and when one spins the block again, the same kinetic energy is produced by the stress σ_2.

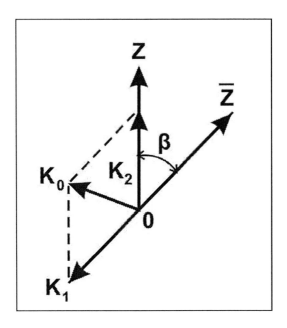

Figure 1. Two systems of coordinates (Z, $\overline{Z}$) rotated about their common axis through angle β. Z axis is parallel to axis of body rotation and directed from its southern to northern pole.

Let the volume V be a uniform sphere with its moment of inertia I independent of the choice of the rotation axis. Then the equality of the kinetic and potential energy drives at $|\mathbf{K}_1| = |\mathbf{K}_2|$. The difference of these vectors is the sought moment of force $\mathbf{K}_0$ produced by rotation of V in a noninertial system: $\mathbf{K}_0 = \mathbf{K}_2 - \mathbf{K}_1$. From the cosine theorem,

$$|K_0| = 2|K_1| \sin \beta/2. \tag{2}$$

2. The sought field of elastic strain $\mathbf{U}$ is known (Feynman, Leighton, Sands, 1964; Landau and Lifshits, 2003) to satisfy the equation of elastic equilibrium

$$graddiv U - arotrot U = 0 \tag{3}$$

with zero infinity conditions

$$|U| \to 0 \text{ at } r = (x_1^2 + x_2^2 + x_3^2)^{1/2} \to \infty \tag{4}$$

zero force applied to V:

$$F_i = \int \sigma_{ij} dS_i = 0 \qquad (5)$$

and a moment of force independent of the block size:

$$K_i = \int x_k e_{ikl} \sigma_{lj} dS_j \neq f(R_0), \qquad (6)$$

where $a = (1-2v)/2(1-v)$ is Poisson's ratio, R_0 is the radius of the sphere V, and e_{ikl} is the Levi-Civita symbol.

The solution to (3) – (6) in the spherical coordinates (r, θ, φ) with the origin $r = 0$ at the center of V at $r \geq R_0 r$ consists in the strain (**U**) and stress (σ) fields:

$$U_r = U_\theta = 0, \ U_\varphi = Ar^{-2} \sin\theta, \qquad (7)$$

$$\sigma_{r\varphi} = \sigma_{\varphi r} = 3/2 AGr^{-3} \sin\theta \qquad (8)$$

where G is the shear modulus, A is constant (defined below). The other stress tensor components are zero.

Substituting (8) into (6) for the moment of force of the elastic field gives

$$K_{1z} = \int_0^\pi \int_0^{2\pi} \sigma_{r\varphi} r^3 \sin\theta d\theta d\varphi = 3\pi^2 AG. \qquad (9)$$

The other moment components are zero: $K_{1x} = K_{1y} = 0$ (Figure 1).

3. Integrating the density of the elastic strain energy $W = \sum\{\lambda/2(\varepsilon_{ij}\delta_{ij})^2 + G\varepsilon_{ij}^2\}$, where λ is the modulus of dilation, ε_{ij} is the strain and δ_{ij} is the Kronecker symbol, over the body (Earth) volume, assumed to be incompressible, gives the elastic energy produced by $\mathbf{K}_1$:

$$W = 9/2 A^2 G \int_{R_0}^\infty \int_0^\pi \int_0^{2\pi} r^{-4} \sin\theta dr d\theta d\varphi = 4\pi A^2 G R_0^{-3}. \qquad (10)$$

Equating it to kinetic energy (1) and taking into account that the sphere moment of inertia is $I = 8/15 \pi \rho R_0^5$, where ρ is the density of the body (Earth) medium, gives A as

$$A = \Omega R_0^4 \sqrt{\frac{\rho}{15G}} \qquad (11)$$

With regard to (2), the moment of force of the elastic field around the block orthogonal to the block rotation plane is

$$K = -6\pi^2 \Omega R_0^4 \sqrt{\frac{\rho G}{15}} \sin \beta / 2. \qquad (12)$$

The elastic energy is

$$W = 16/15 \pi \rho \Omega^2 R_0^5 \sin^2 \beta / 2, \qquad (13)$$

the strain is

$$U_r = U_\theta = 0, \ U_\varphi = \Omega r^{-2} R_0^4 \sqrt{\frac{\rho}{15G}} \sin \theta \sin \beta / 2, \ r \geq R_0 \qquad (14)$$

and the stress is

$$\sigma_{r\varphi} = \sigma_{\varphi r} = 3/2 \Omega r^{-3} R_0^4 \sin \theta \sin \beta / 2. \qquad (15)$$

The other stress components are zero.

Note that if the block has an elliptic geometry, the final equations remain almost the same and acquire coefficients close to 1 (Vikulin and Ivanchin, 1998).

Estimates. With the model parameters $\rho = 3$ g/cm^3, $G = 10^{11}$ N/m^2, $\Omega = 7.3 \cdot 10^{-5}$ rad/s, $R_0 = 100$ km corresponding to crustal blocks of $M \approx 8$ earthquake sources, (12) – (15) give $U_0 \approx 10$ m, $\sigma_0 \approx 100$ bar, $W_0 \approx 10^{16-18}$ J, $K_0 \approx 10^{21-23}$ N·m. These estimates approach, to the order of magnitude, the real values recorded in great earthquakes for slip, stress release, elastic energy, and seismic moment, respectively, corresponding to a rotation angle of the block (earthquake source) of $\beta_0 \approx U_0 / R_0 = 10^{-4}$ rad. At the duration of a seismic cycle (recurrence time of great earthquakes at a site) of 100 to 1000 years, the mechanic (model) estimate of the block rotation velocity is $10^{-(4-6)}$ deg/year, which is close to the geological rotation rates of Iceland (Melekestsev, 1979), the Nazca and Juan Fernandez plates (Uyeda, 2003), and other plates and crustal blocks (Zonenshain and Savostin, 1979).

ACTION-AT-A-DISTANCE OF THE ROTATION ELASTIC FIELD

The interaction energy of rotating blocks can be estimated using the known rule that elastic energy (within the limits of Hooke law) is proportional to square strain. Then, writing the strain of some part of the body as a sum of strains produced there separately by each block gives the equation in which the cross terms define the interaction energies.

In the **model of two blocks** for which the total elastic energy is

$$W = G \int (a+b)^2 dV = G\{\int a^2 dV + \int b^2 dV + 2\int ab dV\},$$

where a and b are the elastic strain tensors produced by the rotation of the first and second blocks, respectively, integration is over the total body volume. First two terms in the right-hand side of the elastic energy equation are the inherent elastic energies, each found by (13). The third term defines the interaction energy of the two blocks:

$$W_{int} = 2G \int ab dV . \qquad (16)$$

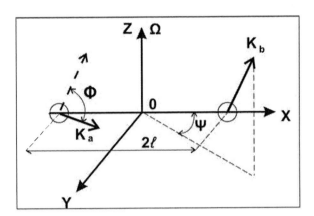

Figure 2. Relative orientation of moments of force K_a and K_b in model of two blocks in Cartesian coordinates XYZ. Z axis is parallel to Earth's rotation axis and directed from South pole to North pole. Ω is angular velocity of Earth's rotation; ϕ and ψ are angles that define directions K_a and K_b, respectively; $2l$ is spacing between gravity centers of blocks.

The interaction energy (Figure 2) is found as follows. The gravity centers of the blocks lie on the X axis being spaced at $2l$, and the origin of the coordinates (Figure 2) is equidistant from the gravity centers. The Z axis is chosen such that the vector K_a lie in the plane XY and Z be orthogonal to XY. The Z direction corresponds to the right-hand coordinate system.

The direction K_b relative to K_a is defined by the angles (Figure 2), which can be found from $\cos\phi = \dfrac{(K_a \cdot K_b)}{|K_a||K_b|}$, and ψ, which is the angle between the projection of K_b onto XY and the X axis. The matrix of the K_b rotation toward K_a is

$$Q = \begin{Vmatrix} \cos\psi\cos\varphi & -\sin\psi & \cos\psi\sin\varphi \\ \sin\psi\cos\varphi & \cos\psi & \sin\psi\sin\varphi \\ -\sin\varphi & 0 & \cos\varphi \end{Vmatrix}.$$

Then the tensor b in the coordinates defined by $\mathbf{K}_b$ is $b = QbQ'$, where Q' is the transposed matrix.

Assuming the blocks to be spheres of the radius sizes R_{0a} and R_{0b}, expressing the tensors a and b via the respective strains (7), substituting them into (1) with regard to (11), and calculating the respective integrals in bipolar coordinates gives the interaction energy of two blocks rotating at the same velocity as

$$W_{\text{int}} = 3/2\pi\rho\Omega^2 R_{0a}^4 R_{0b}^4 l^{-3} \cos\phi. \quad (17)$$

The interaction energy is localized outside the blocks where, according to (16), both strain tensors are nonzero:

$$W_{\text{int}} \neq 0, \ r \geq R_{0a}, \ r \geq R_{ob} \quad (18)$$

The derivation was made on the assumption that the blocks were spaced at the distance l far greater than their sizes. Then, one may neglect the block size and apply integration for points. This assumption is, however, not critical because $l \geq 2R_0$, and, with the finite block size included, the correction is second-order vanishing: $(R_{0a\,(0b)}/l)^2$.

The moment of force due to the interaction energy is found by differentiating (17) with respect to ϕ:

$$K_{\text{int}} = -3/2\pi\rho\Omega^2 R_{0a}^4 R_{0b}^4 l^{-3} \sin\phi. \quad (19)$$

Moment (19) is applied, from the elastic field, to the surface of each block and is directed in a way to reduce the interaction energy. It has the same absolute value but opposite directions for the two blocks.

Estimates. Assuming that the two interacting blocks have equal sizes ($R_{0a} = R_{0b} = R_0$), from (12) and (19), we obtain

$$\frac{K_{\text{int}}}{K} = \frac{\Omega R_0}{v_S}\left(\frac{R_0}{l}\right)^3, \quad (20)$$

where $v_S = (G/\rho)^{1/2}$ is the shear-wave velocity. It is clear from (20) that the inertial interaction effects related to the rotation of blocks inside a rotating body become ever more significant with growing rotation velocity Ω of the latter and with the block size R_0.

The ratio of the interaction energy W_{int} to the block inherent energy W, by (13) and (17), is

$$\frac{W_{int}}{W} = \frac{45}{32} \frac{(R_0/l)^3 \cos\phi}{(\sin\beta/2)^2} = \delta. \qquad (21)$$

Therefore, the maximum distance ($\cos\phi = 1$), at which the interaction energy is of the same order of magnitude as the block energy ($\delta = 1$), is given by

$$l_0 \approx 2\beta^{-2/3} R_0 = (10^2 - 10^3) R_0. \qquad (22)$$

According to (22) the elastic fields that arise around the blocks inside a rotating body, are actually of action at a distance. The numerical estimation was with $\beta \approx 10^{-4}$ rad in (22), which corresponds to that in a great earthquake.

Thus, the action-at-a-distance effect of the rotation elastic field is that great earthquakes in closely spaced blocks with parallel ($\phi = 0$) (or antiparallel, $\phi = \pi$) moments occur simultaneously (or do not occur at all). Indeed, if one block has accumulated enough energy for a great earthquake, the other interacting blocks should have at least the same energy, as a result of parallel ($\phi = 0$) interaction. In the case of antiparallel ($\phi = \pi$) moments of the blocks, their interaction energy, on the contrary, cancels the energy of the nucleating earthquake. This inference is supported by the available seismological evidence (Vikulin, 2009) of earthquake doublets and pairs, as well as the absence of great earthquakes for hundreds of years in seismic areas.

EARTHQUAKE MIGRATION

Earthquakes occur within narrow belts that extend through the whole globe. The Pacific seismic belt is the most active one where earthquakes release up to 80–85% of total seismic energy and include almost all great events of the world. The belt width is 100–200 km and the length reaches $4 \cdot 10^5$ km. The upper 50–100 km section of the zone consists of 100–300 km long seismic blocks (an average of L_0 = 200 km). The block sizes correspond to the sources of great earthquakes ($M \approx 8$ and more). The sources of some great events involve several blocks and amount to 10^3 km or more: Aleutian, 1957, M = 9, L = 1200 km; Sumatra, 2004, M = 9, L = 1000 km; etc, (Vikulin, 2009). The host island arcs and Pacific continental margins are as long as $4 \cdot 10^3$ km (e.g., the Aleutian arc).

Uniform chain of blocks. Consider a one-dimensional chain of rotating interacting blocks located inside a body which rotates at the angular velocity Ω and assume that all blocks are spheres with the same radius R_0 ($\approx 1/2L_0$).

Let all blocks in the chain be in uniform motion. Then, according to the above results, the equation of motion for a block in the chain is

$$I\frac{d^2\beta}{dt^2} = K_1 + K_2, \qquad (23)$$

where β is the angle through which the block turns as a result of earthquake nucleation, $I = 8/15\pi\rho R_0^5$ is its moment of inertia, $\mathbf{K}_1$ is the moment of force of elastic stress around the block produced by its rotation, with its value defined by (12), and $\mathbf{K}_2$ is the moment of force responsible for interaction of the block with other blocks in the chain.

The moment $\mathbf{K}_2$ is obviously proportional to both the elastic energy of a certain block ($Vd^2\beta/dx^2$) and to the elastic energy corresponding to all other blocks in the chain. Let the latter energy be selected as a value equal to the mean linear density of the elastic energy of the chain w, where $V = 4/3\pi R_0^3$ is the block volume and x is the coordinate along the chain. Thus, the moment responsible for the interaction of the block with the other blocks is

$$K_2 = \varsigma w V d^2\beta/dx^2, \qquad (24)$$

where ς is the dimensionless coefficient that characterizes the chain homogeneity. This coefficient may be assumed $\varsigma = 1$ for the Pacific belt, which is quite homogeneous.

Finally, equation of motion (23) for a block with the coordinate x at the time t, with regard to (12) and (24), in dimensionless coordinates $\xi = k_0 x$ and $\eta = c_0 k_0 t$, becomes

$$\partial^2\vartheta/\partial\xi^2 - \partial^2\vartheta/\partial\eta^2 = \sin\vartheta \qquad ..(25)$$

where $\vartheta = \beta/2$. The wave number and the velocity that represent the seismic process, are found, respectively, as

$$k_0^2 = \frac{3\pi\Omega}{wV}\left(\frac{3V}{4\pi}\right)^{4/3}\sqrt{\frac{\rho G}{15}}, \qquad (26)$$

$$c_0^2 = \frac{wV}{I}. \qquad (27)$$

Equation (25) is the sine-Gordon (SG) nonlinear wave equation.

Strongly nonlinear equations that have solutions in the form of solitons, including the SG equation, are currently of broad use in technological, physical, and geophysical applications (Bykov, 2000; Nikolaevskiy, 1996; Scott, Chu, McLaughlin, 1973). SG equation (25) differs from other equations of this kind as its constants c_0 and k_0 defined by the nonlinear properties of the medium are dependent on its rotation angular velocity. Below it is shown that wave motion observed in geophysics can be really modeled within the limits of problem (23)–(27).

Properties of SG solutions. The SG equation of motion for a chain of seismic blocks has many solutions (Scott, Chu, McLaughlin, 1973). In technological, physical and geophysical applications, these solutions are often in the form of solitons (sol), or localized (solitary) waves (Bykov, 2000, 2005, 2008), and/or excitons (ex) (Davydov, 1982) in the case of long chains in which the effect of chain ends is negligible (as in Earth's seismic belts and, specifically, in the Pacific belt).

The qualitative dependence of the excitation energy E on the velocity of solitons (sol, I) and excitons (ex, II) is shown in Figure 3 borrowed from (Davydov, 1982). The relationships for the excitation energy of solitons and excitons satisfy the conditions

$$E_{sol} \approx V_{sol}^n, \; 0 \leq V_{sol} \leq V_{01}; \; E_{ex} \geq E_0 > 0; \; E_{ex} \approx V_{ex}^p, \; V_{01} \leq V_{ex} \leq V_{02}; \; n > p, \quad (28)$$

where V_{01} is the characteristic rate of the process.

In the quasi-linear approximation, when a linearized SG equation is applicable to describe the process, the dispersion law for exciton solutions (Davydov, 1982; Scott, Chu, McLaughlin, 1973) is

$$\omega^2 = \omega_0^2 (1 + \lambda_0^2 / \lambda^2), \; \lambda_0 = 2\pi c_0 / \omega_0, \quad (29)$$

where ω and λ are the exciton frequency and wavelength, respectively, ω_0 is the own frequency of the block rotation, and λ_0 is the corresponding wavelength.

The primary feature of dispersion law (29) is its link with the nonlinearity of the chain of blocks (material that fills the seismic focal volume) rather than with its discrete structure.

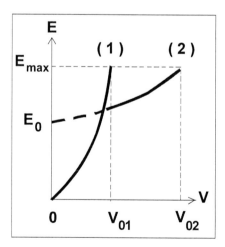

Figure 3. SG wave solutions, after (Davydov, 1982). 1 – solitons, 2 – exitons. V_{01} and V_{02} are characteristic rates of process, corresponded "limit" ($E \to E_{max}$) solition ($0 \leq E_{sol} \leq E_{max}$; $0 \leq V_{sol} \leq V_{01}$) and exiton ($0 < E_0 \leq E_{ex} \leq E_{max}$; $0 \leq V_{ex} \leq V_{02}$) solutions. E_{max} is limit energy, corresponded strongest earthquake.

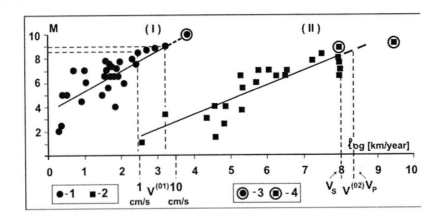

Figure 4. $M_{1,2}(\log V_{1,2})$ are the dependences to Pacific earthquakes, after (Vikulin, 2009). 1 – velocities of earthquake foci migration along Pacific outskirts – global migration (I) with limit velocity $V^{(01)} = (1 \div 10)$ cm/s; 2 – migration velocities of aftershocks and foreshocks in sources strong earthquakes – local migration (II) with limit velocity $V^{(02)} = (V_S \div V_P) \approx (4 \div 8)$ km/s. 3 – migration velocity of strongest Pacific earthquake-doubles with M = 8,1-8,7 in 1897-1901. 4 – velocities, corresponded to migration of main shocks in earthquake-doubles 4.XI.1952, M = 9,0, Kamchatka and 13.X.1963, M = 8,7, Kuril Islands. (I), (II) – global ($M_{01}(\log V_1)$) and local ($M_{02}(\log V_2)$) relationships. M_{max} = 8,3 is limit foreshock and aftershock magnitude, corresponded foreshock of strongest Chili, 1960, M = 9,5 earthquake (Duda, 1963). $V_P \approx 8$ km/s and $V_S \approx 4$ km/s – longitudinal and shear waves.

Another feature of (29) is that the frequency of waves that propagate along the chain of blocks is always higher than ω_0. It is physically obvious that ω_0 corresponds to a long wavelength ($\lambda \to \infty$ in the limit) when all blocks in the chain move as a single whole, without chain deformation. This case of zero exciton state corresponds to extrapolation of the exciton relationship $E_{ex}(V)$ in (28) into the range $V_{ex} < V_{01}$:

$$V_{ex} = 0, \; E_{ex} = E_{min} > 0. \qquad (30)$$

Waves of earthquake migration. See Figure 4 for the available data on velocities of seismicity migration along the Pacific seismic belt (Vikulin, 2003, 2009). The total velocity field (Figure 4) splits into two distinct zones (I, II) divided by

$$V^{(01)} = (10^{2,5} \div 10^{3,5}) = \text{km/year} \approx (1 \div 10) \text{ cm/s}. \qquad (31)$$

The zone of $V_2 > V^{(01)}$ includes the migration velocities of foreshocks and aftershocks, or local migration and the zone of $V_1 < V^{(01)}$ is for the velocities that govern migration on a large scale in time and space, or global migration.

The data of Figure 4 were used to calculate the least square earthquake magnitudes within each zone as

$$M_{01} \approx 2\log V_1 (km/year), \ 0 < V_1 \leq V^{(01)}, \quad (32)$$

$$M_{02} \approx \log V_2 (km/year), \ V^{(01)} \leq V_2 \leq V^{(02)}, \quad (33)$$

where M_0 are the magnitudes of migrating earthquakes in each sample. These relationships are almost the same as those we obtained earlier (Vikulin and Ivanchin, 1998) using a twice smaller data set. Therefore, migration relationships (32) and (33) appear rather reliable being derived from a sufficiently large data collection, and further extension of the latter will cause no change to the results.

Assuming that the magnitudes M of earthquakes are related to the elastic energy they release as log E (J) = $1.8M + 4.3$, according to the Gutenberg-Richter law, and using the global relationship between the magnitude of an earthquake and its source size L (log L (km) = $0.4 M - 1.0$), (32) and (33) become, respectively

$$E_1 = V_1^{4-5}, \ \log L_1 (km) \approx \log V_1 (km/year), \ 0 < V_1 \leq V^0, \quad (32.1)$$

$$E_2 = V_2^{2-3}, \ \log L_2 (km) = 0{,}5 \cdot \log V_2 (km/year), \ V^{(01)} \leq V_2 \leq V^{(02)}, \quad (33.1)$$

where V_1 and V_2 are the velocities corresponding to global and local migration relationships (I) and (II) (Figure 4).

An idea of what are the earthquake migration waves comes from quite a large amount of data on surface strain waves recorded in earthquake sources and in their surroundings for the foreshock and aftershock periods. According to geodetic instrumental measurements (Bakhtiyarov and Levin, 1993), strain waves are solitary peaks or troughs (solitons) propagating along the Earth's surface at velocities in a range including most earthquake migration velocities (Figure 4). Modeling results suggest that strain waves can be solutions of the equations of motion which are strongly nonlinear equations, most often of SG type (Bykov, 2000, 2008).

Thus, seismological, geophysical, and modeling data provide explicit evidence that earthquake migration waves are solitons and exitons with their nature being controlled by the nonlinearity of the lithosphere.

Zones (I) and (II) in the earthquake migration data of Figure 4, the corresponding first relationships in (32.1) and (33.1), and the dividing position of the point $V^{(01)}$ show a qualitative correlation with the properties of SG solutions (23), namely, the relationships (1) and (2) in Figure 3, relationships (28), and the same position of the points V_{01} and V_{02} (Figure 3). Therefore, the obtained relationships correspond to soliton ((32) and (32.1)) and exciton ((33) and (33.1)) SG solutions with $V^{(01)}$ as the characteristic rate of the process.

By analogy with the common elastic waves (the case of tectonic approximation (Nikolaevsky, 1996)), assuming the exciton wavelength λ_0 to equal the seismic block size

$$\lambda_0 \approx R_\oplus, k_0 = 2\pi / R_\oplus. \qquad (34)$$

the characteristic rate of the process c_0 is given by

$$c_0^2 = \frac{3\sqrt{15}}{8\pi^2} \Omega R_0 \sqrt{\frac{G}{\rho}} \approx V_R V_S, \qquad (35)$$

or, with the above model parameters,

$$c_0 = (1-10) \text{ cm/s}. \qquad (36)$$

Thus, the characteristic rate c_0 of the model seismic process ((23) – (27)) turns out to approach V_{01} in (31), i.e., the empirical migration relationships (32), (32.1), and (33), (33.1) are, respectively, the soliton and exciton solutions to (23) – (27).

This inference is confirmed by the proof that the own frequency ω_0 (29) can be identified, within the limits of problem (23) – (27), with the frequency of nutation (Chandler wobble) which is actually a zero exciton solution corresponding to state (30). Furthermore, it becomes possible, within the limits of the same problem, to predict and explain splitting of the nutation (Chandler) frequency into two components (Vikulin and Krolevets, 2001, 2002): The migration velocity along the latitude differs from that along the longitude for the Doppler velocity due to the Earth's rotation (Vikulin, 2009).

According to (35), c_0 can be expressed, to the numerical multiplier, as a geometric mean product of the centrifugal velocity $V_R = \Omega R_0$ and the shear-wave velocity $V_S = (G/\rho)^{1/2}$. That is the reason why we called model (23) – (27) the rotation model (Vikulin and Ivanchin, 1998).

CONCLUSIONS

1. The reported study includes the formulation and an analytical solution of the problem for the elastic stress field which arises around a crustal block subject to internally driven rotation. The only physical assumption of the model is that the Earth's rotation changes the direction of the block angular momentum and this change produces an elastic field with a moment of force around the block. For a block with the size R $\approx$ 100 km rotating through an angle of $\beta \approx 10^{-4}$ rad, the predicted (model) values of elastic energy, seismic moment, stress, and strain are similar to the respective values known for great earthquakes (M $\approx$ 8 and more). This rotation angle, in turn, is consistent with the inferred rotation velocity of lithospheric blocks and plates, under reasonable assumptions. Thus, the model is applicable to describe the processes in earthquake sources.

   Elastic stress fields of this kind being of action at a distance, the seismic process in global seismic belts can be simulated in terms of a chain of blocks (earthquake sources). A problem for motion of a chain of related blocks has been formulated and solved in the phenomenological form. The soliton and exciton (according to Davydov (1982)) solutions to this problem allow describing the full scope of known velocities of earthquake migration. The soliton solutions, which are global earthquake migration waves or slow tectonic waves according to Bykov (2005, 2008), are characterized by maximum velocities of $\approx (1 \div 10)$ cm/s. The exciton solutions (fast tectonic waves) represent waves of local migration (Bykov, 2005) of foreshocks and aftershocks in earthquake sources by maximum velocities of $V_P \div V_S \approx (4 \div 8)$ km/s (Figure 4).

In the tectonic approximation (Nikolaevsky, 1996), assuming that the exciton wavelength equals the size of a seismic block (a source of a great earthquake), the characteristic model velocity is

$$c_0^2 \approx V_R V_S,$$

and is the geometrically mean product of the centrifugal V_R and shear-wave V_S velocities.

Thus, the reported study concerns with a new type of elastic rotation waves associated with motion of seismic blocks in a rotating Earth, which control earthquake migration. Theoretical (Bulletin, 2009; Milanovsky, 2007; Teisseyre, Nagahama, Majewski, 2008; Teisseyre, Takeo, Majewski, 2006; Vikulin, 2003, 2004) and instrumental (Bulletin, 2009; Huang, 2003; Lee, Igel, Trifunac, 2009; Teisseyre, Takeo, Majewski, 2006) rotary seismology, intensive developed at last time, is allowed to corroborate it possible the existence of rotation waves and it is proved in practice a migration character such waves.

2. One can believe that in this way we can approach the forecast of the strength of the medium. The possibility for quantitative description of the foreshock and aftershock stages of the seismic cycle (Vikulin, 2003, 2006, 2009) allows us to create a kind of earthquake prediction theory, containing the timing of events and their foci locations, which is more deterministic than statistic in character.

As we can see, the rotary wave model, explaining remote action effect between earthquake's foci, may open new ways for the solution to the earthquake prediction problem. To this end, it would be necessary, firstly, to improve the presented theoretical approach and, secondly, to develop a network of stations recording rotation waves, in place of the existing single stations of such a character (Bulletin, 2009; Huang, 2003; Lee, Igel, Trifunac, 2009).

In the frame of the rotary model, the nature of interaction between seismicity and the solar activity cycles could be studied, as the seismicity and its relation to the moment of momentum of blocks become relate to the "complex dynamics of all solar system" (Vikulin, 2009).

3. The rotary wave model, utilizing the concept of stresses and force moments, is constructed in the frame of classical elasticity theory with symmetric stress tensor (Vikulin, 2003, 2006, 2009; Vikulin, Ivanchin, 1998). In our approach, we did not need the concept of elasticity theories (Bulletin, 2009; Nowacki, 1975; Teisseyre, Nagahama, Majewski, 2008) based on Cosserat's continuum (Cosserat and Cosserat,

1909) and it similar: Le Roux (1911), Mindlin (1963), ets. It can be noted the Cosserat's elasticity theory, in contrast to the classic elasticity theory (Feynman, Leighton, and Sands, 1964; Landau, Lifshits, 2003), is purely mathematical (is not physical). We think that this can be of consequence for only one reason: the geophysical space-time (in sense A. Einstein) must consider as spin continuum (Vikulin, 2009; Vikulin, Ivanchin, 2002).

4. In turns out the idea of the earth crust block which might rotate due to the internal sources, brings some new productive tools. In the frame of our and other models it possible to get quantitative description of the wide spectrum geophysics and geodynamics phenomena (Lee, 1928; Melekestsev, 1979; Kurlenya, Oparin, 2000; Teisseyre, Takeo, Majwski, 2006; Tveritinova, Vikulin, 2005; Vikulin, 2003, 2009; Vikulin, Krolevets, 2001, 2002; Vikulin, Tveritinova, 2007; Xie Xin-sheng, 2004; ets.). Thus, the basis of a new school, "the vortex dynamics of the lithosphere" (Milanovsky, 2007; Vikulin, 2004; Vikulin, Tveritinova, 2008), is founded.

The relations between tectonic vortex movement intensity and planet's rotation, as analyzed above, might be extended onto a search for similar relations in the domain of atmosphere vortex movements, like cyclones, on the Earth and planets.

Finally, we believe that nature of vortex movements is connected with some more "deep" parameters related to the fine matter structure; such relations might become recognized along with the development of seismic technology of sufficiently high precision to study elastic nonlinear waves of spinning polarization (Vikulin, 2009).

ACNOWLEDGMENTS

The author is grateful to professors S. Duda and R. Teisseyre and W. Lee for help and discussions on the problems of the paper.

REFERENCES

[1] Bakhtiarov, V. F. & Levin, V. E. (1993). The Use of Optical Range Finders at the Mishennaya Geodetic Observatory (Kamchatka) for Recording Ground Motion [in Russian]. *Optichesky Zhurnal, 10,* 82-85.

[2] Bulletin Seismological Society of America. Special Issue: "*Rotational Seismology and Engineering Applications*", 2009, 99, No. 2B, 945-1485.

[3] Bykov, V. G. (2000). *Nonlinear Wave Processes in Rocks* [in Russian], Dalnauka, Vladivostok.

[4] Bykov, V. G. (2005). Strain Waves in the Earth: Theory, Field Data, and Models. *Geologiya i Geofizika (Russian Geology and Geophysics) 46(11),* 1176-1190 (1158-1172).

[5] Bykov, V. G. (2008). Stick-slip and Strain Waves in the Physics of Earthquake Rupture: Experiments and Models. *Acta Geophysica,* 56, 270-285.

[6] Chao, B. F. & Gross, R. S. (1995). Changes in the Earth's Rotational Energy Induced by Earthquakes. *Geophys. J. Int., 122,* 776-783.

[7] Cosserat, E. & Cosserat, F. (1909). *Theorie des Corps Deformables.* Librairie Scientifique A. Hermann et Fils, Paris.

[8] Davydov, A. S. (1982). Solitons in Quasi-one-Dimensional Molecular Structures [in Russian]. *Uspekhi Fizicheskikh Nauk, 138(4)*, 603-643.

[9] Duda, S. J. (1963). Strain Release in the Circum-Pacific Belt, Chile 1960. *J. Geophys. Res., 68*, 5531-5544.

[10] Feynman, R. P., Leighton, R. B. & Sands, M. (1964). The Feynman Lectures on Physics. V. 2. Addison-Wesley Publishing Co., *Inc. Reading*, Massachusetts & Palo Alto & London.

[11] Goldin, S. V. (2003). Physics of the Alive Earth, in: A. V. Nikolaev, ed., *Problems of Geophysics in the 21st Century* [in Russian]. Book 1, Nauka, Moscow, 17-36.

[12] Huang, B. S. (2003). Ground Rotational Motion of the 1991 Chi-Chi, Taiwan Earthquake as Interred from Dence array Observations, *Geophys. Res. Letters, 30*, No. 6, 1307-1310.

[13] V. A., Karzhakin, ed., 2002. *Lithospheric Tectonics and Geophysics.* Proc. XXXV Tectonic Workshop [in Russian]. GEOS, Moscow.

[14] Kurlenya, M. V. & Oparin, V. N. (2000). Problems of Nonlinear Geomechanics [in Russian]. Part II. *FTPRPI, 4*, 3-26.

[15] Landau, L. D. & Lifshits, E. M. (2003). *The Theory of Elasticity* [in Russian]. Nauka, Moscow.

[16] Le Roux, 1911. Etude Geometrique de la Torsion et de la Flexion. *Ann. Scient. De L'Ecole Normale Sup.*, 28, Paris.

[17] Lee, J. S. (1928). Some Characteristic Structural Types in Eastern Asia and their Bearing upon the Problems of Continental Movements, *Geol. Mag., LXVI*, 422-435.

[18] Lee, W. H. K., Igel, H. & Trifunac, M. D. (2009). Recent Advances in Rotational Seismology. *Seismological Research Letters, 80*, No. 3, 479-490.

[19] Lukiyanov, A. V. (1999). Nonlinear Effects in Models of Tectonic Genesis, in: A. V. Lukiyanov, ed., *Problems of Lithospheric Geodynamics* [in Russian]. Nauka, Moscow, 253-287.

[20] Melekestsev, I. V. (1979). The Vortex Volcanic Hypothesis: Some Application Prospects, in: V. S. Sobolev, ed., *Problems of Magmatism* [in Russian]. Nauka, Moscow, 125-155.

[21] E. E. Milanovsky, (Ed.), 2007. *Rotation Processes in Geology and Physics* [in Russian]. Dom Kniga, Moscow.

[22] Mindlin, R. D. (1963). Influence of Couple-Stress on Stress Concentrations. *Exp. Mech., 3*, 1-7.

[23] A. V., Nikolaev, ed., (2003). *Problems of Geophysics in the 21st Century* [in Russian]. Books 1, 2, Nauka, Moscow.

[24] Nikolaevsky, V. N. (1996). Geomechanics and Fluidodynamics. Kluwer, *Academic Publishers*, Dordecht & Boston & London.

[25] Nowacki, W. (1975). *Theoria Sprezystosci.* Warszawa: Panstwowe Wydawnictwo Naukowe.

[26] Ostrovsky, L. A. (2005). Non-Classical Nonlinear Acoustics, in: A. V. Gaponov-Grechov, & V. I. Nekorkin, eds., *Nonlinear Waves* [in Russian]. IPF RAN, Nizhni Novgorod, 109-124.

[27] Panin, V. E. (2002). Plastic Strain and Fracture of Solids as an Evolution of Loss of

their Shear Stability at Different Scale Levels [in Russian]. *Probl. Materials Science*, *1(29)*, 34-49.

[28] Peive, A. V. (1961). Tectonics and Magmatism [in Russian]. *Izv. AN SSSR, Ser Geol.*, *3*, 36-54.

[29] Ponomarev, B. C. (2008). *Energy Saturation of Geological Medium* [in Russian]. *Nauka*, Moscow.

[30] *Problem Materials Science*. Special Issue: "Mesostructure. Proc. International Workshop, St. Petersburg, 4-7 December 2001" [in Russian], 2002. *1(29)*.

[31] Sadovsky, M. A. (2004). The Living Earth, in: A. V. Nikolaev, (Ed.), Mikhail Aleksandrovich Sadovsky. *Essays. Memoirs. Documents* [in Russian]. Nauka, Moscow, 242-245.

[32] Scott, C., Chu, F. Y. F. & McLaughlin, D. W. (1973). The Soliton: a New Concept in Applied Science, *Proceedings IEEE, 61*, No. 10, 79-123.

[33] Sedov, L. I. (1973). *Continuum Mechanics* [in Russian]. Nauka, Moscow.

[34] Saffman, P. G. (1992). *Vortex Dynamics*. Cambridge University Press, Cambridge.

[35] Sleznak, O. I. (1972). *Lithospheric Vortex Systems and Precambrian Structures* [in Russian]. Naukova Dumka, Kiev.

[36] R., Teisseyre, H. Nagahama, & E. Majewski, eds., (2008). *Physics of Asymmetric Continua: Extreme and Fracture Processes: Earthquake Rotation and Soliton Waves*. Berlin & Heidelberg: Springer-Verlag.

[37] R., Teisseyre, M. Takeo, & E. Majewski, eds., (2006). Earthquake Source Asymmetry, *Structural Media and Rotation Effects*. Berlin: Springer.

[38] Tveritinova, T. Yu. & Vikulin, A. V. (2005). Geological and Geophysical Signature of Vortex Structures in the Crust [in Russian]. Vestn. KRAUNC, Ser. Nauk o Zemle, 5, 59-77. www.kscnet.ru.

[39] S. Uyeda, (Ed.), (2003). International Geological-Geophysical Atlas of the Pacific Ocean [in Russian]. *Intergovernmental Oceanographic Commission*, Moscow & St. Petersburg.

[40] Ustinova, V. N., Vyltsan, I. A. & Ustinov, V. G. (2005). Space and Time Evolution of Cyclic Events in the Earth, from Geophysical Data [in Russian]. *Geofizika*, *3*, 65-71.

[41] Vikulin, A. V. (2003). Physics of the Wave Seismic Process [in Russian]. KGPU, *Petropavlovsk-Kamchatsky*. www.kscnet.ru

[42] A. V. Vikulin, ed., (2004). Vortices in Geological Processes [in Russian]. IVGiG DVO RAN; KGPU, *Petropavlovsk-Kamchatsky*. www.kscnet.ru.

[43] Vikulin, A. V. (2006). Earth Rotation, Elasticity and Geodynamics: Earthquake Wave Rotary Model, in: R., Teisseyre, M. Takeo, & E. Majewski, eds., *Earthquake Source Asymmetry*, Structural Media and Rotation Effects. Berlin & Heidelberg: Springer-Verlag, 273-289.

[44] Vikulin, A. V. (2009). *Earth's Physics and Geodynamics* [in Russian]. KamGU, Petropavlovsk-Kamchatsky. www.kscnet.ru.

[45] Vikulin, A. V. & Ivanchin, A. G. (1998). A Rotation Model of the Seismic Process. *Tikhookeanskaya Geologiya, (Russian Geology of the Pacific Ocean)*, *15(6)*, 1225-1240.

[46] Vikulin, A. V. & Ivanchin, A. G. (2002). Rotation and Elasticity [in Russian]. *Probl. Materials Science*, *1(29)*, 435-441.

[47] Vikulin, A. V. & Krolevets, A. N. (2001). Chandler Wobble: Seismotectonic

Implications. *Geologiya i Geofizika (Russian Geology and Geophysics)*, *42(6)*, 996-1009 (947-958).

[48] Vikulin, A. V. & Krolevets, A. N. (2002). Seismotectonic Processes and the Chandler Oscillation. *Acta Geophys. Pol.*, *50(3)*, 395-411.

[49] Vikulin, A. V. & Tveritinova, T. Yu. (2007). Energy of Tectonic Process and Vortex Geological Structures [in Russian]. *Dokl. Earth Sci.*, *413, A(3)*, 336.

[50] Vikulin, A. V. & Tveritinova, T. Yu. (2008). Momentum-Wave Nature of Geological Medium. *Moscow Univ. Geology Bull.*, *63(6)*, 368-371.

[51] Vladimirov, V. I. & Romanov, A. E. (1986). *Disclinations in Crystals* [in Russian]. Nauka, Leningrad.

[52] Xie Xin-Sheng, (2004). Discussion on Rotational Tectonics Stress Field and the Genesis of Circum-Ordos Landmass Fault System. *Acta Seismol. Sinica*, *17(4)*, 464-472.

[53] Zonenshain, L. P. & Savostin, L. A. (1979). *Introduction into Geodynamics* [in Russian]. Nauka, Moscow.

In: Exciton Quasiparticles
Editor: Randy M. Bergin

ISBN: 978-1-61122-318-7
© 2011 Nova Science Publishers, Inc.

Chapter 6

EXCITON DYNAMICS STUDY OF InAs/GaAs QUANTUM DOT HETEROSTRUCTURES

*Ya-Fen Wu[1], Jiunn-Chyi Lee[2], Jen-Cheng Wang[3] and Tzer-En Nee[3]**

[1]Department of Electronic Engineering, Ming Chyi University of Technology, Taiwan
[2]Department of Electrical Engineering, Technology and Science Institute of Northern Taiwan
[3]Department of Electronic Engineering, Chang Gung University, Taiwan

ABSTRACT

The elementary excitation dynamics differ qualitatively from those in higher-dimensional systems, since the density of states in the zero-dimensional quantum dot (QD) systems is a series of δ-functions. Many unique phenomena, including electronic, optical, magnetic, and thermal characteristics, have been observed. As far as the optical properties of the semiconductor QDs are concerned, the excitonic process has attracted a lot of investigations because it is expected to realize very high-efficiency photonic devices due to the Bosonic character of excitons. A key issue is to attain a profound understanding of the corresponding dynamics to facilitate the research for innovative heterodevice architectures. In this work, a steady-state thermal model taking into account the dot size distribution, the random population of density of states, and all of the important mechanisms of exciton dynamics, including radiative and nonradiative recombination, thermal escaping and relaxing, and state filling effects is proposed. These mathematical analyses successfully explain the abnormality of the exciton-related emissions observed in the low dimensional nanostructures. Not only the temperature- and excitation-dependent luminescence measurement systems, but also the metal-organic chemical vapor epitaxy is systematically discussed.

INTRODUCTION

Particularly impressive results have arisen from optical studies of quantum wells and superlattices. In 1974 Dingle *et al.* [1] directly observed the step-like characteristic of the absorption spectrum related to the two-dimensional characteristic of the density of states (DOS) in quantum wells. A decrease in the GaAs layer thickness resulted in a shift of the steps toward higher photon energies. Optical studies also underlined the tremendously increased role of excitonic effects [2,3] and demonstrated zero-dimensional properties of excitons localized in potential fluctuations of quantum wells. By the end of 1980s the main properties of quantum wells and superlattices were well understood and the interest of researchers shifted toward structures with further reduced dimensionality to quantum wires and quantum dots (QDs) [4]. The carriers are localized in all three dimensions and breakdown the classical band structure model of a continuous dispersion of energy as a function of momentum. The strong localization of the electronic wave function leads to a δ-function-like DOS. A profound size-dependent change of all macroscopic material properties as compared to the bulk structures occurs.

The QDs have been successfully produced using self-organization effects, e.g. produced during growth of strained heterostructures. These effects are also called self-assembly. Self-assembled semiconductor QDs are presently of much interest both in the quest to understand the basic physics of quasi-zero- dimensional nanostructures and for their important optoelectronic applications. The requirement for high performance optoelectronic devices has spurred much experimental effort directed toward understanding and exploiting the electronic and optical properties of QDs [5–9]. The δ-function-like electron DOS are expected to be useful for imparting optoelectronic devices with improved parameters. Evidence of the high quality of QD structures is given by the demonstration of QD-based detectors [10], charge storage devices [11], and injection lasers with low threshold current densities and predicted high-temperature stability [12–14].

The temperature dependence of the PL from QD system has been a subject of extensive studies for clarifying the processes of energy relaxation and energy transfer in the QD ensemble [15–17]. The measured temperature-dependent PL spectra of QDs usually exhibit two main features. First, the integrated PL intensity quenches at high temperature. It is attributed to thermal escape of carriers from the QDs into the wetting layer or barrier material, where they are lost through, for example, nonradiative recombination centers [18]. The second feature is the decrease of the full width at half maximum (FWHM), together with a redshift of the emission wavelength, are observed in the mid-temperature range. The fast redshift of the PL peak energy and the anomalous decrease of the linewidth with temperature have been explained by enhanced carrier redistribution among dots due to carrier thermionic emission and carrier transport through the wetting layer [19,20]. These thermal effects are not expected in an ideal QD structure in which all the dots are decoupled from each other. They have been attributed to a thermal escape occurring at lower temperatures for high-energy dots and carriers being recaptured by dots emitting on the low-energy side of the distribution through the wetting layer [21–23]. As the temperature increases further, the anomalous phenomena become weakened and the FWHM increases with temperature because of the increasing electron-phonon scattering [24–26].

The effect of temperature T on the optical properties of self-assembled QDs is of great importance for their use in devices working at room temperature. Various papers have discussed the temperature dependence of photoluminescence (PL) properties of QDs in InAs/GaAs [15–17], InGaAs/GaAs [27,28], and InAlAs/AlGaAs [29,30] heterostructures. Similar studies were also performed on laser structures based on $In_{0.5}Ga_{0.5}As$ QDs embedded in (AlGa)As barriers. The relaxation dynamics in the zero-dimensional QD systems differ qualitatively from that in higher-dimensional systems, since the density of states is a series of δ-functions. The limited number of states available for carriers impairs carrier relaxation toward the ground state ("phonon bottleneck" effect) [31,32]. In addition, the finite degeneracy of each QD state is expected to result in the state filling effect due to exclusion principle taking effect when only a few carriers populate the lower states. Both effects possibly lead to the intersublevel relaxation rate being comparable to the radiative recombination rate and have been used to explain observed PL from excited states transition at low excitation density [33–35]. Moreover, self-assembled QDs suffer size fluctuations of ~10% [36,37] as well as shape non-uniformities, which average out the otherwise discrete density of states. The non-uniformity of self-assembled QDs and the restricted number of available energy states make QDs very interesting systems for studying carrier transport under optical excitation.

In this chapter, we describe our study of the temperature dependence of the PL spectra measured from samples having different dot densities and size uniformities. We propose a theoretical thermal model that considers the QD size distribution and random population effects. It is based on a set of rate equations at the steady state that connect the carrier dynamics in the intersublevels of the QDs, the wetting layer, and the GaAs barrier states. All of the relevant thermalization and quenching processes that are active in QDs are taken into account. Furthermore, we also present a detailed investigation into the dependence of the PL upon the excitation power intensity (P_{in}) at different temperatures. The experimental data obtained from these PL measurements fit this model well. From a quantitative analysis of the thermal redistribution and state filling effects, we discuss the dependence of the PL spectra on the temperature and excitation power intensity.

EXPERIMENTS

The self-assembled InAs/GaAs QD heterostructure studied in this work was created by using a metal-organic chemical vapor epitaxy system. Two samples were grown on (100) 2°–tilted toward (111)A GaAs substrate. The heterostructure included a 400 nm Si-doped GaAs buffer layer, an InAs QD active region of 3 monolayers (MLs), and a 100 nm undoped GaAs capping layer. The growth rate was 0.1 ML/s, and the V/III ratio during the growth of InAs layer was 6.36. In order to obtain different dot densities and size uniformities for these two samples, the growth interruption (GI) times during dot formation were set to 6 and 15 s, respectively. In order to investigate the average dot size distribution and shape, images of these samples were taken by high-resolution JEOL-JEM2010 transmission electron microscopy (HRTEM) operating at 200 keV. The temperature and excitation dependent PL spectra were measured using a He-Cd laser with a wavelength of 325 nm and the average excitation intensity was in the range from 5 to 40 mW. The samples were mounted in a closed-cycle He cryostat where the temperature was varied from 15 to 280 K. The luminescence was dispersed in a 0.5 m monochrometer and detected with a Ge photodiode using a standard lock-in technique.

Figure 1 shows the plan-view TEM images for samples A and B. The quantitative data on size distribution of the samples have been obtained from the TEM images, the average dot density of samples A and B are 2.4×10^{10} cm^{-2} and 1.2×10^{10} cm^{-2}, and the average dot diameters of the samples are 15 nm and 18 nm, respectively. The normalized PL spectra of the samples recorded at T=15 K and P_{in}=20 mW are presented in Figure 2. All of these spectra exhibit a double-like feature and can be decomposed into two Gaussian peaks; we attribute these two main spectral features of the QDs to the ground state and the excited state emissions. Sample A possesses the larger ground state and excited state transmission energies, i.e., 1.05 eV and 1.11 eV; and the values are 1.01 eV and 1.09 eV for sample B. The values of FWHMs of the ground state and the excited state emissions are 27.1 meV and 88.3 meV for sample A, and 26.8 meV and 79.6 meV for sample B.

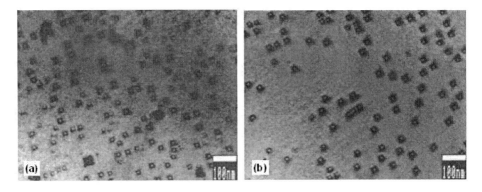

Figure 1. Plan-view TEM images of the InAs quantum dots of (a) sample A, and (b) sample B.

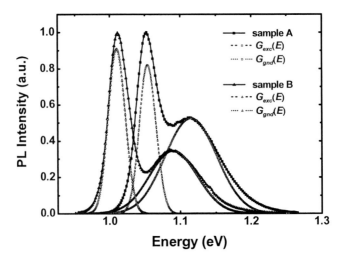

Figure 2. PL spectra of sample A and sample B at T=15 K and P_{in}=20 mW. The Gaussian line fit to the experimental data is also shown. $G_{exc}(E)$ and $G_{gnd}(E)$ represent the fitted distributions of excited state and ground state, respectively.

Generally, the application of GI time results in the formation of larger sized QDs with a regular size distribution [36], as can be seen from Figure 1(a) and Figure 1(b). Considering the quantum-size effect on the peak energies, we believe that the excitons localized in smaller dots will contribute to higher peak energies [38]. As a result, the higher peak energy of sample A (GI=6 s) is attributed to the smaller size of the QDs and the smaller energy separation of it is attributed to the increasing strain of QDs that resulted from its higher dot density [39]. The PL linewidth is mainly determined by the inhomogeneous broadening of InAs islands resulted from size fluctuation of the dot size at low temperature [22,40], the measured data for sample B are consistent with its better size uniformity.

RESULTS AND DISCUSSION

Figure 3(a) displays the temperature dependent FWHMs of PL spectra of sample A and sample B, both the ground state and the excited state are included. Observing the FWHMs, they stay constant up to 75 K and 110 K. As the temperature further increases, the FWHMs decrease and the minimal FWHMs of excited state are found to be around 69 meV at 200 K for both samples. When the temperature is higher than 200 K, the PL linewidths start to increase with temperature.

At low temperature, the PL linewidth of quantum dots is determined by the inhomogeneous broadening resulted from size fluctuation of the dot size. With elevated temperature, the excitons locate at shallow potential minima are thermally activated outside the dots and into the wetting layer and then preferentially relax into large dots with relatively low-localized energy states. Thus, the size distribution of the QDs taking part in the emission becomes narrower and the FWHMs follow the same trend: it decreases as the temperature increases. At the same time, because the emitting QDs are characterized by energetically lower states, as can be seen in Figure 3(b), the peak energies experience a faster peak energy decrease than that of the InAs bulk band gap which has been calculated by the Varshni law:

$$E_g(T) = E_g(0) - \frac{\alpha T^2}{T + \beta}, \qquad (1)$$

where $E_g(T)$ is the band gap energy at temperature T, $E_g(0)$ the band gap energy at 0 K, α and β the material-related constants. For InAs, the values of α and β are 0.276 meV/K and 93 K, respectively.

When the temperature increases further, the electron-phonon scattering becomes important; this factor induces an increase in the FWHMs of the PL band upon increasing the temperature [19,20,41]. It is noticeable that the thermal action described above is observed to be milder for sample B. Thanks to the lower density and better uniformity of dots with this sample, the behavior can be explained by considering a weaker thermal redistribution effect of carriers among dots, which we will discuss later.

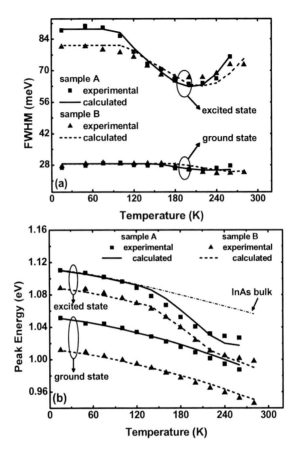

Figure 3. Experimental and calculated temperature dependent (a) FWHMs, and (b) peak energies of the ground state and the excited state for sample A and sample B. The dash-dotted line in (b) is calculated according to the Varshni law using the parameters of InAs and shifted along the energy axis.

$$E_g(T) = E_g(0) - \frac{\alpha T^2}{T+\beta}, \quad (1)$$

To analyze the carrier dynamics of the QD system, we develop a theoretical model that takes into account the QDs size distribution, the state filling effect, and all of the important carrier transport processes, including the carrier capture and relaxation, thermal emission and retrapping, and radiative and nonradiative recombination. Referring to the model of the QD system described schematically in Figure 4, the processes included in this model for the QD system consist of four discrete quantum levels, i.e., the ground state and excited state in the dots, the wetting layer, and the GaAs barrier. Because the process of quantum dot population is intrinsically random, we assume that the DOS of both the ground state $[n_{gnd}(E)]$ and the excited state $[n_{exc}(E)]$ are proportional to the Gaussian distribution, with parameters chosen to match the peak energies and linewidths of the lowest temperature PL spectra [23,42]. If the dot density of the sample is n_{QD} and we take the spin degeneracy into consideration, then

$$\int n_{gnd}(E)\,dE = \alpha \int G_{gnd}(E)\,dE \propto 2 \times n_{QD}, \quad (2)$$

$$\int n_{exc}(E)dE = \alpha \int G_{exc}(E)dE \propto 4 \times n_{QD}, \qquad (3)$$

where $G_{gnd}(E)$ and $G_{exc}(E)$ are the Gaussian distributions of the ground and excited states, respectively, obtained from the PL spectra at $T=15$ K and α is the proportionality constant. Besides,

$$n_{gnd}(E) = nf_{gnd}(E) + ne_{gnd}(E), \qquad (4)$$

$$n_{exc}(E) = nf_{exc}(E) + ne_{exc}(E). \qquad (5)$$

The terms $nf_{gnd}(E)$ and $nf_{exc}(E)$ are the numbers of filled ground and excited states, respectively, of the QDs and $ne_{gnd}(E)$ and $ne_{exc}(E)$ represent the number of empty ground and excited states, respectively.

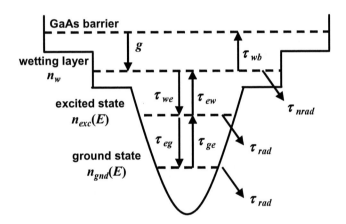

Figure 4. Schematic representation of the processes taken into account in the rate equation model.

The exciton dynamics taken into account in this model are described as follows. First, the coupling among those four carrier reservoirs is treated as a relaxation ladder process from each energy level to its lower level neighbor. We denote the capturing time constant from the wetting layer into the QDs as τ_{we} and the relaxing time constant from excited state to ground state as τ_{eg}. Secondly, thermal emission of the excitons toward an adjacent higher energy level arises when the temperature is sufficiently high. The coefficients corresponding to emitting from the ground state to the excited state, the excited state to the wetting layer, and the wetting layer to the GaAs barrier are given by τ_{ge}, τ_{ew}, and τ_{wb}, respectively. The third type of carrier dynamics considered in this system is the radiative recombination. We have neglected any recombination from the second excited state of the dots, since no PL is observed at energies possible for the second excited state, and assumed that only two discrete electron levels exist inside a quantum dot, i.e., the ground state and the first excited state. The radiative recombination lifetime τ_{rad} is assumed to be the same for both of the states in all of the QDs and is constant with respect to T.

Suppose an external excitation source injects the excitons into the GaAs barrier at a constant rate G and they are transferred into the wetting layer at rate g, which acts as a channel for the carriers transferring among dots. The excitons (n_w) in this layer are captured into the QDs, emit into the GaAs barrier, or are lost at a rate τ_{nrad}^{-1} through the nonradiative recombination centers. The last two terms are usually taken to be responsible for the thermal quenching of the integrated PL intensity of QDs at high temperature. Once excitons emit into the GaAs barrier, they are assumed to be irreversibly lost [41]. Then, we can write a set of rate equations that describes the carriers transferring among each energy level [18,23,43].

For wetting layer:

$$\frac{dn_w}{dt} = g - \int \frac{n_w}{\tau_{we}(E)} dE + \int \frac{nf_{exc}(E)}{\tau_{ew}} dE - \frac{n_w}{\tau_{wb}} - \frac{n_w}{\tau_{nrad}}. \quad (6)$$

For excited state:

$$\frac{d[nf_{exc}(E)]}{dt} = \frac{n_w}{\tau_{we}(E)} - \frac{nf_{exc}(E)}{\tau_{eg}(E)} + \frac{nf_{gnd}(E)}{\tau_{ge}(E)} - \frac{nf_{exc}(E)}{\tau_{ew}} - \frac{nf_{exc}(E)}{\tau_{rad}}. \quad (7)$$

For ground state:

$$\frac{d[nf_{gnd}(E)]}{dt} = \frac{nf_{exc}(E)}{\tau_{eg}(E)} - \frac{nf_{gnd}(E)}{\tau_{ge}(E)} - \frac{nf_{gnd}(E)}{\tau_{rad}}. \quad (8)$$

The state filling effect is essentially significant in the QD system because of the reduced density of states and should be taken into account [43,44]. This effect implies that the relaxation and thermal escaping processes in the QDs are proportional to the number of empty dot states where the excitons arrive. Thus the relaxation and thermal emission time constants of QDs: τ_{we}, τ_{eg}, and τ_{ge} are written as

$$\tau_{we}(E) = \tau_{we0} \times \frac{n_{exc}(E)}{ne_{exc}(E)}, \quad (9)$$

$$\tau_{eg}(E) = \tau_{eg0} \times \frac{n_{gnd}(E)}{ne_{gnd}(E)}, \quad (10)$$

$$\tau_{ge}(E) = \tau_{ge0} \times \frac{n_{exc}(E)}{ne_{exc}(E)}, \quad (11)$$

where τ_{we0} and τ_{ge0} are defined as the intrinsic time constants while the excited states is empty and τ_{eg0} is defined as the intrinsic time constant while the ground state is empty.

Prior to the description of the simulation process, we must discuss the parameters used in the model. The intrinsic exciton lifetime in the QDs (τ_{QD}) at low temperature is estimated in terms of the exciton lifetime in a corresponding quantum well (τ_{WL}) [45,46]:

$$\tau_{QD} = \frac{3}{2}\tau_{WL}\left(\frac{\eta}{k_{ex}}\right)^2. \qquad (12)$$

Here, $k_{ex}=2\pi n/\lambda_{PL}$ is the reciprocal wavelength of the emitted light in the quantum dot material, with the refractive index of InAs, and η is a measure of the lateral dot size. Using values of (1/15) nm^{-1}, 3.6, and 1181 nm for η, n, and λ_{PL}, respectively, and an exciton lifetime τ_{WL} of 25 ps, the radiative recombination lifetime τ_{rad} is calculated to be approximately 500 ps.

To rewrite (6), (7), and (8) under the steady state at $T=15$ K, where the thermal emission can be eliminated, those three equations can be given by

$$g - \int \frac{n_w}{\tau_{we0}} \times \frac{ne_{exc}(E)}{n_{exc}(E)} dE - \frac{n_w}{\tau_{nrad}} = 0, \qquad (13)$$

$$\frac{n_w}{\tau_{we0}} \times \frac{ne_{exc}(E)}{n_{exc}(E)} - \frac{nf_{exc}(E)}{\tau_{eg0}} \times \frac{ne_{gnd}(E)}{n_{gnd}(E)} - \frac{nf_{exc}(E)}{\tau_{rad}} = 0, \qquad (14)$$

$$\frac{nf_{exc}(E)}{\tau_{eg0}} \times \frac{ne_{gnd}(E)}{n_{gnd}(E)} - \frac{nf_{gnd}(E)}{\tau_{rad}} = 0. \qquad (15)$$

Then (13), (14) and (15) are substituted into the following relation

$$\frac{nf_{gnd}(E_{pk})}{nf_{exc}(E_{pk})} \propto \frac{I_{gnd}(E_{pk})}{I_{exc}(E_{pk})}, \qquad (16)$$

where I_{gnd} and I_{exc} are the PL peak intensities of ground state and excited states, respectively, obtained from the measured PL spectra, and E_{pk} represents the peak energy. The value of τ_{eg0} is then determined by using (15). Combining (14) and (15) and using the value of 30 ps for the intrinsic exciton capture time by QDs, yields the value of τ_{we0}. Following the similar procedure, we can obtain the value of g. The lifetime of nonradiative loss in the wetting layer (τ_{nrad}) depends on the quality of the material. Upon increasing the temperature, the rate should increase and saturate around 150 K at a constant value of 100 ps, which is the value that we used in this model [18].

The thermal emission time are derived by assuming that the system reaches quasi-Fermi equilibrium [47]:

$$\tau_{wb} = g \times \exp[(E_{ba} - E_{wl})/k_B T], \qquad (17)$$

$$\tau_{ew} = \tau_{we} \times \exp\{[E_{wl} - E_{exc}(E)]/k_B T\}, \qquad (18)$$

$$\tau_{ge} = \tau_{eg} \times \exp\{[E_{exc}(E) - E_{gnd}(E)]/k_B T\}. \qquad (19)$$

E_{ba}, E_{wl}, $E_{exc}(E)$, and $E_{gnd}(E)$ are the level energies of the GaAs barrier, wetting layer, excited state, and ground state, respectively. Luminescence from the wetting layer is not observed in the spectra, implies the fast capture rate of QDs. The localization energy of QDs with respect to the wetting layer is about 300 meV [19], thus we use the value of 1.31 eV to be the wetting layer energy.

The coupled rate-equation set (6)–(8) are solved numerically under a steady state assumption by fitting the integrated PL intensity. Once the carrier distribution functions $nf_{gnd}(E)$ and $nf_{exc}(E)$ are determined, the PL spectra of ground state (PL_{gnd}) and excited state (PL_{exc}) can be expressed as

$$PL_{gnd}(E) = \beta \times nf_{gnd}(E)/\tau_{rad}, \qquad (20)$$

$$PL_{exc}(E) = \beta \times nf_{exc}(E)/\tau_{rad}, \qquad (21)$$

where β is a normalizing factor. From (20) and (21), the measured temperature and incident-power dependent PL spectra are reproduced.

Under various temperatures and various incident-powers, the PL spectra were measured and studied quantitatively by the rate-equation model. The values of the temperature dependent FWHMs and peak energies for both samples at P_{in}=20 mW are obtained by the rate-equation model demonstrated above and the fitting parameters are listed in Table 1, and shown in Figure 4 compared to their experimental data. Despite the simplicity of this model, the calculated values are in good agreement with the experimental data, demonstrating that the model is indeed realistic. In order to analyze the values of the temperature dependent FWHMs and peak energies of the PL spectra, more detailed information about carrier dynamics is obtained by calculating the excitons changing rate of the discrete energy level in the QD system. We compared the rates of excitons that thermally emit from the excited state (n_{ew}) and the wetting layer (n_{wb}), and those that relax from the wetting layer to the excited state (n_{we}). The rates can be written as

$$n_{ew} = \int \frac{nf_{exc}(E)}{\tau_{ew}} dE = \int \frac{nf_{exc}(E)}{\tau_{we} \times \exp[(E_{wl} - E_{exc})/k_B T]} dE, \qquad (22)$$

$$n_{wb} = \frac{n_w}{\tau_{wb}} = \frac{n_w}{g \times \exp[(E_{ba} - E_{wl})/k_B T]}, \qquad (23)$$

Exciton Dynamics Study of InAs/GaAs Quantum Dot Heterostructures 265

$$n_{we} = \int \frac{n_w}{\tau_{we0}} \times \frac{ne_{exc}(E)}{n_{exc}(E)} dE. \qquad (24)$$

The results calculated for sample A are shown in Figure 5(a). When T goes below 100 K, the value of n_{ew} increases upon increasing the temperature. However, the increment is too small to affect the total excitons existing in the excited state, and the FWHMs exhibit no significant changes. In the temperature range of 100–200 K, the excitons in the excited state that have higher energies begin to emit to the wetting layer and this process leads to a remarkable rise in the value of n_{ew}. Some of these excitons are retrapped by the QDs, giving rise to the lateral transferring of excitons among dots, while others are lost through the nonradiative recombination centers in the wetting layer.

Table 1. Parameters used to calculate temperature dependent PL spectra

Sample	g (s^{-1})	τ_{we0} (ps)	τ_{eg0} (ps)	τ_{rad} (ps)
A	1×10^{21}	17	69	500
B			32	

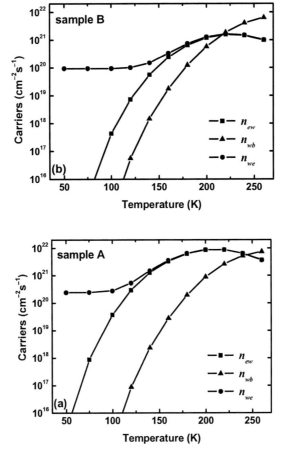

Figure 5. Carrier relaxing and thermal emission of the energy stages in the QD system for (a) sample A and (b) sample B.

More detailed information can be obtained by observing the trend of n_{we} in Figure 5(a). Evidently, both n_{ew} and n_{we} are increased with temperature in the range of 100 K<T<200 K, indicating the fact of thermal redistribution and lateral transition of carriers. Therefore, the decrease in the values of FWHMs and the fast redshift of the peak energies within this temperature range are both explained by considering that the thermal carriers transferred among the quantum dots via the wetting layer. On the other side, the thermal escape from the wetting layer to the barrier is not remarkable yet because the temperature is not high enough. When the temperature exceeds 200 K, the value of n_{wb} increased significantly. This behavior demonstrates that the number of carriers emitted into the GaAs barrier becomes large enough to lead to a reduction in the value of n_{ew}, which in turn suppress the population redistribution effect. Consequently, the degrading of FWHMs gradually decelerates upon the increasing temperature.

A similar thermal action was conducted for sample B. Observing the plot shown in Figure 5(b), the amounts of variety in n_{ew} and n_{wb} with temperature are all less than those of sample A. Besides, the tendency of n_{ew} is in accordance with the curve shown in Figure 5(a), but the turning temperature is even higher. Owing to the lower dot density and higher dot uniformity of this sample, there is a narrower energy spreading for it. Furthermore, inspecting the PL spectra at T=15K for both samples, the energy difference between the ground state and the excited state of sample B is larger than that of sample A. It is reasonable to conclude that the thermal emission in sample B is weaker than that in sample A. The inference is consistent with the consequence examined from Figure 5(b). As a result, the QDs interband transition rate for sample B will be higher than that of sample A due to the weaker thermal population effect of the excited state and better uniformity [48].

The FWHMs and peak energies of the temperature dependent PL spectra of excited state for sample A at various excitation intensities are displayed in the left panels of Figure 6(a) and Figure 6(b). Referring to the curves in the figure, the thermal redistribution effect in the PL spectra degrades upon the increasing incident-power intensity. Because the excitons are distributed randomly into the QD states at low temperature, the spectra exhibit no significant changes with respect to the different incident powers. When the temperature is increased, the processes of thermal emission and retrapping of the excitons via the wetting layer become marked. Consequently, once the incident power is increased at the same temperature, the enhanced filled portion in the energy states leads to a smaller reduction of the FWHMs and a smaller redshift for PL emission energy results from the reduced influence of the thermally redistributed carriers in the excited state [34,49]. As a result, we observe the increment of FWHMs and a slight overall blueshift of the peak energy upon increasing P_{in} at a fixed temperature. A much milder dependence of the action on the incident power is observed for sample B shown in the right panels of Figure 6(a) and Figure 6(b). Interestingly, the temperature dependent values of the FWHMs and the peak energies remain almost unchanged in response to the different incident-power intensities. Because of the better uniformity of this sample, this behavior can be explained by considering the suppressed thermal redistribution effect.

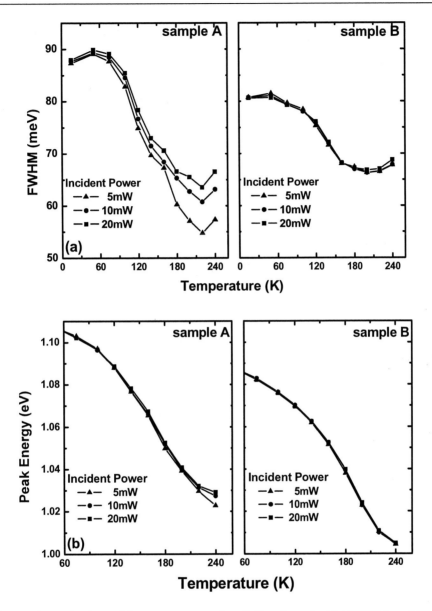

Figure 6. Experimental values of the temperature and excitation dependent (a) FWHMs and (b) peak energies of the excited states of sample A and sample B.

The mechanism of thermal redistribution of excitons activated in the temperature dependent PL spectra also possesses power dependent characteristic. The simulated values of the temperature dependent FWHMs and peak energies for sample A at various incident powers P_{in} are displayed in Figure 7, the experimental PL spectra are reproduced well by the presented model. Simulated values of the power dependent FWHMs and peak energies for sample A and sample B are shown in Figure 8(a) and Figure 8(b), respectively, which are consistent with the experimental results presented in Figure 6. In the data calculated for sample A, the values of the FWHMs and the peak energies undergo less variation at temperatures below 100 K than they do above it. If we consider that the injected excitons are distributed randomly into the QDs when the temperature is low, it is clear that an increasing value of P_{in} raises the PL intensity. But, on the other side, the changes of excitons that result from the increasing temperature will have a less influence on the values of either the FWHMs or the emission energies. To confirm this assumption, we computed the thermal escaping rates of the carriers (n_{ew}) at various values of P_{in} for each value of temperature. Moreover, we also determined the value of N_{ew}, the ratio of n_{ew} with respect to the filled fraction of excited state (f_{exc}):

$$N_{ew} = \frac{n_{ew}}{f_{exc}}, \tag{25}$$

$$f_{exc} = \frac{\int n f_{exc}(E)\, dE}{\int n_{exc}(E)\, dE}, \tag{26}$$

The values of n_{ew} and N_{ew} calculated for sample A that shown in Figure 9(a) support our explanation above. There is only a slight change in N_{ew} upon increasing the value of P_{in} at temperatures below 100 K. When the temperature is higher than 100 K, thermal redistribution becomes active in the QDs; the value of n_{ew} increases upon increasing the value of P_{in}, but the value of N_{ew} decreases accordingly. From the comparison of the values n_{ew} and N_{ew} under different incident powers, we observe that the most distinct difference occurred at the lowest value of P_{in}. Because fewer excitons are injected into the QD system when the value of P_{in} is low, it is provided with a lower portion of filled energy states. This finding implies that the state filling effect plays an important role in the population redistribution process of the QDs. Thus, we predict quite reasonably, that the PL spectra should display a weaker thermal response upon increasing the incident power. Consequently, the values of the FWHMs increase and the peak energies shift slightly toward the shorter wavelength as P_{in} is increasing. Moreover, the decrease of FWHMs in the mid-temperature range is somewhat compensated for by the increase in the incident power because the thermal redistribution effect is depressed at higher incident powers. This deviation is clearly evident in the experimental data presented in Figure 6(a).

Figure 8 also provides the results of performing analogous calculations of the power-dependent behavior of sample B. The power dependent PL spectra remain almost unchanged within the measured power range. The better size uniformity induces a limited energy broadening for the excited state; this result implies that the energies of the states available for the excitons are confined within a narrower range. Thus, the change in f_{exc} that results from the increasing P_{in} becomes less apparent. The deduction is consistent with the lower slope of the power dependent N_{ew} presented in Figure 9(b), which results in relatively stable values of FWHMs and peak energies upon increasing the value of P_{in}, in accordance with our experimental data.

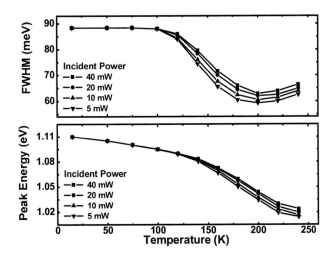

Figure 7. Calculated values of the temperature dependent (a) FWHMs and (b) peak energies of the excited state under different incident power intensity of sample A.

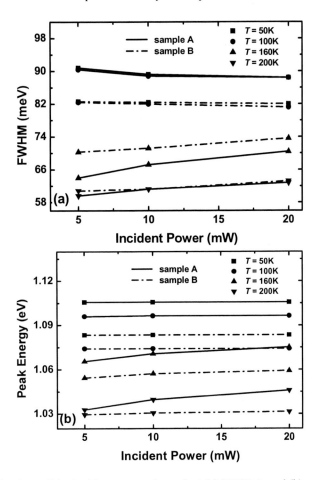

Figure 8. Calculated values of the incident-power dependent (a) FWHMs and (b) peak energies of sample A and sample B at different temperatures.

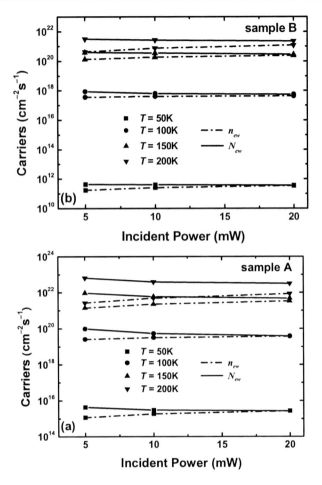

Figure 9. Power dependent thermal escaping carriers for (a) sample A and (b) sample B at different temperatures.

CONCLUSION

In this chapter, we have systematically investigated the exciton dynamics of a QD system both experimentally and theoretically. A thermal model for demonstrating the exciton dynamics of InAs/GaAs QD heterostructures is proposed. The model is based on a set of rate equations which connect the ground state, the excited state, the wetting layer, and the GaAs barrier in the QD system. All of the important mechanisms for explaining the unique evolution of quantum dot PL spectra are taken into account, involving the QDs inhomogeneous broadening, random population of the density of states, thermal escaping and relaxing, radiative and nonradiative recombination, and the state filling effect. The simulated results exhibit a good agreement to the experimental data measured from samples with different dot densities and size uniformities for temperatures ranging from 15 to 280 K and at incident power intensities from 5 to 40 mW. Quantitative discussion of the excitons which thermally excited and relax between the excited state and the wetting layer provides an explicit proof of the thermal redistribution and lateral transition of excitons via the wetting layer. The variation of FWHMs and peak energies with temperature is well explained by the separate calculation for the thermal emission from every stage in the QD system. Moreover, comparison of different samples demonstrates that samples with higher uniformity possess a milder dependence on temperature in PL line shape.

In the quantitative discussions of incident-power dependent excitons transferring mechanisms, we also find that there is an increase in the linewidth and a slight blueshift of the PL peaks upon increasing the value of P_{in}. We attribute this finding to the state filling effect. Our mathematical study of the influence induced by the excitation intensity at different temperatures reveals that the state filling fraction plays an important role in the PL spectra for temperatures above 100 K. The larger portion of filled energy states at higher incident power intensities suppresses this thermal redistribution effect. Therefore, the variety in the PL spectra that arises upon increasing the temperature is compensated partially upon increasing the incident power. The lineshape from sample with better uniformity is almost independent of the incident power. The detailed investigation of exciton dynamics in QD system is of particular significance to the design of QD structures and is expected to lead to improved in optoelectronic devices.

ACKNOWLEDGMENT

This work was supported by the National Science Council of the Republic of China under Contract no. NSC 97-2112-M-182-002-MY3 and NSC 98-2112-M-131-001-MY2.

REFERENCES

[1] Dingle, R; Wiegmann, W; Henry, CH. *Phys. Rev. Lett*, 1974, 33, 827-830.
[2] Christen, J; Bimberg, D. *Phys. Rev. B*, 1990, 42, 7213-7218.
[3] Christen, J; Krahl, M; Bimberg, D. *Superlattice. Microst*, 1990, 7, 1-4.
[4] Kapon, E; Hwang, DM; Bhat, R. *Phys. Rev. Lett*, 1989, 63, 430-433.
[5] Arkawa, Y; Sakaki, H. *Appl. Phys. Lett*, 1982, 40, 939-941.
[6] Brus, L. *IEEE J. Quantum Electron*, 1986, 22, 1909-1914.
[7] Chemla, DS; Miller, DAB. *Opt. Lett*, 1986, 11, 522-524.
[8] Leonard, D; Krishnamurthy, M; Reaves, CM; Denbaars, SP; Petroff, PM. *Appl. Phys. Lett*, 1993, 63, 3203-3205.
[9] Cusack, MA; Briddon, PR; Jaros, M. *Phys. Rev. B*, 1996, 54, R2300-R2303.
[10] Sauvage, S; Boucaud, P; Julien, FH; Gerard, JM; Thierry-Mieg, V. *Appl. Phys. Lett*, 1997, 71, 2785-2787.
[11] Yusa, G; Sakaki, H. *Appl. Phys. Lett*, 1997, 70, 345-347.
[12] Huang, X; Stintz, A; Hains, CP; Liu, GT; Cheng, J; Malloy, KJ. *IEEE Photonic. Tech. Lett*, 2000, 12, 227-229.
[13] Shchekin, OB; Deppe, DG. *Appl. Phys. Lett*, 2002, 80, 3277-3279.
[14] Bhattacharya, P; Ghosh, S. *Appl. Phys. Lett*, 2002, 80, 3482-3484.
[15] Lee, H; Lowe-Webb, R; Johnson, TJ; Yang, W; Sercel, PC. *Appl. Phys. Lett*, 1998, 73, 3556-3558.

[16] Polimeni, A; Patanè, A; Henini, M; Eaves, L; Main, PC. *Phys. Rev. B*, 1999, 59, 5064-5068.
[17] Lobo, C; Leon, R; Marcinkevičius, S; Yang, W; Sercel, PC; Liao, XZ; Zou, J; Cockayne, DJH. *Phys. Rev. B*, 1999, 60, 16647-16651.
[18] Ru, ECL; Siverns, PD; Murray, R. *Appl. Phys. Lett*, 2000, 77, 2446-2448.
[19] Dai, YT; Fan, JC; Chen, YF; Lin, RM; Lee, SC; Lin, HH. *J. Appl. Phys.*, 1997, 82, 4489-4492.
[20] Mazur, YI; Wang, X; Wang, ZM; Salamo, GJ; Xiao, M; Kissel, H. *Appl. Phys. Lett*, 2002, 81, 2469-2471.
[21] Lubyshev, DI; Gonzalez, PP; Mareda, E; Petitprez, E; Lascala, N; Basmaji, P. *Appl. Phys. Lett*, 1996, 68, 205-207.
[22] Brusaferri, L; Sanguinetti, S; Grilli, E; Guzzi, M; Bignazzi, A; Bogani, F; Carraresi, L; Colocci, M; Bosacchi, A; Frigeri, P; Franchi, S. *Appl. Phys. Lett*, 1996, 69, 3354-3357.
[23] Yang, W; Lowe-Webb, RR; Lee, H; Sercel, PC. *Phys. Rev. B*, 1997, 56, 13314-13320.
[24] Xu, ZY; Lu, ZD; Yang, XP; Yuan, ZL; Zheng, BZ; Xu, JZ. *Phys. Rev. B*, 1996, 54, 11528-11531.
[25] Jiang, WH; Ye, XL; Xu, B; Xu, HZ; Ding, D; Liang, JB; Wang, ZG. *J. Appl. Phys.*, 2000, 88, 2529-2532.
[26] Gammon, D; Rudin, S; Reinecke, TL; Katzer, DS; Kyono, CS. *Phys. Rev. B*, 1995, 51, 16785-16789.
[27] Fafard, S; Leonard, D; Merz, JL; Petroff, PM. *Appl. Phys. Lett*, 1994, 65, 1388-1390.
[28] Mukai, K; Ohtsuka, N; Sugawara, M. *Jpn. J. Appl. Phys.*, 1996, 35, L262-L265.
[29] Fafard, S; Leon, R; Leonard, D; Merz, JL; Petroff, PM. *Phys. Rev. B*, 1990, 52, 5752-5761.
[30] Leon, R; Fafard, S; Leonard, D; Merz, JL; Petroff, PM. *Appl. Phys. Lett*, 1995, 67, 521-523.
[31] Bockelmann, U; Bastard, G. *Phys. Rev. B*, 1990, 42, 8947-8951.
[32] Benisty, H. *Phys. Rev. B*, 1995, 51, 13281-13293.
[33] Raymond, S; Fafard, S; Poole, PJ; Wojs, A; Hawrylak, P; Charbonneau, S; Leonard, D; Leon, R; Petroff, PM; Merz, JL. *Phys. Rev. B*, 1996, 54, 11548-11554.
[34] Grosse, S; Sandmann, JHH; Plessen, GV; Feldmann, J. *Phys. Rev. B*, 1997, 55, 4473-4476.
[35] Fafard, S; Leon, R; Leonard, D; Merz, JL; Petroff, PM. *Phys. Rev. B*, 1994, 50, 8086-8089.
[36] Tarasov, GG; Mazur, YI; Zhuchenko, ZY; Maabdorf, A; Nickel, D; Tomm, JW; Kissel, H; Walther, C; Masselink, JT. *J. Appl. Phys.*, 2000, 88, 7162-7170.
[37] Heitz, R; Kalburge, A; Xie, Q; Grundmann M; Chen P; Hoffmann A; Madhukar A; Bimberg D. *Phys. Rev. B*, 1997, 56, 10435-10445.
[38] Cheng, WQ; Xie, XG; Zhong, ZY; Cai, LH; Huang, Q; Zhou, JM. *Thin Solid Films*, 1998, 312, 287-290.
[39] Marcinkevičius, S; Leon, R. *Appl. Phys. Lett*, 2000, 76, 2406-2408.
[40] Zhang, XQ; Ganapathy, S; Kumano, H; Uesugi, K; Suemene, I. *J. Appl. Phys.*, 2002, 92, 6813-6818.
[41] Zhang, YC; Huang, CJ; Liu, FQ; Xu, B; Wu, J; Chen, YH; Ding, D; Jiang, WH; Ye, XL; Wang, ZG. *J. Appl. Phys.*, 2001, 90, 1973-1976.
[42] Lee, H; Yang, W; Sercel, PC. *Phys. Rev. B*, 1997, 55, 9757-9762.

[43] Mukai, K; Ohtsuka, N; Shoji, H; Sugawara, M. *Phys. Rev. B*, 1996, 54, R5243-R5246.
[44] Mukai, K; Ohtsuka, N; Shoji, H; Sugawara, M. *Appl. Phys. Lett*, 1996, 68, 3013-3015.
[45] Adler, F; Geiger, M; Bauknecht, A; Scholz, F; Schweizer, H; Pilkuhn, MH; Ohnesorge, B; Forchel, A. *J. Appl. Phys.*, 1996, 80, 4019-4026.
[46] Malik, S; Ru, ECL; Childs, D; Murray, R. *Phys. Rev. B*, 2001, 63, 155313.
[47] Jiang, H; Singh J. *J. Appl. Phys.*, 1999, 85, 7438-7442.
[48] Nee, TE; Wu, YF; Lin, RM. *J. Vac. Sci. Technol.*, 2005, 23, 954-958.
[49] Motlan, Goldys, EM. *Appl. Phys. Lett*, 2001, 79, 2976-2978.

In: Exciton Quasiparticles
Editor: Randy M. Bergin

ISBN: 978-1-61122-318-7
© 2011 Nova Science Publishers, Inc.

Chapter 7

EXCITON DIFFUSION LENGTH IN TITANYL PHTHALOCYANINE THIN FILMS AS DETERMINED BY THE SURFACE PHOTOVOLTAGE METHOD

Jiří Toušek,[1,] Jana Toušková,[1] Martin Drábik,[1] Zdeněk Remeš,[2] Jan Hanuš,[1] Věra Cimrová,[1,3] Danka Slavinská,[1] Hynek Biederman,[1] Adam Zachary[4] and Luke Hanley[4]*

[1] Charles University, Faculty of Mathematics and Physics, Department of Macromolecular Physics, V Holešovičkách 2, 18000 Prague 8, Czech Republic
[2] Institute of Physics of the Academy of Sciences of the Czech Republic, Cukrovarnická 10, 162 53 Prague, Czech Republic
[3] Institute of Macromolecular Chemistry, Academy of Sciences of the Czech Republic, Heyrovského nám. 2, 162 06 Prague 6, Czech Republic
[4] University of Illinois at Chicago, Department of Chemistry, Chicago, IL, USA

ABSTRACT

Exciton diffusion length was measured by the generalized surface photovoltage (SPV) method. The experiment needs no junction; it uses a spontaneously created space charge region (SCR) at the surface. The measurement is contactless and non-destructive. The SPV signal comes partly from excitons diffusing from neutral bulk towards the interface with the SCR, partly by excitons generated directly in the SCR. Both are separated in its electric field and subsequently generate the photovoltage. The SPV technique was modified to be applicable to arbitrary thickness of the layers and to samples with arbitrary thickness of the space charge region at the surface. Theoretical

---

[*] Corresponding author: E-mail: jiri.tousek@mff.cuni.cz Fax: (+420) 2-2191-2350.

calculations of the photocurrents from the SCR and from the bulk of the layers were carried out and illustrated to show how the different parameters influence the form and relative size of the photogenerated signal. Experimental spectra were compared with the theoretical ones allowing determination of the exciton diffusion length and the thickness of the SCR. We studied titanyl phthalocyanine (TiOPc) thin films prepared using evaporation and surface polymerization by ion-assisted deposition (SPIAD). Bilayer (TiO$_2$/TiOPc) thin films were also prepared, where the TiO$_2$ layer was sputtered from TiO$_2$ target. All films were characterized by the surface photovoltage method using absorption coefficients evaluated from measurement of the optical transmission and reflection. We found that the drift lengths of the charge carriers were shorter than the SCR thickness, which means recombination in this depletion region. Typically, the thickness of the SCR was higher than that of the bulk and the diffusion length of excitons was ~15 nm.

INTRODUCTION

Organic semiconductors have attracted enormous recent interest due to their potential use in organic light emitting diodes, solar cells, gas sensors, and field effect transistors. Oligomers, conjugated polymers and dye-sensitized wide band-gap semiconductors represent the structures that are researched with the goal of practical use. Preparation of organic materials is less expensive and consumes less energy compared with inorganic materials. Organic materials can also be safely disposed of after transformation into non-toxic components by thermal recycling.

Most of the dyes used as sensitizing substances are based on the phthalocyanines. Besides the metal free phthalocyanine H$_2$Pc there are organometallic complexes with a central metal ion of Cu, Pb, Ni, Mg, Co, Fe. Titanyl phthalocyanin (TiOPc) belongs to the group of oxometal phthalocyanines having a metal oxide in the complex center. Phthalocyanines (Pc) show chemical stability and strong absorbance in the visible region, making them well suited for use in photodetectors and solar energy conversion. Furthermore, the excellent luminescent properties of titanyl phthalocyanine enable construction of low cost power saving lasers. This is why this compound has suppressed classical selenium photoreceptors. The dye-sensitizing effect of TiOPc films on n-type titanium dioxide (TiO$_2$) is also well known [1], [2]. Under illumination the electronic structure of an n-TiO$_2$/TiOPc bilayer facilitates transport of the electrons excited in dye molecules from TiOPc to TiO$_2$, which is required for solid state dye-sensitized solar cells [2].

The diffusion length of photogenerated excitons is one of the most important parameters influencing the efficiency of organic solar cells. This parameter is connected with the mobility and lifetime of excitons. The diffusion length is usually determined by the method of photoluminescence quenching or by fitting of the theoretical external quantum efficiency of solar cells to experimental data [3]. On the other hand, the non-destructive surface photovoltage (SPV) method has been used for evaluation of minority charge carriers diffusion length in thick samples of inorganic materials [4]. We have modified this technique so that it would be applicable to arbitrary thickness of the layers and to samples with arbitrary thickness of the space charge region spontaneously formed at the surface. Recently, we have applied the SPV to determine of exciton diffusion length in organic layers of fluorenethiophene copolymers [5], and here we present this method used for exciton diffusion length evaluation in TiOPc layers and TiO$_2$/TiOPc bilayers.

PRINCIPLES OF THE SURFACE PHOTOVOLTAGE METHOD

Semiconducting materials often display a multitude of states at the surface. The majority carriers from the bulk supply the charge localized in the states, and a space charge region (SCR) is formed at the surface. In organic materials free excitons are created in the bulk and in the SCR by illumination. Our model considers the bulk/SCR interface as the place where the charge separation occurs. The minority electrons enter into the SCR while the majority holes remain in the bulk (usual case of a p-type material is considered). This represents one contribution to the surface photovoltage, with the other contribution coming from the carriers generated in the SCR where possible recombination is taken into account. The total surface photovoltage is a sum of both contributions.

The unit area of the illuminated surface is considered below. The total photogenerated current J is a sum of the diffusion bulk current J_b and the drift current J_{scr} from the SCR (Figure 1). The current J_b is independent of recombination in the SCR, as was shown in Ref.[6], meaning that the photons absorbed in the bulk can create a higher signal than those absorbed in the SCR.

In the case of the free surface illumination, the component from the neutral bulk is calculated by solving the usual diffusion equation for x>w

$$d^2\Delta p/dx^2 - \Delta p/L^2 = (-\alpha I/D)\exp(-\alpha[w+x]), \qquad (1)$$

where Δp is the concentration of photogenerated free excitons, L is their diffusion length, I is the intensity of radiation entering into the bulk with the absorption coefficient α. The boundary condition characterizing the sweep of charges at the bulk-SCR boundary is

$$\Delta p(w) = 0. \qquad (2)$$

An additional condition requires recombination to occur at the bottom contact:

$$D\, d\Delta p/dx|_{x=w+d} = -s\Delta p(w+d), \qquad (3)$$

where D is the exciton diffusion coefficient and s is the surface recombination velocity. Instead of s a dimensionless parameter $S = sL/D$ can be used. The diffusion current J_b from the bulk can be simply obtained as J

$$J_b = eD\, d\Delta p/dx|_{x=w}. \qquad (4)$$

In the case of recombination in the SCR the carriers photogenerated in the SCR contribute to the current J_{scr} by a "gain factor" $G(G\leq 1)$, which is the ratio of their drift length and the thickness of the SCR. Consequently, the current J_{scr} is proportional to the integral of the generation rate $g(x)$ over the SCR of thickness w multiplied by the "gain factor" $G(x)$.

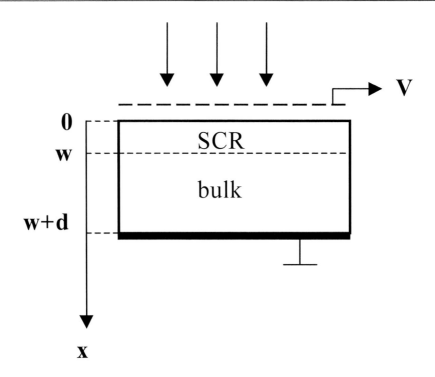

Figure 1. Schematic arrangement of the SPV measurement. The chopped light generates alternating voltage in the sample. Voltage V is induced on the transparent capacitively coupled electrode.

$$J_{scr} = e \int_0^w g(x)G(x)dx \qquad (5)$$

where $g(x) = \alpha I \exp(-\alpha x)$ and according to Shockley–Ramo theorem [7] $G(x) = (l_n(x) + l_p(x))/w = (\mu_n \tau_n E(x) + \mu_p \tau_p E(x))/w$ characterizes the contribution of each photogenerated carrier to the current J_{scr}. $E(x)$ is the electric field in the SCR while $l_{n,p}$, $\mu_{n,p}$, $\tau_{n,p}$ are drift length, mobility and carrier lifetime of electrons and holes, respectively. In fact, the gain factor $G(x)$ accounts for recombination in the SCR.

If the electric field is nonlinear function of coordinate the drift lengths are not proportional to the field. In our case, exponential dependence of l_n, l_p versus x agreed with experiment. The drift lengths at the surface are characterized by l_{n0}, l_{p0} values (Appendix) and together with the SCR thickness w determine the drift lengths at arbitrary point in the SCR. Final equations for the currents J_b and J_{scr} are shown in the Appendix. In [6] was demonstrated that the currents are independent each other and can be added, giving the total photocurrent. Experimentally verified linear relation between the photovoltage and the light intensity leads to proportionality between the photovoltage V and the photogenerated current $J : V \sim J$. Similar calculation can be made if illumination comes from the substrate side (Appendix).

EXPERIMENT

Preparation of the Layers

The TiOPc layers were prepared by the method of evaporation and surface polymerization by ion-assisted deposition (SPIAD). SPIAD films were produced via deposition of TiOPc monomers simultaneously with hyperthermal acetylene ions. Fifty and 100 eV acetylene ions at a 1:1 ion to neutral ratio were used. An organic doser was employed for thermal evaporation of TiOPc. The temperature of the doser for evaporation was about 600 K, which gave a total fluence of ~6.8 x 10^{15} neutrals cm^{-2}. Details about the preparation are described in our former paper [8]. Titania (TiO_2) thin films were prepared by DC magnetron sputtering of a ceramic TiO_2 target. A method using a mixture of argon and oxygen as a working gas (17% of O_2) was chosen for the thin films preparation based on previous experience. Working gas pressure and its total flow rate were held constant at 5 Pa and 6cm^3_{STP}/min, respectively. The magnetron with TiO_2 target was powered from an Advanced Energy 1.5 DC power supply. The DC power supply was operated in a "constant current" mode at 0.2 A, which corresponds to 100 W power. The TiO_2 layers were amorphous with small amount of anatase phase.

The Transmittance and Reflectance Spectra Measurements Optimized for Thin Films

The characterization of the optical losses in the thin films is usually complicated by the presence of interference fringes. A dual beam spectrometer optimized for the precise measurement of the optical properties of thin films, including multilayers, has been developed [9]. Prior the measurements of the transmittance and reflectance spectra the spectrum of the incident light intensity (the baseline) is measured without the sample as the ratio of the sum of the photocurrents from the photodiodes mounted to the integrating sphere and the photocurrent from the auxiliary photodiode. The transmittance spectrum normalized on the baseline is calculated from the ratio of the sum of the photocurrents from the photodiodes mounted to the integrating sphere placed behind the sample and the photocurrent from the auxiliary photodiode. The reflectance is measured with the integrating sphere placed in front of the sample. The sample is slightly tilted towards the incident light and it is not moved during the transmittance and reflectance measurements. Thus, the interference fringes in transmittance (T) and reflectance (R) spectra show the same spectral position of the corresponding minima and maxima and the spectral function 100%-T-R gives the optical absorptance for the smooth, non-scattering thin films. The index of refraction, absorption coefficient and the film thickness are calculated from the T and R spectra using the commercial software FilmWizard. [10].

Absorption spectra of TiOPc evaporated and TiOPc SPIAD samples are shown in Figure 2. The TiO_2, TiOPc layers and the TiO_2/TiOPc bilayers were deposited on quartz substrates. No absorption of the TiO_2 and quartz was found for the wavelengths above 350 and 250 nm, respectively.

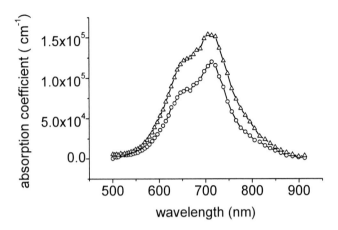

Figure 2. Optical absorption spectra of evaporated TiOPc (triangles) and SPIAD TiOPc (circles) thin films.

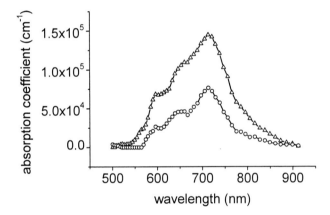

Figure 3. Optical absorption spectra of evaporated TiO_2/TiOPc (triangles) and SPIAD TiO_2/TiOPc (circles) thin films.

The Q-band absorption feature in phthalocyanines appearing at 500 – 800 nm derives from excitation across the HOMO - LUMO junction [11-15]. Our curves show this band with indication of a splitting. The main maximum is at 715 nm for both the evaporated and SPIAD samples.

Figure 3 shows absorption spectra of TiO_2/TiOPc evaporated and TiO_2/TiOPc SPIAD samples.

The shape of the spectra is similar as in Figure 2 but new peaks appear at 600 nm.

Surface Photovoltage Spectra

The surface photovoltage spectra were measured in the sample configuration shown in Figure 4.

The alternating voltage was generated in TiOPc or TiO$_2$/TiOPc layers by low-intensity monochromatic light chopped at a low frequency of 11 Hz. The form of the SPV pulses can be affected by several relaxation processes in the layers [16]. The used frequency was sufficiently low to obtain saturated pulses not influenced by relaxations. The voltage was measured between a transparent conductive electrode capacitively coupled to the sample and an ITO back electrode. The SPV spectra were taken at constant impinging photon flux density and subsequently corrected for the transparency of the glass covered with ITO and of the Mylar sheet. All measurements were carried out at room temperature.

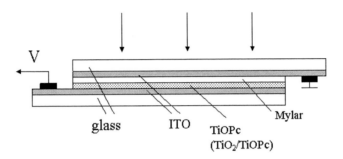

Figure 4. Experimental set up for SPV measurement of TiOPc and TiO$_2$/TiOPc samples.

RESULTS

The forms of absorption coefficients measured on the evaporated and the SPIAD samples were close to each other (see Figure 2). We have used the SPIAD TiOPc absorption coefficients for calculation of theoretical spectral dependences of the photovoltage. Figures 5 – 7 display the theoretical spectra of the layers illuminated from the free surface side as in Figure 1. The SPV spectrum under illumination from the substrate side is shown in Figure 8. All these spectra are normalized to their value in the maximum.

The diffusion length considerably influences the SPV spectrum if illumination comes from the free surface side and if contribution from the SCR is small. Such situation is favorable for extraction of the diffusion length.

Figure 5 demonstrates change in the SPV spectra with the diffusion length in the case of high losses in the SCR. Curve *a* shows that strongly absorbed photons with penetration depth shorter than the SCR thickness generate a signal only in the SCR. But, there is a considerable recombination in this region, which lowers the generated current. The minimum corresponds to absorption of the photons with energy near to the absorption maximum (Figure 2) and is consequence of losses in the SCR. A higher signal is generated by the photons penetrating as far as the bulk as the diffusion length is quite long. The two peaks in the curve *a* are connected with the current created by these weakly absorbed photons. Figure 5, curve *b* depicts the comparable contributions from the bulk and from the SCR. In this case the diffusion length is short.

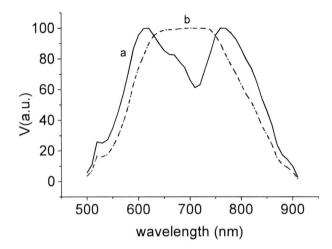

Figure 5. Normalized theoretical SPV spectra of TiOPc layer illuminated from the free surface side for $d = 50$ nm, $w = 200$ nm, $l_n = l_p = 1$ nm. Curve a: $L = 40$ nm, curve b: $L = 2$ nm. Curve b is multiplied by 6.1.

Further SPV spectra show the cases of coincidence with the course of absorption, which is not suitable for the diffusion length evaluation. The SPV spectrum follows the absorption spectrum when the prevailing contribution from only one part of the device (either SCR or bulk).

This is the case in Figure 6, curve c where the main signal comes from the SCR.

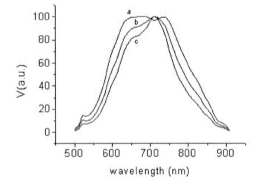

Curve a: $l_n = l_p = 1$ nm
Curve b: $l_n = l_p = 10$ nm
Curve c: $l_n = l_p = 100$ nm
Curves a and b are multiplied by 9.4, and 4.7, respectively.

Figure 6. Normalized theoretical SPV spectra of TiOPc layer illuminated from the free surface side for $d = 100$ nm, $w = 100$ nm, $L = 20$ nm.

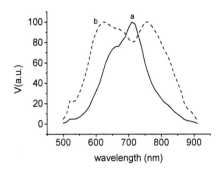

Curve a: w = 10 nm,
Curve b: w = 200 nm.
Curve b is multiplied by 5.

Figure 7. Normalized theoretical SPV spectra of TiOPc layer illuminated from the free surface side for d = 10 nm, L = 20 nm, $l_n = l_p$ = 1 nm.

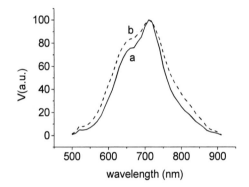

Curve a: L = 20 nm,
Curve b: L = 1 nm.
Curve b is multiplied by 13.8.

Figure 8. Theoretical SPV spectra of TiOPc layer illuminated from the substrate side for d = 20 nm, w = 200 nm, $l_n = l_p$ = 1 nm.

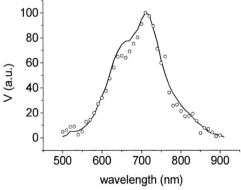

Points: experiment, full line: theory for d = 5 nm, w = 20 nm, $l_n = l_p$ = 1 nm.

Figure 9. Normalized SPV spectrum of evaporated TiOPc layer illuminated from the free surface side.

The second case with prevailing signal from the bulk part is given in Figure 7, curve *a*.

However, the SPV spectrum is qualitatively different if absorption in the front part (SCR) cannot be neglected, as shown in Figure 7, curve *b*. Absorption in the thick SCR leads to a weak signal from the bulk that no more prevails.

Illumination from the substrate side generates the SPV spectra usually weakly influenced by the diffusion length. Theoretical SPV spectra of samples illuminated from this side are shown in Figure 8. Curve *a* follows the absorption spectrum, since signal from the bulk dominates being proportional to the photogenerated charge. A short diffusion length in the bulk and, consequently relative increase of contribution of the SCR influence the spectrum in the way as shown by curve *b*. Comparing the front (Figure 5) and substrate side illumination we see that in the latter case the decrease of the bulk contribution lowers the SPV signal but only weakly changes the shape of the spectra.

Evaluation of the parameters influencing experimental spectra was obtained by fitting the SPV theoretical curves to experimental points. Figures 9, 10 and 12 display SPV experimental spectral dependence of the samples illuminated from the free surface side while Figure 11 is the SPV spectrum measured under illumination from the substrate side. Our samples showed a low value of the surface recombination velocity (S ÷ 1), which practically did not influence the form of the spectra.

The SPV spectrum in Figure 9 follows the absorption coefficient spectral dependence. This is typical for very thin samples and it corresponds to prevailing contribution from the SCR. In such a case, the diffusion length cannot be exactly determined since the thickness of the bulk limits its value (L > 5 nm).

Figure 10 shows the SPV spectrum influenced not only by absorption but also by diffusion processes in the bulk and recombination in the SCR.

Figure 11 shows coincidence of the form of the SPV spectrum with that of absorption (Figure 2). A sufficiently long diffusion length and large losses in the SCR cause the signal from the bulk to dominate the spectrum. This configuration leads to a higher signal compared with illumination from the free surface side.

Diffusion processes in the bulk and contribution from the SCR result in broadening of the SPV spectrum in Figure 12. A peak at 600 nm is distinguishable both in theoretical and experimental curves. The $TiO_2/TiOPc$ bilayer, illuminated from the substrate side, shows the SPV spectrum (not shown) that follows absorption coefficients in Figure 3.

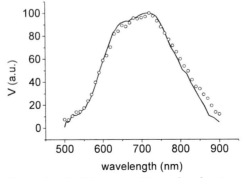

Points: experiment, full line: theory for $d = 25$ nm, $w = 48$ nm, $l_n = l_p = 1$ nm, $L = 14$ nm.

Figure 10. Normalized SPV spectrum of SPIAD TiOPc layer illuminated from the free surface side.

Exciton Diffusion Length in Titanyl Phthalocyanine Thin Films... 285

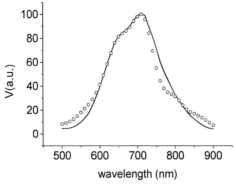

Points: experiment, full line: theory for $d = 25$ nm, $w = 180$ nm, $l_n = l_p = 1$ nm, $L = 15$ nm.

Figure 11. Normalized SPV spectrum of evaporated TiOPc layer illuminated from the substrate side.

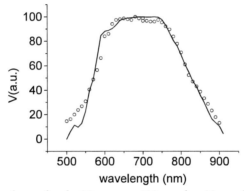

Points: experiment, full line: theory for $d = 20$ nm, $w = 120$ nm, $l_n = 10$ nm, $l_p = 1$ nm, $L = 16$ nm.

Figure 12. Normalized SPV spectrum of evaporated TiO$_2$/TiOPc bilayer illuminated from the free surface side.

DISCUSSION

The different preparation methods of the titanyl phthalocyanine layers lead to higher values of the absorption coefficients for the evaporated samples as compared with those for the SPIAD samples. Prior work found that the SPIAD layers showed the formation of covalently bound dimers whereas monomers were present in the films prepared by thermal evaporation [8]. Both freshly evaporated and SPIAD layers were amorphous. Other work found that amorphous films display a non-planar molecular structure leading to deformation of the TiOPc molecules by intermolecular interaction that results in two optical absorption peaks in the Q-band [12]. Freshly prepared samples displayed a Q-band maximum in the SPIAD films was blue-shifted ~ 20 nm to 714 nm compared to the evaporated films [8]. However, here the absorption spectra with the peak at 715 nm and a shoulder at ~ 660 nm (Figures 2 and 3) of both the evaporated and SPIAD samples do not show any shift. The present measurements were performed 20 months after the sample preparation, during which time, the Q-band maximum of the evaporated samples successively approached that of the

SPIAD samples. The aging-induced change in the absorption spectra of the evaporated layers may have occurred by diffusion of H_2O molecules into the bulk of TiOPc films [17], leading to chemical interaction and modification of the electronic structure and HOMO energy of TiOPc. Even a change in crystal structure can occur by diffusion of water molecules [17], although the effect of oxygen on TiOPc layers is also well known [18], [19]. In any case, it appears that the SPIAD method of TiOPc deposition leads to a more stable dimer structure which does not change in time whereas the films created by simple evaporation technique change their properties during the aging process. As a possible explanation of the additional peak at 600 nm in the absorption spectra of TiO_2/TiOPc bilayers, we consider that during the TiO_2 deposition, oxygen present in the working gas may incorporate into these layers and then penetrate in TiOPc. Recently, it was pointed out that also amorphous TiO_2 films show a photocatalytic activity [20]. We cannot exclude another possibility, namely that water vapor or O_2 from air could penetrate into TiO_2 through the overlaying porous TiOPc layer. Photocatalytic reaction could lead to strongly active O_2^- that could diffuse into TiOPc and as a result, influencing the spectra. However, discussion of the mechanism of oxygen interaction with TiOPc is beyond the scope of this paper as it would require further investigation.

As shown in Figures 10-12 the exciton diffusion lengths were found to be independent of both TiOPc preparation method and the presence of a TiO_2 layer. We can conclude that TiO_2 layer did not influence the photoelectric properties of the phthalocyanine films.

Values of the diffusion lengths can be reliably determined if the thickness of the bulk is not too small as compared with the diffusion length. The simple model applied here assumes existence of one neutral bulk at the substrate and one space charge region created at the free surface. As it was shown in Ref. [6], photocurrents generated in the both regions are fully independent and can be added. As a consequence, beside the diffusion length, the thickness of the SCR was evaluated by fitting the theory to the SPV experimental data. Thicker SCRs of evaporated samples point out their higher resistivity compared to SPIAD samples. A much thinner bulk region compared to the SCR was found in all cases.

A lower SPV signal is generated when illuminating the free surface side compared with illumination from the substrate side of the samples. This difference indicates that the lifetime of the carriers generated in the SCR is shorter than that from the bulk. This agrees with the short drift lengths evaluated from the SPV spectra.

It is also shown (Figure 9) that the shapes of the SPV spectra of very thin samples track the optical absorption spectra. The same holds for the substrate side illumination if the bulk is very thin and the contribution from the SCR is low (Figure 11).

CONCLUSION

The phthalocyanine layers prepared by evaporation as well as by SPIAD show a space charge region at the free surface allowing measurement of the surface photovoltage. Our model of the surface photovoltage effect enables evaluation of the photovoltage spectra and yields thickness of the space charge region and exciton diffusion length. The measured and calculated SPV spectra are in good agreement, which justifies the model used here.

In the majority of samples, the ~15 nm diffusion length was to be found independent of the film preparation method. Optical absorption spectra of evaporated TiOPc and SPIAD

TiOPc thin films differed only by the magnitude of their absorption coefficients. While the absorption spectra of SPIAD samples did not change in time, the maximum of the absorption spectra of the evaporated TiOPc films successively shifted towards the lower wavelengths due to an aging process. The positions of the Q-bands in both spectra correspond to HOMO-LUMO transitions in the SPIAD samples. The same holds for the optical absorption spectra of evaporated TiO_2/TiOPc and SPIAD TiO_2/TiOPc bilayers. The photovoltage spectra match well the solar spectrum, which suggests a possibility to use the titania/ titanyl phthalocyanine combination in photovoltaic solar cells.

APPENDIX

Calculations for the case of p-type organic layer are performed.

The illumination from the free surface side (Figure 1) generates the current $J = J_{b1} + J_{s1}$ where J_{b1} is the bulk current, ($S = sL/D$).

$$J_{b1} = \frac{e l \alpha L}{1-\alpha^2 L^2} \frac{\exp(-\alpha w)}{\{(1-S)\exp(-d/L)+(1+S)\exp(d/L)\}} [(1+S)\exp(d/L)-(1-S)\exp(-d/L)+ \quad (A1)$$
$$+ 2\exp(-\alpha d)(\alpha L - S)] - el\exp(-\alpha w)\frac{\alpha^2 L^2}{1-\alpha^2 L^2}$$

Assuming exponential dependence of the drift lengths versus coordinate x in the form $l_{n,p} = l_{n0,p0} [\exp[-\gamma(x-w)]-1]/[\exp(\gamma w)-1]$ as in [21] then the current from the SCR is

$$J_{s1\,exp} = \frac{e\alpha l}{w} \frac{l_{p0}}{\exp(\gamma w)-1} \int_0^w [\exp(-\gamma(x-w))-1]\exp(-\alpha x)dx +$$
$$\frac{e\alpha l}{w} \frac{l_{n0}}{\exp(\gamma w)-1} \int_0^{l_{n1}} [\exp(-\gamma(x-w))-1]\exp(-\alpha x)dx +$$
$$\frac{e\alpha l}{w} \int_{l_{n1}}^w x\exp(-\alpha x)dx = \frac{el}{w}\left\{\frac{1}{\alpha}[1-\exp(-\alpha l_{n1})] - l_{n1}\exp(-\alpha l_{n1})\right\} \quad (A2)$$
$$+ \frac{el}{w}\frac{l_{n0}}{\exp(\gamma w)-1}\exp(\gamma w)\frac{\alpha}{\gamma+\alpha}\left\{\exp[-(\gamma+\alpha)l_{n1}]-\exp[-(\gamma+\alpha)w]\right\}$$
$$+ \frac{el}{w}\frac{l_{n0}}{\exp(\gamma w)-1}[\exp(-\alpha w)-\exp(-\alpha l_{n1})]$$
$$+ \frac{el}{w}\frac{\alpha}{\alpha+\gamma}\frac{l_{p0}}{\exp(\gamma w)-1}[\exp(\gamma w)-\exp(-\alpha w)]$$
$$- \frac{el}{w}\frac{l_{p0}}{\exp(\gamma w)-1}[1-\exp(-\alpha w)]$$

where γ is a parameter; l_{no}, l_{po} are the drift lengths corresponding to the maximum electrical field in the SCR and l_{n1} is the value of l_n equal to the distance from the surface. For $x < l_{n1}$ it holds $l_n = x$.

If the illumination comes through a transparent contact into the bulk side of the sample the generated current is $J = J_{b2} + J_{s2}$ where the bulk current is

$$J_{b2} = \frac{e I \alpha L}{1-\alpha^2 L^2} \frac{1}{\{(1-S)\exp(-d/L)+(1+S)\exp(d/L)\}} \{2(\alpha L + S) + \\ + \exp(-\alpha d)[(1-S)\exp(-d/L)-(1+S)\exp(d/L)]\} - eI\exp(-\alpha d)\frac{\alpha^2 L^2}{1-\alpha^2 L^2} \quad (A3)$$

Putting origin of coordinates at the bulk/SCR boundary the drift lengths can be written as the following exponential function of coordinate

$l_{n,p} = l_{n0,p0} (\exp[\gamma x] - 1)/ (\exp[\gamma w] - 1)$ and the current from the SCR is

$$J_{s2\,exp} = \frac{e\alpha I \exp(-\alpha d)}{w} \frac{1}{\exp(\gamma w)-1} \\ \left\{ l_{p0} \int_0^w [\exp(\gamma x)-1]\exp(-\alpha x) dx + l_{n0} \int_0^{l_{n1}} [\exp(\gamma x)-1]\exp(-\alpha x) dx \right\} \\ + \frac{e\alpha I \exp(-\alpha d)}{w} \int_{l_{n1}}^w (w-x)\exp(-\alpha x) dx = \\ = \frac{eI\exp(-\alpha d)}{w} \frac{1}{\exp(\gamma w)-1} \left\{ l_{p0}[\frac{\alpha}{\gamma-\alpha}(\exp(\gamma-\alpha)w)-1) + (\exp(-\alpha w)-1)] \right\} \\ + \frac{eI\exp(-\alpha d)}{w} \frac{1}{\exp(\gamma w)-1} \left\{ l_{n0}[\frac{\alpha}{\gamma-\alpha}(\exp(\gamma-\alpha)l_{n1})-1) + (\exp(-\alpha l_{n1})-1)] \right\} \\ + \frac{eI\exp(-\alpha d)}{w} \left\{ w\exp(-\alpha l_{n1}) - l_{n1}\exp(-\alpha l_{n1}) + \frac{1}{\alpha}[\exp(-\alpha w)-\exp(-\alpha l_{n1})] \right\} \quad (A4)$$

ACKNOWLEDGMENT

This work is a part of the research plan MSM 0021620834 that is financed by the Ministry of Education, Youth and Sports of the Czech Republic. We also acknowledge the support of the Grant Agency of the Czech Republic (Grant No. 202/09/1206). The U.S. National Science Foundation also provided support of this work through grant CHE-0241425.

REFERENCES

[1] Yanagi, H; Chen, S; Lee, PA; Nebesny, KW; Armstrong, NR; Fujishima, A. *J. Phys. Chem.*, 1996, 100, 5447- 5451.
[2] Günes, S. *PhD-Thesis*, Johannes Kepler Universitat, *Linz*, 2006.
[3] Rim, SB; Fink, RF; Schöneboom, JC; Erk, P; Peumans, P. *Appl. Phys. Lett*, 2007, 91, 173504- 173504-3 .
[4] Kronik, L; Shapira, Y. *Surf. Sci. Reports*, 1999, 37, 1 -206.
[5] Tousek, J; Touskova, J; Krivka, I; Pavlackova, P; Vyprachticky, D; Cimrova, V. *Org.*

Electronics, 2010, 11, 50-56.
[6] Tousek, J; Touskova, J. *Sol.Energy Mater. Solar Cells*, 2008, 92, 1020- 1024.
[7] He, Z. *Nucl. Instrum. Meth*, 2001, A463, 250-267.
[8] Zachary, AM; Drabik, M; Choi, Y; Bolotin, IL; Biederman, H; Hanley, L. *J. Vac. Sci. Technol*, 2008, A26, 212-218.
[9] Remeš, Z. PhD-Thesis, Charles University Prague, F*aculty of Mathematics and Physics*, 1999.
[10] Film Wizard Professional, *Scientific Computing Int*. USA, http:\\sci-soft.com.
[11] Zhang, O; Wang, D; Xu, J; Cao, J; Sun, J; Wang, M. *Mater. Chem. Phys.*, 2003, 82, 525-528.
[12] Ma, G; Guo, L; Mi, J; Liu, Y; Qian, S; Pan, D; Huahg, Y. *Thin Solid Films*, 2002, 410, 205-211.
[13] Kumar, GA; Thomas, J; Unnikrishnan, NV; Nampoori, VPN; Vallabhan, CPG. M*ater. Res. Bull*, 2001, 36, issues 1-2, 1-8.
[14] Lüer, L; Egelhaaf, HJ; Oelkrug, D; Winter, G; Hanack, M; Weber, A; Bertagnolli, H. *Synth. Metals*, 2003, 138, 305-310.
[15] Mizuguchi, J; Rihs, G; Karfunkel, HR. *J. Phys. Chem.*, 1995, 99, 16217-16227.
[16] Zidon, Y; Shapira, Y; Ditrich, D. *Appl. Phys. Lett*, 2007, 90, 142103-1-142103-3.
[17] Honda, M; Kanai, K; Komatsu, K; Ouchi, Y; Ishii, H; Seki, K. *J. Appl. Phys.*, 2007, 102, 103704- 1-103704-10.
[18] Nishi, T; Kanai K; Ouchi, Y; Willis, MR; Seki, K. *Chem. Phys.*, 2006, 325, 121-128.
[19] Morishige, K; Tomoyasu, S; Iwano, G. *Lang*, 1997, 13, 5184-5188.
[20] Eufinger, K; Poelman, D; Poelman, H; De Gryse, R; Marin, GB. *Appl. Surf. Sci.*, 2007, 254 148-152.
[21] Tousek, J. *phys. stat. sol*, (a) 1991, 128, 531-538.

In: Exciton Quasiparticles
Editor: Randy M. Bergin

ISBN: 978-1-61122-318-7
© 2011 Nova Science Publishers, Inc.

Chapter 8

ACCURACY OF THE COHERENT POTENTIAL APPROXIMATION FOR FRENKEL EXCITONS IN ONE-DIMENSIONAL ARRAYS WITH GAUSSIAN DIAGONAL DISORDER AND NEAREST-NEIGHBOR TRANSFER

I. Avgin[1] and D. L. Huber[2]

[1]Department of Electronics and Electrical Engineering, Ege University,
Bornova 35100, Izmir, Turkey
[2]Physics Department, University of Wisconsin-Madison, Madison, WI, USA

ABSTRACT

This chapter is a report of the results of an assessment of the accuracy of the coherent potential approximation (CPA) when applied to a one-dimensional Frenkel exciton array with Gaussian diagonal disorder and nearest-neighbor transfer. The integrated density of states, the density of states, the inverse localization length, and the optical absorption are compared with data obtained using mode-counting techniques (integrated density of states, density of states, and inverse localization length) and matrix diagonalization (optical absorption) applied to large arrays. The CPA is in excellent agreement with the numerical data thus providing a rare example of a theory that yields accurate results for a realistic model of a disordered system.

INTRODUCTION

The Frenkel exciton model provides a useful characterization of the optical properties of materials consisting of an array of weakly coupled, optically active centers or chromophores. In describing the transfer of excitation between chromophores, there are two limiting models, corresponding to coherent and incoherent transfer. In the incoherent limit, the localized excitation randomly hops from site to site in a thermally activated process. In the coherent limit, the relevant excited states are Frenkel excitons that, under ideal conditions, propagate as coherent waves. In this chapter, we focus on the latter. Our interest is in the effect of static disorder on the excitons in one-dimensional arrays. It is important to note that the analysis is limited to one dimensional systems with nearest-neighbor coupling. In the Frenkel exciton picture, this assumption is appropriate for systems where the coupling between sites arises from wave function overlap (*e.g.* exchange or tunneling)) rather than long-range, multipole interactions. Recently, significant progress has been made in understanding the effect of the disorder on the spatial extent and distribution of the exciton states in disordered, one-dimensional arrays with nearest-neighbor interactions [1]. Numerical techniques based on mode-counting algorithms, can be used to determine both the distribution and spatial extent of the exciton modes in nearest-neighbor systems to a very high level of accuracy. Complementing the numerical work, a theory based on the coherent potential approximation has been shown to give results that are in excellent agreement with the mode-counting findings for the physically relevant case of Gaussian disorder in the single chromophore transition energy. In this chapter we discus and extend the results reported in Ref. 1.

Although the focus of this chapter is on 'macroscopically disordered' arrays, we begin by reviewing earlier results associated with a single impurity in an otherwise ideal array. The Hamiltonian of the one-dimensional Frenkel exciton array in the site representation takes the form

$$H = \sum_n V_n |n\rangle\langle n| + \sum_n t_{n,n+1}(|n\rangle\langle n+1| + |n+1\rangle\langle n|) \qquad (1)$$

In this equation, V_n denotes the energy difference between the ground and excited states of the chromophore on site n and $t_{n,n+1}$ is the transfer integral between nearest-neighbor sites n and $n+1$. The state vector $|n\rangle$ characterizes a state where the chromophore on site n is excited and all other chromophores are in their ground state. In analyzing the effects of a single impurity, we take the Hamiltonian of the ideal array to be

$$H_0 = -\sum_n (|n\rangle\langle n+1| + |n+1\rangle\langle n|) \qquad (2)$$

so that with periodic boundary conditions, the exciton energies are equal to $-2\cos(k)$, with $-\pi \le k \le \pi$. The impurity chromophore, located at site 0, is assumed to have the same transfer integrals as the host array, but the energy difference between the ground and excited states is shifted by an amount 2ε. The complete Hamiltonian has the form

$$H = 2\varepsilon |0\rangle\langle 0| - \sum_n (|n\rangle\langle n+1| + |n+1\rangle\langle n|) \qquad (3)$$

As pointed out by Koster and Slater [2], the diagonal perturbation leads to a state localized at the impurity with energy

$$E = \pm 2(1+\varepsilon^2)^{1/2} \qquad (4)$$

where the + and − signs refer to the cases $\varepsilon > 0$ and $\varepsilon < 0$, respectively. The amplitude of the localized state, a_n, falls off exponentially relative to the site of the impurity according to the equations [3, 4]

$$a_n = C(-1)^n \exp[|n| \ln\{E/2)[1-(1-4/E^2)^{1/2}]\}] \quad \varepsilon > 0 \qquad (5a)$$

$$a_n = C \exp[|n| \ln\{|E|/2)[1-(1-4/E^2)^{1/2}]\}] \quad \varepsilon < 0 \qquad (5b)$$

where C is a normalization constant and E denotes the localized state energy. As pointed out in Ref. 4, the oscillating behavior for localized states above the band edge reflects the fact that they are largely made up of a linear combination of unperturbed states with $k \approx \pi$, whereas the states below the band edge are largely made up from unperturbed states with $k \approx 0$. The rate of decay of the amplitude in units of the reciprocal of the lattice constant is set by the negative of the factor multiplying the factor $|n|$ appearing in the exponential:

$$ILL(E) = -\ln\{|E|/2)[1-(1-4/E^2)^{1/2}]\} \qquad (6)$$

which, for future reference, we have defined as the inverse localization length or $ILL(E)$. We note the limiting behavior: for $|E| \gg 2$, $ILL(E) \sim \ln(|E|)$, so the rate of decay of the amplitude slowly increases with $|E|$, whereas for $|E|$ close to the band edge, we have $ILL(E) \sim (|E|-2)^{1/2}$, indicating that localized states near the band edges have slowly decaying amplitudes.

Although localized states above the upper band edge and below the lower band edge show similar exponential decay in their amplitudes, their optical properties are quite different. To understand this difference, we investigate the $EOS(E)$ or effective oscillator strength [4] which is related to the square of the sum of the amplitudes, divided by a normalization factor:

$$EOS(E) = |\sum_n a_n|^2 / \sum_n |a_n|^2 \qquad (7)$$

The effective oscillator strength is a dimensionless transition dipole moment associated with the optical absorption or emission by the localized state. It is based on the assumption that the local transition dipole moment associated with the perturbed site is the same as the local transition dipole moment of the unperturbed sites. A direct evaluation of the sums in Eq. (7) leads to the result

$$EOS(E) = (1+x)^3 / [(1-x)(1+x^2)] \qquad (8)$$

where

$$x = (E/2)[(1-4/E^2)^{1/2} - 1] \qquad (9)$$

For $|E| \gg 1$, we have $x \to 0$, and the effective oscillator strengths for states above and below the exciton band both approach unity. This behavior is a consequence of the fact the states are increasingly localized at the impurity site which has the same optical properties as the unperturbed sites. For states below the band, when $E \to -2$, we have $x \to 1$, and the *EOS* diverges as $4/[-E-2]^{1/2}$, whereas for states above the band, when $E \to 2$, we have $x \to -1$, and the *EOS* approaches zero as $4^{-1}(E-2)^{3/2}$. It should be noted that the enhanced optical activity of localized states with energies near the $k = 0$ band edge was originally predicted for Mott-Wannier excitons by Rashba and Gurgenishvili [5] and subsequently observed in CdS by Henry and Nassau [6].

The analysis of the localized states associated with a single impurity in an otherwise perfect array was greatly simplified by the two assumptions: nearest-neighbor transfer and the disorder limited to a shift in the energy splitting, with the transfer integrals unchanged. Studies of localized states in systems with nearest-neighbor transfer and arbitrary values for the energy splitting and the impurity-host transfer integrals were reported in Ref. 4. Localized states in systems with infinite-range transfer where the perturbation only involved a shift in the energy splitting were investigated in Ref 7.

MACROSCOPIC DISORDER AND THE COHERENT POTENTIAL APPROXIMATION

The focus of the preceding section was on the localized states associated with a single impurity. In this section, we consider the consequences of 'macroscopic' diagonal disorder associated with a random distribution of the single chromophore transition energies. We again take the transfer integrals to have the value -1 so that the Hamiltonian becomes

$$H = \sum_n V_n |n><n| - \sum_n (|n><n+1| + |n+1><n|) \qquad (10)$$

We assume the V_n have a Gaussian distribution characterized by zero mean and variance σ^2 with no correlation between different sites. Recent studies have shown that the coherent potential approximation or CPA provides an unusually accurate description of the distribution and localization of the exciton modes in the system characterized by Eq. (10) [1]. In this section, we will present and extend the results reported in that reference.

Since its introduction more than forty years ago [8,9], the coherent potential approximation has proven to be a successful method of determining the distribution and properties of states in disordered materials. Due to its success, efforts have been made to determine if there are models for which the CPA gives exact results. Up to this point, the CPA has proven to be exact only for the Lloyd model where there is a Lorentzian distribution of the fluctuations in the diagonal terms in the Hamiltonian [10,11]. Early applications of the CPA to one-dimensional arrays having a Gaussian distribution of the diagonal fluctuations suggest that the approximation is unusually effective for this model as well [12,13]. In our studies, we found excellent agreement between the CPA predictions for the density of states and the inverse localization length and the corresponding results obtained by mode-counting in arrays of $10^7 - 10^8$ sites. As a by-product of this study, we investigate the asymptotic behavior of the inverse localization length and show that it is the characteristic of the decay rate of a wave function localized at a strongly perturbed site in an otherwise perfect array.

Although the focus of this paper is on Frenkel excitons, the results of the analysis are also applicable to the tight binding model for one-electron states in one dimensional arrays.

In the coherent potential approximation, as applied to the system characterized by Eq.(10), the Green's function, $G_0^{CPA}(E)$ is expressed as

$$G_0^{CPA}(E) = \pi^{-1} \int_0^\pi dk [E + 2\cos(k) - V_c(E)]^{-1} \quad (11)$$

where the coherent potential, $V_c(E)$, satisfies the equation

$$\int dV P(V)[V - V_c(E)]/[1 - (V - V_c(E))G_0^{CPA}(E)] = 0 \quad (12)$$

with $P(V)$ having the Gaussian form.

The main focus of this paper is on assessing the accuracy of the coherent potential approximation for the exciton Green's function by comparing the integral of $G_0^{CPA}(E)$ with accurate numerical results obtained from large arrays. A straightforward way to do this would be to diagonalize the Hamiltonian and calculate the Green's function directly from the eigenvalues. This approach is limited, however, by number of sites in the array (typically, $N \sim 10^3 - 10^4$). In one dimensional arrays with nearest-neighbor interactions, one can make use of mode-counting techniques to establish accuracy of the CPA in far larger arrays. It should be noted that our use of the mode-counting approach involves the assumption that the results obtained from mode-counting with a very large but finite array are essentially equivalent to taking the $N \to \infty$ limit, a property referred to as 'self-averaging'.

Information about the real and imaginary parts of the Green's function can be obtained from an analysis of the tridiagonal Hamiltonian matrix [14] and, in particular, the amplitude ratios, $R_n = a_n/a_{n-1}$, associated with the eigenvectors of the Hamiltonian [15],

$$R_n(E) = E - V_n - 1/R_{n-1}(E) \quad (13)$$

With N being the number of sites, the number of sign changes in the sequence $R_1(E)$ (= $E - V_1$), $R_2(E)$, ..., $R_N(E)$ corresponds to the number of modes with energies less than E. When $N \gg 1$, which is the case here, we identify the number of sign changes, divided by N, with the integrated density of states per site, $IDOS(E)$. The integrated density of states can also be obtained from the imaginary part of the Green's function using the equation

$$IDOS(E) - IDOS(0) = \pi^{-1} \int_0^E dE' \operatorname{Im} G_0(E') \quad (14)$$

Information about the spatial extent of the wave functions follows from a consideration of the logarithm of $|R_n(E)|$ [15,16]. The inverse localization length, $ILL(E)$, which is identified with the reciprocal of the average fall-off distance for eigenstates with energy E, is expressed as [15].

$$ILL(E) = -N^{-1} <\ln|a_N/a_1|>_E$$

$$= -N^{-1}\ln\left|\prod_{n=1}^{N} R_n(E)\right| = -N^{-1}\sum_{n=1}^{N}\ln|R_n(E)| \quad N \to \infty \quad (15)$$

The equivalent expression for $ILL(E)$ involving the Green's function utilizes the real part of $G_0(E)$. It can be derived by integrating the real part of the spectral representation for $G_0(E)$ and making use of the expression for the inverse localization length derived in Ref. 15

$$ILL(E) = \int dx\, DOS(x)\ln|E-x| \quad (16)$$

in which $DOS(x)$ denotes the density of states. The equation takes the form

$$ILL(E) - ILL(0) = \int_0^E dE'\, \text{Re}\, G_0(E') \quad (17)$$

In assessing the accuracy of the CPA, we focus on $IDOS(E) - IDOS(0)$ and $ILL(E) - ILL(0)$ whose derivatives with respect to E yield the real and imaginary parts of $G_0(E)$. In Figures 1 and 2, we compare the mode-counting and CPA results for the integrals of the imaginary (Figure 1 and Eq. (14)) and real (Figure 2 and Eq. (17)) parts of the Green's functions calculated with $\sigma^2 = 0.25$, 1.0, and 4.0. The numerical results for the $IDOS$ involved counting the number of sign changes in the sequence, $R_1, R_2,\ldots, R_N$, divided by N, whereas the values for the ILL were obtained from Eq. (15). In both cases, $N = 4\times10^7$. The values of $IDOS(0)$ and $ILL(0)$ were 0.50002 and 0.02821 for $\sigma^2 = 0.25$, 0.49999 and 0.10880 for $\sigma^2 = 1.0$, and 0.50012 and 0.35731 for $\sigma^2 = 4.0$. The CPA results were obtained from by solving Eqs. (11) and (12) for $G_0^{CPA}(E)$ which is then used in the evaluation of the right hand side of Eqs. (14) and (17). Standard MatLab programs in double precision were used in the solution of the self-consistent equation for the coherent potential and in the evaluation of the integrals.

We note that the $IDOS(E) - IDOS(0)$ curves approach 1/2 as $E \to \infty$, consistent with the fact that the number of modes is equal to the number of sites. In the limit as $N \to \infty$, $IDOS(0) = \frac{1}{2}$ since there are equal numbers of positive and negative energy modes. By differentiating $IDOS(E)$, we obtain the density of states, $DOS(E)$, which is expressed as

$$DOS(E) = \pi^{-1}\,\text{Im}\, G_0(E) \quad (18)$$

In Figure 3, we compare the CPA results for the density of states with the results obtained by numerically differentiating the integrated density obtained by mode counting. Not surprisingly, there is again good agreement between the results obtained from the two approaches.

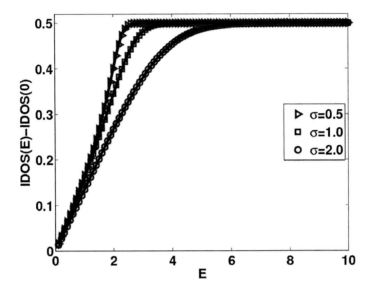

Figure 1. $IDOS(E) - IDOS(0)$ vs E (Ref. 1). The symbols are data points obtained by mode-counting for an array of 4×10^7 sites. The solid lines denote values obtained from the CPA. $\sigma = 0.5$, 1.0, and 2.0.

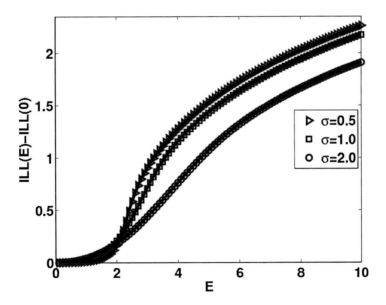

Figure 2. $ILL(E) - ILL(0)$ vs E (Ref. 1). The symbols are data points obtained by mode-counting for an array of 4×10^7 sites. The solid lines denote the values obtained from the CPA. $\sigma = 0.5$, 1.0, and 2.0.

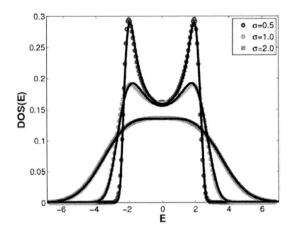

Figure 3. *DOS(E)* vs *E*. The symbols are data points obtained by differentiating the integrated density of states of an array of 5×10^7 sites. The solid lines denote the values obtained from the CPA. $\sigma = 0.5$, 1.0, and 2.0.

In contrast to *IDOS*(0), there appears to be no simple analytical expression for *ILL*(0). The asymptotic behavior of the *ILL* is more interesting. In the limit of large $|E|$, *ILL(E)* approaches the reciprocal of the decay length of a localized state with energy E associated with a perturbed site in an otherwise unperturbed lattice discussed previously. The connection between the energy of the localized state and the energy difference, V_0, is given by Eq. (4) with $\varepsilon = 2V_0$:

$$E = \pm 2(1+(V_0/2)^2)^{1/2} \qquad (19)$$

where the sign corresponds to the sign of V_0. The corresponding inverse localization length is given by Eq. 6. In Figure 4 we compare the mode-counting values of the *ILL(E)* with the results obtained by approximating *ILL(E)* by Eq. (6), which we refer to as the 'single-site approximation'.

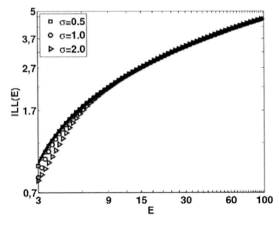

Figure 4. *ILL(E)* vs *E* (Ref. 1). Comparison between mode-counting results obtained from an array of 4×10^7 sites, shown as symbols, and the single-site approximation, Eq. (6), shown as a solid curve. $\sigma = 0.5$, 1.0, and 2.0.

The results displayed in Figure 4 indicate that the single-site approximation works well when the energy of the localized state lies in a region where the probability of the corresponding chromophore energy difference, $V_0 = \pm 2[(E/2)^2 - 1]^{1/2}$, expressed as

$$P(V_0) = (2\pi\sigma^2)^{-1/2} \exp\{-2[(E/2)^2 - 1]/\sigma^2\} \approx (2\pi\sigma^2)^{-1/2} \exp[-E^2/2\sigma^2] \quad (20)$$

is extremely small, i.e. $P(V_0)/P(0) \approx \exp[-E^2/2\sigma^2] \ll 1$. It should be noted that the single-site approximation is not limited to the Gaussian distribution, but is applicable to other distributions in the large-E limit, e.g. the Lorentzian distribution, as can be seen from the exact expression for the *ILL* of the Lorentzian distribution given in Ref. 15. The results obtained from Eq. (6) are also in good agreement with the results for the inverse of the penetration depth obtained from direct diagonalization of the exciton Hamiltonian by Vlaming et al. [18].

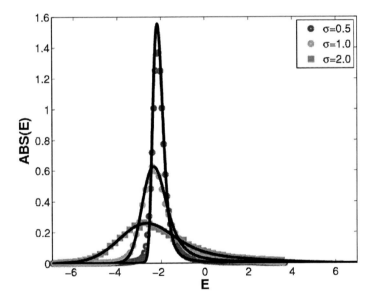

Figure 5. Dimensionless absorption, *ABS(E)*, vs *E*. The solid lines denote the values obtained from the CPA. The symbols are data points obtained by diagonalizing the Hamiltonian of an array of 1000 sites and averaging over 50 configurations. $\sigma = 0.5$, 1.0, and 2.0. A Lorentzian weight factor was employed having a half-width at half-height equal to 0.05.

Unlike the CPA, the mode-counting analysis does not give direct information about the optical absorption. In order to test the accuracy of the CPA for the absorption, it is necessary to calculate the absorption directly from the eigenvectors and eigenvalues of the Hamiltonian by first calculating the *EOS* of each eigenstate, ν, then summing over the eigenstates each weighted by a Lorentzian or similar weight factor replacing the delta function $\delta(E - E_\nu)$. The corresponding expression in the CPA takes the form

$$ABS(E) = \pi^{-1} \operatorname{Im}[E + 2 - V_c(E)]^{-1} \quad (21)$$

where $V_c(E)$ is obtained from Eq. (12). Strictly speaking, Eq. 21 is more properly regarded as an 'effective' dimensionless absorption that is proportional to the experimental absorption when all of the chromophores have the same transition dipole moment and the energy difference between the center of the exciton band and the ground state is much less than the exciton band width (so that the difference is approximately constant over the absorption band).

In Figure 5, we compare the results of the CPA with those obtained by direct diagonalization of the Hamiltonian of an array of 1000 sites, averaged over 50 configurations. As with the density of states and the inverse localization length, there is excellent agreement between the CPA and the numerical data obtained from finite arrays.

CONCLUSION

The main conclusion from the analysis presented above is that the coherent potential approximation gives extremely accurate results for the distribution, localization, and optical absorption of the modes of a one-dimensional Frenkel exciton array with Gaussian diagonal (V_n) disorder and nearest-neighbor transfer. It thus is a rare example of a theory that yields excellent agreement a realistic model of a macroscopically disordered system. It should be noted the CPA can also be applied to systems with long-range transfer [19]. Unfortunately, the same can't be said for the mode-counting approach, which is limited to one-dimensional arrays with nearest-neighbor interactions.

Although the focus in this chapter has been on diagonal disorder, we note that the mode-counting approach is also useful for arbitrary diagonal and off-diagonal ($t_{n,n+1}$) disorder as long as the transfer is limited to nearest-neighbors.

REFERENCES

[1] Avgin, I; Boukahil, A; Huber, DL. *Physica E*, (to be published).
[2] Koster, GF; Slater, JC. *Phys. Rev.*, 1954, 95, 1167, 96, 1208.
[3] Wolfram, T; Callaway, J. *Phys. Rev.*, 1963, 130, 2207.
[4] Avgin, I; Huber, DL; Lumin, J. 2009, 129, 1916.
[5] Rashba, EI; Gurgenishvili, GE. Fiz. Tverd. *Tela*, 1962, 4, 10, *Eng. Trans. Sov. Phys.*, – Solid State, 1962, 4, 759.
[6] Henry, CH; Nassau, K; Lumin, J. 1970, 1-2, 299.
[7] Avgin, I; Huber, DL. *J. Phys. Chem. B*, 2009, 113, 14112.
[8] Soven, P. *Phys. Rev.*, 1967, 156, 809.
[9] Taylor, DW. *Phys. Rev.*, 1967, 156, 1017.
[10] Lloyd, P. J. *Phys. C: Solid St. Phys.*, 1969, 2, 1717.
[11] Neu, P; Speicher, R. *J. Stat. Phys.*, 1995, 80, 1279.
[12] Ishida, I; Matsubara, T. *J. Phys. Soc.*, (Japan), 1975, 38, 534.
[13] Boukahil, A; Huber, DL. *J. Lumin.*, 1990, 45, 13.
[14] Dean, P. Proc. *Phys. Soc.*, 1959, (London) 73, 413.

[15] Thouless, DJ. *J. Phys. C: Solid State Phys.*, 1972, 5, 77. Note that Thouless' function λ is equivalent to the *ILL* used in this chapter.
[16] Eggarter, TP. *Phys. Rev. B*, 1973, 7, 1727.
[17] Avgin, I; Huber, DL. *J. Lumin.*, 2009, 129, 1916.
[18] Vlaming, SM; Malyshev, VA; Knoester, J. *Phys. Rev. B*, 2009, 79, 205121. The results mentioned are displayed in Figure 2 of the Vlaming paper.
[19] Balagurov, DB; Malyshev, VA; Dominquez Adame, F. *Phys. Rev. B*, 2004, 69, 104204.

In: Exciton Quasiparticles
Editor: Randy M. Bergin

ISBN: 978-1-61122-318-7
© 2011 Nova Science Publishers, Inc.

Chapter 9

EXCITON FANO RESONANCE IN SEMICONDUCTOR HETEROSTRUCTURES AND ITS RELATED PHENOMENA

Ken-ichi Hino[*], *Muneaki Hase and Nobuya Maeshima*
Graduate School of Pure and Applied Sciences, University of Tsukuba,
1-1-1 Tennodai, Tsukuba, Ibaraki 305-8573, Japan

Abstract

Excitonic Fano resonance (EX-FR) effects in semiconductor heterostructures, such as quantum wells, superlattices, and biased superlattices are highlighted. Here, these effects are classified into three types of linear EX-FR, non-linear EX-FR, and dynamic EX-FR. In the first two effects, a built-in interaction such as Coulomb coupling and hole-subband mixing cause Fano interference between discrete and continuum states. The linear EX-FR represents a Fano effect probed by means of conventional linear spectroscopy, while in the non-linear EX-FR, this effect is investigated from the viewpoint of non-linear spectroscopy. On the other hand, in the third effect, a Fano coupling is created dynamically by laser irradiation, without which this effect turns off. Therefore, both of the linear and non-linear EX-FRs are attributable to the static interactions inherent to the systems concerned, whereas the dynamic EX-FR literally has the dynamic origin. Moreover, other non-excitonic dynamic FR phenomena, somewhat relevant to the dynamic EX-FR are also presented. One effect for FR are the dynamic localization states realized in laser-driven biased superlattices, and the other FR result was observed during the initial stage of coherent phonon generation; where both of these two FRs are caused by an inter-subband ac-Zener tunneling and a laser-induced electron-phonon interaction, respectively. Because a continuum state plays a key role in FR, the previously mentioned types of FRs are understood in a unified manner based on the framework of the multichannel scattering theory; a conventional variational method seems incorrect since there is no account for the significance of the continuum. The R-matrix theory is considered as a powerful numerical method for solving such a FR problem in both accuracy and efficiency. Applying it to the FR phenomena concerned here, reveals a great number of intriguing characters.

---

[*]E-mail address: hino@bk.tsukuba.ac.jp

PACS 2010 73.21.-b, 78.40.Fy, 78.47.-p, 78.67.-n, 78.67.Pt, 63.20.Kr, 42.50.Hz, 42.50.Md, 32.80.Zb

Keywords: Fano resonance, semiconductor, exciton, terahertz wave, laser, coherent phonon

1. Introduction

Quantum interference of continuum states with embedded bound states, known as Fano resonance (FR) [1, 2], is one of the most fundamental phenomena occurring in diverse physical systems such as nuclei, atoms, molecules, and semiconductors: FR is also termed Feshbach resonance. Representative FR phenomena in atomic and molecular systems seem autoionization of a few-electron atom and molecular predissociation, which are caused by an electron-electron interaction and a vibronic interaction, respectively. Aside from these, FR plays cardinal roles in the phenomena of a hollow atom [3] and a superexcited molecules [4]. A well known FR in a semiconductor is found in a heavily doped p-type silicon crystal, that is generated by an electron-phonon interaction [5]. Further, FR is also found in a semiconductor quantum dot embedded in an Aharonov-Bohm ring [6].

This previously mentioned FR is generated by *static* interactions such as electron-correlation. Differing from this, the concept of laser-induced continuum structure (LICS) was introduced in the study of laser-atom interactions, in which atomic continuum states are structured due to this *dynamic* interaction by embedding a bound state into an originally unstructured continuum [7]. Here, the strength of a FR-coupling can be tuned by altering the strength of an applied laser field. The concept of LICS has been recently applied to strongly interacting Bose-Einstein condensates in an ultracold atomic system as an optical Feshbach effect [8, 9, 10]. In addition, a non-linear Fano effect has been recently discussed in semiconductor quantum dots [11, 12], where FR arising from *static* couplings was probed in an optically non-linear regime at high power. This is somewhat similar to the laser-induced autoionization (LIA) in the study of laser-atom interactions [13, 14], rather than to the LICS.

By the way, it is known that excitons (EXs) in various types of quantum-confined systems of semiconductor form FR states. This is generated by interactions between a discrete state (belonging to one subband pair of an electron and a hole) and a set of degenerate continuum states (belonging to other subband pairs). The resulting pronounced asymmetric profiles were investigated in a great number of articles [15, 16, 17, 18, 19, 20, 21, 22, 23, 24, 25, 27, 28, 29, 30, 31, 32, 33, 34]. It is usually understood that the FR is mediated by a Coulomb coupling and a hole-subband (HS) mixing. Recently, a LIA-type FR of quantum well (QW)-EXs was investigated using four-wave mixing (FWM) spectroscopy [35, 36]. A LICS-type FR of QW-EXs was also examined in a study of the Rabi splitting, where the FR arises from coherent mixing of subband pairs by an infrared laser field [37]. This chapter will be devoted to the review of our recent theoretical studies on excitonic FR in semiconductor heterostructure such as QWs, superlattices (SLs), and biased superlattices that are termed as Wannier-Stark ladders (WSLs). Further, the related phenomena of LICS-type FR in semiconductors will be also presented. A multichannel scattering (MCS) problem [38] has to be tackled to understand FR properly, because FR is formed in more than one continuum state coexisting with bound states at the same energy. The R-matrix theory [39], which is explained later in detail, is applied throughout this chapter as a powerful numerical method to solve the MCS problem of FR.

Bellow, the FR of confined EXs manifested in interband linear absorption spectra of the semiconductor heterostructures such as QWs and WSLs are considered, and a comparison is made with the corresponding high-resolution spectroscopy experiments [23, 24, 25, 26]. Next, the LIA-type excitonic FR induced by non-linear optical transitions is considered, and it is shown that this FR is characterized by asymmetric Autler-Townes (AT) doublet [40] in degenerated FWM spectroscopy [11]. The many-body Coulomb effect on the non-linear FR spectra is also mentioned. Further, anomalous EX-FR states (Floquet EX states) [41, 42] formed in dynamic WSLs (DWSLs) are introduced, where the DWSLs means photo-dressed SLs or WSLs resulting from inter-subband couplings induced by a periodic THz wave [43, 44]. As intensity of the THz wave increases, quasi-energetically degenerated Floquet states are strongly coupled, and thus, excitonic FR is generated, which is caused by the ac-Zener tunneling (ac-ZT) as well as the usual Coulomb interaction. Then, such FR attributable to the optical coupling, similar to the LICS mentioned earlier, is termed here as dynamic FR (DFR) in order to distinguish this from the conventional FR exclusively arising from static interactions.

Moreover, the concept of DFR is applied to other related FR phenomena in semiconductors, which arise from non-excitonic origins. One is electronic dynamic localization (DL) states of DWSL [45], and the other is coherent phonon states generated by intense and ultra-short laser-pulse irradiation [46]. The FR states formed in these systems are considered DFR, because both of the two phenomena originate from the laser-mediated interactions, namely, the inter-subband ac-ZT and the laser-induced electron-phonon interaction. The strength of these DFR-couplings can be tuned by altering either intensity or frequency of an applied laser field, similarly to the LICS. Thus, such tunability would provide a new possibility of coherent control of both spectral intensity and profile in these systems.

In Section 2., the theoretical framework of the MCS theory for FR is presented on a basis of the R-Matrix theory. In Section 3., this is applied to several excitonic FR phenomena occurring in semiconductor heterostructures. In particular, three different FR phenomena are discussed, that is, the linear excitonic FR, non-linear excitonic FR, and dynamic excitonic FR. In Section 4., non-excitonic DFR phenomena are mentioned as related topics of the excitonic FR in the preceding section. Section 5. is a conclusion. All acronyms defined in this article are summarized in Table 1. The atomic units are used throughout this chapter unless otherwise stated.

2. Theoretical Framework

The R-Matrix theory is one of the generalized frameworks for dealing with MCS problems including a many-body correlation effect [39]. The FR problem of concern pertains to this MCS problem. The purpose of this section is to present a theoretical framework tackling the FR problem based on the R-matrix theory. In order to describe the point of the theory without unnecessary generalization, the FR problem of single-QW EXs with no HS mixing will be taken up here as a concrete example. Generalization to more complicated FR systems

Table 1. Summary of acronyms used in text in alphabetical order and corresponding meanings.

acronyms	meanings
ADQW	Asymmetric double QW
AT	Autler-Townes
CWT	Continuous wavelet transformation
DFR	Dynamic Fano resonance
DL	Dynamic localization
DOS	Density of state
DVR	Discrete variable representation
DWSL	Dynamic WSL
EX	Exciton
FR	Fano resonance
FWHM	Full width at half maximum
FWM	Four-wave mixing
HF	Hatree-Fock
HS	Hole-subband
ISW	Incoming scattering wave
KH	Kramers-Henneberger
LIA	Laser-induced autoionization
LICS	Laser-induced continuum structure
MCS	Multichannel scattering
NNTB	Nearest-neighbor tight-binding
PAT	Photon-assisted tunneling
PSB	Photon sideband
QW	Quantum well
RFT	R-matrix Floquet theory
SBE	Semiconductor Bloch equation
SL	Superlattice
SR	Shape resonance
SRFWM	Spectrally-resolved FWM
SSQW	Shallow single QW
THz	terahertz
WSL	Wannier-Stark ladder
ZT	Zener tunneling

is easily made by a slight change of the expression of a system Hamiltonians under consideration. The present framework is divided into the following three stages: an adiabatic expansion (Section 2.1.), an application of the R-matrix propagation technique (Section 2.2.), and the extraction of time-delay (Section 2.3.). This numerical method allows one to implement large-scale MCS calculations without any empirical broadening parameter, thus ensuring high accuracy as well as numerical efficiency and stability.

2.1. Adiabatic Expansion

For the system of the QW-EX, a set of coordinates $\{\boldsymbol{\rho}, z_e, z_h\}$ is defined, where $\boldsymbol{\rho} = (\rho, \phi)$ represents an in-plane relative vector between an electron (e) and a hole (h), with ϕ representing an in-plane angular coordinate, and z_e and z_h represent the z-coordinates of e and h, respectively. Here, the center-of-mass motion of an EX in the plane of the layer is removed, and only the s-radial symmetry of a heavy-hole EX is taken into account for the sake of simplicity. The effective-mass Hamiltonian H_{ex} for the EX is expressed as

$$H_{ex}(\boldsymbol{\rho}, z_e, z_h) = -\frac{1}{2m_\parallel}\nabla_\rho^2 + V(\rho, z_e - z_h) + h_{qw}(z_e, z_h), \qquad (1)$$

where V is a Coulomb potential between e and h, expressed as

$$V(\rho, z_e - z_h) = -\frac{1}{\epsilon\sqrt{\rho^2 + (z_e - z_h)^2}} \qquad (2)$$

with ϵ a static dielectric constant. In addition, h_{qw} represents the Hamiltonian of QW for a combined subband of e and h, and is defined as

$$h_{qw}(z_e, z_h) = \sum_{k=e,h} h_{qw}^{(k)}(z_k). \qquad (3)$$

Here,

$$h_{qw}^{(k)}(z_k) = -\frac{\partial}{\partial z_k}\left(\frac{1}{2m_z^{(k)}}\right)\frac{\partial}{\partial z_k} + u_k(z_k), \qquad (4)$$

where $u_{e/h}$ is the QW potential for the particle e/h, $m_z^{(e/h)}$ is the mass of e/h in the z-direction, and $m_\parallel$ is the in-plane reduced mass of e and h. It should be noted that $m_z^{(e/h)}$ and $m_\parallel$ are the functions of z_e and z_h.

An EX envelope-function Ψ following the effective-mass model satisfies the Wannier equation:

$$(H_{ex} - E)\Psi(\rho, \Omega) = 0, \qquad (5)$$

where E is the given energy of the FR-EX, and Ω has been defined as a lump of the coordinates $\Omega = (\phi, z_e, z_h)$ for convenience of presentation. The νth solution of Ψ_ν is expanded with respect to a set of adiabatic channel functions $\{\Phi_\mu\}$ given by the following equation:

$$\Psi_\nu(\rho, \Omega) = \frac{1}{\sqrt{\rho}}\sum_\mu \Phi_\mu(\rho; \Omega) F_{\mu\nu}(\rho). \qquad (6)$$

$\{F_{\mu\nu}\}$ ($\mu, \nu = 1 \sim N$) is a set of radial functions to be determined, where the number of channels incorporated in Eq. (6) is denoted as N. This set of functions satisfies the following coupled equation:

$$\sum_\mu \langle \Phi_\mu(\rho; \Omega)|H_{ex}|\Phi_{\mu'}(\rho; \Omega)\rangle_\Omega F_{\mu'\nu}(\rho) = EF_{\mu\nu}(\rho), \qquad (7)$$

where pointed brackets $\langle \cdots \rangle_\Omega$ mean an integration over Ω. Here, Φ_μ is an eigenfunction of the Schrödinger equation

$$(h_{qw} + V)\Phi_\mu(\rho; \Omega) = U_\mu(\rho)\Phi_\mu(\rho; \Omega), \tag{8}$$

where the in-plane radius of the EX, ρ, is fixed as an adiabatic parameter. The eigenvalue of $U_\mu(\rho)$ represents the adiabatic potential, which is identical to the μth QW subband energy associated with h_{qw} at the limit $\rho \to \infty$. In the cases dealt with in Sections 2. and 3., $U_\mu(\infty)$ is always finite and $E \geq U_1(\infty)$. Thus, Eq. (7) is considered as a MCS equation; otherwise, this is simply an equation for a bound-state problem. An open [closed] channel μ is defined as an adiabatic channel satisfying $E > U_\mu(\infty)$ [$E < U_\mu(\infty)$]. A radial wave function $F_{\mu\nu}$ is provided by the use of the R-matrix propagation technique to be given here.

2.2. R-Matrix Propagation

Let the whole ρ-space $[0, \rho_{as}]$ be divided into N_R sectors, namely, $S_n \equiv [\rho_n, \rho_{n+1}]$ ($n = 1 \sim N_R$) with $\rho_1 \equiv 0$ and $\rho_{N_R+1} \equiv \rho_{as}$. The Schrödinger equation for Ψ_ν is recast in S_n into

$$\Psi_\nu = [H_L - E]^{-1} L \Psi_\nu, \tag{9}$$

where $H_L(\rho, \Omega) = H(\rho, \Omega) + L(\rho, \Omega)$. Here, the Bloch operator L is defined by

$$L(\rho, \Omega) = \delta(\rho - \rho_{n+1})\mathcal{L}(\rho_{n+1}, \Omega) - \delta(\rho - \rho_n)\mathcal{L}(\rho_n, \Omega) \tag{10}$$

with

$$\mathcal{L}(\rho, \Omega) = a(\Omega)\frac{\partial}{\partial\rho}, \tag{11}$$

and $a(\Omega) = -1/2m_\parallel$. This operator L has been introduced in order to enforce hermiticity on H_L in S_n, while H is no longer hermite there [39]. A set of wavefunctions $\{\psi_k\}$ with the associated eigenvalues $\{\mathcal{E}_k\}$ are provided by solving $[\mathcal{E}_k - H_L]\psi_k = 0$ in S_n, where ψ_k is expanded as

$$\psi_k(\rho, \Omega) = \frac{1}{\sqrt{\rho}}\sum_q \Phi_\mu(\rho_i^{(n)}; \Omega)\mathcal{F}_{qk}(\rho) \tag{12}$$

with $q = (\mu, i)$. $\mathcal{F}_{qk}$ is defined by

$$\mathcal{F}_{qk}(\rho) = \rho^{1/2}\varphi_i(\rho)c_{qk}. \tag{13}$$

Here, $\varphi_i(\rho)$ is the ith point-wise basis function in the discrete variable representation (DVR) associated with the Gauss-Legendre quadrature. c_{qk} is a coefficient to be determined in a variational way. $\{\rho_i^{(n)}\}$ is a set of abscissas associated with this quadrature defined in S_n. See Ref. [47] for more details of DVR.

Putting the complete set $\{\psi_k\}$ into the midst of the right-hand side of Eq. (9) with taking Eq. (6) into consideration yields the following relation for $F_{\mu\nu}$:

$$F_{\mu\nu}(\rho) = \sum_\xi \left[G_{\mu\xi}(\rho, \rho_{n+1})F_{\xi\nu}^d(\rho_{n+1}) - G_{\mu\xi}(\rho, \rho_n)F_{\xi\nu}^d(\rho_n)\right]. \tag{14}$$

Here, a propagator $G_{\mu\nu}$ has been defined as

$$G_{\mu\nu}(\rho,\rho') = \sum_{kqq'} O_{\mu q}(\rho) \frac{\mathcal{F}_{qk}(\rho)\mathcal{F}^{\dagger}_{kq'}(\rho')}{E - \mathcal{E}_k} O^{\dagger}_{q'\nu}(\rho'), \tag{15}$$

with $O_{\mu q'}(\rho) = \left\langle \Phi_\mu(\rho;\Omega) \mid \Phi_{\mu'}(\rho_i^{(n)};\Omega) \right\rangle_\Omega$ and $q' = (\mu', i)$. $F_{\mu\nu}^d$ is a derivative matrix given by

$$F_{\mu\nu}^d(\rho) = \langle \Phi_\mu(\rho;\Omega) \mid \mathcal{L}(\rho,\Omega) \mid \Psi_\nu(\rho,\Omega) \rangle_\Omega. \tag{16}$$

Introducing into Eq. (14) the R-matrix, defined by

$$R_{\mu\nu}(\rho) = \frac{1}{\rho} \sum_\xi F_{\mu\xi}(\rho) \left[F^d(\rho)\right]^{-1}_{\xi\nu}, \tag{17}$$

provides

$$R(\rho_{n+1}) = \tilde{G}(\rho_{n+1},\rho_{n+1}) - \tilde{G}(\rho_{n+1},\rho_n) \frac{1}{R(\rho_n) - \tilde{G}(\rho_n,\rho_n)} \tilde{G}(\rho_n,\rho_{n+1}), \tag{18}$$

where $\tilde{G}(\rho,\rho') = \rho^{-1/2} G(\rho,\rho') \rho'^{-1/2}$, and matrix notations have been used. This formula gives a forward propagation of the R-matrix from the one edge-point ρ_n to the other one ρ_{n+1} within S_n [48]. Thus, the R-matrix $R(\rho_{N_R+1})$ at the asymptotic distance is obtained by iterative application of Eq. (18) from the origin ρ_1 up to ρ_{N_R+1}.

The radial function $F(\rho_1)$ satisfies $[E - h_o] F = 0$, where h_o is a $N \times N$ Hamiltonian derived from Eq. (1) in the vicinity of the origin. The initial value of $R(\rho_1)$ for the R-matrix propagation is determined by putting the resulting solutions $F(\rho_1)$ into Eq. (17) in view of Eq. (16). The radial function $F(\rho_{as})$ is represented by a linear combination of two independent basis functions $\mathcal{E}^{(\pm)}$:

$$F_{\mu\nu}(\rho_{as}) = \mathcal{E}^{(+)}_{\mu\nu}(\rho_{as}) - \sum_{\xi=1}^{N} \mathcal{E}^{(-)}_{\mu\xi}(\rho_{as}) \Sigma_{\xi\nu}(E). \tag{19}$$

$\mathcal{E}^{(\pm)}_{\mu\nu}$ correspond to the regular and irregular solutions of $[E - h_{as}] \mathcal{E} = 0$, where h_{as} is a $N \times N$ asymptotic Hamiltonian derived from Eq. (1):

$$h_{as}(\rho) = A(\rho) \frac{d^2}{d\rho^2} + U(\rho). \tag{20}$$

with $A_{\mu\xi}(\rho) = \langle \Phi_\mu(\rho;\Omega) \mid a(\Omega) \mid \Phi_\xi(\rho;\Omega) \rangle_\Omega$. Both A and U become constant at $\rho \simeq \rho_{as}$, since the adiabatic potential $U_\nu(\infty)$ is equal to the νth subband energy. Hence, $\mathcal{E}$ is of the form: $\mathcal{E}_{\mu\nu}(\rho) = \chi_{\mu\nu} \exp(ik_\nu\rho) \mathcal{N}_\nu$, where χ satisfies an algebraic equation

$$[Ak^2 - U + E]\chi = 0. \tag{21}$$

There are N_o solutions with k real and positive (negative) corresponding to outgoing (incoming) matching functions $\mathcal{E}^{(+)}$ ($\mathcal{E}^{(-)}$) for open channels ($E > U_\nu(\infty)$) and N_c solutions with Im$[k]$ positive (negative) corresponding to decaying (growing) matching function

$\mathcal{E}^{(+)}$ ($\mathcal{E}^{(-)}$) for closed channels ($E < U_\nu(\infty)$). $\mathcal{N}_\nu$ is a normalization constant. A matrix Σ is evaluated by inserting Eq. (19) into the R-matrix $R(\rho_{as})$ obtained by the forward propagation.

There are N_o independent solutions for the MCS problem concerned here at a given $E(> U_1(\infty))$. Letting the solutions be denoted as $\{F_{\mu\alpha}^{(-)}\}$ ($\mu = 1 \sim N, \alpha = 1 \sim N_o$), the incoming scattering wave (ISW) boundary condition reads [38]

$$F_{\mu\alpha}^{(-)}(\rho_{as}) = \mathcal{E}_{\mu\alpha}^{(+)}(\rho_{as}) - \sum_{\xi=1}^{N_o} \mathcal{E}_{\mu\xi}^{(-)}(\rho_{as}) S_{\xi\alpha}^{(-)}(E) \tag{22}$$

for $\mu = 1 \sim N_o$, and $F_{\mu\alpha}^{(-)}(\rho_{as}) = 0$ for $\mu = N_o+1 \sim N(= N_o+N_c)$. $S^{(-)}$ is a scattering matrix (S-matrix). $F^{(-)}$ is expressed as a linear combination of N solutions of Eq. (19) given by $F^{(-)} = FV$, where V is a $N \times N_o$ matrix. V and further $S^{(-)}$ are evaluated in the following. For the sake of convenience, the matrices Σ and V are expressed in the form:

$$\Sigma = \begin{pmatrix} \Sigma_{oo} & \Sigma_{oc} \\ \Sigma_{co} & \Sigma_{cc} \end{pmatrix}, \quad V = \begin{pmatrix} V_o \\ V_c \end{pmatrix}, \tag{23}$$

where Σ_{co} and V_c are block matrices consisting of N_c rows and N_o columns, and so on. V is determined so that $F^{(-)}$ satisfies the ISW boundary conditions and is provided as:

$$V_o = 1, \quad V_c = -\frac{1}{\Sigma_{cc}}\Sigma_{co}. \tag{24}$$

This result yields

$$S^{(-)} = \Sigma_{oo} - \Sigma_{oc}\frac{1}{\Sigma_{cc}}\Sigma_{co}. \tag{25}$$

The normalization constant $\mathcal{N}$ for a matching wavefunction $\mathcal{E}^{(\pm)}$ is determined so as to be energy-normalized, that is,

$$\mathcal{N}_\alpha(E) = \left[\frac{1}{2\pi}\left(\frac{dk_\alpha(E)}{dE}\right)\right]^{1/2}. \tag{26}$$

Such a choice is consistent with a flux conservation designated by a unitarity condition $[S^{(-)}]^\dagger S^{(-)} = S^{(-)}[S^{(-)}]^\dagger = 1$.

Once the radial wavefunction, $\tilde{F}^{(-)}(\rho_{as})$, is determined, $\tilde{F}^{(-)}(0)$ is provided immediately by projecting a $N \times N$ backward propagation matrix T on $\tilde{F}^{(-)}(\rho_{as})$, namely, $\tilde{F}^{(-)}(0) = T\,\tilde{F}^{(-)}(\rho_{as})$, where $\tilde{F}^{(-)}(\rho) = F^{(-)}(\rho)/\sqrt{\rho}$. This projection matrix is given as $T = R(0)\,P\,[R(\rho_{as})]^{-1}$ from Eq. (18), where

$$P = \prod_{n=1}^{N_R} \left[\tilde{G}(\rho_{n+1}, \rho_n)\right]^{-1} \left[\tilde{G}(\rho_{n+1}, \rho_{n+1}) - R(\rho_{n+1})\right]. \tag{27}$$

So far the present method consists of two stages. One is the adiabatic expansion employed in Eqs. (6) and (12). This method allows us to minimize the number of the adiabatic channel functions, since these are eigen functions of the adiabatic Hamiltonian $\mathcal{H}$ with appropriate physical meanings. The other is the R-matrix propagation technique presented here. The DVR basis set $\{\varphi_i\}$ employed in Eq. (13) is optimal based on the Gauss-Legendre quadrature, and a size of each sector S_n is chosen arbitrarily small. Therefore, it is possible to provide accurately and stably the R-matrix basis set $\{\psi_k\}$ large enough to satisfy the closure relation, and further the associated eigen values $\{\varepsilon_k\}$ extending to extremely high energies. In addition, it is noted that EX spectra for "all" given E's are obtained at a stroke just by a single calculation of R in the whole ρ-space, since the propagator of Eq. (15) at every E is represented by the common R-matrix basis set. This feature plays a decisive role to reduce a great deal of computational burden entailed for FR profiles.

2.3. Time Delay

The lifetime of an EX state is determined in terms of a time-delay matrix $\tau(E)$ [49] obtained by $S^{(-)}(E)$, using the expression

$$\tau(E) = -i[S^{(-)}(E)]^{-1}\frac{dS^{(-)}(E)}{dE}. \tag{28}$$

Following this, the lifetime of the concerned state with E is given by

$$T(E) = \frac{\text{Tr}[\tau(E)]}{N_o}. \tag{29}$$

Moreover, $\tau(E)$ is related with an excess density of state (DOS) [49, 50, 51, 52] defined by

$$\rho^{(ex)}(E) = \text{Tr}[\tau(E)]. \tag{30}$$

The excess DOS is also expressed as

$$\rho^{(ex)}(E) = \rho(E) - \rho^{(as)}(E), \tag{31}$$

where $\rho(E)$ and $\rho^{(as)}(E)$ represent the DOS of the concerned EX system and that of a field-free asymptotic state $\mathcal{E}^{(\pm)}(z)$ of Eq. (20). Since $\rho^{(as)}(E)$ shows just structureless continuum, the FR structure observed in $\rho^{(ex)}(E)$ is considered almost similar to that in $\rho(E)$.

3. Excitonic FR in Semiconductor Heterostructures

This section is devoted to the presentation of the following three different types of excitonic FR phenomena: linear excitonic FR (Section 3.1.), non-linear Excitonic FR (Section 3.2.), and dynamic excitonic FR (Section 3.3.). In Section 3.1., the FR is examined from the viewpoint of a linear optical response as initial characterization. On the other hand, in Section 3.2., this is analyzed from the viewpoint of a non-linear optical response by means of degenerate FWM spectroscopy: such non-linear FR would be relevant to the LIA mentioned in Section 1. In Section 3.3., one explores a novel effect of FR under irradiation of an intense THz-wave, where excitonic Floquet states are formed by dressing an interaction between the EX and the THz wave, and the FR arises from ac-ZT as well as a Coulomb coupling between the Floquet photon sidebands (PSBs): this phenomenon would appear similar to the LICS mentioned in Section 1.

3.1. Linear Excitonic FR

3.1.1. QWs

The latest progress of tailoring semiconductor heterostructures of high quality has made it possible to observe detailed spectral profiles of FR and even weak transitions attributable to optically-inactive odd-parity EXs [31, 32, 33, 34]. High-resolution FR spectra are obtained in the asymmetric double QW (ADQW) of 165Å-GaAs/14Å-AlAs/48Å-GaAs [31] and in the shallow single QW (SSQW) of GaAs/Al$_{0.045}$Ga$_{0.955}$As with a 200Å-thick well [32]. Here, the sophisticated approach presented in the preceding section is applied to these systems based on the excitonic 4×4-Luttinger Hamiltonian in place of Eq. (1), where HS mixing is incorporated on an equal footing with Coulomb coupling [23, 24]. Owing to this method, high resolution calculations are feasible with numerical stability and a great deal of reduction of computational time, and one is enabled to obtain Fano profiles having "natural spectral widths." On the other hand, in most conventional methods [15, 16, 17, 18, 19, 20, 21, 22], an empirical broadening parameter is inevitably required in order to compensate for drawbacks built in these methods, which lead to fairly crude evaluation of FR and hamper detailed analysis of it due to numerical instability. To be specific, spectra obtained by the methods are approximated simply by a set of pseudo-continuum states having discrete eigenvalues and being independent of a given energy E of a system concerned. Furthermore, the correct ISW boundary conditions are disregarded from the beginning. The natural spectral widths are intrinsic to FR and indispensable for scrutinizing it, while the homogeneous broadening is usually considered extrinsic and it would be taken into account, if necessary, by convoluting the natural spectra with a suitable Lorentzian profile. In addition, the HS mixing is not taken into account [18, 19, 20, 21, 22]. The present method enables one to overcome almost all of these drawbacks inherent in the conventional methods.

In this section, following conventions, a subband pair, termed as channel, of a heavy-hole (hh) [light-hole (lh)] EX composed of the n_eth electron-subband and the n_hth hole-subband is represented as $en_e - hh(lh)n_h$. It is understood that the subband quantum number relevant to the wide QW of ADQW is denoted as n_e or n_h, while that relevant to the narrow QW is as a primed number n'_e or n'_h. Notations of channels for SSQW follow those of the wide QW in ADQW.

Figures 1(a)-1(c) show calculated FR profiles of EXs in the ADQW [25]. It is remarked that inversion symmetry in the crystal growth direction of ADQW is broken down. However, parity with respect to this symmetry is still considered as an approximately good quantum number in each of the wide and narrow QWs. Thus, the transition with even (odd) parity; namely, with $|n_e - n_h|$ being even (odd) is assumed to be approximately optically allowed (forbidden) under the condition of no HS mixing. In Figure 1(a), there are a plethora of the absorption lines of calculated FR spectra with natural resonance widths. These spectra are composed of a sum of contributions from the Γ_7 and Γ_6 irreducible representations of the point group C_{4v} for a HS state at the zone center. Spectra seen in Figures 1(b) and 1(c) are those convoluted with Lorentzian functions having line widths of 0.5 meV and 4 meV, respectively. Figure 1(d) is the experimental data of photoluminescence excitation (PLE) obtained by Oberli, et al [31]. The homogeneous line width of Figure 1(b) is introduced so as to mimic the peak labeled 1 in Figure 1(d), while the homogeneous line width of Figure 1(c) is for mimicking the spectral profile labeled 3 in Figure 1(d). In Figure 1, excitonic spectra for continuum states staying above the absorption edge (1.538 eV) are depicted and those assigned merely to bound states lying below it are absent. In each of Figures 1(a)-1(c), the calculated results without HS mixing are also shown for comparison with the spectra given by the full calculations.

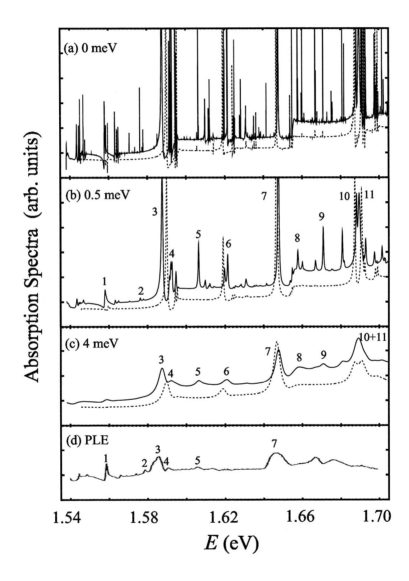

Figure 1. Calculated absorption spectra of exciton in ADQW of 165Å-GaAs/14Å-AlAs/48Å-GaAs versus energy of photon (eV), with (a) natural spectral profiles without convolution, (b) the associated convoluted spectra by the line width of 0.5 meV, and (c) those by the line width of 4 meV. Panel (d) is PLE spectra of Ref. [31]. Labels of peaks are assigned as follows. 1: e1-hh3(1s), 2: e1-lh2, 3: e2-hh2(1s), 4: e2-lh1, 5: e1-hh1', 6: e2-lh2(1s), 7: $e1' - hh1'(1s)$, 8: e1'-hh3, 9: e1'-lh2, 10: e3-hh3(1s), 11: $e1' - lh1'(1s)$. Calculated results without HS mixing are also depicted by dotted lines in panels (a)-(c).

A great number of complicated FR spectra are shown in Figures 1(a) and 1(b), because HS mixing as well as Coulomb coupling is stronger with energy spacing denser [24]. The counterpart spectra without HS mixing appear somewhat less in variety. In the PLE spectra of Figure 1(d), most of fine structures are smeared out due to poorer resolution than that below 1.570 eV, and the contributions from background continua are partially lost. The respective peaks of 3, 6, and 10 are assigned to strong parity-allowed EXs of $e2 - hh2(1s)$, $e2 - lh2(1s)$, and $e3 - hh3(1s)$ intrinsic to the wide QW. The absorption edge relevant to the narrow QW is located at 1.655 eV, and this appears as the abrupt step-like-increase of the spectra in Figure 1(b). The prominent peaks of 7 and 11 are ascribable to parity-allowed EXs of $e1' - hh1'(1s)$ and $e1' - lh1'(1s)$, respectively, intrinsic to the narrow QW.

Here, one focuses on the peaks of 5, 8, 9, 2, and 4. The first three peaks result obviously from relatively strong interplay between the wide QW and the narrow QW, because these peak positions are successfully assigned to the subband energies of $e1 - hh1'$, $e1' - hh3$, and $e1' - lh2$, respectively. Furthermore, it is surprising that these EXs are exclusively enhanced by HS mixing, as is evident from comparison with the counterpart spectra excluding this effect. Actually, the overlaps between the electron-hole subband wavefunctions of the channels pertinent to these EXs are just 0.02 to 0.07. Thus, it is speculated that these anomalous EXs are generated by relatively strong couplings between the different QWs mediated by HS mixing. That is, the oblique transition of electron and hole beyond the barrier is amplified by HS mixing. Because the position of the peak 5 in Figure 1(d) appears almost identical to that in Figure 1(c), this intriguing transition was already observed by an experiment, though overlooked in the original study [31]. The same hybridized EXs of $e1' - hh3$ and $e1' - lh2$ as that of $e1 - hh1'$ are also discerned in Figure 1(c), however, blurred to a certain extent. These are absent from Figure 1(d), since the PLE spectra are largely smeared out with concomitant loss of the background continuum. With regard to the last two peaks of 2 and 4, these are assigned to parity-forbidden lh-EXs supported by the channels of $e1 - lh2$ and $e2 - lh1$, respectively. These peaks also appear in Figure 1(a), whereas these vanish in the spectra excluding HS mixing in Figure 1(b). Hence, it is found that these parity-forbidden lh-EXs are exclusively caused by HS mixing. Both peaks of 2 and 4 are still discerned in the PLE.

Figure 2 shows FR profiles of EXs in the SSQW [25]. Calculated FR spectra with natural resonance widths are given in Figure 2(a). Here, a confining potential is so shallow as to support just a couple of bound subband states. Continuum subband states are evaluated in a variational way, similarly to these bound subband states, simply by introducing infinite potential barriers located far enough not to disturb the bound subband states. The spectra concerned in this SSQW are composed of a sum of contributions from the $\Gamma_7^\pm$ and $\Gamma_6^\pm$ irreducible representations of the point group D_{4h} for a HS state at the zone center, though contributions from Γ_7^- and Γ_6^- are negligibly small in a single photon absorption. The confining potentials of an electron, a hh and a lh support two, three, and two bound states, respectively. The spectra seen in Figures 2(b) and 2(c) are convoluted with Lorentzian functions having the line widths of 0.4 meV and 1.4 meV to mimic the peaks labeled as 1 and 3, respectively, in the PLE spectra of Figure 2(d) obtained by Simmonds, *et al* [32]. In Figures 2(a)-2(c), the excitonic spectra for continuum states staying above the absorption edge of 1.528 eV are depicted, and in addition, the calculated results without HS mixing are also shown as the spectra of reference. In addition, it would be favorable to shift all calculated FR spectra to the higher-energy side by at most $\delta \approx 1$ meV to bring these spectral lines into better agreement with the experiment.

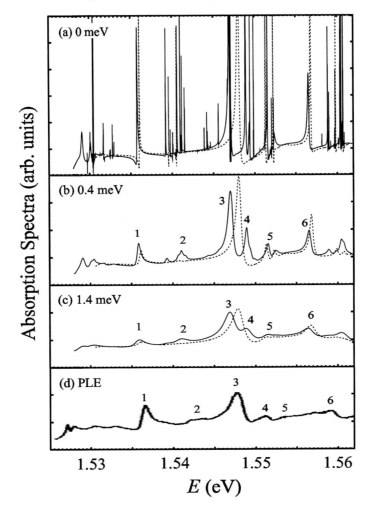

Figure 2. Calculated absorption spectra of exciton in 200Å-thick SSQW of GaAs/Al$_{0.045}$G$_{0.955}$ As versus energy of photon (eV), with (a) natural spectral profiles without convolution, (b) the associated convoluted spectra by the line width of 0.4 meV, and (c) those by the line width of 1.4 meV. Panel (d) is PLE spectra of Ref. [32]. Labels of peaks are assigned as follows. 1: e1-hh3(1s), 2: e1-lh2, 3: e2-hh2(1s), 4: e2-lh1, 5: e2-hh2(ns) ($n \geq 2$), 6: e2-lh2(1s). Calculated results without HS mixing are also depicted by dotted lines in panels (a)-(c).

The spectra of the peak 1 of Figure 2(b), attributable to $e1 - hh3(1s)$, reproduce well the asymmetry pattern of the corresponding FR in the PLE spectra. The FR peaks of 3, 5, and 6 are assigned to $e2 - hh2(1s)$, $e2 - hh2(ns)$ ($n \geq 2$), and $e2 - lh2(1s)$, respectively, and the associated spectra depicted in Figure 2(c) also reproduce the PLE spectra above 1.540 eV. These four EXs are mostly dominated by Coulomb coupling. Next, the relatively small humps of 2 and 4 are attributable to the parity-forbidden lh-EXs of $e1 - lh2$ and

$e2 - lh1$, respectively. It is noticed that the main components of FR profiles pertain to the Γ_7^+ irreducible representation, and hence these FR states arise dominantly from HS mixing. Such observation is also evident from the comparison with the spectra of reference in Figures 2(a)- 2(c), though significance of HS mixing was overlooked in the experiment [32].

3.1.2. WSLs

Biased semiconductor SLs are usually termed as WSLs, where a static electric field (F) is applied along the direction of crystal growth of SLs. On a basis of a single-miniband Kane approximation (corresponding to the Kane wave function [53] and the Houston wave function [54]), an electron in the WSL has localized wave functions and discrete energy spectra with equal spacing, expressed as $\Omega_B = Fd$, where Ω_B and d represent the Bloch frequency and a SL lattice constant, respectively. The equidistant WSL energy spectra were measured by photoluminescence and photocurrent spectroscopy [55, 56], and the related Bloch oscillation was measured by time-resolved FWM spectroscopy [57].

However, due to an inter-subband mixing effect caused by dc-ZT [58], it is proved that the energy spectra are always continuous and the wave functions are neither localized nor square integrable. Thus, the Kane approximation and the existence of equidistant energy spectra can be considered accurate only if spectral broadening due to Zener breakdown is much smaller than Ω_B for a weak bias of F. Resonant Zener tunneling between energetically-aligned neighboring WSL-subbands is termed here as Zener resonance or WSL resonance. Delocalized electron states in coupled WSL induced by the Zener resonance were first observed by photocurrent spectroscopy along with the measurement of the current-voltage characteristics [59], where the delocalization of electrons and the anticrossing behavior attributable to the nearest- and the second-nearest-neighbor resonant couplings were observed.

Regarding excitonic states for WSL, strictly speaking, all of them show FR since dc-ZT causes a coupling of an excitonic discrete level with energetically degenerate continua belonging to neighboring WSL subbands, which results in FR. Then, there are no excitonic bound states in the WSL. In this section, it is shown that FR spectra are strongly modulated due to dc-ZT; when the Zener resonance causes strong anticrossing, namely, repulsion of subband energies [26]. The following interesting effects are observed: (i) The absorption tail edge shifts noticeably toward the lower energy side, and the magnitude of the tail varies with respect to F in an oscillating manner. (ii) Both the intensity and the position of the FR spectral peak are altered in an irregular manner in the higher energy side, as F traverses the anticrossing region.

We begin with FR spectra of a WSL-EX in the lower bias region. The Hamiltonian of Eq. (1) should be replaced by the associated WSL Hamiltonian for this purpose. Here, the calculated spectra are compared with the experimental spectra of the undoped [001] SLs of 67/17-Å GaAs/Ga$_{0.7}$Al$_{0.3}$As with $F = 13.3$ kV/cm given in Ref. [34]. The calculated spectra convoluted by a Lorentzian function with the FWHM of $\Gamma = 0.25$ meV are shown in Figure 3(a), where the contributions of lh EXs are also included. In this case, each peak of an $\mathcal{N}s$-state pertaining to the $hh(n)$ ($lh(n)$) channel is labeled by $hh(n)[\mathcal{N}]$ ($lh(n)[\mathcal{N}]$). This label of $\mathcal{N}$ is omitted for the sake of simplicity, in the case that there is no confusion. The same spectra as those of Figure 3(a) but with $\Gamma = 1.5$ meV are shown in Figure 3(b), and the reported experimental spectra – labeled E – to be compared with these are cited in Figure 3(c). The reported theoretical spectra [60] are also included in Figure 3(c) and labeled as T. The FWHM of 1.5 meV has been employed in Figure 3(b) for a good reproduction of the experimental width of the main peak of $[hh(-1)[1] + lh(-3)]$ in the E spectra. Further, the overall spectral intensity of Figure 3(b) is also normalized to the height of this peak. Although a slight red-shift by about 2 meV is observed at every peak position of the calculated spectra shown in Figure 3(b) in comparison with the E spectra, it is observed that the peak positions, the spectral profiles, and the height of the background continuum are well reproduced in the calculated spectra. When the T spectra are compared with the present spectra, the height of the background continuum is too small.

Next, we address the effect of the issue (i) mentioned earlier regarding the strong modulation of excitonic FR spectra in the higher bias region. The sample of undoped [001] SLs employed for calculations is GaAs/Ga$_{0.75}$Al$_{0.25}$As of 35/11 monolayers (ML) [1 ML=2.83 Å] for the well and barrier thickness. Here, a set of quantum numbers of a joint WSL subband state of an electron and a hole is given by $j \equiv n(b_e, b_h)$, which is termed as a channel, where b_e [b_h] and n_e [n_h] are a miniband index and a WSL index of the electron [the hole], respectively, and $n = n_h - n_e$ is used to represent the joint WSL subband state of this pair. These quantum numbers, based on the Kane representation, are adopted for the convenience of designating the quantum state, though, in actuality, meaningless for strongly coupled WSL. The calculated spectra with F ranging from 90 to 110 kV/cm are shown in Figure 4; the traces, represented by solid, dotted, and chain lines, indicate the FR spectra with coupled WSL channels, the FR spectra with non-interaction WSL channels (under the Kane approximation), and the free spectra without EX effects. A peak of the EX 1s-state is labeled as $n(b_e, b_h)$ in terms of the channel by which this is supported. Both these FR spectra are manipulated by convoluting the original natural spectra by employing $\Gamma = 2$ meV. The free spectra, however, remain unconvoluted and they have a stepped shape. The reference spectra obtained under the Kane approximation for the purpose of comparison are shown only for 100 kV/cm; apart from minor changes, in fact, the spectra remain almost unaltered above $F = 60$ kV/cm with respect to the variance in F.

The FR spectral profiles are seen to be strongly modulated with respect to the change of F, differing considerably from the FR spectra under the Kane approximation. It is noteworthy that the red-shift of the absorption tail edge manifests itself noticeably in the vicinity of $F = 100$ kV/cm, whereas this disappears at $F = 90$ and 110 kV/cm; that is, the dominant absorption begins again with the 0(1,1) transition. It is speculated that this effect arises from the Zener resonance. Further, the red-shift of concern always accompanies intensity reduction in the higher energy region, as a result of the conservation rule of optical oscillator strength, i.e., the sum rule.

Figure 5 (a) shows the energy-fan diagram of coupled WSL for electron as a function of bias F, in which energies calculated under the Kane approximation are also shown just for comparison. Three dominant anticrossings represented by a dashed circle each appear around $F = 95$, 100, and 105 kV/cm. Figure 5 (b) shows the change in absorption intensity at the photon energy of $E = 1.47$ eV as a function of F from 90 to 112 kV/cm, where the calculated result is indicated by filled circles connected by solid lines in order to aid the presentation. It is evident that the intensity of the tail region of the spectra changes in a

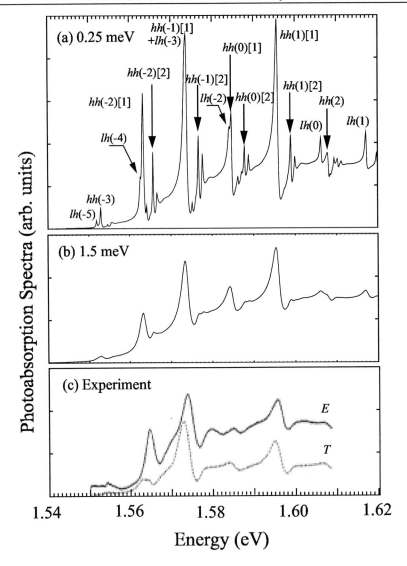

Figure 3. FR spectra of 67/17-Å GaAs/Ga$_{0.7}$Al$_{0.3}$As with $F = 13.3$ kV/cm as a function of photon energy E. (a) The calculated spectra with $\Gamma = 0.25$ meV. Consult the text for the meanings of the labels. (b) The same as panel (a) but with $\Gamma = 1.5$ meV. (c) Experimental spectra reported in Ref. [34], labeled by E. For the purpose of comparison, the calculated spectra included in this reference are also shown with label T.

complex manner accompanying oscillations rather than in a monotonic manner as a whole. It is seen that there are three maxima of intensity at the positions of $F = 95$, 100, and 104 kV/cm. These biases coincide perfectly with each location of the anticrossing shown by the dashed circle in Figure 5(a). Therefore, this observation leads to the speculation that the Zener resonance plays a decisive role in causing the red-shift of the absorption tail edge. Moreover, this effect is quite localized in the proximity of each anticrossing. For instance, the first peak begins at $F = 93$ kV/cm and ends at $F = 96$ kV/cm, followed by

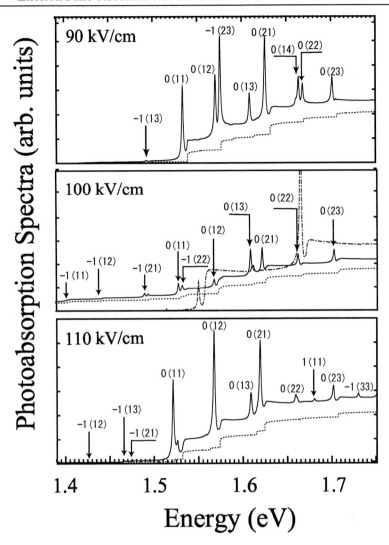

Figure 4. Calculated FR spectra of the present WSL with $F = 90$–110 kV/cm as a function of photon energy E. The solid traces are obtained from the full calculations, including $N_b = 6$, $N_{site} = 20$, and $N = 70$–90, while the dotted traces and the chain trace for $F = 100$ kV/cm indicate free spectra without exciton effects and reference spectra given by the Kane approximation, respectively, for the purpose of comparison. The spectra indicated by the solid and chain lines are convoluted by the Lorentzian function with $\Gamma = 2$ meV, while the free spectra is given with no convolution. In the label of the $n(b_e b_h)$ state, the comma separating b_e and b_h is removed just for a typographical reasons; this should read as $n(b_e, b_h)$. Refer to the text for the meaning of this label.

the onset of the next rise. This succession of the localized peaks and dips relevant to the anticrossings are thus understood to give rise to the appearance of the oscillation structure in the intensity. Mention to the red-shift of the absorption tail edge was first made in Ref. [61]. Nevertheless, the amount of the shift toward the lower energy side is very small, and the significant role of anticrossings in conjunction with the Zener resonance was not addressed.

Finally, the issue (ii) as mentioned is examined. As seen from Figure 4, dc-ZT also causes the spectral modulation of FR spectra regarding this issue. In order to trace the modulation pattern in more detail, Figure 6 shows the spectra in the vicinity of the arrowed peak positions of $0(1, 1)$ for $F = 10 - 110$ kV/cm, with $\Gamma = 2$ meV (solid lines) along with $\Gamma = 0.3$ meV (dotted lines) to indicate shapes of the corresponding natural spectra. Both the peak position and height vary in an irregular manner without monotonic changes. These irregular changes manifest themselves at approximately $F = 70$ and 100 kV/cm, in particular, possibly due to the formation of anticrossings between the electron subbands 1–2 and 1–3, respectively. Thus, it is speculated that these FR spectra are greatly modulated by Zener resonance.

As already discussed in Ref. [62], the absorption spectra of the $0(1, 1)$ state rapidly disappear due to field-induced delocalization with an increase in F to 40 kV/cm; however, this is not the case for larger values of F as seen in Figure 6. This figure also indicates the monotonic red-shift similar to the character of the quantum-confined Stark effect [63, 64]. Furthermore, in the reported experiment [34], the change of Fano's q-value in FR spectra with respect to F measured to approximately 30 kV/cm. This indicates that by increasing F, $|q|$ increases monotonically without a change in its sign, i.e., the asymmetric FR spectra become symmetric. This was conducted for relatively weak F-fields up to a maximum of approximately 30 kV/cm, where the Zener tunneling is still insignificant. Indeed, in Figure 6, the spectra for $F = 10 - 40$ kV/cm with strong peaks and $q < 0$ appear to follow the monotonic change similar to this experiment. However, the spectrum exhibits an anomalous change in q above this; in particular, q appears positive at $F = 50$ kV/cm, and can be brought back to a negative value at $F = 60$–80 kV/cm. In this figure, this is more clearly seen in the dotted curves than in the solid ones. Such spectral irregularity is also observed in the states of $0(1, 2), 0(2, 2), 0(1, 3),$ and $0(2, 1)$, though not shown here.

3.2. Non-Linear Excitonic FR

3.2.1. Introduction

Our next concern is the non-linear spectra of excitonic FR states for WSLs in transient FWM spectroscopy, where the EX is induced by *resonant* pumping [11]; this problem concern is relevant to LIA. Since a FR wavefunction is usually complex because of its nature is ascribable to a MCS problem, sharply energy-dependent phaseshifts associated with Fano couplings make all associated FR transition matrices phase-sensitive complex numbers. It is thus speculated that the change of the FR phaseshifts largely modulates coherence of induced polarizations, and the modulation pattern of the spectra depends on a way of the Fano interference. As is seen later, the spectrally-resolved FWM (SRFWM) signals in relatively strong excitation exhibit an AT-like doublet [40, 65, 66] with the asymmetric amplitudes of two sidebands. This present dressed excitonic-FR problem is distinguished from an excitonic Rabi problem pertaining to pure *bound*-states [67, 68, 69, 70, 71, 72, 73, 74, 75, 76], because in the latter problem, the associated wavefunction can always be set real, and thus, the resulting AT doublet appears symmetric. In the excitonic Rabi flopping, a Rabi energy is saliently enhanced due to internal renormalized-fields arising from the many-body Coulomb effect negligible in the atomic Rabi system [68, 69, 70, 71]. This effect also plays

Exciton Fano Resonance in Semiconductor Heterostructures ...   321

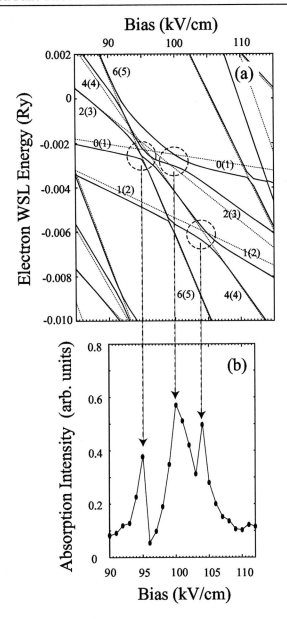

Figure 5. (a) Energy-fan diagram of coupled WSL for electron as a function of bias F. Energies calculated under the Kane approximation are also shown by dotted lines. Approximate Kane quantum numbers $n_e(b_e)$ participating in anticrossing regions are indicated. Dominant anticrossings are also represented by a dashed circle each. (b) Variance of the spectral intensity at $E = 1.47$ eV in the present WSL as a function of F in the regions where anticrossings in the vicinity of $F = 100$ kV/cm are dominant. Here, filled circles represent calculated FR. These circles are connected by a solid line in order to aid the presentation. Downward chain arrows crossing over panels (a) and (b) represent the correspondence between the anticrossings and the peak positions of FR spectra.

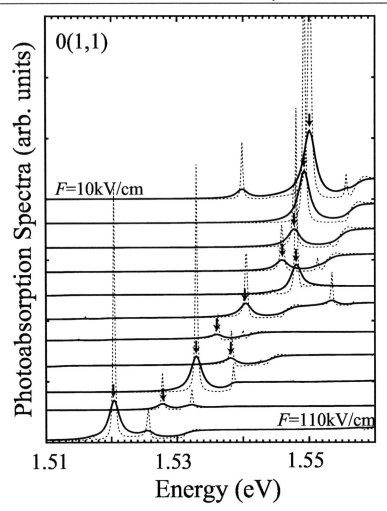

Figure 6. FR spectra of the present WSL in the vicinity of the $1s$ state pertaining to $0(1,1)$. Traces are shifted vertically to enhance the presentation, represent the spectra at F=10 – 110 kV/cm in an ascending order of F from the top to the bottom. Peak positions of the $0(1,1)$ FR state are indicated by arrows. Spectra with the Lorentzian broadening of $\Gamma = 2$ and 0.3 meV are represented by the solid and dotted lines, respectively.

a decisive role in the dressed FR problem.

The semiconductor Bloch equations (SBEs) are applied to the present problem for the excitonic WSL within the Hartree-Fock (HF) approximation [77], by expanding a microscopic polarization and a carrier population density in terms of an excitonic FR basis set. However, it is pointed out that the SBEs in the HF model tend to give incorrect results in that these underestimate excitonic effects on the intraband dynamics [78], and overestimate contributions from continuum due to partial lack of excitonic correlation [73, 76]. Moreover, theories beyond the HF approximation have been presented. For instance, in the method of dynamic controlled truncation of the hierarchy of density matrices [78] and the quasibosonic-EX model [79], indicating significance of the EX-EX correlation in intraband dynamics such as terahertz (THz) radiation [80, 81, 82]. On the other hand, incoherent scattering relevant to excitation-induced dephasing [83, 84, 85, 86, 87] is drastically reduced in the regime of strong and resonant pumping of EXs, leading to little alteration from the HF dynamics at the more sophisticated dressed second-Born level [88]. Discussion of such Coulomb correlation effects is beyond the scope of this section. To our knowledge, investigations toward the many-body Coulomb effect (including the HF effect) on non-linear dynamics of FR excitons have still remained unexplored but a few experimental works [35, 36] suggesting its significance. Therefore, it is considered worth showing the importance of the Coulomb collisions in the dressed FR system concerned here even within the mean-field approximation.

3.2.2. Theory

The Hamiltonian of the system concerned here reads

$$H = \sum_{\substack{l,k,\\i=(c,v)}} \varepsilon_{lk}^{(i)} a_{lk}^{(i)\dagger} a_{lk}^{(i)} + \frac{1}{2} \sum_{\substack{l_1,l_2,l_3,l_4,\\k,k',q(\neq 0),\\i,j=(c,v)}} V_{l_1 l_2 l_3 l_4}(q) a_{l_1 k+q}^{(i)\dagger} a_{l_2 k'-q}^{(j)\dagger} a_{l_3 k'}^{(j)} a_{l_4 k}^{(i)}$$

$$- \sum_{l,l',k} F(t) \left\{ a_{lk}^{(c)\dagger} a_{l'k}^{(v)} d_{ll'} + (h.c.) \right\}. \quad (32)$$

Here $a_{lk}^{(i)\dagger}$ ($a_{lk}^{(i)}$) is a creation (an annihilation) operator of the carrier i (either a conduction electron c or a valence-band electron v) at a site l with a crystal momentum $\mathbf{k}$ in the layer plane, satisfying anti-commutation relations. The first term of Eq. (32) represents a sum of an energy of a WSL subband and a kinetic energy of the in-plane motion. The second term stands for electron-electron interactions accompanied by momentum transfer $\mathbf{q}$. The third term means a dipole interaction of the carrier with an external electric field of $F(t)$ with a dipole moment $d_{ll'}$, where k-dependence on $d_{ll'}$ is neglected as usual, and $(h.c.)$ is meant by taking a hermition conjugate of the first term in the curl brackets. The laser field $F(t)$ is given within the rotating-wave approximation by

$$F(t) = \frac{1}{2} \left(\sum_{j=1,2} F_j(t) \exp\left(i[\mathbf{K}_j \cdot \mathbf{R} - \omega_j t]\right) \right), \quad (33)$$

where F_j is an envelope of the jth pulse ($j = 1$ for a probe beam and $j = 2$ for a pump one) with $\mathbf{K}_j$ and ω_j a momentum and a center frequency of light, respectively, and $\mathbf{R}$ being position conjugate to $\mathbf{K}_j$. Each pulse is considered Gaussian, having a temporal width, σ_j, and the maximum amplitude, F_{j0}. That is,

$$F_j(t) = F_{j0} g_j(t), \quad (34)$$

where

$$g_1(t) = \exp\left(-\frac{(t+\tau)^2}{\sigma_1^2}\right), \quad g_2(t) = \exp\left(-\frac{t^2}{\sigma_2^2}\right), \quad (35)$$

with τ a delayed time.

A density matrix is defined by

$$\rho_{ll'\mathbf{k}}^{(ij)} = \langle a_{l\mathbf{k}}^{(i)\dagger} a_{l'\mathbf{k}}^{(j)} \rangle, \tag{36}$$

where $\langle \cdots \rangle$ has been meant by taking an expectation value. Following a convention, it is convenient to rewrite the density matrix in the electron-hole representation by introducing operators $a_{l\mathbf{k}} \equiv a_{l\mathbf{k}}^{(c)}$ for the electron and $b_{l-\mathbf{k}} \equiv a_{l\mathbf{k}}^{(v)\dagger}$ for the hole. Thus the density matrices are recast into

$$\begin{aligned}
\rho_{ll'\mathbf{k}}^{(vc)} &= \langle b_{l-\mathbf{k}} a_{l'\mathbf{k}} \rangle \equiv P_{ll'\mathbf{k}}, \\
\rho_{ll'\mathbf{k}}^{(cc)} &= \langle a_{l\mathbf{k}}^{\dagger} a_{l'\mathbf{k}} \rangle \equiv N_{ll'\mathbf{k}}^{(e)}, \\
\rho_{ll'\mathbf{k}}^{(vv)} &= \langle b_{l-\mathbf{k}} b_{l'-\mathbf{k}}^{\dagger} \rangle \equiv \delta_{ll'} - N_{ll'-\mathbf{k}}^{(h)},
\end{aligned} \tag{37}$$

where $N_{ll'-\mathbf{k}}^{(h)} = \langle b_{l'-\mathbf{k}}^{\dagger} b_{l-\mathbf{k}} \rangle$. $P_{ll'\mathbf{k}}$ indicates a microscopic polarization of an interband transition, a diagonal element of $N_{ll\mathbf{k}}^{(e/h)}$ stands for a population density of e/h, and an off-diagonal element $N_{ll'\mathbf{k}}^{(e/h)}$ for $l \neq l'$ represents a microscopic polarization of an intraband transition between WSL sites l and l' of e/h.

The Liouville equation for the operators associated with $P_{ll'\mathbf{k}}$ and $N_{ll'\mathbf{k}}^{(e/h)}$ yields the following set of SBE within the random phase approximation [77]

$$i\left(\frac{d}{dt} + \frac{1}{T_2}\right) P_{l_h l_e \mathbf{k}} = \sum_n \left[\epsilon_{nl_e\mathbf{k}}^{(e)} P_{l_h n\mathbf{k}} + \epsilon_{l_h n\mathbf{k}}^{(h)} P_{nl_e\mathbf{k}} \right. \\
\left. + \Omega_{nl_h\mathbf{k}} N_{nl_e\mathbf{k}}^{(e)} + \Omega_{l_e n\mathbf{k}} N_{l_h n-\mathbf{k}}^{(h)} - F(t) d_{l_e n} \delta_{l_h n} \right], \tag{38}$$

$$i\left(\frac{d}{dt} + \frac{1}{T_1^{(e)}}\right) \Delta N_{l_e l'_e \mathbf{k}}^{(e)} = \sum_n \left[\epsilon_{nl'_e\mathbf{k}}^{(e)} N_{l_e n\mathbf{k}}^{(e)} - \epsilon_{l_e n\mathbf{k}}^{(e)} N_{nl'_e\mathbf{k}}^{(e)} \right. \\
\left. + \Omega_{nl_e\mathbf{k}}^{\dagger} P_{nl'_e\mathbf{k}} - \Omega_{l'_e n\mathbf{k}} P_{l_e n\mathbf{k}}^{\dagger} \right], \tag{39}$$

$$i\left(\frac{d}{dt} + \frac{1}{T_1^{(h)}}\right) \Delta N_{l_h l'_h -\mathbf{k}}^{(h)} = \sum_n \left[\epsilon_{l_h n\mathbf{k}}^{(h)} N_{nl'_h -\mathbf{k}}^{(h)} - \epsilon_{nl'_h\mathbf{k}}^{(h)} N_{l_h n-\mathbf{k}}^{(h)} \right. \\
\left. + \Omega_{l'_h n\mathbf{k}}^{\dagger} P_{l_h n\mathbf{k}} - \Omega_{nl_h\mathbf{k}} P_{nl'_h\mathbf{k}}^{\dagger} \right]. \tag{40}$$

Here, an effective EX energy $\epsilon_{ll'\mathbf{k}}^{(e/h)}$ and an effective Rabi energy $\Omega_{ll'\mathbf{k}}$ have been defined that

$$\epsilon_{l_e l'_e \mathbf{k}}^{(e)} = \left(\frac{\mathbf{k}^2}{2m_{e\|}} + \varepsilon_{l_e}^{(e)} + E_g\right) \delta_{l_e l'_e} + \Sigma_{l_e l'_e \mathbf{k}}^{(e)}, \tag{41}$$

$$\epsilon_{l_h l'_h \mathbf{k}}^{(h)} = \left(\frac{\mathbf{k}^2}{2m_{h\|}} + \varepsilon_{l_h}^{(h)}\right) \delta_{l_h l'_h} + \Sigma_{l_h l'_h \mathbf{k}}^{(h)} + \Sigma_{l_h l'_h}^{(ch)}, \tag{42}$$

$$\Omega_{l_e l_h \mathbf{k}} = F(t) d_{l_e l_h} + \Pi_{l_e l_h \mathbf{k}}, \tag{43}$$

with $m_{e(h)\|}$ and $\varepsilon_l^{(e(h))}$ an in-plane mass and a WSL subband energy of $e(h)$, respectively, E_g a band gap of the concerned bulk semiconductor, and $\Sigma_{l_h l'_h}^{(ch)}$ a Coulomb-hole self-energy. Moreover an exchange self-energy $\Sigma_{ll'\mathbf{k}}^{(e/h)}$ and an internal renormalized field-energy $\Pi_{l_e l_h \mathbf{k}}$ have been given by

$$\Sigma_{l_e l'_e \mathbf{k}}^{(e)} = -\sum_{n_e n'_e \mathbf{q}} V_{l'_e l_e n_e n'_e}(\mathbf{q}) N_{n_e n'_e \mathbf{k}-\mathbf{q}}^{(e)}, \qquad (44)$$

$$\Sigma_{l_h l'_h \mathbf{k}}^{(h)} = -\sum_{n_h n'_h \mathbf{q}} V_{l'_h l_h n_h n'_h}(\mathbf{q}) N_{n_h n'_h -(\mathbf{k}-\mathbf{q})}^{(h)}, \qquad (45)$$

$$\Pi_{l_e l_h \mathbf{k}} = \sum_{n_e n_h \mathbf{q}} V_{l_e l_h n_h n_e}(\mathbf{q}) P_{n_h n_e \mathbf{k}-\mathbf{q}}. \qquad (46)$$

In Eq. (38), $d_{l_e l_h}$ is a dipole moment of a transition between the l_eth site of e and the l_hth site of h. In Eqs. (39) and (40), $\Delta N_{ll'\mathbf{k}}^{(e/h)} = N_{ll'\mathbf{k}}^{(e/h)} - N_{ll'\mathbf{k}}^{(e/h,eq)}$ with $N_{ll'\mathbf{k}}^{(e/h,eq)}$ a quasi-equilibrium density. In Eqs. (38)-(40), a phenomenological time of population relaxation $T_1^{(e/h)}$ and a dephasing time T_2 have been introduced.

Since FR wavefunctions of the WSL-EX are prepared in the ρ-space based on the theoretical framework of Section 2., it is convenient to take the Fourier-transformation of $P_{ll'\mathbf{k}}$ and $N_{ll'\mathbf{k}}^{(e/h)}$ from the k-space into the ρ-space, that is

$$Z_{\mathbf{k}} = (2\pi)^{-2} \int d\boldsymbol{\rho} \exp(-i\mathbf{k}\cdot\boldsymbol{\rho}) \tilde{Z}_{\boldsymbol{\rho}} \qquad (47)$$

for $Z_{\mathbf{k}} = P_{ll'\mathbf{k}}$ or $N_{ll'\mathbf{k}}^{(e/h)}$. Furthermore the transforms of $\tilde{P}_{ll'\boldsymbol{\rho}}$ and $\tilde{N}_{ll'\boldsymbol{\rho}}^{(e/h)}$ are expanded with respect to a complete set $\{\varphi_{ll'\alpha}^{E}(\boldsymbol{\rho})\}$ as:

$$\tilde{P}_{ll'\boldsymbol{\rho}}(t) = \sum_{\alpha E} \varphi_{ll'\alpha}^{E}(\boldsymbol{\rho}) p_{\alpha E}(t), \qquad (48)$$

$$\tilde{N}_{ll'\boldsymbol{\rho}}^{(e/h)}(t) = \sum_{\alpha E} \varphi_{ll'\alpha}^{E}(\boldsymbol{\rho}) n_{\alpha E}^{(e/h)}(t). \qquad (49)$$

Here $\varphi_{ll'\alpha}^{E}$ is a projection of the αth excitonic-FR envelope-function Ψ_α of Eq. (6) for a given energy E onto the ρ-space, defined by

$$\varphi_{ll'\alpha}^{E}(\boldsymbol{\rho}) = \langle\langle \phi_l^{(h)}(z_h) \phi_{l'}^{(e)}(z_e) | \Psi_\alpha(|\boldsymbol{\rho}|, \Omega) \rangle\rangle_{z_e, z_h}, \qquad (50)$$

with $\langle\langle \cdots \rangle\rangle_{z_e, z_h}$ representing integrations over two variables of z_e and z_h. Here, this wavefunction satisfies the closure relation

$$\sum_{\alpha E} |\varphi_{ll'\alpha}^{E}(\boldsymbol{\rho})\rangle\langle\varphi_{ll'\alpha}^{E}(\boldsymbol{\rho}')| = \delta(\boldsymbol{\rho}-\boldsymbol{\rho}'). \qquad (51)$$

It should be noted that employing the EX basis set of Eq. (51) used in the expansion of Eqs. (48) and (49) has a merit that the FR effect and the relevant Coulomb interaction between e and h within a single EX is automatically included in theory; however, a part of this effect is conventionally missing in the SBE with the HF model using the free-particle basis set for each of e and h.

To evaluate FWM signals, first one takes spatial Fourier-transformations of $p_{\alpha E}$ and $n_{\alpha E}$ as follows:

$$p_{\alpha E} = \sum_l \exp\left(i[(\mathbf{K} + l\mathbf{\Delta K}) \cdot \mathbf{R}]\right) p_{\alpha E}^{(l)} \qquad (52)$$

and

$$n_{\alpha E} = \sum_l \exp\left(il\mathbf{\Delta K} \cdot \mathbf{R}\right) n_{\alpha E}^{(l)}, \qquad (53)$$

respectively, where $\mathbf{K} = (\mathbf{K}_1 + \mathbf{K}_2)/2$ and $\mathbf{\Delta K} = (\mathbf{K}_1 - \mathbf{K}_2)/2$. Putting these partial-wave expressions into the SBEs corresponding to $p_{\alpha E}$ and $n_{\alpha E}$ yields a set of coupled equations for $p_{\alpha E}^{(l)}$ and $n_{\alpha E}^{(l)}$. A macroscopic polarization P corresponding to a time-resolved FWM signal, diffracted in the $2\mathbf{K}_2 - \mathbf{K}_1$ direction in the time domain, is cast into the form:

$$P(t, \tau) = \sum_{\alpha E} \mu_\alpha(E)\, p_{\alpha E}^{(-3)}(t), \qquad (54)$$

where a dipole moment $\mu_\alpha(E)$ is an interband transition to a FR state in the αth open-channel at a given E, and this is provided by

$$\mu_\alpha(E) = \mu_0 \int dz\, [\Psi_\alpha(|\boldsymbol{\rho}| = 0, \Omega)]_{z \equiv z_e = z_h}, \qquad (55)$$

with μ_0 meaning a dipole moment of an interband transition of the bulk crystal concerned here. The macroscopic polarization in the frequency domain and a SRFWM signal are given by

$$\tilde{P}(\omega, \tau) = \sum_{\alpha E} \mu_\alpha(E)\, \tilde{p}_{\alpha E}^{(-3)}(\omega), \qquad (56)$$

and $|\tilde{P}(\omega, \tau)|^2$, respectively, where $p_{\alpha E}^{(-3)}(t) = \int d\omega\, \exp(-i\omega t)\, \tilde{p}_{\alpha E}^{(-3)}(\omega)$.

3.2.3. Analytic Model without Coulomb Exchange

For qualitative understanding of the underlying physics, it is worth obtaining an analytic expression of $\tilde{p}_{\alpha E}^{(-3)}$ under the following approximations. First, the many-body effects relevant to the second term of Eq. (32) are neglected in the SBEs. Second, since the highest-lying open-channel and the lowest-lying closed channel are usually dominant compared with other distant channels, just this pair of channels at every E is kept and others are neglected. Hereafter, this is termed as the dominant-channel approximation. For the sake of further simplicity, it is assumed that $F_{10} \ll F_{20}$, $\sigma_2 \gg 1$ and $T_1 = T_2 \equiv 2/\Gamma$, where the first equation shows a weak probe limit, and the second one is justified in the narrow band limit of the pumping laser, which makes valid the approximation of $F_2(t) \approx F_{20}$. Under these approximations with a couple of other subsidiary approximations, the analytic expression of $\tilde{p}_{\alpha E}^{(-3)}(\omega)$ is obtained as follows:

$$\tilde{p}_{\alpha E}^{(-3)}(\omega) = \frac{8\pi^2 [\mu'_\alpha(E)]^*}{(\omega - \varepsilon_+ + i\Gamma/2)(\omega - \varepsilon_- + i\Gamma/2)} \mathcal{F}(\omega), \qquad (57)$$

where
$$\mathcal{F}(\omega) \approx \tilde{\Omega}_1^*(2\omega_2 - \omega)\Omega_{20}^2 \frac{(\omega - \omega_2)\exp(-i\theta) + \Gamma\sin\theta}{(\varepsilon - \omega_2 - i\Gamma/2)(\omega - \omega_2 + i\Gamma/2)}, \tag{58}$$

and $\varepsilon_\pm = (\varepsilon + \omega_2 \pm \Omega_0)/2$ with $\Omega_0 = \sqrt{(\varepsilon - \omega_2)^2 + 8\Omega_{20}^2}$ and $\varepsilon = E + \Sigma^{(ch)}$. Here, $\tilde{\Omega}_j(\eta)$ has been defined as

$$\tilde{\Omega}_j(\eta) = \frac{1}{2\pi}\int d\eta \, \exp(i\eta t)\Omega_j(t), \tag{59}$$

where $\Omega_j(t) = \Omega_{j0}g_j(t)\exp(-i\omega_j t)$ with $\Omega_{j0} = \mu_0 F_{j0}/2$ for $j = 1, 2$, and $\tilde{\Omega}_2(\eta) = \Omega_{20}\delta(\eta - \omega_2)$. Further, by use of the Fano model for the case of one-open and one-closed channels [1], $\mu_\alpha(E)$ is expressed by $\mu_\alpha(E) = \mu_{bg}\exp(i\theta)\mu'_\alpha(E)$, where

$$\mu'_\alpha(E) = \frac{\epsilon + q}{\epsilon - i}. \tag{60}$$

Here, q is a Fano parameter, μ_{bg} means a dipole moment due to a background continuum being almost energy-independent, $\epsilon = (\varepsilon - \varepsilon_r)/(\gamma/2)$ with ε_r a FR energy position and γ a natural FWHM of FR, and $\theta = -\arctan(1/q) + \theta_0$ with θ_0 an overall phase constant. $\mathcal{F}(\omega)$ depends linearly on Ω_{10}, while the effect of Ω_{20} is treated nonperturbatively. $\tilde{p}_{\alpha E}^{(-3)}$ obviously has two poles at $\omega = \varepsilon_\pm - i\Gamma/2$, which is reminiscent of the AT doublet [40, 65] because of $|\varepsilon_+ - \varepsilon_-| = \Omega_0$ if Ω_0 is considered the Rabi frequency.

Linear absorption spectra of FR given by the Fano model is shown in Figure 7(a) with assuming $\theta_0 = 0$. For $q = 0$ and -1000, the profiles are symmetric. The former case ($q = 0$) corresponds to the extreme FR limit having an obvious transparent window. On the other hand, the latter ($q = -1000$) corresponds to the weak FR limit with a Lorentzian shape, however, no discernible window. The case of $|q| \gg 1$ is thus considered a pure bound-state rather than the FR, aside from a finite spectral width. Profiles of other q-values are asymmetric.

In Figure 7(b), the SRFWM signals, $|\tilde{P}(\omega,\tau)|^2$ obtained from Eqs. (56) and (57) are shown for some q-values in resonant excitations that $\omega_1 = \omega_2 = \varepsilon_r$, where $\theta_0 = 0$ is assumed again. The signals exhibit an AT-like doublet and the relative intensities of both lobes depend strongly upon q, resulting in asymmetric spectral profiles except for $q = 0$ and -1000, in which the signals are symmetric. Specifically, for $q = -1000$ corresponding to a bound state transition, these symmetric ac-Stark sidebands are common to a two-level atomic system [65]. It is found that q-values rendering the linear absorption spectra symmetric (asymmetric) also induce symmetric (asymmetric) shapes of SRFWM signals. The spectral pattern is determined just by the numerator of Eq. (58). Because the q-value indicates an interference between contributions of an open channel and a closed channel [1], the asymmetric doublet manifested here arises from phase modulation in polarization coherence $\tilde{p}_{\alpha E}^{(-3)}$ due to this Fano coupling through the q-value. Therefore, it is concluded that the asymmetry of the AT-doublet arises from interference due to the phase of θ.

3.2.4. SBE Results with Coulomb Exchange

We will proceed to SRFWM signals, provided by a fully numerical calculation based on the SBEs in Section 3.2.2. As a sample for this calculation, we used 34Å-GaAs/17Å-Al$_{0.3}$Ga$_{0.7}$

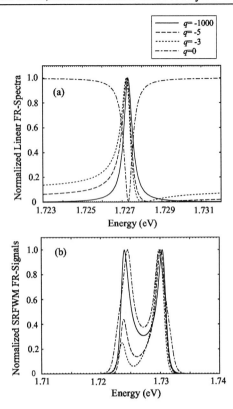

Figure 7. (a) Normalized linear FR-spectra, $|\mu_\alpha(\omega)|^2$, obtained by the Fano model of Eq. (60) versus incident photon energy ω for $q = $ -1000, -5, -3 and 0. (b) Normalized SRFWM FR-signals, $|\tilde{P}(\omega,\tau)|^2$, obtained by the analytic model of Eq. (57) versus ω for the same q-values as those of panel (a) in resonant excitation of $\omega_1 = \omega_2 = \varepsilon_r = 1.7273$ eV. Other setup parameters are: $F_{10} = 1$ kV/cm, $F_{20} = 100$ kV/cm, $\sigma_1 = \sigma_2 = 0.5$ ps, $T_1 = T_2 = 1$ ps, $\tau = 0.8$ ps, and the FR width (FWHM) 0.5 meV.

as superlattices with a bias field of 25 kV/cm along the crystal growth direction. Photoabsorption spectra of the WSL-FR states in the present system are calculated based on the theory in Section 2., and shown in Figure 8, where each peak is labeled by $n(ks)$ with n a WSL index and k a hydrogenic principle quantum number of EX. For the calculation of the SRFWM signals, the whole effects of the HF Coulomb exchange are incorporated without assuming any additional imposition upon σ_j and F_{j0}. The dominant-channel approximation is still adopted, which is valid in the present sample. A frequency distribution of the pulse 2, with $\sigma_2 = 500$ fs, tuned resonantly to $\varepsilon_r = 1.7273$ eV is also depicted in Figure 8. The pump field coherently excites most parts of the FR state of $0(1s)$, and slightly overlaps with adjacent FR peaks such as $-1(3s)$ and $0(2s)$.

The resulting SRFWM signals are shown in Figure 9, where the solid and dashed curves represent traces with and without the many-body effects, respectively, for three delayed times of τ=-0.4, 0, and 0.4 ps, with $T_2 = 1$ ps. Figures 9(a)-9(c) show the results for $F_{10} = 1$ kV/cm and $F_{20} = 50$ kV/cm. The profiles without the many-body effect are almost

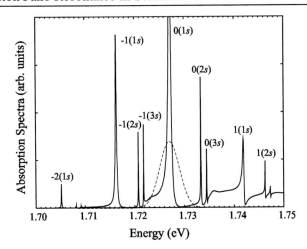

Figure 8. Linear absorption spectra of the sample under consideration versus photon energy. Peaks are denoted by $n(ks)$ with n the WSL index and k an exciton quantum number. The dashed curve stands for a frequency distribution of the pumping pulse with $\sigma_2 = 0.5$ ps and tuned to $\varepsilon_r = 1.7273$ eV.

Lorentzian, whereas those with this effect are modulated to some extent and slight doublet structures are discerned, especially, for $\tau=0.4$ ps. For greater T_2 than 1 ps, comparable to or smaller than the FR width of nearly 0.5 meV of the peak $0(1s)$ in Figure 8, it would be likely that a SRFWM signal reflects a generic pattern of the Fano profile of the linear spectra. Otherwise, a detailed asymmetry pattern seems to be smeared out by more rapid dephasing.

As seen in Figures 9(d)-9(f), the many-body effect is more dominant with an increase in pumping intensity, and this results in marked modification of the spectra, compared with Figures 9(a)-9(c), Here, the signals denoted by the solid and dashed lines of Figure 9(f) are cited as signals, S and S', respectively. The overall shape of S' appears similar to that of the signal of $q = -5$ in Figure 7(b), aside from a slight variance of a relative intensity of the two sidebands, due presumably to the fact that the model calculations described in Section 7 are ensured for $\sigma_2 \gg 1$. The following three characteristics of the dressed FR are found. First, the main profile of S is shifted toward the high-energy side, namely, blue-shifted. Second, the relative intensity of its two lobes is reversed from that of S'; the right lobe is stronger than the left one in S, and vice versa in S'. Third, such asymmetry reversal also accompanies prominent changes of spectral widths; the right lobe is still narrower than the left one, whereas the widths of the two sidebands in S' are almost equal.

The two characters of asymmetry reversal and spectral narrowing are unequivocally attributed to interplays between the phase-modulation induced by FR couplings and the many-body effects. The phase modulation also contributes to the formation of the asymmetry in the AT-like doublet as was discussed in Section 3.2.3. When regarding the blue shift, a similar effect was already observed in the *nonresonant* ac-Stark effect [89, 90, 91], and hence, this would be due mostly to temporal evolution of the repulsive EX-EX interactions.

For still stronger dephasing ($T_2 = 0.2$ ps) of polarizations, SRFWM signals no longer

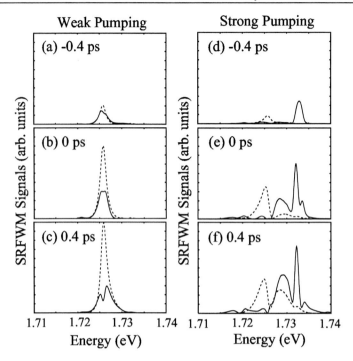

Figure 9. SRFWM spectra, $|\tilde{P}(\omega,\tau)|^2$, obtained by the fully numerical calculations versus ω for different τ's in resonant excitations of $\omega_1 = \omega_2 = \varepsilon_r = 1.7273$ eV. Panels (a)-(c) for weak pumping with $F_{20} = 50$ kV/cm, and panels (d)-(f) for strong pumping with $F_{20} = 100$ kV/cm. Solid (dashed) curves represent spectra with (without) the many-body effect, where (a) and (d) are for $\tau = -0.4$ ps (the pulse 2 precedes the pulse 1), (b) and (e) for $\tau = 0$ ps, and (c) and (f) for $\tau = 0.4$ ps (the pulse 1 precedes the pulse 2), respectively. For more detail, consult the text.

exhibits the AT-like doublet even for $F_{20} = 100$ kV/cm (though not shown here). Increasing the pumping strength to $F_{20} = 400$ kV/cm, the doublet structure is retrieved if the many-body effect is taken into account. For the splitting to manifest itself, it is necessary that an effective Rabi energy, namely, an external pump-field plus an internal renormalized-field, is greater than an inverse of dephasing time.

3.3. Dynamic Excitonic FR

3.3.1. Introduction

The advent of affordable intense THz light sources such as molecular THz lasers and free-electron lasers has enabled innovative research in the areas of high-power THz excitation of semiconductors and the relevant coherent control of quantum dynamics [92, 93]. It has been found that the irradiation of a periodically oscillating THz wave causes photon-assisted tunneling (PAT) in QWs, SLs, and WSLs. In particular, the application of an appropriately controlled periodic drive to these systems brings quantum transport and diffusion to an almost complete standstill [94]. In THz-driven SLs and THz-driven WSLs, such spatial localization is called dynamic localization (DL). Dynamic localization(DL) in THz-driven SLs was first discussed by Dunlap and Kenkre [43], who pointed out that the localization occurs when the special relation of $J_0(x) = 0$ is satisfied, where $x = Fd/\omega$, F is the strength of the THz-field, and ω is its frequency; $J_n(x)$ is the nth-order Bessel function of the first kind. Holthaus [44] also showed that the quasi-energy as a function of x is of repeated gourd-shape, and band reformation and band collapse are accompanied by dynamic delocalization and DL, respectively. Furthermore, Zak [95] pointed out that a similar property also holds in the THz-driven WSL and that DL occurs when $J_n(x) = 0$, where n is an integer equal to the electric matching ratio Ω_B/ω, where $\Omega_B = F_0 d$ is the Bloch frequency of the system concerned

Physics underlying THz-driven SLs and THz-driven WSLs is enriched by additional complexity of an EX effect [41] and interminiband interactions caused by ac-ZT due to the THz wave [96, 97]. In this section, we focus on Floquet EX-states in THz-driven SLs, in which the EX state indicates a characteristic FR effect absent from an EX state in non-THz-driven SLs, namely, original SLs. It should be noted that in the latter system, the lowest joint-miniband of the SLs supports nothing but EX *bound*-states; though the higher joint-minibands can support EX-FR states. On the other hand, in the former system, the THz drive allows this original bound state to be energetically degenerate with EX continuum-states of photon sidebands (PSBs) pertaining to the higher-minibands, resulting in anomalous EX-FR states. As shown in Figure 10, this FR is mediated by ac-ZT, in addition to Coulomb coupling common to the conventional FR presented in Sections 3.1. and 3.2. It should be noted that such FR is not induced without ac-ZT . Here, PSBs mean replicas of the original quantum states formed by absorbing/emitting photons of the THz wave. The resulting FR differs from the conventional FR in that the FR coupling concerned here is enabled to be tuned by modulating the intensity and the frequency of the applied THz field through the ac-ZT. In this sense of the *dynamic* coupling, the present FR is called DFR, contrasted with the conventional FR mediated by a *static* coupling. The DFR coupling becomes more effective with an increase in intensity of the THz field. To understand this DFR effect on Floquet EX-states, linear absorption spectra of the THz-driven SLs are evaluated by use of the Liouvílle equation in the range from a low THz-field region to a relatively strong THz -field region in which DL likely occurs.

3.3.2. Formulation

The total Hamiltonian of the system concerned comprises the joint-miniband SL Hamiltonian composed of field-free Hamiltonians of both of a conduction (c)-band and a valence (v)-band, a Coulomb interaction between electrons, an intersubband interaction caused by a driving THz wave $F(t)$, and an interband interaction invoked by a probe laser $f_p(t)$, where $F(t) = F_{ac} \cos(\omega t)$ and $f_p(t) = f_{p0} \cos(\omega_p t)$ with F_{ac} and f_{p0} as amplitudes, ω and ω_p as frequencies, and t as time. The microscopic polarization of an interband transition defined by $p_{\lambda\lambda'\mathbf{K}_\parallel}(t) \equiv \langle a^{(v)\dagger}_{\lambda\mathbf{K}_\parallel} a^{(c)}_{\lambda'\mathbf{K}_\parallel} \rangle$ is seeked by solving the Liouvílle equation [77], based on the same theoretical framework as made in Section 3.2.2. [11, 42]. Here, $\lambda^{(\prime)}$ represents the lump of a SL miniband index $b^{(\prime)}$ and a SL lattice site $l^{(\prime)}$, namely, $\lambda^{(\prime)} = (b^{(\prime)}, l^{(\prime)})$, and $\mathbf{K}_\parallel$ is the in-plane momentum of a pair of electrons of c and v bands, where this is associated with a relative motion of these two electrons in the plane normal to the direction of crystal

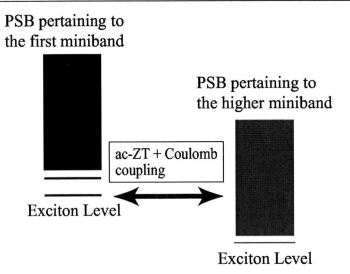

Figure 10. Scheme of EX-DFR mechanism mediated by ac-ZT and Coulomb coupling. This shows that an EX discrete-level pertaining to one PSB relevant to the first SL miniband is embedded in an EX-continuum pertaining to the other PSB relevant to the higher SL miniband.

growth (the z-axis). Further, $a^{(s)\dagger}_{\lambda \mathbf{K}_\parallel}$ and $a^{(s)}_{\lambda \mathbf{K}_\parallel}$ represent creation and annihilation operators of electron with λ and $\mathbf{K}_\parallel$ in band s, respectively, which satisfy usual anti-commutation relations. $\langle \cdots \rangle$ is meant by taking an expectation value.

The main approximations made here are as follows. (i) The nearest-neighbor tight-binding (NNTB) model for c- and v-band SL Hamiltonians is employed. (ii) A Coulomb interaction just for EX composed of a single electron-hole pair is retained, whereas the many-body Coulomb correlation effect is neglected. (iii) It is assumed that a probe laser is weak enough to satisfy the relation $F_{ac} \gg f_{p0}$ and that ω_p is much greater than ω, namely, $\omega \ll \omega_p$. Thus, $F(t)$ does not contribute to interband transitions and $f_p(t)$ does not contribute to intersubband transitions. The resulting absorption coefficient is linear in $f_p(t)$, however non-linear in $F(t)$.

The exact expression of the absorption coefficient $\alpha^{(ex)}_{abs}(\omega_p; \omega)$ for the interband optical transition between "time-dependent" Floquet states is not trivial a priori, differing from that between usual time-independent steady states. For the purpose of deriving it in an ab initio way, we begin with the Liouvílle equation under the approximation (ii). The equation to be solved is reduced to

$$i\left(\frac{d}{dt} + \gamma - i\omega_p\right)\bar{p}(\boldsymbol{\rho}, z_v, z_c, t) + (2\pi)^2 e^{i\omega_p t} f_p^{(+)}(t) d_0^{(vc)} \delta(\boldsymbol{\rho})\delta(z_v - z_c)$$
$$= \int dz \left[\bar{p}(\boldsymbol{\rho}, z_v, z, t) H^{(c)}(z, z_c, t) - H^{(v)}(z_v, z, t)\bar{p}(\boldsymbol{\rho}, z, z_c, t)\right]$$
$$+ \mathcal{H}(\boldsymbol{\rho}, z_v, z_c)\bar{p}(\boldsymbol{\rho}, z_v, z_c, t), \tag{61}$$

where in place of $p_{\lambda\lambda'\mathbf{K}_\parallel}(t)$, the positional representation of a microscopic polarization,

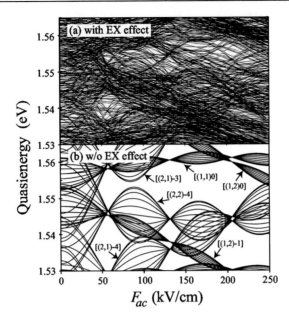

Figure 11. Quasienergy E_ν (eV) of THz-driven SL state as a function of F_{ac} (kV/cm). Panels (a) and (b) show results with and without EX effect, respectively. In panel (b), the main component of Houston state is denoted as $[(b_c, b_v)J]$ in every tilted gourd-shaped quasienergy band.

$\bar{p}(\boldsymbol{\rho}, z_v, z_c, t)$, has been used, which is defined as

$$\bar{p}(\boldsymbol{\rho}, z_v, z_c, t) = e^{i\omega_p t} \sum_{\lambda,\lambda'} \int d\mathbf{K}_\| e^{i\mathbf{K}_\| \cdot \boldsymbol{\rho}} \langle z_v|\lambda\rangle p_{\lambda\lambda'\mathbf{K}_\|}(t)\langle\lambda'|z_c\rangle. \quad (62)$$

$\boldsymbol{\rho}$ is a radial coordinate vector conjugate to $\mathbf{K}_\|$, and $\langle\lambda|z_{c/v}\rangle$ represents a Wannier state of the lth site of SL miniband b at a position $z_{c/v}$ for an electron in c/v band. In Eq. (61), a phenomenological homogeneous broadening γ has been introduced, $d_0^{(vc)}$ is an interband dipole transition matrix element of a bulk material, and $f_p^{(+)}(t) \equiv (f_{p0}/2)e^{-i\omega_p t}$. Further, the Hamiltonian, given by

$$\mathcal{H}(\boldsymbol{\rho}, z_v, z_c) = -\frac{\nabla_\rho^2}{2m_\|} - \frac{1}{\epsilon\sqrt{\rho^2 + (z_c - z_v)^2}}, \quad (63)$$

governs the Coulombic motion of an electron-hole pair in the ρ-direction (the layer plane), with $m_\|$ and ϵ as an in-plane reduced mass of the pair and a static dielectric constant, respectively. In addition, $H^{(s)}(z, z', t)$ has been given by $\langle z|\hat{H}^{(s)}(t)|z'\rangle$, with the operator

$\hat{H}^{(s)}(t)$ defined as

$$\hat{H}^{(s)}(t) = \sum_{\lambda=(l,b)} \left[(-1)^{b+\sigma^{(s)}} \frac{\Delta_b^{(s)}}{4} (|l,b\rangle\langle l+1,b| + |l+1,b\rangle\langle l,b|) + \epsilon_{0b}^{(s)}|\lambda\rangle\langle\lambda| \right]$$
$$- F(t) \frac{1}{2} \sum_{\lambda,\lambda'} \left[|\lambda\rangle Z_{\lambda\lambda'}^{(s)}\langle\lambda'| + |\lambda'\rangle Z_{\lambda'\lambda}^{(s)*}\langle\lambda| \right]. \qquad (64)$$

The first term represents the Hamiltonian of SLs for band s by means of the NNTB model, where $\Delta_b^{(s)}$ is width of miniband b in band s, and $\epsilon_{0b}^{(c/v)}$ is the center of miniband b reckoned from the bottom/top of the c/v band, with $\sigma^{(c)} = 0$ and $\sigma^{(v)} = 1$. Further, the second term stands for a dipole interaction with $F(t)$, and a dipole matrix element is given by $Z_{\lambda\lambda'}^{(s)} = ld\delta_{\lambda\lambda'} + X_{bb'}^{(s)}\delta_{ll'}(1 - \delta_{bb'})$, where it is noted that $X_{bb'}^{(s)}$ causes the ac-ZT to be stressed herein.

Equation (61) is an inhomogeneous equation corresponding to the homogeneous equation for a THz-driven Floquet EX-wavefunction $\psi_\nu(\boldsymbol{\rho}, z_v, z_c, t)$, given by

$$\left(i\frac{d}{dt} + E_\nu \right) \psi_\nu(\boldsymbol{\rho}, z_v, z_c, t)$$
$$= \int dz \left[\psi_\nu(\boldsymbol{\rho}, z_v, z, t) H^{(c)}(z, z_c, t) - H^{(v)}(z_v, z, t)\psi_\nu(\boldsymbol{\rho}, z, z_c, t) \right]$$
$$+ \mathcal{H}(\boldsymbol{\rho}, z_v, z_c)\psi_\nu(\boldsymbol{\rho}, z_v, z_c, t), \qquad (65)$$

where E_ν is a quasienergy and there is temporal periodicity of $\psi_\nu(\boldsymbol{\rho}, z_v, z_c, t+T) = \psi_\nu(\boldsymbol{\rho}, z_v, z_c, t)$ with $T = 2\pi/\omega$. Therefore, it is plausible to expand $\bar{p}(\boldsymbol{\rho}, z_v, z_c, t)$ as $\bar{p}(\boldsymbol{\rho}, z_v, z_c, t) = \sum_\nu \psi_\nu(\boldsymbol{\rho}, z_v, z_c, t) a_\nu$ in terms of the basis set $\{\psi_\nu(\boldsymbol{\rho}, z_v, z_c, t)\}$. Using the orthonormality relation of this set, one obtains

$$a_\nu = \frac{1}{E_\nu - \omega_p - i\gamma} \frac{(2\pi)^2 d_0^{(vc)*}}{T} \int_0^T dt\, e^{i\omega_p t} f_p^{(+)}(t) \bar{\psi}_\nu(t)^*, \qquad (66)$$

where $\bar{\psi}_\nu(t) \equiv \int dz\, [\psi_\nu(\boldsymbol{0}, z, z, t)]$. The price to be paid for obtaining such a simple form of a_ν leads to requirement of solving a rather involved evaluation of Eq. (65).

Employing the Fourier expansion of $\psi_\nu(\boldsymbol{\rho}, z_v, z_c, t)$ in view of its temporal periodicity, and further, making an expansion with respect to a joint-miniband state, one obtains MCS equations for the expansion components, the expression of which is equivalent to that of Eq. (7). Thus, the present EX-DFR problem is properly treated within the framework described in Section 2. (For a more detailed relation between the Floquet theory and the R-matrix theory, see Section 4.1.2..) A scattering channel for the resulting MCS equations is defined as a lump of the indexes of B, k, and J, where these indexes represent the labels of joint-miniband, joint Bloch momentum belonging to B, and PSB, respectively. It should be noted that with an increase in F_{ac}, that is, with an increase in the maximum value of $|J|$, the number of the scattering channels to be required for obtaining convergent results rapidly increases. Such a situation seems to entail a heavy computational burden, and make practical calculations infeasible.

Therefore, in place of the MCS calculation, a variational calculation is adopted here just for the purpose of the initial characterization of the EX-DFR effect. To this end, ψ_ν is represented by

$$\psi_\nu(\rho, z_v, z_c, t) = \sum_{k,B,J,i} \Phi^*_{[BJ]}(k; z_v, z_c, t)\varphi_i(\rho)C_{kBJi,\nu}. \tag{67}$$

$\Phi_{[BJ]}$ is the ZT-free and non-EX joint-miniband Floquet wavefunction defined as $\Phi^*_{[BJ]}(k; z_v, z_c, t) = \phi^{(v)}_{b_v j_v}(k; z_v, t)\phi^{(c)*}_{b_c j_c}(k; z_c, t)$, where $B \equiv (b_c, b_v)$, $J \equiv j_c - j_v$, and $\phi^{(s)}_{bj}(k; z, t)$ is a Houston wavefunction [54] for PSB j and miniband b of band s at k. Moreover, $\varphi_i(\rho)$ stands for the DVR basis function at grid i in the ρ-direction [47]. Equation (65) can be solved by resorting to the standard diagonalization procedure, and hence, a set of the coefficients $\{C_{kBJi,\nu}\}$ is obtained.

Once Eq. (61) is solved, a linear optical susceptibility $\chi(t)$ with respect to $f_p^{(+)}(t)$ is provided by

$$\chi(t) = \frac{1}{\epsilon_0}|d_0^{(vc)}|^2 \sum_\nu \frac{\mathcal{O}_\nu(t)}{E_\nu - \omega_p - i\gamma}, \tag{68}$$

where $\mathcal{O}_\nu(t) \equiv \bar{\psi}_\nu(t)T^{-1}\int_0^T dt'\bar{\psi}^*_\nu(t')$, and ϵ_0 is the dielectricity of vacuum. Accordingly, the absorption coefficient becomes of the form [42]:

$$\alpha^{(ex)}_{abs}(\omega_p; \omega) = \frac{\omega_p}{c}\sum_J \text{Im}\,\chi_J(\omega_p; \omega), \tag{69}$$

where $\chi(t) \equiv \sum_J e^{iJ\omega t}\chi_J(\omega_p; \omega)$, and c is the speed of light. Note that $\text{Im}\,\chi_0(\omega_p; \omega) \geq 0$.

3.3.3. Dynamic EX-FR Spectra

SLs of GaAs/Ga$_{0.75}$Al$_{0.25}$ as of 35/11 monolayers (ML) [1 ML=2.83 Å] for the well and barrier thickness are used for practical calculations. ω is set equal to $[\epsilon^{(c)}_{02} - \epsilon^{(c)}_{01}]/3$, namely, ω =31 meV so that the band $[(1, b_v)J]$ is resonant with the band $[(2, b_v)J]$ by three photon absorption. Quasienergies, E_ν, obtained by solving Eq. (65) are shown in Figure 11 (a) as a function of F_{ac}, while in Figure 11 (b) is shown non-EX results for the purpose of comparison with Figure 11 (a). In Figure 11 (b), the main component of the Houston state $[BJ]$ is designated as $[(b_c, b_v)J]$ in the respective tilted gourd-shaped quasienergy bands. Anticrossings arise from ac-ZT between states with different Bs and Js, however, with the same k; for instance, bands of $[(1, 1)0]$ and $[(1, 2) - 1]$ are repelled strongly. On the other hand, the structure of E_ν of Figure 11 (a) appears more intricate than that of Figure 11 (b) due to the EX effect causing entangled anticrossings among all Houston states. The sequence of Floquet-EX bound states is located right below lower-edges of every tilted gourd-shaped quasienergy band seen in Figure 11 (b), while EX (pseudo-)continuum states spread over all energies above these band edges. This situation implies that a Floquet EX-bound state is embedded in a Floquet EX-continuum state, bringing about DFR due to inter-PSB mixing via ac-ZT, as seen in Figure 10. In addition, the DL seen clearly in Figure 11 (b) is not discernible in Figure 11 (a) any longer. This observation would suggest that EX spectra are blurred to some extent due to the DFR effect at the DL, if the ac-ZT is significant. This appears contrasted with the reported results [41, 98], in which spectral intensity of THz-driven EXs at DL is more enhanced than or comparable to that of EXs under no THz drive.

Figure 12 shows Floquet-EX spectra, $\alpha_{abs}^{(ex)}(\omega_p; \omega)$, as a function of ω_p at several F_{ac}'s ranging from 0 to 132.5 kV/cm, where the spectra of Figures 12(a) and 12 (b) are results with and without the ac-ZT, respectively. The salient peaks seen around ω_p=1.53eV are assigned to the 1s Floquet EX-state pertaining to the $[(1, 1)0]$ band. As F_{ac} becomes large, the position of this peak [denoted as P_0 in Figure 12 (a)] shows red shift, accompanying a decrease in intensity. Further, line-width broadening is observed with an increase in F_{ac}, which would presumably arise from DFR. Note that the DL takes place at $F_{ac} = 57.5$ and 132.5 kV/cm, as seen in Figure 11 (b). However, no particular effect due to the DL is found in Figure 12 (a).

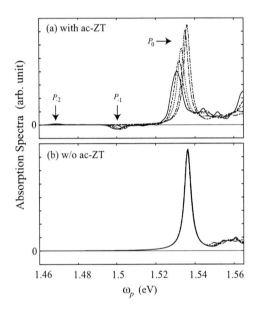

Figure 12. Absorption spectra $\alpha_{abs}^{(ex)}(\omega_p; \omega)$ with γ=2 (meV) as a function of ω_p (eV) at several values of F_{ac} (kV/cm): F_{ac} =132.5 (solid), 112.5 (dashed), 92.5 (dotted), 57.5 (chain), and 0 (double chain). Panels (a) and (b) show results with and without ac-ZT, respectively. In panel (a), the main peak of the 1s Floquet EX-state, the dip of the first replica, and the small peak of the second replica are denoted as P_0, P_{-1}, and P_{-2}, respectively.

Moreover, the additional dip [denoted as P_{-1}] and peak [denoted as P_{-2}] become pronounced around ω_p =1.5 and 1.47 eV, respectively, and move toward the lower-energy side coincidently with the red shift of P_0. Because the energy difference of P_0 from P_{-n} is approximately identical to $n\omega$ with $n = 1, 2$, it is understood that both P_{-1} and P_{-2} are manifested as the replica bands of the parent band P_0, resulting from relatively strong ac-ZT presumably between bands of $[(1, 1)0]$ and $[(1, 2) - 1]$. This tendency is much contrasted with that seen in Figure 12 (b), where the position of the main peak remains almost unaltered without any additional structure. It was also noted that P_{-1} exhibits negative absorption, namely, optical gain. Obviously, this anomaly is caused by the interplay of EX effect and ac-ZT, both of which are indispensable for its manifestation. The spectra without the EX effect show no such signal [97].

Finally, let us mention the recent study of THz-driven EX spectra in multiple QWs [99]. According to this, the excitonic resonance redshifts, and the absorption linewidth broadens due to the dynamic Franz-Keldysh effect. In this report, the polarization direction of THz wave is normal to the direction of crystal growth; and thus, neither DL nor ac-ZT are expected, differing from the present study, in which the polarization is along this direction.

4. Related Phenomena of Dynamic FR

Thus far, the focus has been on FR and DFR relevant to EX effects. In this section, novel DFR phenomena observed in semiconductors without the excitonic origin are presented. In Section 4.1., the DFR effect manifested in DL states of DWSLs [45] is presented exclusively from the viewpoint of a theoretical study based on the framework in Section 2. In Section 4.2., an experimental study on the DFR effect manifested in coherent phonon states of bulk materials under intense and ultra-short laser-pulse irradiation [46] is presented with a follow-up theoretical analysis; the theoretical framework of this analysis does not rely on that in Section 2., differing from the other sections in this chapter.

4.1. Dynamic FR in Dynamic Wannier-Stark Ladder

4.1.1. Introduction

Dynamic Localization (DL) characterizing a THz-driven WSL, called dynamic WSL (DWSL), shows the intriguing effects of quasienergy miniband collapse and complete stop of quantum transport and diffusion, as mentioned in Section 3.3. Further, it has been found that DL in the DWSL leads to the formation of sharp peaks in the optical spectra, and these are therefore more enhanced than or comparable to those in the corresponding WSL (without THz driving) [41, 98]. In addition, DL has also been extensively studied from the viewpoint of driven quantum tunneling and coherent control in systems such as optical SLs [100], atomic hyperfine and Zeeman-level structures [101, 102, 103], spin systems [104], and quantum chaos [105]. The concept of DL has recently been extended to trapped atoms in Bose-Einstein condensates [106], and Cooper pairs in Josephson qubits [107]. In asymmetric double QWs, the phenomenon termed coherent destruction of tunneling has been discussed [108, 109], in which tunneling is destroyed in a manner similar to that by DL, and further its relation with DL is understood from the viewpoint of group theory and the Landau-Zener problem [110].

In this section, we focus on DL from the viewpoint of its stability in the intense region of the THz field. According to some existing theoretical studies based on the lowest two-miniband tight-binding model [96, 111], even if the strength of the THz wave (F_{ac}) increases, miniband collapse is still observed, although ac-ZT across PSBs corresponding to different minibands becomes more significant, and this results in pronounced anticrossings. Therefore, it is considered that DL would retain a stable Floquet state and would be only minimally affected by ac-ZT. However, as F_{ac} is enhanced to the intense region of the order of hundreds of kV/cm, it can be hypothesized that a large number of PSBs contribute to this coupling, invalidating the conventional interpretation mentioned earlier. Further, state-of-the-art THz light sources that can realize such intensities have already been produced [112, 113].

Based on the ac-ZT-free Houston-Floquet picture [96, 111], quasienergies of a single-electron DWSL form a manifold structure labeled as (b, j), where b and j represent the indices of a miniband and a PSB, respectively. In particular, for the concerned DL, the (b, j) level is usually considered to be discrete. In fact, because of dc-ZT arising from an applied bias (with strength F_0), these DL levels would more or less incur shape resonance (SR) decay, although only the $(1, j)$ level can still be approximately considered to be discrete; the SR width broadens as b increases. Figure 15 shows the manifold structure of the DL levels, where each level is shaded, indicating the SR continuum. Here, it should be noted that because of ac-ZT coupling, the parent DL band of $(1, 0)$ with quasienergy E likely interacts with degenerate (SR-broadened) replicas of $(b > 1, j < 0)$, leading to instability. According to a simple perturbation picture, the parent band with $j = 0$ is believed to be predominantly coupled with an adjacent PSB with $|j| = 1$. This decay mechanism is similar to the conventional Fano effect [1] in the sense that the approximately discrete DL band is embedded in the SR-broadened DL bands, both of which couple each other by ac-ZT. Further, the strength of the ac-ZT coupling can be tuned by changing F_{ac} and the THz frequency. Therefore, this new mechanism can be classified as DFR, similar to EX-DFR presented in Section 3.3.

Considering the DFR, the DWSL problem reduces to a MCS problem, and the MCS states are numerically solved based on the R-matrix Floquet theory (RFT) [114]. By evaluating the excess DOS of Eq. (30) of DWSL, it is demonstrated that the DL band $(1, 0)$ is really unstable in a certain region of THz intensity, as shown later. The excess DOS is an important physical quantity because it is associated with the lifetime of the concerned resonance state, and further, it provides the initial characterization for understanding a more complicated problem, for instance, the problem of the transient interband coherent dynamics of a THz-driven semiconductors.

4.1.2. R-Matrix Floquet Theory

The RFT is introduced to the present DWSL problem. This consists of the following two stages. One is the stage of deriving MCS equations by applying to the concerned problem the Floquet expansion based on the Kramers-Henneberger (KH) transformation [115]. The other is the stage of applying the R-matrix propagation technique similar to Section 2.2. to these equations for obtaining the excess DOS.

Let us begin with the DWSL Hamiltonian

$$H(z, t) = \left[p_z + \frac{1}{c} A(t)\right] \frac{1}{m(z)} \left[p_z + \frac{1}{c} A(t)\right] + V(z), \tag{70}$$

where $V(z)$, $m(z)$, and p_z represent the SL confining potential, effective mass of an electron, and momentum operator along the crystal growth direction z, respectively, and $A(t)$ is a vector potential at time t for an applied electric field $F(t) = -\dot{A}(t)/c$, where c is the speed of light. The application of a gauge transformation and the KH transformation [115] to the DWSL wavefunction satisfying

$$\left[H(z,t) - i\frac{\partial}{\partial t}\right]\Psi(z,t) = 0 \tag{71}$$

yields the equation

$$\left[\mathcal{H}(z,t) - i\frac{\partial}{\partial t}\right]\Phi(z,t) = 0, \tag{72}$$

where

$$\mathcal{H}(z,t) = p_z\left[\frac{1}{m(z+a(t))}\right]p_z + V(z+a(t)) + F_0 z + v(z,t), \tag{73}$$

and $v(z,t)$ represents the residual part that is relatively small; its explicit expression is unnecessary, and thus, not given here. The successive transformation from Eq. (71) into Eq. (72) is explicitly expressed as

$$\Psi(z,t) = \exp\left[-if(z,t)\right]\exp\left[-ia(t)p_z\right]\Phi(z,t). \tag{74}$$

In Eq. (74), $f(z,t)$ for the gauge transformation and $a(t)$ for the KH transformation are given by

$$f(z,t) = \frac{A_0(t)}{c}(z+a(t)) + \frac{1}{2m_\infty}\int^t\left(\frac{A_1(t)}{c}\right)^2 dt \tag{75}$$

and

$$a(t) = \frac{1}{m_\infty}\int^t \frac{A_1(t)}{c}dt, \tag{76}$$

respectively. Here, $A(t) = A_0(t) + A_1(t)$, and a dc-electric field and a THz field are defined as $F_0 = -\dot{A}_0(t)/c$ and $F_1(t) = -\dot{A}_1(t)/c$, respectively. It should be noted that $\mathcal{H}(z,t)$ becomes

$$\mathcal{H}_{as}(z) = \frac{p_z^2}{2m_\infty} + F_0 z + V_\infty, \tag{77}$$

and $v(z,t)$ vanishes in the asymptotic region of $|z+a(t)| \gg 1$, where it has been assumed that $V(z+a(t))$ and $m(z+a(t))$ become the constant values of V_∞ and m_∞, respectively.

$\mathcal{H}(z,t)$ ensures the Floquet theorem, because we are concerned with monochromatic THz driving with $F_1(t) = F_{ac}\cos\omega t$, where ω is a frequency. Therefore, $\Phi(z,t)$ is expressed as

$$\Phi(z,t) = \exp(-iEt)\sum_{\nu=-\infty}^{\infty}\exp(i\nu\omega t)\psi_\nu(z), \tag{78}$$

and Eq. (72) is recast into the coupled equations

$$\sum_\nu [L_{\mu\nu}(z) - E\delta_{\mu\nu}]\psi_\nu(z) = 0, \tag{79}$$

where $L_{\mu\nu}(z)$ is given by

$$L_{\mu\nu}(z) = \mathcal{H}_{\mu\nu}(z) + \mu\omega\delta_{\mu\nu} \tag{80}$$

with E representing a quasienergy. Hereafter, it is understood that the time average of an arbitrary function $X(z,t)$ has been defined as

$$X_{\mu\nu}(z) = \frac{1}{T}\int_0^T \exp\left[-i(\mu-\nu)\omega t\right] X(z,t), \tag{81}$$

where $T = 2\pi/\omega$. It should be noted that $L_{\mu\nu}(z)$ becomes $L_\mu^{(as)}(z)\delta_{\mu\nu}$, where

$$L_\mu^{(as)}(z) = \mathcal{H}_{as}(z) + \mu\omega \tag{82}$$

in the region of $|z+\alpha| \gg 1$. Here, α, defined as

$$\alpha = \frac{F_{ac}}{m_\infty \omega^2}, \tag{83}$$

is called the ponderomotive radius corresponding to the excursion amplitude of a classical electron traveling under $F_1(t)$.

For any E, open boundary conditions are imposed on $\{\psi_\nu(z)\}$ at $z = z_{as} < 0$ with $|z_{as}| \gg 1$ because of $F_0 z_{as} + V_\infty + \nu\omega \to -\infty$. Hence, Eq. (79) is regarded as the coupled equations for the MCS problem, where an asymptotic scattering channel is provided by field-free solutions for Eq. (82), and this is designated by a photon index ν. There exist independent solutions, the number of which is the same as that of the channels incorporated in actual calculations, because all channels are open. Thus, $\psi_\nu(z)$ is hereafter written as $\psi_{\nu\beta}(z)$ in order to specify the βth solution.

In view of the same form of Eq. (79) as that of Eq. (7), the present MCS equations can be solved in actual calculations by virtue of the R-matrix propagation technique in Section 2.2. Here, the scattering boundary condition

$$\psi_{\mu\beta}(z) = \chi_\mu^{(+)}(z)\delta_{\mu\beta} - \chi_\mu^{(-)}(z) S_{\mu\beta}^{(-)}(E) \tag{84}$$

is imposed on $\psi_{\mu\beta}(z)$, where $\chi_\mu^{(\pm)}(z)$ is the energy-normalized progressive wave in the direction of $\pm z$, satisfying

$$[L_\mu^{(as)}(z) - E]\chi_\mu^{(\pm)}(z) = 0 : \tag{85}$$

$\chi_\mu^{(\pm)}(z)$ is associated with the Airy function. In terms of the obtained S-matrix $S^{(-)}(E)$, the excess DOS $\rho(E)$ of Eq. (30) and the lifetime $T(E)$ of Eq. (29) of the concerned state with E are provided. It should be noted that there exists the periodicity

$$\rho^{(ex)}(E) = \rho^{(ex)}(E + k\omega), \tag{86}$$

where k is an integer, because of the relation

$$S_{\mu,\beta}(E) = S_{\mu+k,\beta+k}(E + k\omega). \tag{87}$$

4.1.3. Ponderomotive Interactions

The actual calculations are implemented for the SL of 35/11 ML GaAs/Ga$_{0.75}$Al$_{0.25}$ as (1 ML = 2.83 Å) with a lattice constant $d = 246$. The concerned SLs are designed to be composed of ten QWs that are surrounded by Ga$_{0.75}$Al$_{0.25}$ as in the outer regions. Hence, m_∞ and V_∞ are identical to material parameters associated with this material. Hereafter, we consider the case of $\Omega_B = n\omega$, where n is set to 2, and Ω_B is the Bloch frequency, given by $\Omega_B = 5 \times 10^{-3}$ with $F_0 = 104.5$ kV/cm. Further, α is changed from 1 ($F_{ac} = 2.5$ kV/cm) to 108.2 ($F_{ac} = 268.4$ kV/cm).

Defining as

$$U_{\mu-\nu}(z) \equiv V_{\mu\nu}(z) + F_0 z \delta_{\mu\nu} \tag{88}$$

the ponderomotive interaction arising from the renormalization of the THz field to $V(z)$, $L_{\mu\nu}(z)$ is rewritten as

$$L_{\mu\nu}(z) = p_z \left[\frac{1}{m(z + a(t))} \right]_{\mu\nu} p_z + U_{\mu-\nu}(z) + v_{\mu\nu}(z) + \mu\omega\delta_{\mu\nu}. \tag{89}$$

In the diagonal term of $L_{\mu\mu}(z)$, the first term plays the role of a kinetic energy operator, and the remaining terms represent effective interactions. Here, $U_{\eta=0}(z)$ is generally considered to dominate $v_{\mu\mu}(z) = v_{00}(z)$. However, for a large value of F_{ac}, this would not always be the case. The last term of $\mu\omega$ with $\mu \neq 0$ contributes to PSB formation. Further, the off-diagonal term of $L_{\mu\nu(\neq\mu)}(z)$ is governed by $U_{\eta\neq 0}(z)$ for the same reason as the diagonal one. Figure 13 (a) shows the change in the diagonal term, $U_0(z)$, for $\alpha = 1$, 50, and $\alpha_{DL} = 108.2$, where α_{DL} corresponds to the DL associated with the first zero of $J_n(x) = 0$. Here, the erosion of the potential barrier from $V(z)$ appears pronounced for larger values of α; $U_0(z)$ for $\alpha = 1$ is indistinguishable from $V(z)$. In particular, it is seen that the height of $U_0(z)$ at $\alpha = \alpha_{DL}$ is greater in the well region of $V(z)$ than that in the associated barrier region, unlike the height of $U_0(z)$ at $\alpha = 50$. Figure 13 (b) shows the change in the off-diagonal term, $U_{\eta\neq 0}(z)$, for $\alpha = \alpha_{DL}$ within the range of a single QW site. As α increases, $U_{\eta\neq 0}(z)$ contributes more significantly to interactions between different channels even if $|\eta|$ is large, whereas $U_{\eta\neq 0}(z)$ for $\alpha = 1$ almost vanishes; this is not shown here.

4.1.4. Excess DOS

The contribution of a DFR decay caused by ac-ZT can be identified by examining the change in $\rho^{(ex)}(E)$ with respect to α. Without this coupling, the lifetime of a DWSL reflecting on $\rho^{(ex)}(E)$ would be responsible for the SR decay; it is considered that PAT would not affect the lifetime. In this case, the lifetime would be similar to that of a WSL, where SR is caused by dc-ZT because of a combined potential, $V(z) + F_0 z$, independently of α. Therefore, it is believed that the reduction in the lifetime of a DWSL relative to that of the associated WSL is attributable to ac-ZT. In the following, an interaction caused by the ponderomotive interaction $U_{\eta\neq 0}(z)$ is termed as interchannel coupling.

Figure 14 shows the calculated results of $\rho^{(ex)}(E)$ for $\alpha = 1 \sim \alpha_{DL}$. $\rho^{(ex)}(E)$ for $\alpha = 1$ is almost identical to that of the WSL for $\alpha = 0$, where there exist four discernible peaks labeled as $(1, -1), (2, -2), (3, -3)$, and $(1, 0)$; the levels of $b = 1$ and 2 and the

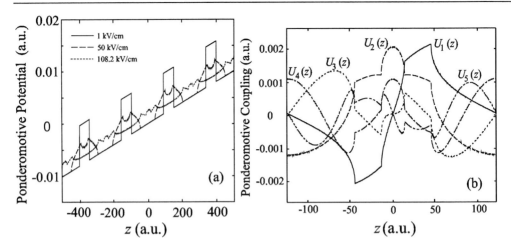

Figure 13. (a) Ponderomotive potential $U_0(z)$ as a function of z-coordinate for $\alpha = 1$, 50, and $108.2(=\alpha_{DL})$ indicated by solid, dashed, and dotted lines, respectively. Here, $F_0 = 104.5$ kV/cm. (b) Ponderomotive couplings $U_\eta(z)$ ($\eta = 1 \sim 5$) as a function of z-coordinate (a.u.) within a single QW site for $\alpha = 108.2(=\alpha_{DL})$. Here, $U_\eta(z)$ with $\eta = 1, 2, 3, 4$, and 5 are indicated by solid, dahsed, dotted, chain, and double-chain lines, respectively.

levels of $b \geq 3$ originate from the original QW levels below and above the barrier height of $V(z)$, respectively. Here, the peaks for $b = 1$ are still stable [$T(E) \approx 4.3$ ps] despite the dc-ZT caused by the relatively large F_0, differing from the other two peaks for $b = 2$ and 3 that appear blurred. This observation of the sharp peak for $b = 1$ is attributed to the fact that a potential drop in a well region of the QW is still smaller than the barrier height of $V(z)$, and therefore, a tilted SL confining potential is capable of supporting the $b = 1$ level as an almost discrete one, namely, a sharp SR one. Moreover, because of this fact, it is also ensured that the DL level of $(1, j \neq 0)$ is assumed to be almost discrete because of the PSB of $(1, 0)$, as mentioned in Section 4.1.1.

The variance of the peak positions for each b with respect to α is shown by connecting these positions by thin solid lines. The pronounced anticrossing behavior between adjacent b's is seen around the specific α's, as indicated by the dotted ovals. Without the ponderomotive coupling, $U_{\eta \neq 0}(z)$, differing from the spectra in Figure 14, no anticrossings manifest themselves, and the obtained peak positions appear only weakly dependent on α; this is not shown here. However, the peak values of the $(1, -1)$ and $(1, 0)$ levels without $U_{\eta \neq 0}(z)$ appear to be similar to those shown in Figure 14. With a further increase in α from around $\alpha = 30$, $\rho^{(ex)}(E)$ is found to decrease rapidly, and the discernible peaks are overlapped and entangled with each other because of strong interchannel interactions; this leads to a difficulty in assigning each peak to an approximate quantum number (b, j). This tendency is pronounced at $\alpha = \alpha_{DL}$, and therefore, the associated $T(E)$'s of the blurred peaks decrease to a decay lifetime of approximately 120 fs. Further, a vestige of the collapse of the quasienergy miniband is no longer seen in the profile of $\rho^{(ex)}(E)$. This result is in sharp contrast to that of WSL at $\alpha = 1$. It is understood that such instability of DL is attributed to the DFR mechanism shown in Figure 15; this is caused by ac-ZT-mediated interchannel

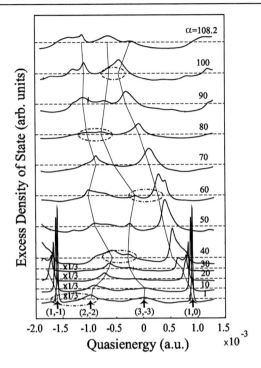

Figure 14. Excess DOS, $\rho^{(ex)}(E)$, as a function of quasienergy E in range of $\alpha = 1 \sim 108.2 (= \alpha_{DL}^{(2,1)})$. Curves are shifted for clarity. Four discernible peaks are assigned to approximate quantum numbers (b, j); each peak attributable to the same (b, j) is connected by thin solid lines and some prominent anticrossings are indicated by chain ovals. Here, $F_0 = 104.5$ kV/cm and $\omega = 2.5 \times 10^{-3}$.

coupling $U_\eta(z)$ between the energetically degenerate DL states, where one is the relatively stable state of $(1, 0)$ and the other is the unstable continuum-like one of $(b > 1, j < 0)$.

It should be noted that the present DFR differs from the conventional Fano effect in that sharp and blurred SR states are coupled via a dynamic interaction (namely, ac-ZT) in the former and discrete and continuum states are coupled via a static interaction in the latter. Strictly speaking, FR is usually considered to be resonance caused by an interaction between closed and open channels, unlike the present DFR, in which there exists no closed channel, as mentioned in Section 4.1.1. The sharp SR state for DL is considered to be a quasi-closed channel. In the study of the autoionization of a negative hydrogen ion (H$^-$), it is known that FR is caused by an interchannel coupling similar to that of DFR, where the lowest doubly excited state of $(2s)(2p)$ $^1P^o$ embedded in an SR continuum plays the role of this quasi-closed channel [116]. Therefore, it is considered that the DFR presented here is a new effect having both the *tunable* dynamic interaction and the *SR-mediated* FR as key roles, unlike conventional FR.

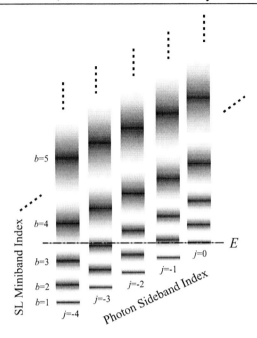

Figure 15. Schematic diagram showing DL manifold and DFR mechanism between DL levels. Each level is shaded depending on degree of magnitude of SR. For more detail, consult text.

4.2. Dynamic FR in Coherent Phonon Generation

The mechanism responsible for the FR in phonon spectra has been debated for decades using conventional light scattering spectroscopies. It has been concluded that both interband or intraband electronic transitions can give rise to the continuum overlapping the optical phonon energy in semiconductors, such as Si [5, 117, 118] and GaAs [119], as well as in semimetals, like Sb [120]. On the other hand, for complex oxides, like ferroelectrics and cuprate superconductors, the continuum originates from broad multi phonon processes [121, 122]. In both cases, the phonon spectra can be fitted to a Fano lineshape, characteristic of a continuous-discrete interference

$$I(\epsilon, q) = \frac{(\epsilon + q)^2}{\epsilon^2 + 1}. \tag{90}$$

in terms of Eq. (60) [1, 5]. Here, parameters γ and q are defined as [5],

$$\gamma = \pi |V_p|^2 \quad ; \quad \pi \gamma q^2 = T_p^2 / T_e^2, \tag{91}$$

where T_p and T_e are the probabilities for one-phonon and electronic (or multiple-phonon) Raman scattering, respectively, and V_p is the matrix element of the interaction between the discrete level and continuum.

In metals there have been theoretical discussions on the Fano-interference effects, in which both interband- and intraband-electronic Raman scattering can contribute to the electronic continuum [123]. However, experimental investigation of the Fano-interference in metals might be inaccessible by conventional time-integrated methods like Raman scattering [124, 125]. This can be attributed to either the very fast dephasing of the electronic continuum or to screening of electron-phonon coupling.

Femtosecond (fs) real-time probes present an experimental alternative to the conventional time-integrated methods. These techniques have an additional advantage to conventional spectroscopy, since they utilize photoexcitation to study both buildup and the decay of the FR in time. Utilizing femtosecond FWM technique Siegner et al. studied the dynamics of FR in GaAs [35, 36]. Recently, transient Fano-type anti-resonance between non-equilibrium carriers and coherent LO phonon in Si was also reported [46], with the lifetime estimated to about 100 fs. Moreover, a possibility of observing a build-up of the FR on a sub-fs time scale in an atom was also recently addressed [126]. Thus, in the photoexcited state the coherence between the discrete state and the continuum is expected to be short-lived, with the relaxation dynamics governed by dephasing [35, 36].

While photoexcited carrier dynamics in metals and semiconductors have been studied by fs pump-probe techniques[127], only recently observation of DFR due to optical phonons and electronic continuum has been reported. Using high sensitivity pump-probe techniques, such as second harmonic generation, and transient reflectivity (see Figure16), coherent optical phonons have been studied in Gd [128, 129], Cd and Zn [130], Si [46, 131, 132]. These materials would be potential candidates for the observation of DFR.

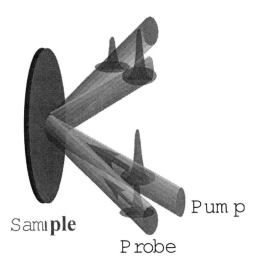

Figure 16. Schematic of the pump-probe reflectivity measurement. The stronger pump pulse generates coherent phonons in solids, while the weaker probe pulse detects the change in the reflectivity through the photodiodes.

Coherent optical phonons of Si revealed the ultrafast formation of renormalized quasi-particles [46]. The anisotropic transient reflectivity of n-doped Si(001) featured the coherent optical phonons with a frequency of 15.3 THz (Figure 17(a)). Rotation of the sample by 45° from $\Gamma_{25'}$ to Γ_{12} configuration led to disappearance of the coherent oscillation (Figure 17(b,d)), which suggests a phase shift (π) of the coherent phonons in the $R_\parallel$ and $R_\perp$ components, resulting in the cancellation of the oscillation with Γ_{12} configuration (Figure

17(d) inset) [46, 133]. A Similar polarization dependence was also observed for the coherent optical phonons of Ge [133] and diamond [134]. In time-dependent spectral amplitude (chronogram) in Figure 17(c), the optical phonon was seen as the horizontal band at 15.3 THz for $t > 0$. The chronogram also revealed the broadband response at $t \simeq 0$ due to the coherent electronic coupling of the pump and probe fields via the third-order susceptibility during coherent Stokes and anti-Stokes Raman scattering.

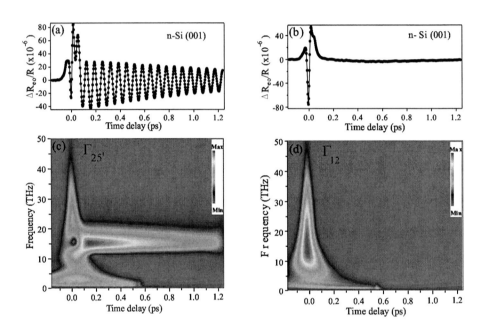

Figure 17. (a) Transient anisotropic reflectivity change for Si(001) in $\Gamma_{25'}$ geometry. (b) Transient anisotropic reflectivity change for Si(001) in Γ_{12} geometry. (c) Continuous wavelet transform of (a). (d) Continuous wavelet transform of (b).

The most intriguing aspect of the chronogram was the anti-resonance at 15.3 THz slightly after $t = 0$. It revealed interference effects leading to the coherent phonon generation and subsequent "dressing" by electron-hole pairs photoexcited near the Γ point and along the Λ direction. Many-body time-dependent approach showed that the deformation potential scattering is the origin of the destructive interference [135]. These results clearly demonstrated the possibility of observing the quantum mechanical manifestations of carrier-phonon interactions in the real time [136], which until now could only be deduced from transport measurements and spectral lineshape analysis.

A similar signature of a Fano-type resonance appears near the time-zero region for coherent E_{2g} phonons in Zn under intense laser excitation, in which both the destructive (anti-resonance) and constructive interference are observed [137].

To account for the observed FR by theoretical mean, starting from the Fröhlich Hamiltonian describing the polar coupling in a semiconductor, one combines it with the light-matter interaction [135];

$$\begin{aligned}\mathcal{H} &= \sum_{\mathbf{k}} \varepsilon_{\mathbf{k}}^v d_{\mathbf{k}}^\dagger d_{\mathbf{k}} + \sum_{\mathbf{k}} \varepsilon_{\mathbf{k}}^c c_{\mathbf{k}}^\dagger c_{\mathbf{k}} + \omega_{\text{LO}} \sum_{\mathbf{q}} a_{\mathbf{q}}^\dagger a_{\mathbf{q}} \\ &+ \sum_{\mathbf{q}} \sum_{\mathbf{k}\sigma} M_{\mathbf{q}} c_{\mathbf{k}}^\dagger c_{\mathbf{k}+\mathbf{q}}(a_{-\mathbf{q}} + a_{\mathbf{q}}^\dagger) \\ &+ \sum_{\mathbf{k}} \mathbf{j}(\mathbf{k}) \cdot \mathbf{A}_0 e^{i\omega\tau} c_{\mathbf{k}}^\dagger d_{-\mathbf{k}}^\dagger \Theta_0(\tau) + \text{h.c.} \\ &+ \sum_{\mathbf{k}} \mathbf{j}(\mathbf{k}) \cdot \mathbf{A}_1 e^{i\omega(\tau-\delta\tau)} c_{\mathbf{k}}^\dagger d_{-\mathbf{k}}^\dagger \Theta_1(\tau - \delta\tau) + \text{h.c.},\end{aligned} \qquad (92)$$

where $d_{\mathbf{k}}^\dagger (d_{\mathbf{k}})$ is the hole operator for the valence band $\varepsilon_{\mathbf{k}}^v$ and $c_{\mathbf{k}}^\dagger (c_{\mathbf{k}})$ is the electron operator for the conduction band $\varepsilon_{\mathbf{k}}^c$, while $a_{\mathbf{q}}^\dagger (a_{\mathbf{q}})$ is the phonon operator for the LO phonon ω_{LO}. The electron in the conduction band interacts with the optical phonon via the coupling matrix $M_{\mathbf{q}}$ ($\propto 1/|\mathbf{q}|$). The last two terms in $\mathcal{H}$ are from the light-matter interaction by two ultrashort pulses. ω is the energy of the laser. The length of pumping pulse is given by Δ_0 in $\Theta_0(\tau) = \Theta(\tau) - \Theta(\tau - \Delta_0)$ ($\Theta(\tau)$: Heaviside step function). In the same way, Δ_1 is the length of the probing pulse. $|\mathbf{A}_0|$ and $|\mathbf{A}_1|$ are intensities of two pulses, respectively and $\mathbf{j}(\mathbf{k})$ is the current. $\delta\tau$ is the time-delay between two pulses.

By exciting a semiconductor with ultra-short pump pulses at $\tau = 0$, the electron-hole pair can be excited across the band-gap and the electron in the conduction band undergoes scattering. Such dynamics can be described by the state of the whole system $|\Psi(\tau)\rangle$ with $|\mathbf{A}_0| \to 0$ and $|\mathbf{A}_1| \to 0$,

$$\begin{aligned}|\Psi(\tau)\rangle &= C(\tau)|0\rangle_c|0\rangle_v|0\rangle_{\text{ph}} + \sum_{\mathbf{q}} C_{\mathbf{q}}(\tau)|0\rangle_c|0\rangle_v|\mathbf{q}\rangle_{\text{ph}} \\ &+ \sum_{\mathbf{k}\mathbf{k}'} C_{\mathbf{k};\mathbf{k}'}(\tau)|\mathbf{k}\rangle_c|\mathbf{k}'\rangle_v|0\rangle_{\text{ph}} \\ &+ \sum_{\mathbf{k}\mathbf{k}'\mathbf{q}} C_{\mathbf{k};\mathbf{k}';\mathbf{q}}(\tau)|\mathbf{k}\rangle_c|\mathbf{k}'\rangle_v|\mathbf{q}\rangle_{\text{ph}}.\end{aligned} \qquad (93)$$

In $|\Psi(\tau)\rangle$, we note $|\mathbf{k}\rangle_c = c_{\mathbf{k}}^\dagger|0\rangle_c$, $|\mathbf{k}\rangle_v = d_{\mathbf{k}}^\dagger|0\rangle_v$, and $|\mathbf{q}\rangle_{\text{ph}} = a_{\mathbf{q}}^\dagger|0\rangle_{\text{ph}}$ for a single electron, hole, and phonon state, respectively. The many-body Hilbert space spanned by $|\Psi(\tau)\rangle$ makes us envisage the one-phonon Raman process. The initial condition $|\Psi(0)\rangle$ should be the ground state of the system before irradiation, therefore we have $|\Psi(0)\rangle = |0\rangle_c|0\rangle_v|0\rangle_{\text{ph}}$. The time-dependent Schrödinger equation $i\partial/\partial\tau|\Psi(\tau)\rangle = \mathcal{H}|\Psi(\tau)\rangle$ now gives infinitely many coupled differential equations for $C(\tau)$, $C_{\mathbf{q}}(\tau)$, $C_{\mathbf{k};\mathbf{k}'}(\tau)$, and $C_{\mathbf{k};\mathbf{k}';\mathbf{q}}(\tau)$. It is the basic scheme of the many-body time-dependent diagonalization to solve those coupled differential equations.

From the solution, we can calculate the time-dependent Raman scattering cross section $P(\tau)$ with $\bar{P} = \sum_{\mathbf{q}} \left[a_{\mathbf{q}} + a_{\mathbf{q}}^\dagger\right]$

$$P(\tau) = \langle\Psi(\tau)|\bar{P}|\Psi(\tau)\rangle. \qquad (94)$$

$P(\tau)$ is equivalent to an usual correlation function for the Raman scattering [135, 138]. $P(\tau)$ is given by $P_0(\tau) + P_1(\tau)$,

$$P_0(\tau) = \sum_{\mathbf{q}} \left[C^*(\tau) C_{\mathbf{q}}(\tau) + C(\tau) C^*_{\mathbf{q}}(\tau) \right], \tag{95}$$

$$P_1(\tau) = \sum_{\mathbf{k}\mathbf{k}'\mathbf{q}} \left[C^*_{\mathbf{k};\mathbf{k}'}(\tau) C_{\mathbf{k};\mathbf{k}';\mathbf{q}}(\tau) + C_{\mathbf{k};\mathbf{k}'}(\tau) C^*_{\mathbf{k};\mathbf{k}';\mathbf{q}}(\tau) \right].$$

An observation of the coherent LO phonon by the transient reflectivity change is governed by the first-order Raman tensor $\partial \chi / \partial Q$, where χ is the linear susceptibility and Q is the phonon coordinate, and therefore, the Raman scattering cross section gives the valid observable [139].

Intriguing insight into the early-time dynamics of the transient Fano interference can be obtained by the continuous wavelet transformation (CWT). In Figure18, we provide the intensity distribution in the frequency-time domain, obtained by CWT. As shown in Figure18 (a), the CWT decomposes the signal into the broad band response (0 ∼ 50 THz) near $\tau = 0$ and the LO phonon response at ≈ 8.7 THz. The broad band response near $\tau = 0$ is mainly from the response of the electron-hole pair and rapidly suppressed by the pair dephasing in the time scale of hundreds of femtoseconds. The LO phonon response is from $P_0(\tau)$. Interestingly, the total response in Figure18 (a) is found to have the obvious dip (or antiresonance) near $\tau = 0$ slightly above the phonon mode frequency. This feature can be understood from the destructive interference between responses of the electron-hole pair and the phonon, that is, the destructive Fano interference. The dip at birth of the electron-hole pair and the phonon (i.e., at $\tau \approx 0$) is found robust as shown in Figure18(a). It is to be noted that the destructive interference (the dip) at ≈ 10 THz actually reflects the case of the negative asymmetry parameter q in the Fano function of Eq. (90) [1, 5].

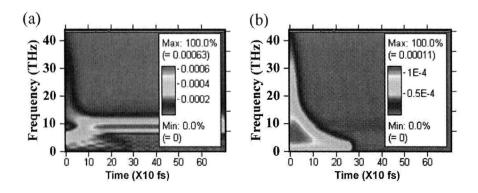

Figure 18. Intensity distribution by continuous wave transformation (CWT) in the frequency-time domain. Left panels are from the total response $P_0(\tau) + P_1(\tau)$, while right panels from the response of only the electron-hole channel $P_1(\tau)$. We have $\omega = 3.81$ eV.

5. Conclusion

Fano Resonance phenomena have been intensively investigated for a couple of decades in condensed matter systems, in particular, in semiconductor systems, as comprehensively reviewed in this chapter. Most of these studies were conventionally directed towards detecting Fano interference in the frequency domain by means of linear optical spectroscopies. Because FR states are very short-lived in general, a real-time observation of this effect appears to be difficult as it stands. However, the pump-probe technique complete with ultra-short pulse lasers has made it possible to explore the FR dynamics in the time domain. Such an approach has opened a new research area of the FR problem, namely, the non-linear FR problem, in which FR is probed by means of non-linear optical spectroscopies, as mentioned in Sections 3.2. and 4.2. Nevertheless, it should be noted that the FR effect to be explored by this method is still caused by *static* interactions inherent in the materials concerned.

In this chapter, we have presented another FR effect, termed DFR, which has the *dynamic* origin attributable to interactions of electron, exciton, and phonon with external ac-fields such as intense THz-waves and high-intensity ultra-short pulse lasers. As shown in Sections 3.3., 4.1., and 4.2., appropriately-tuned external ac-fields are allowed to give rise to the DFR in the material system concerned, even if no FR with a static origin manifests itself in it. Dynamic FR is realized in a dressed material that is formed by renormalizing light-matter interactions to the original material, and this dressed material is regarded as *meta*-material consisting of the original material and the renormalized light field. The DFR depends straightforward on the external light sources through strength, frequency, polarization, and so on, and there would be a new possibility of providing an unexplored manner for coherent control of materials.

Acknowledgments

This work was supported by a Grant-in-Aid for Scientific Research on Innovative Areas "Optical science of dynamically correlated electrons (DYCE)" (No. 21104504) of the Ministry of Education, Culture, Sports, Science and Technology (MEXT), Japan. M. H. acknowledges H. Petek, A. M. Constantinuscu, M. Kitajima, and J. D. Lee for their significant contributions to the studies presented.

References

[1] U. Fano, *Phys. Rev.* **124**, 1866 (1961).

[2] B. W. Shore, *Rev. Mod. Phys.* **39**, 439 (1967).

[3] H. Winter, *J. Phys. Condens. Matter* **8**, 10149 (1996).

[4] Y. Hatano, *Phys. Rep.* **313**, 109 (1999).

[5] F. Cerdeira, T. A. Fjeldly, and M. Cardona, *Phys. Rev. B* **8**, 4734 (1973).

[6] K. Kobayashi, H. Aikawa, S. Katsumoto, and Y. Iye, *Phys. Rev. Lett.* **88**, 256806 (2002).

[7] P. L. Knight, M. A. Lauder, and B. J. Dalton, *Phys. Rep.* **190**, 1 (1990).

[8] J. L. Bohn and P. S. Julienne, *Phys. Rev. A* **56**, 1486 (1997).

[9] R. Ciurylo, E. Tiesinga, and P. S. Julienne, *Phys. Rev. A* **74**, 022710 (2006).

[10] K. Enomoto, K. Kasa, M. Kitagawa, and Y. Takahashi, *Phys. Rev. Lett.* **101**, 203201 (2008).

[11] K. Hino, K. Goto, and N. Toshima, *Phys. Rev. B* **69**, 035322 (2004).

[12] M. Kroner, A. O. Govorov, S. Remi, B. Biedermann, S. Seidl, A. Badolato, P. M. Petroff, W. Zhang, R. Barbour, B. D. Gerardot, R. J. Warburton, and K. Karrai, *Nature* **451**, 311 (2008).

[13] M. V. Fedorov and A. E. Kazakov, *Prog. Quantum Electron.* **13**, 1 (1989).

[14] Z. Deng and J. H. Eberly, *Phys. Rev. A* **36**, 2750 (1987).

[15] G. Wen and Y. C. Chang, *Phys. Rev. B* **45**, 6101 (1992).

[16] C. Y. Chao and S. L. Chuang, *Phys. Rev. B* **48**, 8210 (1993).

[17] R. Winkler, *Phys. Rev. B* **51**, 14395 (1995).

[18] S. Glutsch, D. S. Chemla, and F. Bechstedt, *Phys. Rev. B* **54** 11592 (1996).

[19] S. Glutsch, P. Lefebvre, and D. S. Chemla, *Phys. Rev. B* **55**, 15786 (1997).

[20] S. Glutsch and F. Bechstedt, *Phys. Rev. B* **57**, 11887 (1998).

[21] A. N. Forshaw and D. M. Whittaker, *Phys. Rev. B* **54**, 8794 (1996).

[22] M. Graf, P. Vogl, and A. B. Dzyubenko, *Phys. Rev. B* **54**, 17003 (1996).

[23] K. Hino, *Phys. Rev. B* **62**, R10626 (2000), [*Erratum:* **63**, 119901 (2001)].

[24] K. Hino, *Phys. Rev. B* **64**, 075318 (2001).

[25] K. Hino, *J. Phys. Soc. Jpn.* **71**, 2280 (2002).

[26] K. Hino and N. Toshima, *Phys. Rev. B* **71**, 205326 (2005).

[27] P. Dawson, K. J. Moore, G. Duggan, H. I. Ralph, and C. T. B. Foxon, *Phys. Rev. B* **34**, 6007 (1986).

[28] D. C. Reynolds, K. K. Bajaj, C. Leak, G. Peters, W. Theis, P. W. Yu, K. Alavi, C. Colvard, and I. Shidlovsky, *Phys. Rev. B* **37**, 3117 (1988).

[29] W. Theis, G. D. Sanders, C. E. Leak, D. C. Reynolds, Y. -C. Chang, K. Alavi, C. Colvard, and I. Shidlovsky, *Phys. Rev. B* **39**, 1442 (1989).

[30] Y. Kajikawa, *Phys. Rev. B* **48**, 7935 (1993).

[31] D. Y. Oberli, G. Böhm, G. Weimann and J. A. Brum, *Phys. Rev. B* **49**, 5757 (1994).

[32] P. E. Simmonds, M. J. Birkett, M. S. Skolnick, W. I. E. Tagg, P. Sobkowicz, G. W. Smith and D. M. Whittaker, *Phys. Rev. B.* **50**, 11251 (1994).

[33] S. Bar-Ad, P. Kner, M. V. Marquezini, S. Mukamel, and D. S. Chemla, *Phys. Rev. Lett.* **78**, 1363 (1997).

[34] C. P. Holfeld, F. Löser, M. Sudzius, K. Leo, D. M. Whittaker, and K. Köhler, *Phys. Rev. Lett.* **81**, 874 (1998).

[35] U. Siegner, M. -A. Mycek, S. Glutsch, and D. S. Chemla, *Phys. Rev. Lett.* **74**, 470 (1995).

[36] U. Siegner, M. -A. Mycek, S. Glutsch and D. S. Chemla, *Phys. Rev. B* **51**, 4953 (1995).

[37] S. M. Sadeghi and J. Meyer, *J. Phys. Condens. Matter* **9**, 7685 (1997).

[38] M. J. Seaton, *Rep. Prog. Phys.* **46**, 167 (1983).

[39] P. G. Burke and W. D. Robb, *Adv. At. Mol. Phys.* **11**, 143 (1975).

[40] S. H. Autler and C. H. Townes, *Phys. Rev.* **100**, 703 (1955).

[41] K. Yashima, K. Hino, and N. Toshima, *Phys. Rev. B* **68**, 235325 (2003).

[42] K. Yashima, K. Oka, K. Hino, N. Maeshima, X. M. Tong, *Solid State Communications* **149**, 823 (2009).

[43] D. H. Dunlap and V. M. Kenkre, *Phys. Rev.* **34**, 3625 (1986).

[44] M. Holthaus, *Phys. Rev. Lett.* **69**, 351 (1992).

[45] A. Kukuu, T. Amano, T. Karasawa, N. Maeshima, and K. Hino, *Phys. Rev. B* **82**, 115315 (2010).

[46] M. Hase, M. Kitajima, A. M. Constantinescu, H. Petek, *Nature* **426**, 51 (2003).

[47] K. Hino, *J. Phys. Soc. Jpn.* **67**, 3159 (1998).

[48] J. C. Light, I. P. Hamilton and J. V. Lill, *J. Chem. Phys.* **82**, 1400 (1985).

[49] F. T. Smith, *Phys. Rev.* **118**, 349 (1960).

[50] R. Balian and C. Bloch, *Annals of Physics* **60**, 401 (1970).

[51] J. L. Kinsey, *Chem. Phys. Lett.* **8**, 349 (1971).

[52] M. Desouter-Lecomte and X. Chapuisat, *Phys. Chem. Chem. Phys.* **1**, 2635 (1999).

[53] E. O. Kane, *J. Phys. Chem. Solids* **12**, 181 (1959).

[54] W. V. Houston, *Phys. Rev.* **57**, 184 (1940).

[55] E. E. Mendez, F. Agullo-Rueda, and J. M. Hong, *Phys. Rev. Lett.* **60**, 2426 (1988).

[56] P. Voisin, J. Bleuse, C. Bouche, S. Gaillard, C. Alibert, and A. Regreny, *Phys. Rev. Lett.* **61**, 1639 (1988).

[57] V. G. Lyssenko, G. Valusis, F. Löser, T. Masche, K. Leo, M. M. Dignam, and K. Köhler, *Phys. Rev. Lett.* **79**, 301 (1997).

[58] C. Zener, *Proc. R. Soc. Lond. A* **145**, 523 (1934).

[59] H. Schneider, H. T. Grahn, K. von Klizting, and K. Ploog, *Phys. Rev. Lett.* **65**, 2720 (1990).

[60] D. M. Whittaker, *J. Phys. IV* **5**, 199 (1993); *Europhys. Lett.* **31**, 55 (1995).

[61] S. Glutsch and F. Bechstedt, *Phys. Rev. B* **60**, 16584 (1999).

[62] H. Rosam, D. Meinhold, F. Löser, V. G. Lyssenko, S. Glutsch, F. Bechstedt, F. Rossi, K. Köhler, and K. Leo, *Phys. Rev. Lett.* **86**, 1307 (2001).

[63] D. A. B. Miller, D. S. Chemla, T. C. Damen, A. C. Gossard, W. Wiegmann, T. H. Wood, and C. A. Burrus, *Phys. Rev. Lett.* **53**, 2173 (1984).

[64] K. Hino, Solid State Comm. **128**, 9 (2003).

[65] P. L. Knight and P. W. Milonni, *Phys. Rep.* **66**, 21 (1980).

[66] E. T. Jaynes and F. W. Cummings, *Proc. IEEE* **51**, 89 (1963).

[67] S. T. Cundiff, A. Knor, J. Feldmann, S. W. Koch, E. O. Göbel and H. Nickel, *Phys. Rev. Lett.* **73**, 1178 (1994).

[68] M. Lindberg and R. Binder, *Phys. Rev. Lett.* **75**, 1403 (1995).

[69] H. Giessen, A. Knor, S. Haan, S. W. Koch, S. Linden, J. Kuhl, M. Hetterich, M. Grün and C. Klingshirn, *Phys. Rev. Lett.* **81**, 4260 (1998).

[70] A. Schülzgen, R. Binder, M. E. Donovan, M. Lindberg, K. Wundke, and H. M. Gibbs, *Phys. Rev. Lett.* **82**, 2346 (1999).

[71] R. Binder and M. Lindberg, *Phys. Rev. B* **61**, 2830 (2000).

[72] C. Ciuti and F. Quochi, *Soild State Commun*, **107**, 715 (1998).

[73] M. Saba, F. Quochi, C. Ciuti, D. Martin, J. -L. Staehli, B. Deveaud, A. Mura, and G. Bongiovanni, *Phys. Rev. B* **62**, R16322 (2000).

[74] F. Quochi, G. Bongiovanni, A. Mura, J. L. Staehli, M. A. Dupertuis, R. P. Stanley, U. Oesterle and R. Houdré, *Phys. Rev. Lett.* **80**, 4733 (1998).

[75] F. Quochi, C. Ciuti, G. Bongiovanni, A. Mura, M. Saba, U. Oesterle, M. A. Dupertuis, J. L. Staehli, and B. Deveaud, *Phys. Rev. B* **59**, R15594 (1999).

[76] F. Quochi, M. Saba, C. Ciuti, R. P. Stanley, R. Houdré, U. Oesterle, J. L. Staehli, B. Deveaud, G. Bongiovanni, and A. Mura, *Phys. Rev. B* **61**, R5113 (2000).

[77] H. Haug and S. W. Koch, *Quantum Theory of the Optical and Electronic Properties of Semiconductors* (World Scientific, Singapore, 1993), Chap. 12.

[78] V. M. Axt, G. Bartels, and A. Stahl, *Phys. Rev. Lett.* **76**, 2543 (1996).

[79] M. Hawton and D. Nelson, *Phys. Rev. B* **57**, 4000 (1998).

[80] P. H. Bolivar, F. Wolter, A. Müller, H. G. Roskos, H. Kurz, and K. Köhler, *Phys. Rev. Lett.* **78**, 2232 (1997).

[81] J. M. Lachaine, M. Hawton, J. E. Sipe, and M. M. Dignam, *Phys. Rev. B* **62**, R4829 (2000).

[82] M. Dignam and M. Hawton, *Phys. Rev. B* **67**, 035329 (2003).

[83] L. Schultheis, J. Kuhl, A. Honold, and C. W. Tu, *Phys. Rev. Lett.* **57**, 1635 (1986).

[84] A. Honold, L. Schltheis, J. Kuhl, and C. W. Tu, *Phys. Rev. B* **40**, 6442 (1989).

[85] M. U. Wehner, D. Steinbach, and M. Wegener, *Phys. Rev. B* **54**, R5211 (1996).

[86] D. Birkedal, V. G. Lyssenko, J. M. Hvam, and K. El Sayed, *Phys. Rev. B* **54**, 14250 (1996).

[87] J. M. Shacklette and S. T. Cundiff, *Phys. Rev. B* **66**, 045309 (2002).

[88] C. Ciuti, C. Piermarrocchi, V. Savona, P. E. Selbmann, P. Schwendimann, and A. Quattropani, *Phys. Rev. Lett.* **84**, 1752 (2000).

[89] A. Mysyrowicz, D. Hulin, A. Antonetti, A. Migus, W. T. Masselink, and H. Morkoc, *Phys. Rev. Lett.* **56**, 2748 (1986).

[90] C. Ell, J. F. Müller, K. El Sayed, and H. Haug, *Phys. Rev. Lett.* **62**, 304 (1989).

[91] W. H. Knox, D. S. Chemla, D. A. B. Miller, J. B. Stark, and S. Schmitt-Rink, *Phys. Rev. Lett.* **62**, 1189 (1989).

[92] K. Sakai (Ed.), *Terahertz Optoelectronics*, (Springer-Verlag, Berlin, 2005).

[93] S. D. Ganichev and W. Prettl, *Intense Tetahertz Excitation of Semiconductors*, (Oxford University Press, 2006).

[94] S. Kohler, J. Lehmann, and P. Hänggi, *Phys. Rep.* **406**, 379 (2005).

[95] J. Zak, *Phys. Rev. Lett.* **71**, 2623 (1993).

[96] K. Hino, K. Yashima, and N. Toshima, *Phys. Rev. B* **71**, 115325 (2005).

[97] K. Hino, X. M. Tong, and N. Toshima, *Phys. Rev. B* **77**, 045322 (2008).

[98] R. B. Liu and B. F. Zhu, *Phys. Rev. B* **59**, 5759 (1999).

[99] H. Hirori, M. Nagai, and K. Tanaka, *Phys. Rev. B* **81**, 081305(R) (2010).

[100] K. W. Madison, M. C. Fischer, and M. G. Raizen, *Phys. Rev. A* **60**, R1767 (1999).

[101] S. Haroche, C. Cohen-Tannoudji, C. Audoin, and J. P. Schermann, *Phys. Rev. Lett.* **24**, 861 (1970).

[102] G. Xu and D. J. Heinzen, *Phys. Rev. A* **59**, R922 (1999).

[103] T. Shirahama, X. M. Tong, K. Hino, and N. Toshima, *Phys. Rev. A* **80**, 043414 (2009).

[104] J. Karczmarek, M. Stott, and M. Ivanov, *Phys. Rev. A* **60**, R4225 (1999).

[105] M. Glück, A. R. Kolovsky, and H. J. Korsch, *Phys. Rep.* **366**, 103 (2002).

[106] A. Eckardt, C. Weiss, and M. Holthaus, *Phys. Rev. Lett.* **95**, 260404 (2005).

[107] M. Sillanpää, T. Lehtinen, A. Paila, Y. Makhlin, and P. Hakonen, *Phys. Rev. Lett.* **96**, 187002 (2006).

[108] F. Grossmann, T. Dittrich, P. Jung, and P. Hänggi, *Phys. Rev. Lett.* **67**, 516 (1991).

[109] M. Grifoni and P. Hänggi, *Phys. Rep.* **304**, 229 (1998).

[110] Y. Kayanuma and K. Saito, *Phys. Rev. A* **77**, 010101(R) (2008).

[111] M. Holthaus and D. W. Hone, *Philosophical Magazine B* **74**, 105 (1996).

[112] B. Bartal, I. Z. Kozma, A. G. Stepanov, G. Almási, J. Kuhl, E. Riedle, and J. Hebling, *Appl. Phys. B* **86**, 419 (2007).

[113] N. Karpowicz, J. Dai, X. Lu, Y. Chen, M. Yamaguchi, L. Zhang, C. Zhang, M. Price-Gallagher, C. Fletcher, O. Mamer, A. Lesimple, and K. Johnson, *Appl. Phys. Lett.* **92**, 011131 (2008).

[114] J. Purvis, M. Dörr, M. Terao-Dunseath, C. J. Joachain, P. G. Burke, and C. J. Noble, *Phys. Rev. Lett.* **71**, 3943 (1993).

[115] W. C. Henneberger, *Phys. Rev. Lett.* **21**, 838 (1968).

[116] K. Hino, M. Nagase, H. Okamoto, T. Morishita, M. Matsuzawa, and M. Kimura, *Phys. Rev. A* **49**, 3753 (1994).

[117] M. Chandrasekhar, J. B. Renucci, and M. Cardona, *Phys. Rev. B* **17**, 1623 (1978).

[118] M. Chandrasekhar, H. R. Chandrasekhar, M. Grimsditch, and M. Cardona, *Phys. Rev. B* **22**, 4825 (1980).

[119] L. A. O. Nunes, L. Ioriatti, L. T. Florez, and J. P. Harbison, *Phys. Rev. B* **47**, 13011 (1993).

[120] M. L. Bansal and A. P. Roy, *Phys. Rev. B* **33**, 1526 (1986).

[121] D. L. Rousseau, and S. P. S. Porto, *Phys. Rev. Lett.* **20**, 1354 (1968).

[122] D. Mihailovic, K. F. McCarty, and D. S. Ginley, *Phys. Rev. B* **47**, 8910 (1993).

[123] M. V. Klein, in *Light Scattering in Solids III*, edited by M. Cardona and G. Güntherodt, *Topics in Applied Physics* Vol. 51 (Springer-Verlag, Berlin, 1982), p. 140.

[124] W. B. Grant, H. Schulz, S. Hüfner, and J. Pelzl, *Phys. Stat. Sol. (b)* **60**, 331 (1973).

[125] G. A. Bolotin, Yu. I. Kuz'min, Yu. V. Knyazev, Yu. S. Ponosov, and C. Thomsen, *Phys. Sol. State*, **43**, 1801 (2001).

[126] M. Wickenhauser, J. Burgdörfer, F. Krausz, and M. Drescher, *Phys. Rev. Lett.* **94**, 023002 (2005).

[127] W. S. Fann, R. Storz, H. W. K. Tom, and J. Bokor, *Phys. Rev. Lett.* **68**, 2834 (1992).

[128] A. Melnikov, I. Radu, U. Bovensiepen, O. Krupin, K. Starke, E. Matthias, and M. Wolf, *Phys. Rev. Lett.* **91**, 227403 (2003).

[129] U. Bovensiepen, A. Melnikov, I. Radu, O. Krupin, K. Starke, M. Wolf, and E. Matthias, *Phys. Rev. B* **69**, 235417 (2004).

[130] M. Hase, K. Ishioka, J. Demsar, K. Ushida, and M. Kitajima, *Phys. Rev. B* **71**, 184301 (2005).

[131] D. M. Riffe and A. J. Sabbah, *Phys. Rev. B* **66**, 165217 (2002).

[132] D. M. Riffe and A. J. Sabbah, *Phys. Rev. B* **76**, 085207 (2007).

[133] T. Pfeifer, W. Kütt, H. Kurz, and R. Scholz, *Phys. Rev. Lett.* **69**, 3248 (1992).

[134] K. Ishioka, M. Hase, M. Kitajima, and H. Petek, *Appl. Phys. Lett.* **89**, 231916 (2006).

[135] J. D. Lee, J. Inoue, and M. Hase, *Phys. Rev. Lett.* **97**, 157405 (2006).

[136] R. Huber, F. Tauser, A. Brodschelm, M. Bichler, G. Abstreiter, and A. Leitenstorfer, *Nature* **414**, 286 (2001).

[137] M. Hase, J. Demsar, and M. Kitajima, *Phys. Rev. B* **74**, 212301 (2006).

[138] R. Loudon, *Proc. Roy. Soc. (London)* **A275**, 218 (1963).

[139] T. Dekorsy, G. C. Cho, and H. Kurz, in *Light Scattering in Solids VIII*, edited by M. Cardona and G. Güntherodt (SpringerVerlag, Berlin, 2000).

In: Exciton Quasiparticles
Editor: Randy M. Bergin

ISBN: 978-1-61122-318-7
© 2011 Nova Science Publishers, Inc.

Chapter 10

MAGNETOEXCITON BINDING ENERGY IN POLAR CRYSTALS, QUANTUM WELLS AND GRAPHENE BILAYERS

Z. G. Koinov[*]
Department of Physics and Astronomy,
University of Texas at San Antonio,
San Antonio, TX, USA

Abstract

Strong magnetic fields can dramatically change the dynamical properties of excitons because in this regime the electrons and holes are confined primarily to the lowest Landau Level (LLL), and the Coulomb energy is much smaller than the exciton cyclotron energy. In this Chapter we apply the Bethe-Salpeter (BS) formalism to study magnetoexciton binding energy in bulk polar crystals and quantum wells assuming the existence of Fröhlich interaction between the electrons and the longitudinal optical phonons. In the case of a bulk material the BS equation in the LLL approximation is reduced to a one-dimensional Schrodinger equation which has a nonlocal potential. It is shown that the magnetoexciton binding energy in polar quantum wells has the same form as in the non-polar structures but with an effective dielectric constant which depends on the strength of the magnetic field. Since the unique electronic behaviors of graphene is a result of the unusual quantum-relativistic characteristics of the so-called Dirac fermions, we study magnetoexciton binding energy in graphene bilayers embedded in a dielectric by applying the relativistic BS equation in the LLL approximation. It is shown that in graphene bilayer structures the magnetoexciton mass (binding energy) is four times lower (higher) than the corresponding magnetoexciton mass (binding energy) in coupled quantum wells with parabolic dispersion.

1. Introduction

Magnetoexcitons are bound states between two charged fermions (an electron from the conductive band and a hole from the valence band) in the presence of a magnetic field. The

---
[*]E-mail address: Zlatko.Koinv@utsa.edu

theoretical interest in magnetoexcitons is due to the two main reasons. First, the combination of modern experimental techniques in obtaining extremely high magnetic fields and advanced technologies for making both quantum-well and graphene-based systems opens a broad perspective for creating on this basis functional nanoelectronic devices. Second, the solution of this problem is related to the continued activities to produce a Bose-Einstein condensate (BEC) of excitons in a quantum well and to observe superfluid properties of excitons in a bilayer graphene. The existence of a wide barrier material between electron and hole quantum wells (or graphene bilayers) increases the exciton lifetime, but this also reduces the exciton binding energy, and so decreases the critical temperature for condensation. A possible way to increase the binding energy, and therefore the critical temperature, is to apply magnetic field. In the presence of a magnetic field the exciton binding energy increases due to the two-dimensional confinement of excitons by the magnetic field.

Turning our attention to magnetoexciton dispersion in non-relativistic systems, such as coupled quantum wells (CQW's) with parabolic dispersions ($E_{c,v} = \hbar^2 k^2/2m_{c,v}$) [1], we find that the following Hamiltonian

$$\widehat{H} = -\frac{\hbar^2}{2\mu}\nabla_{\mathbf{r}}^2 + \frac{ie\gamma\hbar}{2\mu c}(\mathbf{B}\times\mathbf{r}).\nabla_{\mathbf{r}} + \frac{e^2 B^2}{8\mu c^2}\mathbf{r}^2 - V(\mathbf{r}+\mathbf{R}_0) \qquad (1)$$

is used to obtain the magnetoexciton dispersion. Here m_c and m_v are the electron and hole effective masses, $\mu = m_c m_v/(m_c + m_v)$ is the exciton reduced mass. The parameter $\gamma = (m_v - m_c)/(m_c + m_v)$ accounts for the difference between the electron and the hole masses, and $\mathbf{R}_0 = R^2 \mathbf{Q}_0$, where $\mathbf{Q}_0 = (-Q_y, Q_x, 0)$, and $R = (\hbar c/eB)^{1/2}$ is the magnetic length. $V(\mathbf{r}) = e^2/(\varepsilon_0\sqrt{|\mathbf{r}|^2 + d^2})$ represents the electron-hole Coulomb attraction screened by the dielectric constant ϵ_0 (d is the barrier thickness). Since the Coulomb term in the Hamiltonian is the only term which depends on the exciton momentum $\mathbf{Q} = (Q_x, Q_y, 0)$, the magnetoexciton dispersion does not depend on the electron and hole masses and the magnetoexciton mass is determined only by Coulomb interaction. In strong magnetic fields one can apply the lowest Landau level (LLL) approximation. In the LLL approximation the binding energy E_{CQW} and the magnetoexciton mass M_{CQW} are as follows:

$$\frac{M_{CQW}}{M_{2D}} = \sqrt{\frac{2}{\pi}}\left[(1+\frac{d^2}{R^2})e^{\left(\frac{d^2}{2R^2}\right)} Erfc\left(\frac{d}{\sqrt{2}R}\right) - \frac{d}{R}\right]^{-1} \qquad (2)$$

$$E_{CQW} = E_b \exp\left(\frac{d^2}{2R^2}\right) Erfc\left(\frac{d}{\sqrt{2}R}\right).$$

Here $Erfc(x)$ is the complementary error function, $E_b = \sqrt{\pi}e^2/(\sqrt{2}\varepsilon_0 R)$ and $M_{2D} = 2^{3/2}\varepsilon_0\hbar^2/(\sqrt{\pi}e^2 R)$ are the two-dimensional magnetoexciton binding energy and magnetoexciton mass, respectively.

Strictly speaking, the excitons are bound states between two charged fermions, and therefore, the appropriate framework for the description of the bound states is the Bethe-Salpeter (BS) formalism [2]. The purpose of this Chapter is to present a nonrelativistic BS approach to the problem of magnetoexcitons in polar crystals and quantum wells [3], as well as the relativistic BS approach to magnetoexcitons in graphene bilayers [4].

2. Magnetoexcitons in Polar Semiconductors

In the polar semiconductors the Fröhlich interaction with the longitudinal optical (LO) phonons dominates as compared with the interaction with phonons of other types. To account for the Fröhlich interaction we have to use two quite different dielectric constants: the static ε_0 and high-frequency dielectric constants ε_∞. It turns out that one has to used a many-body theory to describe the screening of the electron-hole interaction in polar crystals. In the absence of a magnetic field, for example, the electron-hole interaction in polar crystals has been investigated decades ago by applying different many-body techniques, such as a perturbation theory [6], path integrals [7], a variational method [8], a Green's function method [9], and the equation-of-motion method [10]. The common conclusion of all approaches is that the exciton binding energy can be calculated using the Schrödinger equation, but with a more complicated potential instead of the screened Coulomb potential. Each of the above mentioned theoretical methods provides its own potential, usually a non-local one, but regardless of the applied techniques the potential always depends on the static and high-frequency dielectric constants ε_0 and ε_∞, the LO phonon energy $\hbar\omega_0$, and the electron m_c and hole m_v bare masses. Unfortunately, to solve the Schrödinger equation we need the input bare-mass parameters m_c and m_v. But, the experimentally obtained parameter sets should be interpreted as polaron-mass parameters. Thus, we need some additional procedure of converting the polaron-mass parameters into the bare-mass parameters. One may well ask whether this conversion does not affect the accuracy of the numerically calculated exciton binding energy. A possible test of the results provided by the above theoretical approaches is to use the fact that the magnetoexciton binding energy in two dimensions (2D) should depend only on ε_0, ε_∞ and the LO phonon energy $\hbar\omega_0$, but not on the electron and hole masses.

In what follows, we assume the presence of a strong constant magnetic field **B** along the z-axis and we use a symmetric gauge where the vector potential of the magnetic field **A** is defined by $\mathbf{A}(\mathbf{r}) = (1/2)\mathbf{B} \times \mathbf{r}$. As in the case of the absence of a magnetic field, it is common to assume that the electron and the hole, constituting the magnetoexciton, interact individually with the LO phonons. This assumption leads to following Hamiltonian for the magnetoexciton-phonon system:

$$\widehat{H} = \widehat{H}_{exc} + \widehat{H}_{ph} + \widehat{H}_{e-ph} + \widehat{H}_{h-ph}, \qquad (3)$$

where $\widehat{H}_{exc}$ is the exciton part of the total Hamiltonian [11]:

$$\widehat{H}_{exc} = E_g + \frac{\hbar^2 Q_z^2}{2M} - \frac{\hbar^2}{2\mu}\frac{d^2}{dz^2} - \frac{\hbar^2}{2\mu}\nabla_\mathbf{r}^2 + \frac{ie\gamma\hbar}{2\mu c}(\mathbf{B}$$
$$\times \mathbf{r}).\nabla_\mathbf{r} + \frac{e^2 B^2}{8\mu c^2}\mathbf{r}^2 + V_C(\mathbf{r} + R^2\mathbf{Q}_0; z). \qquad (4)$$

Since the interaction with LO phonons is included in (3), the Coulomb attractive interaction between the electron and the hole, $V_C(\mathbf{r}) = -e^2/(\varepsilon_\infty|\mathbf{r}|)$, must be screened by the high-frequency dielectric constant. The Hamiltonian (4) is written using the center-of-mass $\mathbf{R} = \alpha_c \mathbf{r}_c + \alpha_v \mathbf{r}_v$ and the relative $\mathbf{r} = \mathbf{r}_c - \mathbf{r}_v$ coordinates. The coefficients $\alpha_c = (1-\gamma)/2$ and $\alpha_v = (1+\gamma)/2$ are expressed in terms of the parameter γ. $M = m_c + m_v$ is the exciton total mass. The electron and the hole dispersion laws in a 2D crystal are $E_c(\mathbf{k}) = E_g + \hbar^2 \mathbf{k}^2/2m_c$ and $E_v(\mathbf{k}) = \hbar^2 \mathbf{k}^2/2m_v$, respectively. Here E_g is the semiconductor band gap, and $\mathbf{k}$ is 2D vector.

The operators $\widehat{H}_{ph}$ and $\widehat{H}_{e(h)-ph}$ represent the corresponding Hamiltonian for the phonon system and the Fröhlich electron(hole)-LO-phonon interaction.

If we neglect the electron-phonon coupling effects, the magnetoexciton dispersion in 3D crystals is $E(\mathbf{Q}, Q_z) = E_g + \frac{1}{2}\hbar\Omega + \hbar^2 Q_z^2/2M - E_b(\mathbf{Q})$, where $\Omega = eB/\mu c$ is the exciton cyclotron frequency. The magnetoexciton "in-plane" dispersion $E_b(\mathbf{Q})$ has to be calculated by solving the following Schrödinger equation [11]:

$$-E_b(\mathbf{Q})\Psi(z) = -\frac{\hbar^2}{2\mu}\frac{d^2\Psi(z)}{dz^2} - V_\mathbf{Q}(z)\Psi(z), \qquad (5)$$

where in the LLL approximation

$$V_\mathbf{Q}(z) = \frac{e^2}{\varepsilon_\infty}\int d^2\mathbf{r}\frac{\psi_{00}^2(\mathbf{r})}{\sqrt{(\mathbf{r}+R^2\mathbf{Q}_0)^2+z^2}}. \qquad (6)$$

Here, $\psi_{00}(\mathbf{r}) = \exp\left(-r^2/4R^2\right)/\sqrt{2\pi R^2}$ is the LLL wave function. In 2D crystals the exciton dispersion is $E_b(\mathbf{Q}) = E_{2D}\exp(-Q^2 R^2/4)I_0(Q^2 R^2/4)$, where $I_0(x)$ is the modified Bessel function.

The case of magnetoexcitons with zero in-plane pseudomomentum $\mathbf{Q} = 0$ is an analogue to the problem of the polaronic effect on the shallow donors in the presence of a strong magnetic field [12]. The assumption of the absence of a transverse motion of the exciton as a whole simplifies very much the problem, but it neglects the fact that even a small transverse exciton velocity (or small transverse wave vector $\mathbf{Q}$) will induce an electric field in the rest frame of the exciton. This electric field will push the electron and the hole apart, so the binding energy must decrease as the transverse velocity increases. In other words, the magnetic field induces a coupling between the center-of-mass and the relative internal motion. The coupling effect complicates the calculations, so each of the above mentioned techniques need major modifications in order to be applied to the magnetoexciton-LO-phonon problem.

In what follows we shall modify the approach which is based on the assumption that the electron-hole attractive interaction in polar crystals is due to the exchange of the longitudinal photons [13]. Since the photons propagate in the crystal, they interact with the polarization created by the lattice vibrations, and therefore, the system under consideration consists of electrons, photons and phonons. In this approach the interactions between the particles are the electron-photon and phonon-photon interactions. The last interaction leads to the following longitudinal part of the photon Green function $D_\parallel(\mathbf{q},\omega) = \left(2\pi\hbar c^2/\omega^2\right)\varepsilon^{-1}(\mathbf{q},\omega)$ where $\varepsilon^{-1}(\mathbf{q},\omega)$ is the inverse dielectric constant. Let $\omega_\lambda(\mathbf{q})$ are the longitudinal normal modes in the crystal, which can be determined by the solutions of the equation $\varepsilon(\mathbf{q},\omega_\lambda) = 0$. For photon energies closed to the resonance $\hbar\omega_\lambda(\mathbf{q})$ the longitudinal photon Green function assumes the form:

$$D_\parallel(\mathbf{q},\omega) = \frac{2\pi\hbar c^2}{\omega_\lambda^2\left[\partial\varepsilon(\mathbf{q},\omega)/\partial\omega\right]_{\omega=\omega_\lambda(\mathbf{q})}}\frac{1}{\omega-\omega_\lambda(\mathbf{q})+\imath 0^+}.$$

In the absence of a magnetic field, this approach [13] and the equation-of-motion method [10], both provide similar results for the energy gap shift, polaron masses and the exciton binding energy.

As we have already mentioned, the dispersion of magnetoexcitons in the presence of a strong magnetic field has to be studied by the BS equation. In the case of a polar material, the kernel of the BS equation should be screened by the appropriate dielectric function which takes into account the interaction with the LO phonons. In the presence of a strong magnetic field the LLL approximation for the single-particle Green's function (this approximation ignores transitions between Landau levels and considers only the states on the lowest Landau level) should describe qualitatively the main features of the magnetoexcitons in polar crystals. The LLL approximation greatly simplifies the problem because the dynamics of the LLL is essentially $D - 2$-dimensional [14], and only in the LLL approximation we observe a dimensional reduction (from two spacelike coordinates and one timelike coordinate to no spacelike coordinates and one timelike coordinate, i.e. $2 + 1 \to 0 + 1$) in the dynamics of fermion pairing in the presence of a constant magnetic field.

2.1. Bulk Magnetoexcitons

We shall investigate the role of the interaction with LO phonons by applying the BS formalism widely used in quantum field theory for describing the two-fermion bound states. The basic assumption in the BS formalism is that the electron-hole bound states are described by the BS wave function (BS amplitude) $\Psi(1; 2) = \Psi(\mathbf{r}_c, z_c, t_1; \mathbf{r}_v, z_v, t_2)$, where the variables 1 and 2 represent the corresponding coordinates and the time variables. This function determines the probability amplitude to find the electron at the point $(\mathbf{r}_c, z_c)$ at the moment t_1 and the hole at the point $(\mathbf{r}_v, z_v)$ at the moment t_2. The BS amplitude satisfies the following equation:

$$\Psi(1;2) = \int d(1', 2', 1'', 2'') G_c(1; 1') G_v(2'; 2) I \begin{pmatrix} 1' & 1'' \\ 2' & 2'' \end{pmatrix} \Psi(1''; 2''). \qquad (7)$$

Here I is the irreducible BS kernel, and $G_{c,v}$ are the electron and the hole single-particle Green's functions. When the screening effects are taken into account, the irreducible kernel represents the screened Coulomb interaction between electrons and holes that constitute the excitons:

$$V(\mathbf{r}; z; t) = -\int \frac{d^2\mathbf{q}}{(2\pi)^2} \frac{dq_z}{2\pi} \frac{d\omega}{2\pi} \frac{4\pi e^2}{\sqrt{|\mathbf{q}|^2 + q_z^2}} \varepsilon^{-1}(\mathbf{q}, q_z, \omega) \exp\left[\imath(\mathbf{q}\cdot\mathbf{r} + q_z z - \omega t)\right]. \qquad (8)$$

In the case when the screening is due to the interaction with bulk LO phonons with frequency ω_0, the inverse dielectric function $\varepsilon^{-1}(\mathbf{q}, q_z, \omega)$ depends only on ω:

$$\varepsilon^{-1}(\omega) = \frac{1}{\varepsilon_\infty} - \frac{\omega_0}{2\varepsilon^*}\left[\frac{1}{\omega_0 - \omega - \imath 0^+} + \frac{1}{\omega_0 + \omega - \imath 0^+}\right]. \qquad (9)$$

The BS equation in the center-of-mass and reduced coordinates assumes the form:

$$\Psi_{\mathbf{Q},Q_z}(\mathbf{r},\mathbf{R};z,Z;t,t') = \int dz'dZ'd^2\mathbf{r}'d^2\mathbf{R}'dt_1 dt_2$$

$$G_c(\mathbf{R} + \alpha_v\mathbf{r}, \mathbf{R}' + \alpha_v\mathbf{r}'; Z + \frac{m_{vz}}{M_z}z, Z' + \frac{m_{vz}}{M_z}z'; t - t_1)$$
$$G_v(\mathbf{R}' - \alpha_c\mathbf{r}', \mathbf{R} - \alpha_c\mathbf{r}; Z' - \frac{m_{cz}}{M_z}z', Z - \frac{m_{cz}}{M_z}z; t_2 - t')$$
$$V(\mathbf{r}'; z'; t_1 - t_2)\Psi_{\mathbf{Q},Q_z}(\mathbf{r}', \mathbf{R}'; z', Z'; t_1, t_2), \tag{10}$$

where $G_{c,v}$ are the single-particle Green's functions.

The BS amplitude depends on the relative internal time $t - t'$ and on the "center-of-mass" time:

$$\Psi_{\mathbf{Q},Q_z}(\mathbf{r}, \mathbf{R}; z, Z; t, t') = \exp\left(-\frac{\imath E(\mathbf{Q}, Q_z)}{\hbar}(\alpha_c t + \alpha_v t')\right)\psi_{\mathbf{Q},Q_z}(\mathbf{r}, \mathbf{R}; z, Z; t - t'), \tag{11}$$

where $E(\mathbf{Q}, Q_z)$ is the exciton dispersion. Introducing the time Fourier transforms according to the rule $f(t) = \int_{-\infty}^{\infty} f(\omega) \exp(\imath\omega t) \frac{d\omega}{2\pi}$, we transform the above BS equation into the following form:

$$\psi_{\mathbf{Q},Q_z}(\mathbf{r}, \mathbf{R}; z,, Z; \omega) = \int dz' dZ' d^2\mathbf{r}' d^2\mathbf{R}' \frac{d\Omega}{2\pi}$$
$$G_c(\mathbf{R} + \alpha_v\mathbf{r}, \mathbf{R}' + \alpha_v\mathbf{r}'; Z + \alpha_v z, Z' + \alpha_v z'; \hbar\omega + \alpha_c E)$$
$$G_v(\mathbf{R}' - \alpha_c\mathbf{r}', \mathbf{R} - \alpha_c\mathbf{r}; Z' - \alpha_c z', Z - \alpha_c z; \hbar\omega - \alpha_v E)$$
$$V(\mathbf{r}'; z'; \omega - \Omega)\psi_{\mathbf{Q},Q_z}(\mathbf{r}', \mathbf{R}'; z', Z'; \Omega). \tag{12}$$

where $\psi_{\mathbf{Q},Q_z}(\mathbf{r}, \mathbf{R}; z, Z; \Omega)$ is the Fourier transform of $\psi_{\mathbf{Q},Q_z}(\mathbf{r}, \mathbf{R}; z, Z; t)$. Since the translation symmetry is broken by the magnetic field, the single-particle Green's functions can be written as a product of phase factors and translation invariant parts. The phase factor depends on the gauge. In the symmetric gauge we have [15]:

$$G_{c,v}(\mathbf{r}, \mathbf{r}'; z, z'; \omega) = e^{\imath \frac{e}{\hbar c} \mathbf{r} \cdot \mathbf{A}(\mathbf{r}')} \widetilde{G}_{c,v}(\mathbf{r} - \mathbf{r}'; z - z'; \omega). \tag{13}$$

The broken translation symmetry requires a phase factor for the BS amplitude:

$$\psi_{\mathbf{Q},Q_z}(\mathbf{r}, \mathbf{R}; z, Z; \Omega) = e^{\imath \frac{e}{\hbar c} \mathbf{r} \cdot \mathbf{A}(\mathbf{R})} \chi_{\mathbf{Q},Q_z}(\mathbf{r}, \mathbf{R}; z, Z; \Omega). \tag{14}$$

The BS equation (12) admits translation invariant solution of the form:

$$\chi_{\mathbf{Q},Q_z}(\mathbf{r}, \mathbf{R}; z, Z; \omega) = e^{-\imath(\mathbf{Q}\cdot\mathbf{R}+Q_z Z)}\widetilde{\chi}_{\mathbf{Q},Q_z}(\mathbf{r}; z; \omega). \tag{15}$$

The function $\widetilde{\chi}_{\mathbf{Q},Q_z}(\mathbf{r}; z; \omega)$ satisfies the following BS equation:

$$\widetilde{\chi}_{\mathbf{Q},Q_z}(\mathbf{r}; z; \omega) = \int dz' dZ' d^2\mathbf{r}' d^2\mathbf{R}' \frac{d\Omega}{2\pi} \exp\left[\frac{\imath e}{\hbar c}((\mathbf{r}+\mathbf{r}').\mathbf{A}(\mathbf{R}'-\mathbf{R}) + \gamma\mathbf{r}.\mathbf{A}(\mathbf{r}'))\right]$$
$$\widetilde{G}_c(\mathbf{R} - \mathbf{R}' + \alpha_v(\mathbf{r} - \mathbf{r}'); Z - Z' + \alpha_v(z - z'); \hbar\omega + \alpha_c E)$$
$$\widetilde{G}_v(\mathbf{R}' - \mathbf{R} + \alpha_c(\mathbf{r} - \mathbf{r}'); Z' - Z + \alpha_c(z - z'); \hbar\omega - \alpha_v E)$$
$$V(\mathbf{r}'; z'; \omega - \Omega)\widetilde{\chi}_{\mathbf{Q},Q_z}(\mathbf{r}'; z'; \Omega). \tag{16}$$

The substitution $\mathbf{R}' \rightarrow \mathbf{R}' + \mathbf{R} + \gamma\mathbf{r}$ provides the following equation for the Fourier transform of the exciton wave function $\widetilde{\chi}_{\mathbf{Q},Q_z}(\mathbf{k}; k_z; \omega) =$

$\int dz d^2\mathbf{r} \exp\left[-\imath\left(\mathbf{k}.\mathbf{r}+k_z z\right)\right]\widetilde{\chi}_{\mathbf{Q},Q_z}(\mathbf{r};z;\omega)$:

$$\widetilde{\chi}_{\mathbf{Q},Q_z}(\mathbf{k}-\tfrac{\gamma}{2}\mathbf{Q};k_z;\omega) = \int \frac{dp_z}{2\pi} \frac{d^2\mathbf{q}}{(2\pi)^2} \frac{d^2\mathbf{p}}{(2\pi)^2} d^2\mathbf{R} \int_{-\infty}^{\infty} \frac{d\Omega}{2\pi} e^{-\imath(\mathbf{q}+\mathbf{Q}).\mathbf{R}}$$

$$\widetilde{G}_c\left(\tfrac{1}{2}\mathbf{q}+\mathbf{k}-\tfrac{e}{\hbar c}A(\mathbf{R});k_z+\alpha_v Q_z;\hbar\omega+\alpha_c E\right)$$

$$\widetilde{G}_v\left(-\tfrac{1}{2}\mathbf{q}+\mathbf{k}-\tfrac{e}{\hbar c}A(\mathbf{R});k_z-\alpha_c Q_z;\hbar\omega-\alpha_v E\right)$$

$$V\left(\mathbf{p}-\left[\mathbf{k}-\tfrac{2e}{\hbar c}A(\mathbf{R})\right];p_z-k_z;\omega-\Omega\right)\widetilde{\chi}_{\mathbf{Q},Q_z}(\mathbf{p}-\tfrac{\gamma}{2}\mathbf{Q};p_z;\Omega),$$

where $V(\mathbf{k};k_z;\omega) = -4\pi e^2/\left(\mathbf{k}^2+k_z^2\right)\varepsilon^{-1}(\omega)$ and $\widetilde{G}_{c,v}(\mathbf{k};k_z;\hbar\omega)$ are the Fourier transforms of $\widetilde{G}_{c,v}(\mathbf{r};z;\hbar\omega)$.

In the effective-mass approximation the exact fermion Green's functions $G_{c,v}$ are replaced by the corresponding propagator of the free fermions but with renormalized masses. The translation invariant parts $\widetilde{G}_{c,v}$ in the Landau level representation have the following forms:

$$\widetilde{G}_c(\mathbf{r};z;\hbar\omega) = \int \frac{d^2\mathbf{k}}{(2\pi)^2}\frac{dk_z}{2\pi}\widetilde{G}_{c,v}(\mathbf{k};k_z;\hbar\omega)\exp\left[\imath\left(\mathbf{k}.\mathbf{r}+k_z z\right)\right],$$

$$\widetilde{G}_c(\mathbf{k};k_z;\hbar\omega) = 2\sum_{n=0}^{\infty}\frac{(-1)^n\exp\left(-R^2\mathbf{k}^2\right)L_n\left(2R^2\mathbf{k}^2\right)}{\hbar\omega-\left[\hbar^2 k_z^2/2m_c+E_g+\hbar\Omega_c(n+1/2)\right]+\imath 0^+}, \quad (17)$$

$$\widetilde{G}_v(\mathbf{k};k_z;\hbar\omega) = \sum_{n=0}^{\infty}\frac{(-1)^n\exp\left(-R^2\mathbf{k}^2\right)L_n\left(2R^2\mathbf{k}^2\right)}{\hbar\omega+\left[\hbar^2 k_z^2/2m_v+\hbar\Omega_v(n+1/2)\right]-\imath 0^+}.$$

Here, $L_n(x)$ are the Laguerre polynomials, and $\hbar\Omega_{c,v} = \hbar eB/cm_{c,v}$ are the electron and hole cyclotron energies. In strong magnetic fields the probability for transitions to the excited Landau levels due to the Coulomb interaction is small. The resonant condition $\omega_0 \simeq \Omega_{c,v}$ requires to take into account Landau levels with $n \geq 1$. It can be seen, that all Landau levels could be taken into account by rewriting the BS equation (7) in the form $G^{-1}G^{-1}\Psi = I\Psi$ [16], but in this case there is no more dimensional reduction. In what follows we assume that the resonant condition does not hold, and therefore, the contributions to the Green's functions from the excited Landau levels are negligible. Thus, we can apply the LLL approximation, where we keep only $n = 0$ term in the last expressions:

$$\widetilde{G}_c(\mathbf{k};k_z;\hbar\omega) \approx \frac{2\exp\left(-R^2\mathbf{k}^2\right)}{\hbar\omega-\left[E_g+\hbar^2 k_z^2/2m_c+\hbar\Omega_c/2\right]+\imath 0^+},$$

$$\widetilde{G}_v(\mathbf{k};k_z;\hbar\omega) \approx \frac{2\exp\left(-R^2\mathbf{k}^2\right)}{\hbar\omega+\left[\hbar^2 k_z^2/2m_v+\hbar\Omega_v/2\right]-\imath 0^+}. \quad (18)$$

The solution of the BS equation in the LLL approximation can be written in the following form:

$$\widetilde{\chi}_{\mathbf{Q},Q_z}(\mathbf{k};k_z;\omega) = \exp\left[-R^2\left(\mathbf{k}+\tfrac{\gamma}{2}\mathbf{Q}\right)^2-\imath\mathbf{Q}_0.\mathbf{k}R^2\right]\varphi_{Q_z}(k_z;\omega). \quad (19)$$

Thus, the LLL approximation reduces the problem from $3+1$ dimensions to $1+1$ dimensions, and therefore, the functions $\varphi_{Q_z}(k_z;\omega)$ and the energy $E(\mathbf{Q},Q_z)$ can be obtained from the following equation:

$$\varphi_{Q_z}(k_z;\omega) = \int \frac{dp_z}{2\pi}\frac{d\Omega}{2\pi}\frac{d\omega}{2\pi}I_{\mathbf{Q}}(p_z-k_z;\omega-\Omega) \times$$

$$\left[\frac{1}{\hbar\omega+\alpha_c E-\left(E_g+\frac{\hbar^2}{2m_c}(k_z+\alpha_c Q_z)^2+\frac{\hbar\Omega_c}{2}\right)+i0^+}+\frac{1}{\hbar\omega-\alpha_v E+\frac{\hbar^2}{2m_v}(k_z-\alpha_v Q_z)^2+\frac{\hbar\Omega_v}{2}-i0^+}\right]\times$$
$$\varphi_{Q_z}(p_z;\Omega). \qquad (20)$$

In the LLL approximation, the exciton dispersion is determined by the term

$$I_{\mathbf{Q}}(q_z;\omega) = \frac{4\pi e^2}{\varepsilon_\infty}\int\frac{d^2\mathbf{q}}{(2\pi)^2}\int d^2\mathbf{r}\frac{\psi_{00}^2(\mathbf{r})e^{i\mathbf{q}\cdot(\mathbf{r}+R^2\mathbf{Q}_0)}}{q^2+q_z^2}\times$$
$$\left[1-\frac{\omega_0}{2}\frac{\varepsilon_\infty}{\varepsilon^*}\left(\frac{1}{\omega_0-\omega-i0^+}+\frac{1}{\omega_0+\omega-i0^+}\right)\right]. \qquad (21)$$

The solution of (20) can be chosen in the following form:

$$\varphi_{Q_z}(k_z,\omega) = \phi(k_z)\left(\hbar\omega+\alpha_c E-\left[E_g+\frac{\hbar^2}{2m_c}(k_z+\alpha_c Q_z)^2+\frac{\hbar\Omega_c}{2}\right]+i0^+\right)^{-1}$$
$$\left(\hbar\omega-\alpha_v E+\left[\frac{\hbar^2}{2m_v}(k_z-\alpha_v Q_z)^2+\frac{\hbar\Omega_v}{2}\right]-i0^+\right)^{-1} \qquad (22)$$

Integrating both sides of (21) over ω, we find the following equation for the exciton wave function $\Phi_{Q_z}(k_z)$ and exciton energy $E(\mathbf{Q},Q_z) = E_g + \frac{1}{2}\hbar\Omega + \hbar^2 Q_z^2/2M - E_b(\mathbf{Q},Q_z)$:

$$-E_b(\mathbf{Q},Q_z)\Phi_{Q_z}(k_z) = \frac{\hbar^2 k_z^2}{2\mu}\Phi_{Q_z}(k_z)$$
$$-\frac{4\pi e^2}{\varepsilon_\infty}\int\frac{dq_z}{2\pi}\frac{d^2\mathbf{q}}{(2\pi)^2}\int d^2\mathbf{r}\frac{\psi_{00}^2(\mathbf{r})e^{i\mathbf{q}\cdot(\mathbf{r}+R^2\mathbf{Q}_0)}}{q^2+(k_z-q_z)^2}\Phi_{Q_z}(q_z)$$
$$+\frac{2\pi e^2}{\varepsilon^*}\int\frac{dq_z}{2\pi}\frac{d^2\mathbf{q}}{(2\pi)^2}\int d^2\mathbf{r}\frac{\psi_{00}^2(\mathbf{r})e^{i\mathbf{q}\cdot(\mathbf{r}+R^2\mathbf{Q}_0)}}{q^2+(k_z-q_z)^2}\Phi_{Q_z}(q_z)\times$$
$$\left[\frac{\hbar\omega_0}{\hbar\omega_0+E_b(\mathbf{Q},Q_z)+\Delta_{Q_z}(q_z,k_z)}+\frac{\hbar\omega_0}{\hbar\omega_0+E_b(\mathbf{Q},Q_z)+\Delta_{Q_z}(k_z,q_z)}\right],$$

where

$$\Delta_{Q_z}(k_z,q_z) = \frac{\hbar^2 k_z^2}{2m_c}+\frac{\hbar^2 q_z^2}{2m_v}+\frac{\hbar^2 Q_z(k_z-q_z)}{M}.$$

In position representation the last equation assumes the form:

$$-E_b(\mathbf{Q},Q_z)\Phi_{Q_z}(z) = -\frac{\hbar^2}{2\mu}\frac{d^2\Phi_{Q_z}(z)}{dz^2}-V_{\mathbf{Q}}(z)\Phi_{Q_z}(z)$$
$$+\int_{-\infty}^{\infty}dz' U_{\mathbf{Q},Q_z}(z,z';E_b(\mathbf{Q},Q_z))\Phi_{Q_z}(z'). \qquad (23)$$

Thus, we obtain an equation similar to eq. (5), but with two major differences. First, there is an extra non-local potential

$$U_{\mathbf{Q},Q_z}(z,z';E_b(\mathbf{Q},Q_z)) =$$
$$\frac{2\pi e^2}{\varepsilon^*}\int\frac{dk_z}{2\pi}\frac{dq_z}{2\pi}\frac{d^2\mathbf{q}}{(2\pi)^2}\int d^2\mathbf{r}\frac{\psi_{00}^2(\mathbf{r})\exp(i[\mathbf{q}\cdot(\mathbf{r}+R^2\mathbf{Q}_0)+k_z z-q_z z'])}{q^2+(k_z-q_z)^2}\times$$
$$\left[\frac{\hbar\omega_0}{\hbar\omega_0+E_b(\mathbf{Q},Q_z)+\Delta_{Q_z}(q_z,k_z)}+\frac{\hbar\omega_0}{\hbar\omega_0+E_b(\mathbf{Q},Q_z)+\Delta_{Q_z}(k_z,q_z)}\right], \qquad (24)$$

which represents the effect of the exciton-LO phonon interaction. The second difference is that because of the interaction with the LO phonons the in-plane motion and the motion along the z-direction are not independent.

For $\mathbf{Q} = 0$ and $Q_z = 0$, after integrations over $\mathbf{r}$ and $\mathbf{q}$, we find the following equation for the magnetoexciton binding energy $\epsilon = E_b/\hbar\Omega$ and the corresponding wave function $\Phi(z)$ (z is in units l):

$$0 = \left(-\frac{1}{2}\frac{d^2}{dz^2} - V_0(z) + \epsilon\right)\Phi(z) + \int_{-\infty}^{\infty} dz' U_{0,0}(z,z';\epsilon)\Phi(z'). \tag{25}$$

Here, $V_0(z) = \sqrt{\frac{\pi}{2}}(R/a_B) Erfc\left(\frac{|z|}{\sqrt{2}R}\right)\exp\left(\frac{z^2}{2R^2}\right)$, and the non-local potential (in units $\hbar\Omega$) is:

$$U_{0,0}(z,z';\epsilon) = \frac{1}{2}(R/a_B)\beta\left(1 - \frac{\varepsilon_\infty}{\varepsilon_0}\right)$$
$$\int \frac{dk_z}{2\pi}\frac{dq_z}{2\pi}\exp\left(\imath[k_z z + q_z(z-z')]\right)\exp\left(\frac{k_z^2}{2}\right)\Gamma\left(0,\frac{k_z^2}{2}\right)$$
$$\times \left[\left(\beta + \epsilon + \frac{1}{2}[\alpha_c(k_z+q_z)^2 + \alpha_v q_z^2]\right)^{-1}\right.$$
$$\left. + \left(\beta + \epsilon + \frac{1}{2}[\alpha_v(k_z+q_z)^2 + \alpha_c q_z^2]\right)^{-1}\right], \tag{26}$$

where $\beta = \omega_0/\Omega$, and $a_B = \varepsilon_\infty\hbar^2/\mu e$ is the exciton Bohr energy. $Erfc(x)$ and $\Gamma(a,x)$ are the complementary error function and the incomplete gamma function, respectively.

The last term in (25), which is non-local in space and depends on the binding energy and the electron and the hole bare masses. The dependence on the bare masses is due to the $3+1 \to 1+1$ reduction, so one can expect that in a pure 2D case the non-local potential should be mass independent one.

Without the non-local term, Eq. (25) has been studied decades ago by many authors [11]. More recently, it was found (see, e.g. Ref. [17] and references therein) that the eigenvalues can be separated into two distinct classes: the states having no node and the states having node (or notes) in their eigenfunctions. The states having no node in their wave functions are tightly bound while the states having nodes in their wave functions are weakly bound. A complete numerical evaluation of Eq. (25) for arbitrary value of the magnetic field is a complicated problem beyond the main goal of this paper.

2.2. Quantum-well Magnetoexcitons

Our approach can be applied not only to bulk crystals, but to the quasi two-dimensional systems as well. In quantum wells, however, besides the bulklike phonon modes one has to take into account the presence of slab modes [18], interface modes [19], and half-space modes [20]. In quasi two-dimensional systems we still have a dimensional reduction $2+1 \to 0+1$, but each of the above modes will create extra poles in the inverse dielectric function $\varepsilon^{-1}(\mathbf{q},\omega)$. The problem becomes too complicated and cannot be solved analytically.

In what follows we calculate the magnetoexciton dispersion in a quantum well taking into account only the interaction with Fröhlich's bulk LO phonons. In other words, we shall take into account the effect of size-quantization in a quantum well on the electron (hole) spectrum, whereas the phonon spectrum stays the same as in a homogeneous medium, as if the entire space were filled with the quantum well material. Strictly speaking, this approach can provide only an approximate solution of the quasi two-dimensional problem, but its results are exact ones in the pure 2D case.

In a single quantum well (SQW) and coupled quantum wells (CQW) the Fourier transform of the exciton wave function satisfies the following BS equation:

$$\tilde{\chi}_{\mathbf{Q}}(\mathbf{k} - \tfrac{\gamma}{2}\mathbf{Q}; \omega) = \int \frac{d^2\mathbf{q}}{(2\pi)^2} \frac{d^2\mathbf{p}}{(2\pi)^2} d^2\mathbf{R} \int_{-\infty}^{\infty} \frac{d\Omega}{2\pi} \exp\left[-\imath(\mathbf{q} + \mathbf{Q}).\mathbf{R}\right]$$
$$\tilde{G}_c\left(\tfrac{1}{2}\mathbf{q} + \mathbf{k} - \tfrac{e}{\hbar c}\mathbf{A}(\mathbf{R}); \hbar\omega + \alpha_c E\right) \times$$
$$\tilde{G}_v\left(-\tfrac{1}{2}\mathbf{q} + \mathbf{k} - \tfrac{e}{\hbar c}\mathbf{A}(\mathbf{R}); \hbar\omega - \alpha_v E\right)$$
$$V\left(\mathbf{p} - \left[\mathbf{k} - \tfrac{2e}{\hbar c}\mathbf{A}(\mathbf{R})\right]; \omega - \Omega\right) \tilde{\chi}_{\mathbf{Q}}(\mathbf{p} - \tfrac{\gamma}{2}\mathbf{Q}; \Omega), \qquad (27)$$

where $V(\mathbf{k}; \omega) = -(2\pi e^2 f(|\mathbf{k}|)/|\mathbf{k}|) \varepsilon^{-1}(\omega)$, where $f(\mathbf{k})$ is the structure factor:

$$f(|\mathbf{q}|) = f(q) = \int_{-\infty}^{+\infty} dz_c \int_{-\infty}^{+\infty} dz_v \exp\{-q(z_c - z_v)]\} \varphi_{0c}^2(z_c) \phi_{0v}^2(z_v). \qquad (28)$$

In our calculations, we take into account only the first electron E_{0c} and hole E_{0v} confinement levels with wave functions $\varphi_{0c}(z_c)$ and $\phi_{0v}(z_v)$, respectively.

In the LLL approximation the exact fermion Green's functions $G_{c,v}$ are replaced by the corresponding propagator of the free fermions $G_{c,v}^{(0)}$:

$$\tilde{G}_c(\mathbf{k}; \hbar\omega) \approx \frac{2\exp\left(-R^2\mathbf{k}^2\right)}{\hbar\omega - [E_g + E_{0c} + \hbar\Omega_c/2] + \imath 0^+},$$
$$\tilde{G}_v(\mathbf{k}; \hbar\omega) \approx \frac{2\exp\left(-R^2\mathbf{k}^2\right)}{\hbar\omega + E_{0v} + \hbar\Omega_v/2 - \imath 0^+}. \qquad (29)$$

The solution of the BS equation in the LLL approximation can be written in the following form:

$$\tilde{\chi}_{\mathbf{Q}}(\mathbf{k}; \omega) = \exp\left[-R^2\left(\mathbf{k} + \tfrac{\gamma}{2}\mathbf{Q}\right)^2 - \imath\mathbf{Q}_0.\mathbf{k}R^2\right] \varphi_E(\omega). \qquad (30)$$

Thus, the LLL approximation reduces the problem from $2+1$ dimensions to 1-dimension problem for obtaining function $\varphi(\omega)$ energy $E(\mathbf{Q})$ from the following BS equation:

$$\varphi_E(\omega) = -\frac{1}{\left[\hbar\omega + \alpha_c E - E_g - E_{0c} - \frac{\hbar\Omega_c}{2} + \imath 0^+\right]\left[\hbar\omega - \alpha_v E + E_{0v} + \frac{\hbar\Omega_v}{2} - \imath 0^+\right]} \times$$
$$[I(|\mathbf{Q}|, \varepsilon_\infty) \int_{-\infty}^{\infty} \frac{d\Omega}{2\pi} \varphi_E(\Omega)$$
$$- I(|\mathbf{Q}|, \varepsilon^*) \int_{-\infty}^{\infty} \frac{d\Omega}{2\pi} \varphi_E(\Omega) \frac{\omega_0}{2} \left(\frac{1}{\omega_0 - \omega + \Omega - \imath 0^+} + \frac{1}{\omega_0 + \omega - \Omega - \imath 0^+}\right)]. \qquad (31)$$

The exciton dispersion is determined by the term:

$$I(\mathbf{Q}, \varepsilon) = \frac{2\pi e^2}{\varepsilon} \int d^2\mathbf{r} \frac{d^2\mathbf{q}}{(2\pi)^2} \psi_{00}^2(\mathbf{r}) \frac{f(|\mathbf{q}|)e^{\imath\mathbf{q}.(\mathbf{r} + R^2\mathbf{Q}_0)}}{|\mathbf{q}|}. \qquad (32)$$

The solution of (31) can be chosen in the following form:

$$\varphi_E(\omega) = \frac{1}{\left[\hbar\omega + \alpha_c E - E_g - E_{0c} - \frac{\hbar\Omega_c}{2} + \imath 0^+\right]\left[\hbar\omega - \alpha_v E + E_{0v} + \frac{\hbar\Omega_v}{2} - \imath 0^+\right]}. \qquad (33)$$

Thus, by integrating both sides of (32) over ω, we find the following equation for the exciton dispersion $E(|\mathbf{Q}|) = E_g + E_{0c} + E_{0v} + \hbar\Omega/2 - E_b(|\mathbf{Q}|)$:

$$E_b(|\mathbf{Q}|) = I(|\mathbf{Q}|, \varepsilon_\infty) - I(|\mathbf{Q}|, \varepsilon^*)\frac{\hbar\omega_0}{\hbar\omega_0 + E_b(|\mathbf{Q}|)}. \tag{34}$$

Solving for $E_b(|\mathbf{Q}|)$, we obtain:

$$E_b(|\mathbf{Q}|) = \frac{1}{2}\left[I(|\mathbf{Q}|, \varepsilon_\infty) - \hbar\omega_0 + \sqrt{(I(|\mathbf{Q}|, \varepsilon_\infty) - \hbar\omega_0)^2 + 4\frac{\varepsilon_\infty}{\varepsilon_0}I(|\mathbf{Q}|, \varepsilon_\infty)\hbar\omega_0}\right]. \tag{35}$$

When $\mathbf{Q} = 0$ the exciton binding energy $E_b = E_b(\mathbf{Q} = 0)$ is:

$$E_b = \frac{I_\infty}{2}\left[1 - \frac{\hbar\omega_0}{I_\infty} + \sqrt{\left(1 - \frac{\hbar\omega_0}{I_\infty}\right)^2 + 4\frac{\varepsilon_\infty}{\varepsilon_0}\frac{\hbar\omega_0}{I_\infty}}\right],$$

where $I_\infty = I(|\mathbf{Q}| = 0, \varepsilon_\infty)$.

In the pure 2D case the exciton dispersion is determined by the terms $I_{2D}(\mathbf{Q}, \varepsilon_\infty)$ and $I_{2D}(\mathbf{Q}, \varepsilon^*)$, where

$$I_{2D}(\mathbf{Q}, x) = \frac{2\pi e^2}{x}\int d^2\mathbf{r}\frac{d^2\mathbf{q}}{(2\pi)^2}\psi_{00}^2(\mathbf{r})\frac{e^{i\mathbf{q}\cdot(\mathbf{r}+R^2\mathbf{Q}_0)}}{|\mathbf{q}|}.$$

The exciton dispersion $E(|\mathbf{Q}|) = E_g + \hbar\Omega/2 - E_b(|\mathbf{Q}|)$ is:

$$\begin{aligned}E_b(|\mathbf{Q}|) &= \frac{1}{2}[I_{2D}(|\mathbf{Q}|, \varepsilon_\infty) - \hbar\omega_0 \\ &+ \sqrt{(I_{2D}(|\mathbf{Q}|, \varepsilon_\infty) - \hbar\omega_0)^2 + 4\frac{\varepsilon_\infty}{\varepsilon_0}I_{2D}(|\mathbf{Q}|, \varepsilon_\infty)\hbar\omega_0}].\end{aligned}$$

When $\mathbf{Q} = 0$ the magnetoexciton binding energy assumes the form $E_b = \sqrt{\frac{\pi}{2}}e^2/\epsilon(B)R$ but with an effective dielectric constants:

$$\epsilon(B) = \frac{2\varepsilon_\infty}{1 - \frac{\hbar\omega_0}{E_{2D}} + \sqrt{\left(1 - \frac{\hbar\omega_0}{E_{2D}}\right)^2 + 4\frac{\varepsilon_\infty}{\varepsilon_0}\frac{\hbar\omega_0}{E_{2D}}}}, \tag{36}$$

The corresponding exciton mass is:

$$M_{2D} = \frac{2^{5/2}\varepsilon_\infty}{\sqrt{\pi}e^2 R}\left[1 + \frac{1 + \left(2\frac{\varepsilon_\infty}{\varepsilon_0} - 1\right)\frac{\hbar\omega_0}{E_{2D}}}{\sqrt{\left(1 - \frac{\hbar\omega_0}{E_{2D}}\right)^2 + 4\frac{\varepsilon_\infty}{\varepsilon_0}\frac{\hbar\omega_0}{E_{2D}}}}\right]^{-1}$$

3. Magnetoexciton Dispersion in Graphene Bilayers Embedded in a Dielectric

A lot of experimental and theoretical studies in recent years are focusing on the unusual relativisticlike, kinematic properties of the electronic states in graphene predicted theoretically decades ago [21, 22]. A major breakthrough was done in 2004 when a group lead by Novoselov [23, 24] tested and confirmed that the graphitic monolayers have anomalous relativisticlike properties. Because electrons and holes in a graphene behave like massless Dirac particles, there is a number of unusual properties, such as high charge carrier mobility [24], the graphene's conductivity never falls below a minimum value [25, 26], and an anomalous quantum Hall effect [27]. Bilayer graphene systems, where carriers in one layer are electrons and carriers in the other are holes, have been considered as ideal candidates for observing superfluid properties at room temperatures [28, 29, 30].

In what follows we examine how both the magnetoexciton binding energy and magnetoexciton mass in graphene bilayer systems vary with the magnetic field and the separation d between the layers in the LLL approximation. It is worth mentioning that the calculations done by treating the Coulomb interaction as a perturbation [31] provide in the LLL approximation *a number of extra terms which do not exist in the case of CQW's*. From a general point of view, we have to expect that the binding energy is *exactly* four times higher than E_{CQW}, while the magnetoexciton mass is *exactly* four times lower than M_{CQW}. The physical reason for the above statement lies in the fact that in the LLL approximation we have a dimensional reduction in the dynamics of the electron-hole pairing from two space variables plus a time variable to zero space variable and a time variable. Because of this $2+1 \to 0+1$ reduction the results should be insensitive to the type of the band dispersion. The factor four is due to the four-component-spinor description used in the relativistic case.

3.1. Relativistic Bethe-Salpeter Equation

The system under consideration is made from two graphene sheets embedded in a dielectric and separated by distance d. Each of the two graphene layers has two Dirac-like linear dispersion $\hbar v_F k$ bands centered at two non-equivalent points $\mathbf{K}$ and $\mathbf{K'}$, where v_F is the Fermi velocity of electrons in graphene. Since the layers are embedded in a dielectric, there is no hopping of π-electrons between the layers. There is a potential difference $\pm V_g/2$ (gate voltage) applied to each of the two layers which allows us to adjust the charge density in the layers. We assume that the potential difference is chosen in a manner that the electrons are in the top layer (pseudospin index $\tau = 1$) and the same number of holes in the bottom layer ($\tau = 2$).

The unit cell of graphene has two atoms, A and B, each belonging to the different sublattice. The operator $\psi^{(\tau)\dagger}_{\sigma,A,\alpha}(\mathbf{r})$ ($\psi^{(\tau)\dagger}_{\sigma,B,\alpha}(\mathbf{r})$) creates an electron of spin $\sigma = \uparrow, \downarrow$ on the atom A (atom B) of the unit cell in layer τ defined by the position vector $\mathbf{r}$. We introduce four component spinors:

$$\Psi^{(\tau)}_\sigma(\mathbf{r}) = \begin{pmatrix} \psi^{(\tau)}_{\sigma,A,\mathbf{K}}(\mathbf{r}) \\ \psi^{(\tau)}_{\sigma,B,\mathbf{K}}(\mathbf{r}) \\ \psi^{(\tau)}_{\sigma,B,\mathbf{K'}}(\mathbf{r}) \\ \psi^{(\tau)}_{\sigma,A,\mathbf{K'}}(\mathbf{r}) \end{pmatrix}, \quad \overline{\Psi}^{(\tau)}_\sigma(\mathbf{r}) = \Psi^{(\tau)\dagger}_\sigma(\mathbf{r})\gamma^0, \qquad (37)$$

where the following representation of the Dirac matrices is chosen:

$$\gamma^0 = \begin{pmatrix} 1 & 0 & 0 & 0 \\ 0 & -1 & 0 & 0 \\ 0 & 0 & -1 & 0 \\ 0 & 0 & 0 & 1 \end{pmatrix}, \quad \gamma^1 = \begin{pmatrix} 0 & 1 & 0 & 0 \\ -1 & 0 & 0 & 0 \\ 0 & 0 & 0 & -1 \\ 0 & 0 & 1 & 0 \end{pmatrix}, \quad (38)$$

$$\gamma^2 = \begin{pmatrix} 0 & -i & 0 & 0 \\ -i & 0 & 0 & 0 \\ 0 & 0 & 0 & i \\ 0 & 0 & i & 0 \end{pmatrix}. \quad (39)$$

In continuum approximation the non-interacting quasiparticles in the layers are described by the Hamiltonian:

$$H_0 = \sum_{\sigma,\tau} \int d^2\mathbf{r} \overline{\Psi}_\sigma^{(\tau)}(\mathbf{r}) \widehat{H}^{(\tau)} \Psi_\sigma^{(\tau)}(\mathbf{r}), \quad (40)$$

where

$$\widehat{H}^{(\tau)} = v_F \left(\gamma^1 \widehat{p}_x + \gamma^2 \widehat{p}_y\right), \widehat{p}_x = -i\hbar \frac{\partial}{\partial x}, \widehat{p}_y = -i\hbar \frac{\partial}{\partial y}. \quad (41)$$

The action that describes the non-interacting quasiparticles in a layer τ is:

$$S_0^{(\tau)} = \int d^2\mathbf{r} dt \overline{\Psi}_\sigma^{(\tau)}(\mathbf{r},t) \left[\gamma^0 i\hbar \frac{\partial}{\partial t} - v_F\left(\gamma^1 \widehat{p}_x + \gamma^2 \widehat{p}_y\right)\right] \Psi_\sigma^{(\tau)}(\mathbf{r},t) \quad (42)$$

In the presence of a perpendicular magnetic field $\mathbf{B} = (0, 0, B)$ and a potential difference $\pm V_g/2$ (gate voltage) applied to each of the two layers, the action (42) assumes the form:

$$S_0^{(\tau)} = \int d^2\mathbf{r} dt \overline{\Psi}_\sigma^{(\tau)}(\mathbf{r},t) \left[\gamma^0\left(i\hbar \frac{\partial}{\partial t} - V_g^{(\tau)}\right) - v_F\left(\gamma^1 \widehat{\pi}_x + \gamma^2 \widehat{\pi}_y\right)\right] \Psi_\sigma^{(\tau)}(\mathbf{r},t), \quad (43)$$

where $\widehat{\pi}_{x(y)} = \widehat{p}_{x(y)} \mp (e/c)\mathbf{A}_{x(y)}(\mathbf{r})$, and $\mathbf{A}(\mathbf{r}) = (1/2)\mathbf{B} \times \mathbf{r}$ is the vector potential in a symmetric gauge.

In what follows we assume that the interaction between an electron with a position vector $\mathbf{r}_1$ from the top layer ($\tau = 1$) and a hole with a position vector $\mathbf{r}_2$ from the bottom layer ($\tau = 2$) is described by the Coulomb potential $V(\mathbf{r}_1 - \mathbf{r}_2) = e^2/\varepsilon_0 \sqrt{|\mathbf{r}_1 - \mathbf{r}_2|^2 + d^2}$. Instead of two position vectors $\mathbf{r}_1$ and $\mathbf{r}_2$, we introduce the center-of-mass $\mathbf{R} = \alpha(\mathbf{r}_1 + \mathbf{r}_2)$ and the relative $\mathbf{r} = \mathbf{r}_1 - \mathbf{r}_2$ coordinates ($\alpha = 1/2$).

The basic assumption in our BS formalism is that the electron-hole bound states are described by the BS wave function (BS amplitude). This function determines the probability amplitude to find the electron at the point $\mathbf{r}_1$ at the moment t_1 and the hole at the point $\mathbf{r}_2$ at the moment t_2. The BS amplitude depends on the relative internal time $t - t'$ and on the "center-of-mass" time:

$$\Phi^{\mathbf{Q}}(\mathbf{r},\mathbf{R};t,t') = \exp\left(-\frac{iE(\mathbf{Q})\alpha}{\hbar}(t+t')\right) \phi^{\mathbf{Q}}(\mathbf{r},\mathbf{R};t-t'), \quad (44)$$

where $E(\mathbf{Q})$ is the exciton dispersion. The BS equation for the equal-time BS amplitude in the center-of-mass and reduced coordinates is [32]:

$$\Phi^{\mathbf{Q}}(\mathbf{r}, \mathbf{R}; t, t) = \int d^2\mathbf{r}' d^2\mathbf{R}' dt' G^{(1)}(\mathbf{R} + \alpha\mathbf{r}, \mathbf{R}' + \alpha\mathbf{r}'; t - t')\gamma^0 \times$$
$$G^{(2)}(\mathbf{R}' - \alpha\mathbf{r}', \mathbf{R} - \alpha\mathbf{r}; t' - t)\gamma^0 V(\mathbf{r}')\Phi^{\mathbf{Q}}(\mathbf{r}', \mathbf{R}'; t', t'). \qquad (45)$$

The Fourier transforms of the electron and hole propagators $G^{(\tau)}(\mathbf{r}, \mathbf{r}'; t)$ are define in terms of the Dirac four component spinors $\psi^\kappa(\mathbf{r})$ and the corresponding eigenvalues $E_n = \hbar v_F \sqrt{2n}/R$ [32]:

$$G^{(\tau)}(\mathbf{r}, \mathbf{r}'; \omega) = \sum_\kappa \frac{\psi^\kappa(\mathbf{r})\overline{\psi}^\kappa(\mathbf{r}')}{\hbar\omega - E_n \pm \imath 0^+} \qquad (46)$$

Here we keep only the positive energy pole contributions, $n = 0, 1, 2, ...$, and $\kappa = (n, j_z, \sigma)$, where j_z is the z component of the total angular momentum.

By means of the time Fourier-transforms, we transform the BS equation (12) into the following form (repeated indexes $\nu, \mu = 1, 2, 3, 4$ are summed up):

$$\phi^{\mathbf{Q}}_{\nu,\mu}(\mathbf{r}, \mathbf{R}; \omega) = \int' d^2\mathbf{r}' d^2\mathbf{R}' \frac{d\Omega}{2\pi} G^{(1)}_{\nu,\nu'}(\mathbf{R}$$
$$+\alpha\mathbf{r}, \mathbf{R}' + \alpha\mathbf{r}'; \hbar\omega + \alpha(E(\mathbf{Q}) - V_g))\gamma^0_{\nu',\nu''}$$
$$G^{(-1)}_{\mu,\mu'}(\mathbf{R}' - \alpha\mathbf{r}', \mathbf{R} - \alpha\mathbf{r}; \hbar\omega - \alpha(E(\mathbf{Q}) - V_g))\gamma^0_{\mu',\mu''}$$
$$V(\mathbf{r}')\phi^{\mathbf{Q}}_{\nu'',\mu''}(\mathbf{r}', \mathbf{R}'; \Omega). \qquad (47)$$

When the translation symmetry is broken by the magnetic field, the Green's functions can be written as a product of phase factors and translation invariant parts. The phase factor depends on the gauge. In the symmetric gauge the Green's functions are:

$$G^{(\tau)}(\mathbf{r}, \mathbf{r}'; \omega) = \exp\left[\imath \frac{e}{\hbar c}\mathbf{r}.\mathbf{A}(\mathbf{r}')\right] \widetilde{G}^\tau(\mathbf{r} - \mathbf{r}'; \omega). \qquad (48)$$

The broken translation symmetry requires a phase factor for the BS amplitude:

$$\phi^{\mathbf{Q}}(\mathbf{r}, \mathbf{R}; \Omega) = \exp\left[\imath \frac{e}{\hbar c}\mathbf{r}.\mathbf{A}(\mathbf{R})\right] \chi^{\mathbf{Q}}(\mathbf{r}, \mathbf{R}; \Omega). \qquad (49)$$

The BS equation (47) admits translation invariant solution of the form:

$$\chi^{\mathbf{Q}}(\mathbf{r}, \mathbf{R}; \omega) = \exp[-\imath(\mathbf{Q}.\mathbf{R})] \widetilde{\chi}^{\mathbf{Q}}(\mathbf{r}; \omega). \qquad (50)$$

The function $\widetilde{\chi}^{\mathbf{Q}}(\mathbf{r}; \omega)$ satisfies the following BS equation:

$$\widetilde{\chi}^{\mathbf{Q}}(\mathbf{r}; \omega) = \int d^2\mathbf{r}' d^2\mathbf{R}' \frac{d\Omega}{2\pi} \exp\left[\frac{\imath e}{\hbar c}((\mathbf{r} + \mathbf{r}').\mathbf{A}(\mathbf{R}' - \mathbf{R}))\right]$$
$$\widetilde{G}^{(1)}(\mathbf{R} - \mathbf{R}' + \alpha(\mathbf{r} - \mathbf{r}'); \hbar\omega + \alpha(E - V_g))\gamma^0$$
$$\widetilde{G}^{(2)}(\mathbf{R}' - \mathbf{R} + \alpha(\mathbf{r} - \mathbf{r}'); \hbar\omega - \alpha(E - V_g))\gamma^0 V(\mathbf{r}')\widetilde{\chi}^{\mathbf{Q}}(\mathbf{r}'; \Omega). \qquad (51)$$

The substitution $\mathbf{R}' \to \mathbf{R}' + \mathbf{R}$ provides the following equation for the Fourier transform of the exciton wave function $\tilde{\chi}^\mathbf{Q}(\mathbf{k}; \omega) = \int d^2\mathbf{r} \exp(-\imath \mathbf{k}.\mathbf{r}) \tilde{\chi}^\mathbf{Q}(\mathbf{r}; \omega)$ of the exciton wave function:

$$\tilde{\chi}^\mathbf{Q}(\mathbf{k}; \omega) = \int \frac{d^2\mathbf{q}}{(2\pi)^2} \frac{d^2\mathbf{p}}{(2\pi)^2} d^2\mathbf{R} \int_{-\infty}^{\infty} \frac{d\Omega}{2\pi} e^{-\imath(\mathbf{q}+\mathbf{Q}).\mathbf{R}}$$
$$\tilde{G}^{(1)}\left(\tfrac{1}{2}\mathbf{q} + \mathbf{k} - \tfrac{e}{\hbar c}\mathbf{A}(\mathbf{R}); \hbar\omega + \alpha(E - V_g)\right)\gamma^0 \times$$
$$\tilde{G}^{(2)}\left(-\tfrac{1}{2}\mathbf{q} + \mathbf{k} - \tfrac{e}{\hbar c}\mathbf{A}(\mathbf{R}); \hbar\omega - \alpha(E - V_g)\right)\gamma^0$$
$$V\left(\mathbf{p} - \left[\mathbf{k} - \tfrac{2e}{\hbar c}\mathbf{A}(\mathbf{R})\right]\right)\tilde{\chi}_\mathbf{Q}(\mathbf{p}; \Omega), \qquad (52)$$

where $\tilde{G}^{(\tau)}(\mathbf{k}; \hbar\omega)$ are the Fourier transforms of $\tilde{G}^{(\tau)}(\mathbf{r}; \hbar\omega)$.

In the effective-mass approximation the exact fermion Green's functions $G^{(\tau)}$ are replaced by the corresponding propagator of the free fermions. The translation invariant parts of the free fermion propagators can be decomposed over the Landau level poles [33]:

$$\tilde{G}^{(\tau)}(\mathbf{k}; \hbar\omega) = 2\imath \sum_{n=0}^{\infty}(-1)^n e^{-R^2\mathbf{k}^2} \frac{\hbar\omega\gamma^0 f_1(k) + f_2(\mathbf{k})}{\hbar^2\omega^2 - 2n\hbar v_F^2 eB/c},$$
$$f_1(k) = \tfrac{1}{2}(1 - \imath\gamma^1\gamma^2)L_n(2R^2k^2) - \tfrac{1}{2}(1 + \imath\gamma^1\gamma^2) \times$$
$$L_{n-1}(2R^2k^2), \quad f_2(\mathbf{k}) = 2v_F\hbar(k_x\gamma^1 + k_y\gamma^2)L^1_{n-1}(2R^2k^2). \qquad (53)$$

Here $L^1_n(x)$ are the generalized Laguerre polynomials, $L^1_{-1}(x) = L_{-1}(x) = 0$ and $L_n(x)$ are the Laguerre polynomials. In strong magnetic fields the probability for transitions to the excited Landau levels due to the Coulomb interaction is small. Thus, the contributions to the Green's functions from the excited Landau levels is negligible, and therefore, one can apply the LLL approximation, where we keep only $n = 0$ term:

$$\tilde{G}^{(1)}(\mathbf{k}; \hbar\omega) \approx \imath \exp(-R^2\mathbf{k}^2) \frac{\gamma^0(1 - \imath\gamma^1\gamma^2)}{\hbar\omega + \imath 0^+},$$
$$\tilde{G}^{(2)}(\mathbf{k}; \hbar\omega) \approx \imath \exp(-R^2\mathbf{k}^2) \frac{\gamma^0(1 - \imath\gamma^1\gamma^2)}{\hbar\omega - \imath 0^+}. \qquad (54)$$

The solution of the BS equation in the LLL approximation can be written in the following form:

$$\tilde{\chi}^\mathbf{Q}(\mathbf{k}; \omega) = \exp\left[-R^2\mathbf{k}^2 - \imath\mathbf{R}_0.\mathbf{k}\right]\Phi_E(\omega). \qquad (55)$$

Here $\Phi_E(\omega)$ is a 4×4 matrix. Thus, the LLL approximation reduces the problem from $2 + 1$-dimensions to $0 + 1$-dimension problem. The matrix $\Phi_E(\omega)$ and the magnetoexciton dispersion $E(\mathbf{Q})$ are determined by the solutions of the following equation:

$$\Phi_E(\omega) = -I(|\mathbf{Q}|) \int_{-\infty}^{\infty} \frac{d\Omega}{2\pi} \frac{\gamma^0(1 - \imath\gamma^1\gamma^2)\gamma^0\Phi_E(\Omega)\gamma^0(1 - \imath\gamma^1\gamma^2)\gamma^0}{(\hbar\omega + \alpha(E - V_g) + \imath 0^+)(\hbar\omega - \alpha(E - V_g) - \imath 0^+)}. \qquad (56)$$

The solution of (56) is $E(\mathbf{Q}) = V_g - 4I(\mathbf{Q})$, where the function $\Phi_E(\omega)$ is given by:

$$\Phi_E(\omega) = \begin{pmatrix} 0 & 0 & 0 & 0 \\ 0 & 1 & 0 & 1 \\ 0 & 0 & 0 & 0 \\ 0 & 1 & 0 & 1 \end{pmatrix} \frac{1}{(\hbar\omega + \alpha(E - V_g) + \imath 0^+)(\hbar\omega - \alpha(E - V_g) - \imath 0^+)}. \qquad (57)$$

Thus, in the LLL approximation, the magnetoexciton dispersion is determined by the Coulomb interaction term $I(\mathbf{Q}) = \int d^2\mathbf{r}\varphi_{00}^2(r)V(\mathbf{r}+\mathbf{R}_0)$, where $\varphi_{00}(r) = (\sqrt{2\pi}R)^{-1}\exp(-r^2/4R^2)$ is the ground-state wave function of an electron in a magnetic field. For small wave vectors we calculate:

$$E(\mathbf{Q}) \approx V_g - 4E_{CQW} + \frac{\hbar^2 Q^2}{2M(B)}, \qquad \frac{M(B)}{M_{CQW}} = \frac{1}{4}. \qquad (58)$$

In graphene bilayer structures the magnetoexciton mass (binding energy) is four times lower (higher) than the corresponding magnetoexciton mass (binding energy) in coupled quantum wells with parabolic dispersion and the same d, ε_0 and B. In the limit of very small interlayer separation $d \ll R$ the asymptotical values of the binding energy and the effective magnetic mass of magnetoexciton in bilayer graphene are $4E_b$ and $M_{2D}/4$, respectively.

It is worth mentioning that the calculations done by treating the Coulomb interaction as a perturbation [31] provide in the LLL approximation *a number of extra terms which do not exist in the case of CQW's*. From a general point of view, we have to expect that the binding energy is *exactly* four times higher than E_{CQW}, while the magnetoexciton mass is *exactly* four times lower than M_{CQW}. The physical reason for the above statement lies in the fact that in the LLL approximation we have a dimensional reduction in the dynamics of the electron-hole pairing from two space variables plus a time variable to zero space variable and a time variable. Because of this $2 + 1 \to 0 + 1$ reduction the results should be insensitive to the type of the band dispersion. The factor four is due to the four-component-spinor description used in the relativistic case.

4. Conclusion

In this Chapter we have applied the BS formalism to the magnetoexcitons in polar materials and graphene structures. We have obtained analytical results for the binding energy and the exciton mass in the LLL approximation which greatly simplifies the calculations. One may well ask whether the magnetoexciton dispersion will be significantly affected by the contributions from the infinity number of Landau levels above the LLL. Turning our attention to the parabolic band quantum-well structures [34] we find that beyond the LLL approximation, the BS equation contains an extra term (BS term) This term takes into account the transitions to the Landau levels with indexes $n > 1$. The corresponding contributions to the magnetoexciton binding energy and mass can be obtained by applying a variational procedure. The results are as follows. In a strong magnetic field, the ground-state energy is very close to that obtained by means of the Schrödinger equation, but the magnetoexciton dispersion is determined by the BS term rather than the electron-hole Coulomb term in the Schrödinger equation. In the relativistic case, going beyond the LLL approximation is an ambitious task which probably will be a subject of future research (to the best of our knowledge the only paper which goes beyond the LLL approximation is [35]).

References

[1] S. I. Shevchenko, Phys. Rev. B **56**, 10355 (1997); Yu. Lozovik, and A. M. Ruvisky, Zh. Eksp. Teor. Fiz. **112**, 1791 (1997) [Sov.Phys. JETP 85, 979 (1997)]; A. B. Dzyubenko,

JETP Lett. **66**, 617 (1997); S.I. Shevchenko, Phys. Rev. B **57**, 14809 (1998); Yu. E. Lozovik, O. L. Berman, and V. G. Tsvetus, Phys. Rev. B **59**, 5627 (1999); O. L. Berman, Yu. E. Lozovik, D. W. Snoke, and R. D. Coalson, Phys. Rev. B **73**, 235352 (2006); O. L. Berman, R. Ya. Kezerashvili, and Yu. E. Lozovik, Phys. Rev. B **80**, 115302 (2009).

[2] E. E. Salpeter and H. A. Bethe, Phys. Rev. **84**, 1232 (1951); M. Gell-Mann and F. Low, Phys. Rev. **84**, 350 (1951); C. G. Wick, Phys. Rev. **96**, 1124 (1954); R. E. Cutkosky, Phys. Rev. **96**, 1135 (1954).

[3] Z. G. Koinov, Phys. Rev. B **79**, 075409 (2009); Phys. Status Solidi (B) **246**, 397 (2009).

[4] Z. G. Koinov, Phys. Rev. B **79**, 073409 (2009).

[5] I. V. Lerner and Yu. E. Lozovik, Zh. Eksp. Teor. Fiz. **80**, 1488 (1981) [Sov. Phys. JETP **53**, 763 (1981)].

[6] S. Wang and M. Matsuura, Phys. Rev. B **10**, 3330 (1974).

[7] J. Adamowski, B. Gerlach, and H. Leschke, Phys. Rev. B **23**, 2943 (1981).

[8] J. Pollmann and H. Büttner, Phys. Rev. B **16**, 4480 (1977), U. Rössler and H. R. Trebin, Phys. Rev. B. **23**, 1961 (1981).

[9] J. Sak, Phys. Rev. B **6**, 2226 (1972), S. D. Mahanti and C. M. Varma, Phys. Rev. B **6**, 2209 (1972), Z. G. Koinov, J. Phys.: Condens. Matter **2**, 6507 (1990).

[10] A. Oswald and I. Egri, Phys. Rev. B **6**, 3291 (1983).

[11] R. London, Amer. J. Phys. **27**, 649 (1959); R. J. Elliott and R. London, J. Phys. Chem. Solids **15**, 196 (1960); H. Hasegawa and R. E. Howard, *ibid.* **21**, 179 (1961); L. P. Gorkov and I. E. Dzyaloshinskii, Zh. Eksp. Teor. Fiz. **53**, 717 (1967)[Sov. Phys. JETP **26**, 449 (1968).

[12] A. Elagovan and K. Navaneethakrishnan, J. Phys.: Condens. Matter **5**, 4021 (1993); Zhi-jie Wang, Yong-gang Weng, Jun-jie Shi1, Zi-xin Liu and Shao-hua Pan, Z. Phys. B, **104**,227 (1997).

[13] Z. Koinov, J. Phys.: Condens. Matter **2**, 6507 (1990); **3**, 6313 (1991).

[14] V. P. Gusynin, V. A. Miransky, and I. A. Shovkovy, Phys. Rev. Lett. **73**, 3499 (1994); Phys. Rev. D **52**, 4718 (1995).

[15] D. Lehmann, Communications in Math. Phys. **173**, 155 (1995).

[16] E. A. Shabat and V. V. Usov, Phys. Rev. D **73**, 125021 (2006); Z. Koinov, Phys. Rev. B **77**, 165333 (2008).

[17] B. M. Karnakov and V. S. Popov, Zh. Eksp. Teor. Fiz. **124**, 996 (2003) [Sov. Phys. JETP **97**, 890 (2003)].

[18] R. Fuchs and K. L. Kliewer, Phys. Rev. **140**, A2076 (1965); K. L. Kliewer and R. Fuchs, Phys. Rev. **144**, 495 (1966); **150**, 573 (1966); X. X. Liang, Sh. W. Gu, and D. L. Lin, Phys. Rev. B **34**, 2807 (1986).

[19] T. Tsuchiya and T. Ando, Phys. Rev. B **47**, 7240 (1993).

[20] J. J. Licari and R. Evrard, Phys. Rev. B **15**, 2254 (1977); L. Wendler and R. Haupt, Phys. Status Solidi B **143**, 487 (1987); S. N. Klimin, E. P. Pokatilov, and V. M. Fomin, Phys. Status Solidi B **190**, 441 (1995).

[21] G. W. Semenoff, Phys. Rev. Lett. **53**, 2449 (1984).

[22] F. D. M. Haldane, Phys. Rev. Lett. **61**, 2015 (1988).

[23] K. S. Novoselov, A. K. Geim, S. V. Morozov, D. Jiang, Y. Zhang, S. V. Dubonos, I. V. Grigorieva, and A. A. Firsov, Science **306**, 666 (2004).

[24] K. S. Novoselov, A. K. Geim, S. V. Morozov, D. Jiang, M. I. Katsnelson, I. V. Grigorieva, S. V. Dubonos, and A. A. Firsov, Nature (London) **438**, 197 (2005).

[25] K. Ziegler, Phys. Rev. Lett. **97**, 266802 (2006).

[26] K. Nomura and A. H. MacDonald, Phys. Rev. Lett. **98**, 076602 (2007).

[27] Y. Zhang, Y.-W. Tan, H. L. Stormer, and P. Kim, Nature **438**, 201 (2005).

[28] C.-H. Zhang and Y. N. Jorlecar, Phys. Rev. B **77**, 233405 (2008).

[29] H. Min, R. Bistritzer, Jung-Jung Su, and A. H. MacDonald, Rev. B **78**, 121401(R) (2008).

[30] Yu. E. Lozovik and A. A. Sokolik, Pisma Zh. Eksp. Teor. Fiz. **87**, 61 (2008).

[31] O. L. Berman, Yu.E. Lozovik, and G. Gumbs, Phys. Rev. B **77**, 155433 (2008); O. L. Berman, R. Ya. Kezerashvili, and Yu. E. Lozovik, Phys. Rev. B **78**, 035135 (2008).

[32] K. A. Kouzakov and A. I. Studenikin, Phys. Rev. C **72**, 015502 (2005).

[33] E. V. Gorbar, V. P. Gusynin, V. A. Miransky and I. A. Shovkovy, Phys. Rev. B **66**, 045108 (2002).

[34] Z. G. Koinov, Phys. Rev. B **65**, 155332 (2002).

[35] A. E. Shabad and V. V. Usov, Phys. Rev. D **73**, 125021 (2006).

In: Exciton Quasiparticles
Editor: Randy M. Bergin

ISBN: 978-1-61122-318-7
© 2011 Nova Science Publishers, Inc.

Chapter 11

WANNIER- MOTT-FRENKEL HYBRID EXCITON IN SEMICONDUCTOR-ORGANIC SYSTEMS CONTAINING QUANTUM DOTS

Nguyen Que Huong [*]
Marshall University, Huntington WV, USA

Abstract

The chapter will describe the theory of the formation of a hybridization state of Wannier Mott exciton and Frenkel exciton in different hetero-structure configurations involving quantum dot. The hybrid excitons exist at the interfaces of the semiconductors quantum dots and the organic medium, having unique properties and a large optical non-linearity. The coupling at resonance is very strong and tunable by changing the parameters of the systems (dot radius, dot-dot distance, generation of the organic dendrites and the materials of the system etc). Different semiconductor quantum dot-organic material combination systems have been considered such as a semiconductor quantum dot lattice embedded in an organic host, a semiconductor quantum dot at the center of an organic dendrite, a semiconductor quantum dot coated by an organic shell.

The formation and the properties of the organic-semiconductor hybrid excitons have been modulated by electric and magnetic fields. The hybrid excitons are as sensitive to external perturbation as Wannier-Mott excitons. Upon the application of the magnetic and electric fields the coupling term between the two kinds of excitons increases.

The most important feature of this system is, by adjusting the system parameters as well as the external fields and their orientation, one can tune the resonance between the two kinds of excitons to get different regions of mixing to obtain the expected high non-linearity.

PACS 73.21.La, 73.22.Dj, 78.67.Hc **Keywords:** nanocrystals, semiconductors, organic material.

---

[*]E-mail address: nguyenh@marshall.edu

1. Introduction

Development of innovative growth techniques and realization of systems in two-, one- and zero-dimensional confined geometry made it possible to have new organic and inorganic structures with unique properties. Especially, being confined in all three dimensions, quantum dots have the biggest quantum size effects and are very attractive objects for experimental and theoretical investigations [1]- [12]. Growth of organic multilayer structures analogous to semiconductor (inorganic) superlattices has been made largely [13] - [15]. In particular, possibilities of combining organic and inorganic materials in one heterostructure lead to a series of investigation of a new type of hybrid excitation [17]-[24] and open a new field of research for investigation of optical properties of organic-inorganic systems, which is very promising from both technological and background scientific points of view.

Excitons play a fundamental role in optical properties of a solid, especially for the optical processes happening close to or below the band gap. Being an electron excitation wave which does not carry electric current, by definition an exciton is a bound state of an electron and a hole, what is created by light. There are two different kinds of excitons in solid: Frenkel excitons and Wannier- Mott excitons. The Frenkel exciton is the electronic state of a molecular crystal, where the electrons and the holes situate on the same molecule or atom. Lattice constants of molecular crystals are very small, $a \sim 5A$, so the Frenkel exciton can be considered as a tightly bound exciton with the exciton radius equal the crystal lattice constant. In opposite, in semiconductors the typical electron-hole distance is large, and the Wannier-Mott excitons are relatively weakly bound with the Bohr radii $a_B \sim 100A$ in III-V materials and $a_B \sim 30A$ in II-VI materials. Due to their properties, the Frenkel excitons are also called small-radius excitons, while the Wannier Mott excitons are large-radius. Because of their large radius, the interaction between the Wannier- Mott excitons is very large, while the one between Frenkel excitons can be neglected in most cases. The Frenkel excitons have very strong oscillator strength, comparable to the oscillator strength of the molecule, while the oscillator strength of an Wannier-Mott exciton is much weaker. In addition, the small radii of the Frenkel excitons make their wavefunctions very difficult to overlap each other to reach the saturation density and the exciton resonance while the Wannier exciton has a rather low saturation density $n_s \sim 1/\pi a_B^2$.

The differences between Frenkel excitons of organic materials and Wannier-Mott excitons of semiconductors give scientists the idea to form systems having hybrid excitation states with combining properties of both kinds of excitons. One expects to have an hybrid exciton with the large exciton radius of the Wannier exciton while possessing the large oscillator strength of the Frenkel exciton. The hybrid state can have at the same time a large optical resonance nonlinearity and a low exciton saturation density. In other words, one expects to have mixing states with the complimentary properties of both kinds of excitons [17]. To realize this idea, physicists have been proposing systems with different geometrical configurations of organic-semiconductor materials [17] where the Frenkel excitons of organic material and the Wannier-Mott excitons in semiconductor are in resonance of each other and interact with each other by the dipole-dipole interaction at their interface to form the mixing state.

The very first effort in this field is the model of a nanostructure consisting of neighbouring organic and inorganic quantum wells by Agranovich et al. [18]. The resonant interaction of two kinds of excitons at the interface of two 2D quantum wells leads to strong mixing and to the apperance of new states with large exciton radii (typical for Wannier-Mott excitons) and large oscillator strength (typical for the Frenkel excitons). The minimum of the dispersion curve is shifted away from the center of the Brilloin zone, which influences the optical properties. The dipole-dipole coupling decreases quickly with increasing the interwell distance. Another configuration studied in [19] was the excited states in parallel neighboring organic and inorganic semiconductor quantum wires. For this geometry, the hybrid exciton state is different from zero even for zero wavevector. The model of a single semiconductor quantum dot with an organic shell was also proposed in [20] and a strong mixing was found for the weak confinement regime in the limit of dot radius $R_D >> a_B$ (exciton Bohr-radius). For these models, the authors predicted a large enhancement of non-linearities at resonance, in some cases it was about two-orders of magnitude in comparison with the traditional systems. The system of a semiconductor quantum dot array embedded in a medium of organic material [21, 22] and a dendrimeric structure with a dot at the center [23] have been considered by Huong and Birman. Such structures were reported to have been fabricated in several labs [12, 25].

In this chapter we will discuss the formation and properties of Wannier-Mott-Frenkel hybrid exciton in different systems containing quantum dots with the possibilities of tuning the system properties by parameters of the systems.

2. Semiconductor Quantum Dot Array in an Organic Medium

The oscillator strength of a quantum dot and the optical non-linearities increase proportionally to the radius of the dot as long as the quantum size effect of the zero-dimensional quantum dot still works. When the dot radius is too large, the quantum size effect no longer works, the exciton energy becomes continuous as in the bulk semiconductors, and the non-linearities stop increasing with the increasing of the dot size. There exists some size of the quantum dot where the nonlinearities archieve the maximum value, the limitation that the pure dot cannot overcome.

It is already known [26] that when many quantum dots are arranged together in an array, due to the multipole interaction of excitons in different dots, an exciton inside a quantum dot can be considered not localized in that dot, but propagating through the lattice via the mechanism of exciton transfer processes. By that transfer process, the quantum dot array can help to overcome the above limitation and enhance the optical nonlinearities in the system.

While the dot array is embedded in an organic material, due to the interaction of this propagating exciton with the medium, a new hybrid exciton will appear in the system [21, 22]. This hybrid exciton, which is a mixed state of the transfer exciton and the Frenkel exciton of the medium when these are at resonance, has a large exciton radius (due to the large Wannier-Mott exciton radius) and a large oscillator strength (due to the large oscillator strengths of the both Frenkel and transfer Wannier-Mott excitons). The small mass of the transfer exciton leads to a large coherence length of the system. This fact, as well as the hybridization, will give a very large optical non-linearity.

2.1. Total Hamiltonian of the System

Let us consider a nanostructure consisting of a three dimensional array of semiconductor quantum dots placed into some organic material as a host medium [21, 22]. The size of the system should be considerably smaller than the wavelength of light that corresponds to the transition between excited and ground state [26]. For simplicity we use the ideal array with dots of the same radius R and the same dot-dot spacing d. The excitations in inorganic semiconductor quantum dots are the Wannier Mott excitons, that will interact with the Wannier-Mott excitons in other dots through the multidipole interaction. The Frenkel excitons in organic medium can move relatively freely between the sites.

The total Hamiltonian of the system will be taken as follows:

$$H = H_W + H_F + H_{int} \qquad (1)$$

where H_W is the Wannier Mott exciton Hamiltonian, which consists of terms for a free excitons and for the hopping between the excitons in different dots:

$$H_W = \sum_{\vec{n},l} E^W_{\vec{n}l} a^+_{\vec{n}l} a_{\vec{n}l} + \sum_{\vec{n}\vec{n}'ll'} t_{\vec{n}\vec{n}'ll'}(a^+_{\vec{n}l} a_{\vec{n}'l'} + h.c.) \qquad (2)$$

where $a^+_{\vec{n}l}$, $(a_{\vec{n}l})$ are creation (annihilation) operators of Wannier excitons in quantum dots. Index l labels the exciton states and $\vec{n}$ indicates the sites of the dot in the dot lattice. Here we assume that the dots are distributed on sites of a three-dimensional lattice with the position $\vec{n} = (n_x, n_y, n_z)$ of each site in (x,y,z) coordinates, where the distance between the sites (the "lattice" constant) equals d. For simplicity, we assume a cubic array, i.e, the number of dots in each directions N_x, N_y, N_z is the same $N_x = N_y = N_z = N$.

The Frenkel exciton Hamiltonian is written in the following form:

$$H_F = \sum_{\vec{k},m} E^F_m(\vec{k}) b^+_{\vec{k}m} b_{\vec{k}m} \qquad (3)$$

where $b^+_{\vec{k}m}$, $(b_{\vec{k}m})$ are creation (annihilation) operators for the Frenkel exciton in the organic medium with wave vector $\vec{k}$ in the mth-exciton state.

The Hamiltonian interaction between the Wannier Mott excitons in the dots and the Frenkel excitons in the medium has the form:

$$H_{int} = \sum_{ml\vec{n}\vec{k}} g_{lm}(\vec{r}_{\vec{n}}, \vec{k})(a^+_{\vec{n}l} b_{\vec{k}m}) \qquad (4)$$

In (2) and (3) $E^W_{\vec{n}l}$ and $E^F_{\vec{k}m}$ are the excitation energies of Wannier excitons in the dots and the Frenkel exciton in the medium, respectively. For the Wannier excitons confined to a spherical dot, the oscillator strengths are concentrated mainly on the low excited states. To a good approximation only the interaction of the lowest states of excitons(the ground state and the lowest excited state) is taken into account. We also assume that the energy difference between the energy levels E^F and E^W is much smaller than the distance to other bands. $g_{lm}(\vec{r}_{\vec{n}}, \vec{k})$ is the coupling constant of Wannier- Mott and Frenkel excitons, and $t_{\vec{n}\vec{n}'ll'}$ is the hopping constant between Wannier- Mott excitons in the dots. This hopping constant, which has its origin in the multipolar interaction of excitons in different dots, in general, is different in different directions because of the direction of polarization. Here we assume only nearest dots interact with each other, so the hopping constants for the nearest dots in the x, y, z direction are t_x, t_y, t_z, respectively.

Since the model is studied in the regime where the dot radius is of the same order of the ground state exciton Bohr radius, we can assume there exists only one exciton in each dot and omit the exciton-exciton interaction in the same dot. Changing to k-space by Fourier transformation with:

$$a^+_{n_x,n_y,n_z;l} = 1/\sqrt{N^3} \sum_{k_x,k_y,k_z} [\exp i(n_x k_x d + n_y k_y d + n_z k_z d)] a^+_{k_x,k_y,k_z;l} \quad (5)$$

and

$$g_{lm}(\vec{r}_{\vec{n}}, \vec{k}) = \sum_{\vec{k}'} [\exp i(n_x k'_x d + n_y k'_y d + n_z k'_z d)] G_{lm}(\vec{k}', \vec{k}). \quad (6)$$

where d is the distance between dots, $\vec{k} = \{k_x, k_y, k_z\}$ is the wave vector of the exciton in the coupled dots. The Fourier transform of t_x, t_y, t_z will be $t(k_x), t(k_y), t(k_z)$, respectively. Notice here that if one makes the translational transformation with a lattice vector $\vec{L}$, due to the exponential forms of the Frenkel and Wannier exciton state functions, by translational invariance, one will get $\sum_L \exp i(\vec{k} - \vec{k}')\vec{L} = \delta(k - k')$. So the coefficient $G_{lm}(\vec{k}, \vec{k}')$ will be different from zero only if $k = k'$. Then instead of $G_{lm}(\vec{k}, \vec{k}')$ we can write the coupling constant as $G_{lm}(\vec{k})$ and omit the sum over k'. Then we get the total Hamiltonian (1) as:

$$H = \sum_{\vec{k}l} E^W_{tr}(\vec{k}, l) a^+_{l\vec{k}} a_{l\vec{k}} + + \sum_{m\vec{k}} E^F_m(\vec{k}) b^+_{\vec{k}m} b_{\vec{k}m} + \\ + \sum_{ml\vec{k}} G_{lm}(\vec{k})(a^+_{\vec{k}l} b_{\vec{k}m} + a_{\vec{k}l} b^+_{\vec{k}m}) \quad (7)$$

where $E^W_{tr}(\vec{k}, l)$ is the eigenenergy of the Wannier-Mott exciton Hamiltonian component including the hopping transfer process

$$H_W = \sum_{k,l} E^W_{tr}(k) a^+_{kl} a_{kl} + 2 \sum_l \sum_{k_x,k_y,k_z,l} (t(k_x)\cos(k_x d) + t(k_y)\cos(k_y d) \\ + t(k_z)\cos(k_z d)] a^+_{l\vec{k}} a_{l\vec{k}} \quad (8)$$

In the perfect confinement approximation the exciton wave functions must vanish at the boundary of dots, and the energy of the Wannier exciton confined in the spherical quantum dot has discrete values according to the zeros of the Bessel function:

$$E^W_{nl}(k) = E_g - E^b_{ext} + \frac{\hbar^2 \gamma^2_{nl}}{2 M_{ex} R^2}, \quad (9)$$

where E_g is the band gap, E^b_{ext} is the exciton binding energy, and γ_{nl} is the n-th zero of the spherical Bessel function $J_l(x)$ of order l, which depends on the magnitude of the dot radius R. M_{ex} is the effective mass of the exciton. The lowest excitation in the dots will be the state with $l = 0, n = 1$. In writing the expression for $E_{nl}^W(k)$ we are assuming the dots are spherical as a good approximation to the actual shape.

For kd small, the transfer energy of the Wannier excitons of the semiconductor dot array $E_{tr}^W(\vec{k})$ becomes:

$$E_{tr}^W(\vec{k}) = E^W + 2\sum_i t(k_i) - d^2 \sum_i t(k_i)k_i^2, \qquad (10)$$

where $i = \{x, y, z\}$. Besides the exciton energy in single quantum dots, the energy of the Wannier exciton in the quantum dot array also includes the large transfer energy between two nearest quantum dots in the array. We notice here that because of the confinement, the energy and the state of one quantum dot exciton cannot be described by the wave vector. But the energy and the state of the transfer exciton in the quantum dot array do have the k-vector dependence and we will need to include the dispersion relation for the energy of this transfer exciton. This energy strongly depends on the value of the hopping constant $t(\vec{k})$ and the direction of the polarization vector of the exciton, which we will investigate in the next sections. The presence of the transfer exciton allows us to change the energy region of the resonance and also the optical properties of the hybrid exciton.

From equation (7) the system of Wannier- Mott excitons and Frenkel excitons interacting with each other can be interpreted as follows: the Wannier- Mott excitons in quantum dots interact with each other to form a transfer exciton propagating through the lattice. This transfer exciton in its turn is coupled at resonance with the Frenkel excitons in the organic medium to form an hybrid organic-inorganic exciton state.

In our model we consider the exciton-exciton interaction and the hybridization as the principal effect, so here we omit the potential scattering between the dot array and the medium.

II. 3. Coupling coefficients $t(\vec{k})$ and $G(\vec{k})$

Being the transfer energy between two nearest dots, the Wannier-Wannier exciton coupling coefficient $t(\vec{k})$, or the hopping constant, is estimated as the electrostatic interaction between excitons in dots. Each exciton confined in a dot has its transition dipole moment, which interacts with the corresponding moment of another dot when the distance between the two dots is comparable to the dot radius. As mentioned above, the oscillator strength of a dot exciton is concentrated mainly on the lowest excited states, therefore we assume that only the transition dipole moments to the lowest excited states are involved in the interaction for an array. This multipolar interaction is intrinsically strongly short-range, and dependent upon the distance between dots, therefore the nearest neighbor approximation is suitable.

$$t(\vec{k}) = <W_i(\vec{k})|H_{d-d}|W_j(\vec{k})> \qquad (11)$$

where $W_i(\vec{k}), W_j(\vec{k})$ are the exciton wave functions in the two dots,

$$|W_i(\vec{k})> = \frac{1}{V_0} \int \phi(\vec{r}_i^{eh})\varphi(\vec{r}_i)e^{i\vec{k}(\vec{r}_{ei}+\vec{r}_{hi})/2}\Psi_e^+(\vec{r}_{ei})\Psi_h^+(\vec{r}_{hi})d\vec{r}_{ei}d\vec{r}_{hi}|0> \qquad (12)$$

V_0 is the dot volume, $\phi(\vec{r}_i^{eh})$ is the relative electron-hole motion function, $\vec{r}_i^{eh} = \vec{r}_{ei} - \vec{r}_{hi}$, $\vec{r}_{ei}, (\vec{r}_{hi}), \Psi_e^+(\vec{r}_{ei}), (\Psi_h^+(\vec{r}_{hi}))$ are the coordinates and creation operators of electron (hole) in the dot, respectively. $\varphi(\vec{r})$ is the exciton envelope function [1,4]:

$$\varphi(\vec{r}_i)_{nlm} = Y_{lm}(\theta_i, \phi_i) \frac{2^{1/2}}{R^{3/2}} \frac{J_{nl}(\gamma_{nl}\frac{r_i}{R})}{J_{l+1}(\gamma_{nl})} \tag{13}$$

H_{d-d} is the interaction Hamiltonian between two dipole moments in these two dots. In our case, where the distance between two dots is larger than the dot radius $d > R$, the ordinary dipole-dipole interaction can be used. Neglecting the higher orders the interaction between two quantum dots can be written in the form:

$$H_{d-d} = \int \frac{3(\vec{r}.\vec{p}_2)(\vec{r}\vec{p}_1) + (\vec{p}_1.\vec{p}_2)r^2}{r^5} \tag{14}$$

where $\vec{p}_1, \vec{p}_2$ are the polarization vectors of the Wannier Mott excitons inside the two dots

$$p_i = \mu_D^W \Psi_e(r_e)_i \Psi_h(r_h)_i + h.c. \tag{15}$$

μ_D^W is the optical transition dipole moment of the Wannier-Mott exciton. For the transition dipole moment to the excited state $(n, l = 0, m = 0)$ of the spherical quantum dot:

$$\mu^w = \frac{(2)^{3/2}}{n\pi} \phi_{1s}(0) p_{cv} R^{3/2} \tag{16}$$

As the result the hopping coefficient between two spherical quantum dots has the form:

$$t(k) = \phi_{ns}(0)^2 f_{ns}\{(\vec{\mu}_1^w.\vec{\mu}_2^w) - 3(\vec{\mu}_1^w.\hat{n}_{12})(\vec{\mu}_2^w.\hat{n}_{12})\} \tag{17}$$

$\vec{\mu}_{1,2}^w$ are transtion dipole moments to the excited state $(n, l = 0, m = 0)$ for the quantum dot spheres 1 and 2, respectively, $\hat{n}_{12}$ is the unit vector directed along the straight line connecting two excitons, which due to the small dot radius we can approximately treat as directed along the line connecting two dot centers. f_{ns} is the integral taken over the volumes of the two dots :

$$f_{ns} = \int \varphi(\vec{r})_{ns} d^3r \int \frac{\varphi(\vec{r'})_{ns}}{|d+r+r'|^3} d^3r' \tag{18}$$

For the exciton polarization parallel to the direction connecting two dot centers the hopping coefficient is equal to :

$$t_\parallel = -2\phi_{ns}(0)^2 f_{ns}(\mu^w)^2 \tag{19}$$

and for the exciton polarization perpendicular to the direction connecting two dot centers:

$$t_\perp = \phi_{ns}(0)^2 f_{ns}(\mu^w)^2 \tag{20}$$

The hopping constant depends strongly on the polarization direction of excitons, the direction of k-vector and relationship between the k-vector and the polarization mode of the exciton. The longitudinal and transverse modes have different energies.

As we see from (17) - (20), the hopping coefficient t depends on R/d, and one can tune the dot separation d with respect to dot radius R in order to determine the optimum t.

The organic medium can also be described as a lattice with organic molecules occupying every site and the Frenkel excitons moving between the sites. Because of the small "lattice constant", the organic molecular lattice can be considered as a "microscopic" lattice in comparison with the macroscopic size of the dot lattice. The organic lattice constant is of order 5Å, while the dot radius is about 30-100 Å and the dot lattice constant is usually around 60-500 Å. The resonance coupling of Frenkel excitons in the medium and Wannier excitons in the dot array is determined by the interaction parameter [18]

$$G(k) = <F, k|H_{int}|W, k>$$ (21)

where the interaction Hamiltonian is taken similarly to [18]

$$H_{int} = -\sum_n E(r_n)P(r_n)$$ (22)

Here $E(r_n)$ is the operator of the electric field created at point r_n in the organic medium by the excitons in quantum dots, $P(r_n)$ is the transition polarization operator of the Frenkel exciton at molecular site r_n of the organic medium.

The Frenkel exciton wave function is written in the form:

$$F(k) = \frac{1}{N_F^{1/2}} \sum_n e^{ikr_n} \chi_s^f(r_n) b_n^+ |0>$$ (23)

$\chi_s^f(r_n)$ is the excited state of the molecule at site r_n. The obtained expression for the hybridization coefficient of the semiconductor quantum dot and the organic medium has the following form:

$$G(k) = \frac{3\epsilon_1}{2\epsilon_2 + \epsilon_1} \frac{\pi}{2} \frac{\sin\theta}{(NF)^{1/2}} \mu^F \mu^w \phi_{ns}(0) D_{ns}(k)$$ (24)

where θ is the angle between exciton transition dipole moments of the quantum dot and the organic molecule,

$$D_{ns}(k) = \int_{Medium} e^{ikr'} \chi_{ns}^f(r') d^3r' \int_{Dot} \frac{\varphi_{nlm}(r)}{r|r-r'|^3} d^3r$$ (25)

The first integral is taken over the dot and the second one is taken over the volume of the whole medium.

2.2. The Hybrid Exciton State

Consider the case where the energy separation between the Wannier-Mott and and the Frenkel excitons is much less than the distance to other exciton bands and seek for the mixing state only between the two nearest bands. As a basic set we choose the mixing state such that when the Wannier-Mott exciton is excited, the Frenkel exciton is in its ground state, and when the Frenkel exciton is excited, the Wannier-Mott exciton is in its ground state and write the new hybrid excited state as the following:

$$|\Psi(k)> = u_l(k)f^F(0)\Psi_l^W(k) + v_{l'}(k)f^W(0)|\Psi_{l'}^F(k)>$$ (26)

where $\Psi^W(k)$ and $\Psi^F(k)$ are excited states and $f^W(0), f^F(0)$ are ground states of the Wannier-Mott exciton in dot array and Frenkel excitons in medium, respectively. Since we will consider only the lowest excited states of the exciton, so from now on we will omit the indices l and l'. The coefficients u(k) and v(k) have the following form:

$$u(k) = \frac{G(k)}{\{[E^F(k) - E_{tr}^W(k)]^2 + G^2(k)\}^{1/2}}$$
$$v(k) = \frac{E^F(k) - E_{tr}^W(k)}{\{(E^F(k) - E_{tr}^W(k))^2 + G^2(k)\}^{1/2}} \quad (27)$$

Then in term of hybrid operators $\alpha_{\vec{k}}, \alpha_{\vec{k}}^+$, the Hamiltonian (7) can be written:

$$H' = \sum_k E(k)\alpha_{\vec{k}}^+ \alpha_{\vec{k}} \quad (28)$$

with the energy $E(\vec{k})$ of the hybrid state given by the following dispersion relation:

$$E(\vec{k}) = 1/2\{E^F(\vec{k}) + E_{tr}^W(\vec{k})\} \pm 1/2\{[E^F(\vec{k}) - E_{tr}^W(\vec{k})]^2 + 4G^2(\vec{k})\}^{1/2} \quad (29)$$

Due to the weak dependence of the Frenkel exciton energy upon the k- vector, the Frenkel exciton energy may be taken independent of the wave vector k, $E^F(k) = E^F(0)$. We can see from (14) that the existence of the array of dots, which results in the appearance of the transfer exciton energy $E_{tr}^W(\vec{k})$, enhances the coupling between these two kinds of exciton at resonance, i.e., the gap between two hybrid exciton branches becomes large. The coupling is strong when $E^F(\vec{k})$ and $E_{tr}^W(\vec{k})$ are in resonance. The resonance coupling behavior depends strongly on the hopping coefficient $t(\vec{k})$ and the hybridization coefficient $G(\vec{k})$.

We also can obtain the average exciton radius of the new hybrid exciton:

$$a_{hybrid} = |u(k)|^2 a_W + |v(k)|^2 a_F \quad (30)$$

where a_W, a_F are radius of the Wannier-Mott exciton and Frenkel exciton, respectively. As previously mentioned, $a_W \gg a_F$, therefore for strong hybrid state we have a large hybrid exciton radius $a_{hybrid} \sim |u(k)|^2 a_W$. Since $u(k)$ depends on the coefficients G(k) and t(k), the hybrid exciton radius is also dependent on those coefficients.

Figure 1 shows the hybrid exciton dispersion curves plotted for ZnSe dots embedded in a standard organic material. The parameters were taken as $E^F(0) - E^W(0) = 5meV, a_B = 30$ Å, $\mu^F = 5D, N = 5$. In Figure 1 two branches of the hybrid exciton are plotted for an array of dots with radius $R = 40$Å, and the dot lattice constant $d = 80$ Å.

2.3. Nonlinear Optical Response

At resonance, the oscillator strength of the hybrid state is determined by its Frenkel exciton component and its exciton radius is determined by its Wannier component. As the result, the hybrid exciton has both the large exciton radius and large oscillator strength. In addition, as already noted in [26] and can be seen from (10), the Wannier transfer exciton has a rather small translational mass, which depends on the hopping constant and the number of dots. For instance, for the simple case where we assume all $t_x = t_y = t_z = t$, comparing the (10)

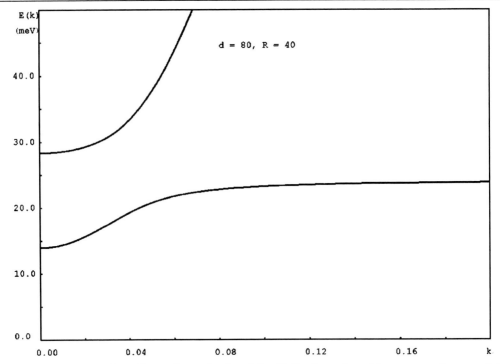

Figure 1. Hybrid exciton dispersion calculated for ZnSe dots placed in an organic material. The dispersion curve is plotted for dot radius R=50 Å and dot spacing d=200 Å. *Huong and Birman* [21]

with the energy formula $E = \hbar^2 k^2/2m$, we have the formula for the so called translational mass

$$M = \frac{\hbar^2}{2t^2 d^2} \tag{31}$$

This translational mass depends on the hopping constant between dots and the dot-dot spacing, and it is a few times smaller than the ordinary reduced electron-hole mass. This small translational mass is one reason for a large coherence length, which is related to the homogeneous linewidth of the excitonic transition.

$$l_c = \left(\frac{3\pi^2}{2^{1/2}}\right)^{1/3} \frac{\hbar}{(M\hbar\Gamma_h)^{1/2}} \tag{32}$$

The increase of the oscillator strength as well as the coherence length leads to a large figure merit of the hybrid exciton. The large radius leads to rather low saturation density, and the existence of large oscillator strength and the low saturation density promises the large non-linearities of the state.

In the presence of the external electric field the third order susceptibility has been calculated using the standard perturbation theory [27]-[30]. Considering only the resonance case and neglecting contributions from the other nonresonant levels, we have the result for

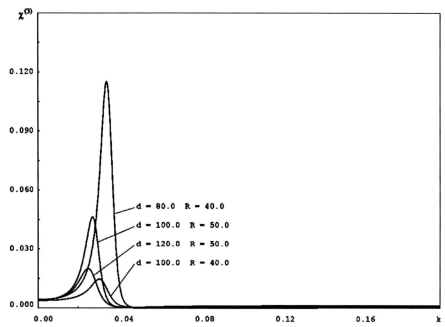

Figure 2. Third-order nonlinear susceptibility for the hybrid exciton state of ZnSe dots in organic material. Plots are for different dot radius R and separation d. *Huong and Birman* [21]

the lowest optical nonlinearity of the hybrid excitons:

$$\chi^{(3)}(w) \approx \frac{\mu_F^4}{V} \frac{(2\sqrt{2})^4}{\pi^2} \left(\frac{V_{Medium}}{V_{cell}}\right)^2 l_c^3 \left(\frac{R}{d}\right)^6 \phi_{1s}^4(0) \times \left\{ \frac{1}{(w - \tilde{w} + i\gamma_\perp)^2 (w - \tilde{w} - i\gamma_\parallel)} \right\} \quad (33)$$

V_{Medium} is the volume of the organic host, and V_{cell} is the volume of one cell in the organic lattice, l_c is the coherence length, $\gamma_\perp$ and $\gamma_\parallel$ are the transverse and longitudinal relaxation constants of the excitonic transition, respectively, R is the dot radius, d is the dot to dot separation, a is the Bohr radius, and $\hbar\tilde{w}$ is the lowest excitation energy of the hybrid exciton.

The value of $\chi^{(3)}(w)$ at resonance may be very large. By changing the number of dots and other parameters of the array, one can control the value of the non-linearity.

Fig. 2 is the numerical results for ZnSe dots [21]. We use here the following typical parameters of organic and semiconductor materials: $v^F = 100$Å, $a_{org} = 5$ Å, $\mu^F = 5D$, $a_{1B} = 30$ Å, $E^F(0) - E^W(0) = 5 meV$. At resonance, e.g. where the hybridization is strongest, we have a very high peak of nonlinear susceptibility with the enhancement of about 5 orders of magnitude in comparison with that of the Wannier exciton. The nonlinear coefficient is larger for smaller dots and closer spacing.

The imaginary part of the susceptibility gives rise to the coefficient for the two-photon absorption a_2. The results for the two-photon absorption a_2 coefficient for the material with hybrid excitons shows that it significantly exceeds that of the other materials in the near infrared spectral region. For example, we obtain the values of $a_2 \sim 5 \times 10^{-10} m/W$ for the array of quantum dots with $R = 100 Å$ and $d = 200$ Å in an organic medium at a wavelength of 1060 nm. This value exceeds a_2 for the other materials in the visible and near infrared spectral regions.

3. Hybrid Exciton State in a Quantum Dot- Dendrite System

In this section, the hybrid exciton formed in the system with a quantum dot at the core of a dendrimer [23] will be discussed. Dendrimers are highly symmetric and perfectly, repeatedly branched macromolecules with controlled structures, which have very strong potential for optimal energy funnels, gene and drug delivery applications. Dendrimers with controlable sizes and structure can be used as light emmiters and can serve as building blocks in nanodevices [31]-[36].

In the quantum dotdendrimer system, a semiconductor dot is placed inside a dendrimer structure during the synthesis process. The Wannier exciton in the dot and the Frenkel exciton in the dendrimer interact with each other through dipole-dipole interaction to form the hybrid excitation. This hybrid exciton plays an important role in optical properties of the system and by changing the size and structure of the nanocrystal dendrimer, one can expect to change the hybrid exciton properties, therefore to control optical processes of the system.

3.1. A Quantum Dot- Dendrite Model

The problem is solved using the Greens function approach [37]. The double-time real space Greens functions with the diagram technique is very useful and important in solving problems with complicated structures. The Greens functions are directly related to properties of the system, so once the Greens functions are calculated, the optical properties such as nonlinearities, fluorescence, scattering... also can be evaluated.

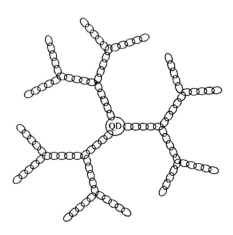

Figure 3. A quantum dotdendron ligand model. *Huong and Birman* [23]

A model of a quantum dot as a spherical core at the center, which is attached to three dendrimeric branches is considered. Each dedrimeric branch consists of tertiary amine groups linked by three-carbon chains. Then each branching point is attached to two protein branches, and so on as in Figure 3. Coupling exists between the quantum dot and the attached protein branches, between the molecules in the same branches as well as between the molecule at a branching point and the protein chain attached to it. But no coupling between molecules of different branches is assumed. Here, again, we assume only nearest neighbor interactions.

The tight-binding Hamiltonian of the system can be written as the following:

$$H = \sum_{i=1}^{N} \mathcal{E}_i a_i^+ a_i + \sum_{i,j} V_{ij} a_i^+ a_j \qquad (34)$$

a_i^+, a_i are exciton creation and annihilation operators, $\mathcal{E}_i$ is the exciton energy at each site, with i labelling the sites of quantum dot and molecules at the end of each generation of the dendrimeric branch. For the quantum dot the energy will be the energy of the exciton confined in the dot[8] and for the molecular sites it will be the Frenkel exciton energy. V_{ij}, where i, j are the sites of the molecules at the branching point, i.e. the end of each generation, is the effective interaction integral between excitons of different dendrimer generations.

The double-time Green's functions are writtten as

$$G_{ij}(t) = \frac{2\pi}{\hbar} << a_i(t), a_j^+(0) >> = -i\frac{2\pi}{\hbar}\Theta(t) < [a_i(t), a_j^+(0)] > \qquad (35)$$

where [,] is a commutator, a(t) is the Heisenberg representation of the operator a, $\Theta(t)$ is Heaviside function, $< ... >$ is the thermal average over a grand canonical ensemble. Using the Fourier transformation to transfer the time Greens functions to the energy variable, the Schwinger-Dyson equations for the Greens functions have the form:

$$(E - \mathcal{E}_i)G_{ij}(E) - \sum V_{ik}G_{kj}(E) = \delta_{ij} \qquad (36)$$

Since the system consists of both Wannier and Frenkel excitons, the Green funtions here are the Greens function of hybridized state and E is the energy of the new excitation.

3.2. Effective Interaction Coefficients Between Dendrimer Generations

The interaction between different generations, called the interaction at a macroscopic scale, has been made from all the interactions between nearest neighboring molecules, called the interaction at the microscopic scale. For example, for a chain in the Figure 4. we want to calculate the interaction between the quantum dot at the center, considered as the zero generation, and the molecule at the end of one of the first molecular chains, which is the first generation. In the chain there is a number of molecules, so the quantum dot and the molecule finishing the first generation may be far apart and do not interact directly with each other if we assume only nearest neighbor interaction. The exciton in the quantum dot interacts with the exciton in the first molecule of the first chain, the first molecule interacts with the second molecule, and so on, the next to the last molecule interacts with the last molecule of the chain. Actually in this process the exciton in the quantum dot interacts with the exciton in the last molecule of the first chain indirectly through the effective superexchange interaction. The problem is very similar to the problem of electron transfer in a chain with one impurity [39, 40].

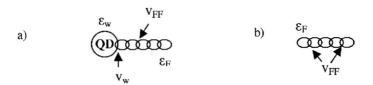

Figure 4. (a) A quantum dotmolecule chain. (b) A moleculemolecule chain. *Huong and Birman* [23]

By solving Dyson equation for Huckel Hamiltonian of the linear chain

$$H = \sum_{\alpha} \epsilon_\alpha a_\alpha^+ a_\alpha + \sum_{\alpha,\beta}(a_\alpha^+ a_\beta + a_\beta^+ a_\alpha) \tag{37}$$

the effective interaction between the dendrimer generations is obtained. The details are given in [23]. As the result, the effective interaction between site 0 and site n, i.e. between the exciton in the quantum dot at the core and the Frenkel exciton of the first dendrimeric generation, has the form

$$V_{WF} = \frac{v_0}{E - vT} T^{n-1} \tag{38}$$

where v_0 is the interaction of Wannier exciton at the core and Frenkel exciton at the nearest molecular site, and v is the interaction between two Frenkel excitons at the nearest sites, and E is the energy of the mixed state in this chain.

$$E = \epsilon_W + \frac{v_0^2}{E - \epsilon_F - vT} \tag{39}$$

Similarly, the effective interaction between the Frenkel excitons in the neighboring generations is obtained as:

$$V_{FF} = \frac{v}{E - vT} T^{n-1} \tag{40}$$

with the energy

$$E = \epsilon_F + \frac{v^2}{E - \epsilon_F - vT} \tag{41}$$

where the transfer function T is given in the form

$$T = \frac{E - \epsilon_F \pm \sqrt{(E - \epsilon_F)^2 - 4v^2}}{2v} \tag{42}$$

Now we can replace the molecular chains with nearest neighbor interaction between molecules by localized sites with effective interaction (38) and (40). The effective interaction coefficients depend on the energies of excitons, the nearest neighbor interaction coefficients, and also the number of molecules in each chain.

3.3. Diagram Techniques for Green Function in Orbital Representation

To solve the Schwinger-Dyson equations for the dendrimer system with many generations, it is convenient to use the electron transfer graph method developed in [37, 38] where the method is used to study electron transfer between localized sites. Then for the Schwinger-Dyson equation (36), every site corresponds to a graph vertex. A nondiagonal term V_{jk} of the Hamiltonian corresponds to an oriented edge originating at vertex j and ending at vertex k with the value of the edge equal to the interaction integral V_{jk}. Diagonal terms $E - \mathcal{E}_{ii}$ of the Hamiltonian corresponds to a loop attached to the vertex i and the value of the loop equals $W_{ii} = E - \mathcal{E}_{ii}$. The details of the diagram technique method can be found in [37, 38]. In this section we just write briefly about definitions and terminology, the main method and results of the authors [37, 38] which will be used in the following sections. In this diagram technique, a path is a sequence of graph edges with succesive edges originating at the point where the previous one ends, and a cycle is a path with the last edge ending at the point where the first one originates. A length of the path is the number of edges in it, where $P_{ii} = 1$. The value O of the cycle is the product of the values of all edges in the cycle, with the sign being negative except for the loop for any cycle with length more than 1. The cyclic term is a set of cycles which have no common vertex and pass through every vertex. The value of cyclic term O is a product of the value of all cycles of the cyclic term and the cyclic value of the graph Θ is the sum of all the cyclic terms.

For a linear chain of sites it is shown in [37, 38] that the Greens function G_{ij} is equal to

$$G_{ij} = \frac{\Theta_{1,i-1} P_{ij} \Theta_{j+1,n}}{\sum O^{\{k\}}} \quad (43)$$

where P_{ij} is the product of the edges along the pathway from vertex i to vertex j

$$P_{ij} = V_{i,i+1} V_{i+1,i+2} ... V{j-1,j} \quad (44)$$

with $\Theta = \sum O^{\{k\}}(i,j)$ being the cyclic value of the graph and $O^{\{k\}}(i,j)$-the cyclic term.

A very important feature of this method is the continuous fraction representation for sites with side groups attached. For the case when there are side groups attached to some site i of a linear path then the graph of the main chain remains the same, only the value of the loop at site i is different. In this case it was proved in [37, 38] that the diagonal element of the Greens function of side graph can be expressed through the Greens function of subgraph in the continued fraction representation.

The value of the loop with the side groups could be expanded into continuous fraction of loops values of the extended side subgraph of the graph. Instead of the original loop value $G_i^{-1} = E - \mathcal{E}$ the loop value of the site with side group becomes

$$\begin{aligned}\tilde{G}_i^{-1} &= (E - \mathcal{E}_i) - \sum_k \frac{V_{ik}^2(i)}{\tilde{\Theta}_{ki}/\Theta_{ki}} \\ &= (E - \mathcal{E}_i) - \sum_k V_{ik}^2(i) G[\tilde{\gamma}_k(i)]_{kk} \end{aligned} \quad (45)$$

For instance for the site chain in the Fig. 5b the value of the loop at site 2 has the form

$$\tilde{W}_{22} = E - \mathcal{E}_2 - \frac{V_{2a} V_{a2}}{E - \mathcal{E}_a - \frac{V_{ab} V_{ba}}{E - \mathcal{E}_b}} \quad (46)$$

3.4. The Quantum Dot-Dendron Ligand

The above real space Green's function method and its diagramatic technique has been used to investigate the Wannier-Frenkel hybrid exciton in different quantum dot-organic dendrimeric configurations [23]. Here we will discuss one of the systems, the nanocrystal-dendron ligand. The priority of the dendron ligand is its perfect symmetry with the dot at the center, and also its closely packed ligand shell.

Let us assume a quantum dot is attached to three molecular branches, then each terminal molecule is attached to other two branches, and so on as in Figure 3. Using the diagram with continous fraction representation we can consider the ligand as a linear chain with sidegroups attached to each site.

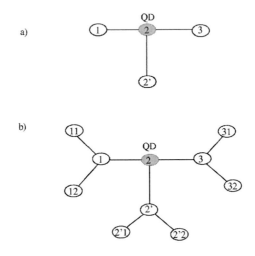

Figure 5. (a) A quantum dotdendron ligand with one generation. (b) A quantum dotdendron ligand with two or more generations. *Huong and Birman* [23]

At first we consider the first generation system with the quantum dot at the center attached to three chains of molecule in Figure 5. Considering this system as a linear chain of chain 1, chain 2 and chain 3, with chain 2' is attached to chain 2. For this case we have:

$$G_{11} = \frac{1}{\Theta}[(E - \tilde{\mathcal{E}}_2)(E - \mathcal{E}_3) - V_{12}V_{21}]$$

$$G_{12} = \frac{1}{\Theta}V_{12}(E - \mathcal{E}_3) = G_{21}$$

$$G_{13} = \frac{1}{\Theta}V_{12}V_{23} = G_{31}$$

$$G_{22} = \frac{1}{\Theta}(E - \mathcal{E}_1)(E - \mathcal{E}_3)$$

$$G_{23} = \frac{1}{\Theta}(E - \mathcal{E}_1)V_{23} = G_{32}$$

$$G_{33} = \frac{1}{\Theta}[(E - \mathcal{E}_1)(E - \tilde{\mathcal{E}}_2) - V_{23}V_{32}] \quad (47)$$

$$\Theta = (E - \mathcal{E}_1)^2(E - \tilde{\mathcal{E}}_2)(E - cal E_3) - V_{12}V_{21}(E - \mathcal{E}_3) - V_{23}V_{32}(E - \mathcal{E}_1) \quad (48)$$

where the value of the loop at site 2 includes the side graph

$$\tilde{\mathcal{E}}_2 = \mathcal{E}_2 - \frac{V_{22'}V_{2'2}}{E - \mathcal{E}'_2} \tag{49}$$

For the dot-dendron of one-generation, the quantum dot is situated at the chain 2, connected to molecules at chains 1, 3 and 2'. So at chain 2 we have Wannier exciton with energy $\mathcal{E}_W$, at chains 1,3 and 2' we have Frenkel excitons with energy $\mathcal{E}_F$. The Green functions of the system have the following form:

$$\begin{aligned}
G_{11} &= \frac{1}{\Theta}[(E - \mathcal{E}_W - \frac{V_{WF}V_{FW}}{E - \mathcal{E}_F})(E - \mathcal{E}_F) - V_{WF}V_{FW}] \\
G_{12} &= \frac{1}{\Theta}V_{FW}(E - \mathcal{E}_F) = G_{21} \\
G_{13} &= \frac{1}{\Theta}V_{FW}V_{WF} = G_{31} \\
G_{22} &= \frac{1}{\Theta}(E - \mathcal{E}_F)^2 \\
G_{23} &= \frac{1}{\Theta}(E - \mathcal{E}_F)V_{WF} = G_{32} \\
G_{33} &= \frac{1}{\Theta}[(E - \mathcal{E}_F)(E - \tilde{\mathcal{E}}_W - \frac{V_{WF}V_{FW}}{E - \mathcal{E}_F}) - V_{WF}V_{FW}]
\end{aligned} \tag{50}$$

$$\Theta = (E - \mathcal{E}_1)^2(E - \tilde{\mathcal{E}}_2)(E - calE_3) - V_{12}V_{21}(E - \mathcal{E}_3) - V_{23}V_{32}(E - \mathcal{E}_1) \tag{51}$$

where the value of the loop at site 2 includes the side graph

$$\tilde{\mathcal{E}}_2 = \mathcal{E}_2 - \frac{V_{22'}V_{2'2}}{E - \mathcal{E}'_2} \tag{52}$$

For the two-generation quantum dotdendron, we can consider the dendron of one generation with side group attached to each of the terminal molecules in Fig. 8b. The loop values $\mathcal{E}_1, \mathcal{E}_2, \mathcal{E}_3$ become $\tilde{\mathcal{E}}_1, \tilde{\mathcal{E}}_2, \tilde{\mathcal{E}}_3$ with the side groups value attached:

$$\begin{aligned}
\tilde{\mathcal{E}}_1 &= \mathcal{E}_1 - \frac{V_{1,11}V_{11,1}}{E - \mathcal{E}_{11}} - \frac{V_{1,12}V_{12,1}}{E - \mathcal{E}_{12}} \\
\tilde{\mathcal{E}}_3 &= \mathcal{E}_3 - \frac{V_{3,31}V_{31,3}}{E - \mathcal{E}_{31}} - \frac{V_{3,32}V_{32,3}}{E - \mathcal{E}_{32}} \\
\tilde{\mathcal{E}}_2 &= \tilde{\mathcal{E}}_2 - \frac{V_{2',2'1}V_{2'1,2'}}{E - \mathcal{E}_{2'1}} - \frac{V_{2,2'2}V_{2'2,2}}{E - \mathcal{E}_{2'2}}
\end{aligned} \tag{53}$$

Using these values we will be able to find Green functions for hybrid exciton in the system of two generation of dendrimer. Continuing this process of side group graphs attached to the chains, we can go farther to any generation just by changing the loop values in Eq. 47. Since every branching point is attached to two branches, in going from n^{th} generation to $n + 1^{th}$ generation, the value of the loop i just needs to be added by the value of the two branches of $(n+1)^{th}$ generation attached to the generation n^{th} of the loop. If we know the energy and Greens functions of a dendrimer with n generations, we always can find ones for the dendrimer with $(n + 1)$ generations. Because of the continuous fraction, from some generation, the effects of $(n+1)^{th}$ generation is small and can be treated as some small perturbations.

The energy of the Wannier-Frenkel hybrid exciton is the pole of the Green functions, or in another words, is obtained from the condition for the zeros of the Greens function determinator Θ.

For the dot-dendron ligand of one generation the Wannier-Mot exciton energy is obtained in the form:

$$E^{WF} = \frac{1}{2}[\mathcal{E}_F + \mathcal{E}_W \pm \sqrt{(\mathcal{E}_F - \mathcal{E}_W)^2 + 8V_{WF}V_{FW}}] \tag{54}$$

And for the energy of the hybrid exciton of the quantum dot-dendron ligand of two generations we find:

$$E^{WF} = \frac{1}{2}\{\mathcal{E}_F + \mathcal{E}_W + \frac{4V_{FF}^2}{E - \mathcal{E}_F} - \frac{2V_{WF}^2}{E - \mathcal{E}_W} \pm [\mathcal{E}_F^2 + \mathcal{E}_W^2 + 6\mathcal{E}_W\mathcal{E}_F$$
$$+ \frac{16V_{WF}^4}{(E - \mathcal{E}_F)^2} - \frac{16V_{FF}^4}{(E - \mathcal{E}_F)^2} + \frac{24V_{FF}^2 V_{WF}^2}{(E - \mathcal{E}_F)^2} + 8V_{WF}^2]^{1/2}\} \tag{55}$$

Notice here that using the continuous fractional diagram technique, each time when we have more generations, the values of $2V_{FF}^2/(E - \mathcal{E}_F)$ are added to each branching point. This means that when one more generation is added, in Eq. 51 only the value of E will change for $2V_{FF}^2/(E - \mathcal{E}_F)$ time some integers. For n large, when it is difficult to solve Eq. 51 analytically, the self-consistent method is used to solve the equations. With this process, for quantum dot-dendron ligands of any generation the expressions for the energy of the hybrid exciton always can be obtained. The Green functions also allow ones to calculate other physical quantities of the hybr

The hybridization effect is strong at resonance of the two kinds of excitonic states, and the structure of the dendrimer (i.e. the number of molecule in one branch and/or the number of the braches, the number of generations...) as well as its material also decides the hybridization. In addition, the homogeneous broadening of the absorption lines could be caused by different reasons such as the surface roughness of the dots and the imperfect symmetry of the structure or the participation of the second or the third-order interaction between molecular sites. It is shown that [23] for the CdSe quantum dot, the exciton linewidth is much smaller than the exciton linewidth for the quantum well, about 0.12 meV [41] which is very small. For Frenkel excitons, there are organic substances with the exciton linewidths about 1 mev or smaller.18 Both these linewidths are smaller in comparison with the hybridization parameter VWF/5 meV.

4. Hybrid Exciton under Electric and Magnetic Fields

Since the hybrid Wannier-Frenkel exciton can be considered among the most promising materials for optical and electronic devices, it is helpful to study the effect of electric and magnetic fields on this mixed state. There are several theoretical works studying the composite organic-inorganic quantum wells in the presence of electric field [43], magnetic field [43], and mutually perpendicular electric and magnetic fields [44]. It was found that by tuning the applied fields, various interesting properties of dispersion laws were obtained

[43, 44]. The quadratic Stark dependence of the hybrid exciton energy on the electric field was found as expected for the confined Stark effect for the hybrid exciton.

We consider a heterostructure of an inorganic semiconductor spherical quantum dot covered by a thin organic layer. The excitations in the quantum dot are Wannier-Mott excitons and in the organic layer the Frenkel excitons are moving between lattice sites. The total Hamiltonian of the system under applied electric field $\vec{E}$ and magnetic field $\vec{B}$ is given by the following:

$$H = H_F + H_W(\vec{E}, \vec{B}) + H_{int}(\vec{E}, \vec{B}) \tag{56}$$

H_{int} is the interaction Hamiltonian between the two kinds of excitons, which leads to a hybridization coupling constant which governs the hybridization:

$$\Gamma = <\Psi_F|H_{int}|\Psi_F> \tag{57}$$

Because of the large Bohr radius of the Wannier-Mott exciton, the semiconductor quantum dot is more affected by the electric and magnetic fields and we can neglect the field effect on the organic layer due to localization of the Frenkel exciton. So in the Hamiltonian (56), the fields changes the second and the third terms only. In this strong confinement limit, the effect of electric and magnetic field has been calculated using perturbation theory, using the Hamiltonian of Wannier exciton for the non-perturbation basic functions to give analytical values of the wavefunctions and energies. It is shown that the perturbation theory works well for the small quantum dots with very little different from numerical calculation. The energies and wavefunctions of the Wannier-Mott excitons under the effects of the electric and magnetic fields then are calculated in the first and second orders of the perturbation.

The interaction H_{int} in (57) then is estimated from the electric field and magnetic field operators produced by the molecular layer in the organic dot under the applied fields and then the coupling constant Γ will be calculated on the basis of the Frenkel exciton wave function and the new Wannier exciton wanvefunction. By this method the Wannier-Mott-Frenkel hybrid exciton in the presence of the electric field is described by the coupling states. The hybridization is strong at the resonance region, and the region depends on the applied field. The resonance region where the Wannier and Frenkel exciton are strongly mixed is shifted by the values and orientation of the fields, so by tuning the field one can change the resonance region and the properties of the hybridization. The electric field reduces the exciton binding energy of the hybrid excitons. The Stark shift of the Wannier exciton energy levels and its wavefunction depending on applied electric field have been obtained [45]. In addition to the spatial confinement of the quantum dot, the magnetic confinement of electron and hole leads to more localization of the electrons and holes. From the results it is shown [45] that the application of the field enhances the coupling between the two kinds of excitons and increases the gap between two branches of the hybrid exciton.

By tuning the magnitude and orientation of the fields, one could obtain different physics properties of the semiconductor-organic systems.

5. Conclusion

We presented in this chapter the possibility of creating systems using semiconductor quantum dots and organic materials which offers a strong resonance coupling of Frenkel and Wannier excitons. The system gives hybrid exciton states with special properties of both kinds of exciton, i.e.having large exciton radius as well as large oscillator strength. The energies and different optical properties of the semiconductor dot-organic systems can be tuned by changing the parameters of the systems such as the number of dots, dot radius, dot-dot spacing, number of molecular branches and number of molecules in each branch.... These results could help to create organic-semiconductor combined systems with expected optical and physical properties to be used in different applications.

References

[1] L. Brus, J. Chem. Phys. **80**, 4403 (1984).

[2] AL. L. Efros, A. L. Efros, Sov. Phys. Semicon. **16**, 772 (1982); A. L. Efros, Phys. Rev B 46, 7448 (1992);A. L. Efros and A. V. Rodina, Phys.Rev B47, 10005 (1993).

[3] A.l. Ekimove, I. A. Kudryavtsev, M. G, Ivanov and Al. L. Efros, Sov. Phys. Solid State, 31, 1385 (1989).

[4] J. B. Xia, Phys. Rev. B 40, 8500(1989).

[5] L. Banyai, S. W. Koch, Semiconductor Quantum Dots, Series on Atomic Molecular and Optical Physics. Vol. 2, World Scientific, Singapore, 1993.

[6] U. Wogon, Optical Properties of Semiconductor Quantum Dots, Springer Verlag, 1996

[7] C. Sercel and K. J. Vahala, Phys. Rev, B42, 3690(1990).

[8] T. Takagahara, Phys. Rev, B47, 4569 (1993).

[9] E. Hanamura, Phys. Rev. B 37, 1273 (1988).

[10] G. Bastard Wave Mechanics Applied to Semiconductor Heterostructures, New York, Wiley, 1988.

[11] S. Nomura, Y. Segawa, T. Kobayashi, Phys. Rev, B 49, 34 (1994).

[12] D. J. Norris, A. Sacra, C. B. Murray, M. G. Bawendi, Phys. Rev. Lett. 72, 2612 (1994); Phys. Rev. B**53**, 16338 (1994).C. Murray, C. Kagan, and M. Bawendi, Annu. Rev. Mater. Sci. 30, 545 (2000); C. Murray, D. Norris, and M. Bawendi, J. Am. Chem. Soc. 115, 8706 (1993); C. Murray, C. Kagan, and M. Bawendi, Science 270, 1335 (1995).

[13] Y. Morgan, C. A. Leatherdale, M. Drndic, M. V. Jarosz, and M. Bawendi, Phys. Rev. B 66, 075339 (2002).

[14] F. F. So and S. R. Forest, Phys. Rev. Lett. 66, 2649 (1991)

[15] Y. Imanishi, S. Hattori, A. Kakuta, S. Numata, Phys. Rev. Lett. 71, 2098 (1993)

[16] A.L. Efros, M. Rosen, M. Kuno, M. Nirmal, D. J. Norris, M. Bawendi, Phys. Rev. B **54**, 4883 (1996); M. Nirmal, D. J. Norris, M. Kuno, M. G. Bawendi, A.L. Efros, M. Rosen, Phys. Rev. Lett.**75**, 3728 (1995).

[17] V. M. Agranovich, V. I. Yudson and P. Reineker, in "Electronic Excitation in Organic Based Nanostructures", Edited by V. M. Agranovich and G. F. Bassani, Elsevier Academic Press 2003

[18] V. Agranovich, Solid State Comm. **92**, 295 (1994).

[19] V. I. Yudson, P. Reineker, V. M. Agranovich, Phy. Rev. B, **52**, R5543 (1995).

[20] A. Engelmann, V. I. Yudson, P. Reineker, Phy. Rev. B, **57**, 1784 (1998).

[21] N.Q. Huong and J.L. Birman, Phys. Rev. B **61**, 13131 (1999); J. L. Birman and Nguyen Que Huong, J.Lumin. **125**, 196 (2007).

[22] Y. Gao, N. Q. Huong and J. L. Birman, M. J. Potasek, J. Appl. Phys. **96** 4839 (2004); Y. Gao, N. Q. Huong, J.L. Birman and M. J. Potasek, Proc. SPIE Int. Soc. Opt. Eng. 5592, 272 (2005).

[23] N.Q. Huong and J.L. Birman, Phys. Rev. B **67**, 075313 (2003).

[24] V. M. Agranovich, D. M. Basko, G. C. La Rocca, F. Bassani, J. Phys. Condens. Matter **10**, 9369 (1998).

[25] S. Facsko, T. Dekorsy, C. Koerdt, C. Trape, H. Kurz, A. Vogt, H. L. Hartnagel, Science **285**, 1551 (1999).

[26] T. Takagahara, Solid State Comm.,**78**,No. 4, 279 (1991); Surface Science **267**, 310 (1992);T. Takagahara and E. Hanamura, Phys. Rev. Lett. **56**, 2533 (1986).

[27] S. Schmitt-Rink, D. A. B. Miller and D. S. Chemla, Phys. Rev. B **35**, 8113 (1987); L. Banyai, Y.Z. Hu, M. Lindberg, S. W. Koch, Phy. Rev. B **38**, 8142 (1988).

[28] Karl W. Boer, "Survey of Semiconductor Physics: Electron and other Particles in Bulk Semiconductors", Van Nostran Reinhold, New York 1990.

[29] T. S. Moss, Handbook on semiconductors, vol.1, p. 23, North-Holland Publishing Company, Amsterdam-NewYork-Oxford 1982.

[30] S.Mukamel, Principle of Nonlinear Optical Spectroscopy, Oxford University Press, New York-Oxford 1995.

[31] D.A. Tomalia, H. Baker, J. Dewald, M. Hall, G. Kallos, S. Martin, J. Roeck, J. Ryder, and P. Smith, Macromolecules **19**, 2466(1986); M. Zhao, L. Sun, and R. Crooks, JACK 120, 487 (1998); K. Sooklai, L. Hanus, P. Harry, and C. Murphy, Chem. Rev. **14**, 10 (1998).

[32] J.P. Majoral and A.M. Caminade, Chem. Rev. **99**, 845 (1999)

[33] M.R. Shortreed, S.F. Swallen, Z.Y. Shi, W. Tan, Z. Xu, C. Devadoss, J.S. Moore, and R. Kopelman, J. Phys. Chem. B **101**, 6318(1997).

[34] W.C.W. Chan and S. Nie, Science **281**, 2016 (1998).

[35] Y.A.Wang, J.J. Li, H. Chen, and X. Peng, J. Am. Chem. Soc. **124**, 2293(2002).

[36] M.L. Steigerwald and L.E. Brus, Acc. Chem. Res. **23**, 183 (1990).

[37] Y. Magarshak, J. Malinsky, and A.D. Joran, J. Chem. Phys. **95**, 418 (1991).

[38] J. Malinsky and Y. Magarshak, Int. J. Quantum Chem. **25**, 183 (1991).

[39] M.A.F. Gomes, A.A.S. Gama, and R. Ferreira, Chem. Phys. Lett. **53**, 499 (1978)

[40] C.P. Melo, H.S. Brandi, and A.A.S. Gama, Theor. Chim. Acta **63**, 1 (1983)

[41] S.A. Empedocles, D.J. Norris, and M.G. Bawendi, Phys. Rev. **27**, 3878 (1996).

[42] V.M. Agranovich, G.C. la Rocca, F. Bassani, and P. Reincker, Pure Appl. Chem. **69**, 1203 (1997).

[43] S. Jarin, S.Jarizi, S.Romdhane, H.Bouchriha, R.Bennaceur, Phys. Let. A **—234**, 141(1997); Phys. Stat. Sol(a) 164,335 (1997).

[44] S. Jarin, H. Abassi, R. Bennaceur, Phys. Letters A 249, 517 (1998).

[45] Huong and Birman, to be published.

In: Exciton Quasiparticles
Editor: Randy M. Bergin

ISBN: 978-1-61122-318-7
© 2011 Nova Science Publishers, Inc.

Short Communication

EXCITONS IN CDO PARABOLIC QUANTUM DOTS

*M. A. Grado-Caffaro** *and M. Grado-Caffaro*
Scientific Consultants, C/ Julio Palacios, Madrid, Spain

ABSTRACT

The main aspects of the physics of excitons confined into a parabolic quantum dot of cadmium oxide are discussed starting from considering in the dot an exciton gas of harmonic oscillators whose Fermi energy coincides with the band-gap shift experienced by cadmium oxide in the visible region. At this point, we must remark that the role of CdO as transparent material in the visible range is highly significant. In particular, the frequency of oscillation of the excitons is determined. In addition, some issues related to the plasma-optical effect are examined.

Keywords: Cadmium oxide; Excitons; Parabolic quantum dots; Band-gap shift; Plasma-optical effect.

1. INTRODUCTION

The role of excitons as relevant elementary excitations is certainly notorious in the physics of condensed matter. In particular, nanosystems as, for example, quantum dots, exhibit a variety of phenomena where excitons participate significantly. In this respect, consider, for instance, quantum dots based on III-V and II-VI compounds. Certainly, semiconductor quantum dots are nanostructures which have been investigated extensively and continue to be investigated given the importance of the potential applications related to quantum dots (see, for example, ref.[1]). However, a number of questions relative to theoretical work remain open. These questions refer mainly to the physics of quantum dots made of some materials as, for instance, quantum dots of oxides transparent in the visible region; at this point, we can mention CdO and ZnO quantum dots.

---
* Corresponding author: Email: caffaro@sapienzastudies.com

In this communication, we will present a brief theoretical analysis upon the behaviour of excitons in CdO parabolic quantum dots. In fact, we shall determine the exciton Fermi velocity by regarding excitons as quantum harmonic oscillators. Within this context, we shall calculate the corresponding frequency of oscillation. On the other hand, a very important fact will be considered; we refer to the band-gap shift experienced by cadmium oxide in the visible range. The above shift depends on the partial pressure of oxygen when the growth of CdO crystals takes place [2,3]. Moreover, charge carrier spatial density depends also on the partial pressure of oxygen. In the light of the plasma-optical effect [4], carrier effective mass relies on the aforementioned carrier density (see, for instance, ref.[5]). This fact will be examined in the following.

2. THEORY

A quantum dot may be conceived as a zero-dimensional structure. Considering a parabolic quantum dot, it is well-known that parabolicity means that the corresponding potential energy of interaction refers to harmonic oscillator so this energy is a quadratic function of distance. Therefore, for excitons in a parabolic quantum dot, the quantized energy in the one quasi-particle approximation is the total energy of a three-dimensional quantum harmonic oscillator and is given by:

$$E_n = \hbar \omega_0 \left(n + \frac{3}{2} \right) \quad (1)$$

where $\hbar$ is the reduced Planck constant, ω_0 is the angular fundamental frequency of oscillation, and $n = 0,1,2,...$ is the corresponding radial (total) quantum number.

On the other hand, we may regard the dot as a spherical quantum box so that we can write (see, for instance, ref.[6]):

$$E_{n+1} - E_n = \frac{h v_F}{4r} \quad (2)$$

where v_F is the magnitude of the involved Fermi velocity (electron-hole velocity), r is the radius of the dot, and h is the Planck constant $h = 2\pi\hbar$.

From formulae (1) and (2), one gets that $v_F = 2r\omega_0/\pi$. Inserting this expression into the following relation for the one-quasi-particle Fermi energy $E_F = m_r^* v_F^2/2$, where m_r^* is the electron-hole reduced effective mass, we have:

$$E_F = \frac{2 m_r^* r^2 \omega_0^2}{\pi^2} \quad (3)$$

On the other hand, one has:

$$E_F = \frac{\hbar^2}{2m_r^*}(3\pi^2 N)^{2/3} \qquad (4)$$

where N is the carrier spatial density (only n-type CdO exists) which, in the one-quasi-particle approximation, reads $N = 1/(4\pi r^3/3)$. Replacing this formula into (4) and equating the resulting expression with (3), it follows:

$$\omega_0 = \frac{h}{4m_r^* r^2}\left(\frac{9\pi}{4}\right)^{1/3} \qquad (5)$$

Looking at eq.(5), we see that the fundamental frequency of oscillation is inversely proportional to m_r^* and proportional to the square of the dot curvature (note that curvature is $1/r$).

Now if we equate (1) to the total harmonic energy $4r^2 m_r^* \omega_0^2/2$, one gets an expression such that, once inserted (5) into this expression, we find $n \approx 5$ which, as expected, corresponds to a highly excited state.

The energy band-gap shift (Burstein-Moss shift) experienced by CdO coincides with the Fermi energy. The shift in question becomes a constant quantity despite the slightly unequal values measured by different workers (see, for instance, ref.[2]). Therefore, taking the right-hand side of (4) as constant, it follows that $m_r^* \propto N^{2/3}$ as carrier-density dependence of the effective mass. This fact is strongly related to the plasma-optical effect (see, for example, refs.[4,5]) by which the plasma resonant frequency plays an important role. This frequency is roughly proportional to $N^{1/6}$ [5] whereas the conjunction of relationship (5) and that $m_r^* \propto N^{2/3}$ gives $\omega_0 \propto N^{-2/3}$.

In the context of CdO thin films prepared by activated reactive evaporation of cadmium in the presence of oxygen, since N depends upon the partial pressure of oxygen [2,3] by eq.(4) then it is clear that the energy band-gap shift depends also on the aforementioned pressure. Therefore, by virtue of our preceding analysis, it is clear that m_r^*, ω_0, v_F, and E_n depend on the partial pressure of oxygen. There is a value of this pressure called optimal pressure because, at this pressure, the transmittance in the visible region is maximum [2].

3. CONCLUSION

Although quantum dots are in reality three-dimensional structures, to a certain extent or at a level of abstraction associated with a strictly theoretical standpoint, one may conceive the structures in question as zero-dimensional. Within this framework, we can consider that

$r \to 0$ in formula (5) so $\omega_0 \to \infty$. In addition, inserting formula (5) into $v_F = 2r\omega_0/\pi$, one sees that the magnitude of the Fermi velocity varies linearly with the curvature of the dot. On the other hand, we wish to remark that the plasma-optical effect together with considerations on optical emission and absorption are the basic ingredients to study the optical properties of CdO parabolic quantum dots. In this respect, let us regard, for instance, optical absorption in CdO parabolic quantum dots for the visible range. The corresponding coefficient of absorption gives an adequate measure; this coefficient, as a function of photon energy, is $\alpha(\omega) \propto \sqrt{\hbar\omega - E_F}/(\hbar\omega)$ [7]. Given that the Fermi energy (energy band-gap shift) depends on the partial pressure of oxygen, then the coefficient of optical absorption in the visible range depends also on the above pressure which, mathematically, acts as a parameter [3].

REFERENCES

[1] *Single semiconductor quantum dots*, P. Michler, (Ed.) (Springer-Verlag, 2009).

[2] Sravani, C; Ramakrisna Reddy, KT; Jayarama Reddy, P. Influence of oxygen partial pressure on the physical behaviour of CdO films prepared by activated reactive evaporation, *Mater. Lett*, 1993, 15, 356-358.

[3] Grado-Caffaro, MA; Grado-Caffaro, M. Dependence on the partial pressure of oxygen of the shift in the energy band-gap of CdO thin films in the visible region, *Active and Passive Elec. Comp.*, 2001, 24, 57-61.

[4] Wooten, F. *Optical properties of solids* (Academic Press, New York, 1972, 52-55).

[5] Grado-Caffaro, MA; Grado-Caffaro, M. Plasma-optical effect in GaAs PIN photodiodes, *Active and Passive Elec. Comp.*, 1993, 15, 63-66.

[6] Grado-Caffaro, MA; Grado-Caffaro, M. A theoretical analysis on the Fermi level in multiwalled carbon nanotubes, *Mod. Phys. Lett, B*, 2004, 18, 501-503.

[7] Pankove, JI. *Optical processes in semiconductors* (Dover, New York, 1971).

INDEX

A

absorption spectra, 11, 15, 60, 62, 66, 69, 83, 84, 141, 144, 145, 158, 191, 201, 282, 287, 288, 307, 315, 317, 322, 329, 331, 333
abstraction, 401
accelerator, 189, 202
acetaldehyde, 189
acetic acid, 199
acetone, 189
acid, 190, 199, 209
active centers, 293
actuality, 319
aggregation, 210
aging process, 288, 289
alanine, 189, 200
aluminium, 198, 199, 200, 202, 205, 211, 212
amine, 389
ammonia, 186
ammonium, 199
amplitude, 16, 17, 27, 28, 29, 31, 121, 152, 155, 176, 177, 218, 295, 297, 325, 342, 348, 363, 364, 371, 372
anatase, 281
anisotropy, vii, 1, 11, 22, 26, 46, 59, 92, 93
annealing, 3, 198, 205, 206
annihilation, 14, 19, 20, 21, 24, 27, 70, 71, 72, 73, 78, 81, 117, 325, 334, 380, 389
aqueous solutions, 184, 189, 198, 200, 205, 208, 209
argon, viii, 2, 74, 83, 85, 86, 88, 281
Asia, 253
assessment, x, 293
asymmetry, 99, 100, 101, 317, 329, 331, 350
atmosphere, 195, 197, 198, 205, 206, 252
atomic positions, 226
atoms, 46, 47, 79, 135, 190, 206, 216, 218, 226, 306, 339, 370
Auger scattering, viii, 133

B

backscattering, 106
band gap, 11, 12, 15, 44, 72, 106, 113, 125, 135, 141, 169, 175, 261, 327, 362, 378, 381
barriers, 135, 259, 316
base, 222
beams, 35, 142
bending, 4
benzene, 219, 226
Bethe-Salpeter (BS), xi
bias, 318, 319, 323, 330, 340
biexciton, 154, 165
binding energies, 184
binding energy, viii, xi, 2, 12, 39, 40, 43, 90, 92, 93, 96, 98, 125, 135, 184, 359, 360, 361, 362, 363, 367, 369, 370, 374, 381, 395
bleaching, 145, 146, 147, 148, 149, 150, 151, 152, 153, 154, 155, 156, 158, 159, 160, 161, 162, 170, 171, 172, 174, 175
blueshift, 268, 273
Boltzmann constant, 70
bonds, 165, 226
Bose-Einstein condensates, 306, 339
boson, 215, 231
Bosonic character of excitons, ix, 257
branching, 389, 393, 394
breakdown, 103, 114, 258, 318
breathing, viii, 133
bubble diagram, ix, 213, 218
bulk materials, ix, 183, 339

C

cadmium, xii, 399, 400, 401
calcination temperature, 192, 193, 199, 202, 204
calculus, 232

candidates, 347, 370
carbon, 190, 226, 389, 402
carbon dioxide, 190
carbon nanotubes, 402
catalysis, 210
cation, 199
CdO parabolic quantum dot, vii, 400, 402
ceramic, 198, 199, 210, 281
ceramic materials, 199
chain deformation, 248
chain scission, 190
charge density, 229, 370
chemical, x, 16, 134, 136, 184, 189, 257, 259, 278, 288
chemical interaction, 288
chemical properties, 136
chemical stability, 278
chemicals, 189
Chicago, 277
Chile, 253
China, 273
chloroform, 138, 143
circularly polarized light, 112
circulation, 239
classes, 219, 367
closure, 230, 313, 327
clusters, 184, 207
coherence, 177, 184, 322, 347, 379, 386, 387
coherent potential approximation (CPA), x, 293
collisions, 16, 19, 24, 325
colloidal core, vii, 138
colloidal nanocrystals (NCs), viii, 133
combustion, 199, 205, 212
communication, 400
competition, 134, 150, 157, 158, 170, 171, 173, 175, 177
complex numbers, 322
complexity, 333
composites, 209
composition, 72, 96, 97, 98, 108, 113, 114, 119, 120, 121, 125, 136, 205, 223
compounds, vii, 32, 35, 46, 74, 97, 101, 103, 104, 106, 184, 198, 199, 205, 399
comprehension, 135
condensed matter, vii, 351, 399
conductance, 189
conduction, 3, 4, 12, 20, 32, 33, 39, 40, 42, 65, 84, 104, 325, 333, 349
conductivity, 370
configuration, 46, 74, 105, 116, 222, 282, 286, 347, 379

confinement, viii, 133, 134, 135, 137, 138, 140, 141, 157, 158, 159, 161, 165, 167, 169, 175, 184, 206, 360, 379, 381, 382, 395
conservation, 25, 26, 70, 71, 72, 137, 225, 239, 312, 319
constant rate, 264
construction, 16, 278
contour, 4, 8, 9, 38, 39, 57, 64, 74, 99, 108, 113, 215, 218, 221, 224, 227, 230, 235
controversial, 177
convention, 219, 326
cooling, 74, 227
cooling process, 74
copolymers, 279
copper, 184, 206, 207
core/shell CdSe nanorods, vii, viii, 133
correlation, 213, 214, 215, 216, 220, 227, 229, 231, 249, 296, 306, 307, 324, 325, 334, 350
correlation function, 213, 214, 215, 216, 227, 228, 229, 350
correlations, 214, 221, 231
cost, 278
Coulomb energy, xi, 359
Coulomb interaction, 3, 134, 137, 141, 307, 327, 333, 360, 363, 365, 370, 373, 374
covering, 126
crossing over, 323
crust, 252
crystal growth, 212, 314, 318, 330, 339, 340
crystal quality, 88
crystal structure, 288
crystalline, 16, 61, 187, 189, 192, 193, 198, 199, 202, 205, 208, 209, 212
crystallization, 186
Cuba, 208, 209, 212
cycles, 251, 391
cyclones, 252
Czech Republic, 183, 277, 290

D

damping, 4, 6, 7, 11, 14, 17, 23, 30, 36, 40, 47, 48, 49, 51, 52, 53, 54, 55, 61, 100, 101, 102, 108, 113, 119, 176, 177, 229
data collection, 249
data set, 249
decay, 21, 137, 138, 141, 147, 148, 149, 150, 151, 152, 154, 155, 156, 160, 161, 162, 164, 165, 166, 167, 168, 169, 171, 172, 173, 174, 177, 184, 185, 197, 198, 206, 212, 233, 295, 296, 300, 340, 343, 344, 347
decay times, 148, 149, 151, 154, 155, 156, 167, 168, 172, 184

decomposition, 186, 192, 195
deduction, 270
defects, viii, 18, 19, 53, 85, 103, 133, 136, 140, 150, 151, 152, 156, 159, 160, 162, 164, 165, 169, 173, 177, 196, 198, 199, 203, 209, 210
deformation, 20, 46, 175, 248, 287
degenerate, 12, 13, 306, 313, 318, 333, 340, 345
degradation, 189, 190
dendrites, xi, 377
density functional theory, ix, 213
density matrices, 324
depolarization, viii, 2, 106, 108, 112, 113, 114
deposition, x, 202, 208, 278, 281, 288
depth, 21, 22, 26, 27, 28, 283, 301
derivatives, 99, 298
destruction, 339
detection, 195
deviation, 96, 97, 140, 211, 270
dielectric constant, xi, 3, 14, 17, 29, 34, 38, 39, 40, 41, 42, 43, 47, 93, 94, 108, 135, 309, 335, 359, 360, 361, 362, 369
dielectric permittivity, 29
dielectrics, 184
differential equations, 349
diffraction, 22, 192
diffusion, vii, x, 1, 16, 19, 20, 21, 22, 24, 27, 232, 277, 278, 279, 283, 284, 286, 288, 332, 339
diffusion process, 286
dilation, 241
dimensionality, 136, 258
diodes, 278
dipole moments, 214, 382, 383, 384
discrete variable, 310
disorder, x, 199, 235, 293, 294, 296, 302
dispersion, vii, xi, 1, 2, 3, 4, 5, 6, 7, 8, 9, 10, 11, 16, 17, 18, 20, 22, 23, 24, 27, 28, 29, 35, 36, 38, 40, 47, 48, 49, 50, 64, 65, 68, 69, 70, 88, 94, 100, 104, 106, 126, 138, 165, 167, 218, 232, 247, 258, 359, 360, 362, 363, 364, 366, 367, 368, 369, 370, 372, 373, 374, 379, 382, 385, 386, 394
displacement, 225, 226, 227, 230
dissociation, 228
distortions, 231
distribution, vii, viii, ix, 1, 16, 19, 21, 22, 23, 24, 25, 26, 27, 70, 71, 72, 74, 76, 78, 80, 133, 134, 140, 159, 164, 165, 170, 172, 185, 206, 211, 214, 215, 221, 227, 231, 257, 258, 259, 260, 261, 262, 266, 294, 296, 301, 302, 350
distribution function, 19, 21, 22, 23, 24, 25, 27, 214, 215, 266
divergence, 238
DOI, 209, 212
donors, 362

dopants, 199
doping, 185, 198, 199
drug delivery, 388
DSC, 208
dyes, 278
dynamic control, 324
dynamical properties, xi, 359

E

earthquakes, ix, 237, 242, 245, 248, 249, 250
elastic stress field, ix, 237, 238, 250
election, 103, 114
electric charge, vii
electric current, 378
electric field, x, xi, 135, 277, 280, 318, 325, 340, 341, 362, 377, 384, 386, 394, 395
electromagnetic, ix, 47, 51, 213, 222, 229
electromagnetic waves, 51
electron diffraction, 193
electron microscopy, 138, 193, 259
electron state, 144, 297, 318
electronic structure, 135, 136, 138, 159, 278, 288
electrons, xi, 33, 40, 90, 101, 102, 105, 134, 135, 141, 165, 185, 189, 191, 202, 205, 206, 207, 215, 216, 218, 223, 226, 278, 279, 280, 318, 333, 351, 359, 362, 363, 370, 378, 395
elementary excitation, vii, ix, 102, 257, 399
elongation, 140
elucidation, 215
energy parameters, 53
energy transfer, 137, 199, 258
environment, 177, 229, 230
equality, 23, 240
equilibrium, 70, 71, 72, 80, 106, 142, 184, 230, 231, 240, 265, 327, 347
equipment, 200
erosion, 343
ethanol, 138, 199
europium, 200
evaporation, x, 17, 278, 281, 288, 401, 402
evidence, 52, 136, 138, 145, 152, 159, 177, 238, 245, 249
evolution, 162, 163, 175, 207, 217, 229, 272, 331
exciton diffusion length, vii, x, 278, 288
exciton polariton dispersion, vii
exciton relaxation dynamics, vii, viii, 133, 177
Exciton resonance spectra, vii, 1
Excitonic Fano resonance (EX-FR), x
exclusion, 259
experimental condition, 21
extraction, 283, 308

F

fabrication, 186, 208, 210
Fabry-Perot interference, vii, 1, 62, 67, 68
fast processes, viii, 133
fast tectonic waves, ix, 237, 251
Femtosecond pump-probe spectroscopy, viii, 133
Fermi level, 402
fermions, xi, 359, 360, 365, 368, 373
ferroelectrics, 346
fiber, 222
fibers, 222
field theory, 231
film thickness, 281
films, x, 186, 199, 207, 208, 212, 278, 281, 282, 287, 288, 289, 401, 402
filters, 143
first generation, 389, 392
fluctuations, 258, 259, 296
fluorescence, 142, 188, 388
fluorine, 187
fluorophore, ix, 213
force, 227, 235, 239, 240, 241, 242, 243, 244, 246, 250, 251
formation, xi, 15, 27, 28, 33, 71, 99, 101, 103, 136, 137, 141, 154, 156, 165, 184, 185, 186, 187, 188, 190, 191, 192, 196, 198, 199, 200, 207, 259, 261, 287, 322, 331, 339, 343, 347, 377, 379
formula, 29, 59, 64, 74, 168, 220, 224, 225, 230, 232, 233, 311, 386, 401, 402
freezing, 12
Frenkel exciton, x, xi, 293, 294, 297, 302, 377, 378, 379, 380, 382, 383, 384, 385, 389, 390, 393, 394, 395
frequency distribution, 231, 330, 331
frequency of waves, 248

G

gain threshold, 137
gallium, 198, 199
gamma radiation, 187
gas sensors, 278
gauge theory, ix, 213
gaussian broadening, ix, 213
Gaussian diagonal disorder, x, 293
gel, 186, 199, 200, 205, 212
geometry, 11, 16, 54, 55, 58, 86, 87, 88, 106, 121, 137, 168, 242, 348, 378, 379
global earthquake migration, ix, 237, 251
glycine, 199, 209, 212
graph, 217, 391, 393

graphene sheet, 370
gravity, 239, 243
growth, 136, 152, 162, 190, 198, 199, 206, 207, 210, 212, 258, 259, 334, 378, 400
growth rate, 259

H

Hamiltonian, 45, 92, 219, 231, 294, 296, 297, 301, 302, 309, 311, 312, 314, 318, 325, 333, 335, 336, 340, 360, 361, 362, 371, 380, 381, 383, 384, 385, 389, 390, 391, 395
harmonic oscillators, xii, 224, 399, 400
heating rate, 204
height, 301, 319, 322, 343, 344
helicity, 215
helium, 72
heterodevice architectures, ix, 257
Hilbert space, 349
homogeneity, 246
hospitality, 178
host, xi, 184, 199, 206, 245, 294, 296, 377, 380, 387
HRTEM, 193, 194, 202, 259
hybrid, xi, 65, 377, 378, 379, 382, 384, 385, 386, 387, 388, 392, 393, 394, 395, 396
hybridization, xi, 44, 46, 377, 379, 382, 384, 385, 387, 394, 395
hydrogen, 11, 12, 16, 33, 34, 38, 39, 43, 83, 84, 90, 92, 126, 186, 189, 190, 199, 205, 206, 345
hydrogen atoms, 206
hydrogen peroxide, 186, 189, 190, 205, 206
hydrolysis, 186
hydroxide, 187
hydroxyl, 185, 189, 206
hypothesis, 232

I

Iceland, 242
illumination, 186, 278, 279, 280, 283, 286, 288, 289, 290
image, 139, 193, 194, 202, 225, 259, 260
implants, 202
improvements, 136
impulsive, 143
impurities, 18, 19, 53, 192
in transition, 145
InAs/GaAs, v, 257, 259
incidence, 11, 16, 17, 18, 20, 27, 29, 30, 31, 32, 47, 48, 49, 50, 52, 54, 55, 56, 57, 58, 117
income, 3, 16, 22, 24, 26, 47
independence, 216

inequality, 30
inertia, 240, 241, 246
ingredients, 402
inhomogeneity, 138
initial state, 70, 79, 231
integration, 219, 243, 244, 310
interaction effect, 244
interaction effects, 244
interdependence, 22
interface, viii, x, 133, 134, 136, 137, 138, 140, 143, 147, 156, 164, 165, 168, 169, 171, 172, 177, 178, 277, 279, 367, 378, 379
interference, vii, x, 1, 36, 51, 60, 61, 62, 63, 64, 65, 66, 67, 68, 69, 102, 143, 281, 305, 306, 322, 329, 346, 348, 350, 351
internal time, 364, 371
inversion, 12, 23, 154, 225, 314
ion implantation, 187, 208
ionizing irradiation, 194
ionizing radiation, 184, 189, 190, 191, 194, 196, 205
ions, viii, 46, 183, 184, 185, 187, 188, 189, 190, 198, 199, 200, 206, 207, 281
IR spectra, 74
irradiation, ix, x, 183, 184, 185, 186, 187, 188, 189, 190, 191, 192, 193, 194, 196, 198, 200, 201, 202, 203, 204, 205, 207, 208, 210, 305, 307, 313, 332, 339, 349
islands, 261
issues, xii, 134, 291, 399
Italy, 133, 143

J

Japan, 130, 302, 305, 351

K

kinetic equations, 19
kinetics, 147, 148, 149, 150, 151, 159, 161, 171, 172, 176, 206
Kinsey, 353
Kuril Islands, 248

L

lanthanum, 189, 197, 198, 205
laser radiation, 88
lasers, 74, 114, 134, 198, 199, 210, 258, 278, 332, 351
lasing threshold, 138
lattices, 126

laws, 70, 362, 394
lead, 13, 14, 72, 104, 151, 169, 215, 259, 268, 273, 287, 314, 369, 378
LED, 198
lifetime, viii, 21, 24, 25, 53, 70, 80, 133, 136, 153, 154, 155, 162, 164, 165, 169, 170, 173, 174, 263, 265, 278, 280, 288, 313, 340, 342, 343, 344, 347, 360
ligand, 209, 388, 392, 394
light, viii, ix, 2, 3, 16, 17, 18, 19, 21, 22, 27, 28, 29, 30, 32, 35, 36, 40, 44, 46, 47, 57, 61, 63, 65, 66, 70, 71, 72, 74, 75, 76, 78, 80, 94, 102, 104, 106, 112, 114, 126, 142, 146, 164, 168, 184, 186, 191, 198, 199, 209, 211, 213, 215, 217, 227, 229, 265, 278, 280, 281, 283, 314, 325, 332, 340, 346, 348, 349, 351, 378, 380, 388, 400
light beam, 35
light emitting diode, 211, 278
light scattering, 346
linear dependence, 15, 169
linear function, 10, 20
liquid phase, 208
lithium, 142
localization, x, xi, 258, 266, 293, 295, 296, 297, 298, 300, 302, 305, 307, 308, 333, 395
Lorentzian profiles, ix, 213
low temperatures, 72, 125
lower (LPB) polariton branches, vii, 1
lowest Landau Level (LLL), xi
luminescence, vii, viii, x, 1, 2, 13, 14, 15, 16, 17, 18, 22, 24, 26, 27, 33, 34, 37, 39, 40, 59, 66, 67, 68, 69, 70, 72, 73, 74, 75, 76, 78, 80, 81, 82, 83, 84, 85, 88, 89, 90, 93, 103, 106, 107, 108, 112, 113, 116, 118, 120, 126, 138, 139, 142, 145, 165, 183, 185, 195, 196, 197, 198, 199, 200, 205, 206, 208, 210, 211, 212, 257, 260

M

macromolecules, 388
magnetic field, xi, 89, 92, 359, 360, 361, 362, 363, 364, 365, 367, 370, 371, 372, 373, 374, 377, 394, 395
magnetic moment, 12
magnitude, 14, 49, 62, 138, 237, 242, 245, 248, 249, 289, 318, 346, 379, 381, 387, 395, 400, 402
majority, 11, 279, 288
manufacturing, 204
mapping, ix, 213, 218, 220, 221, 224, 225, 226, 227, 229, 230, 231
Marx, 178
mass, vii, xi, 1, 3, 4, 10, 11, 14, 16, 17, 34, 36, 38, 39, 40, 41, 42, 43, 44, 46, 56, 57, 59, 60, 62, 65,

70, 90, 92, 93, 96, 100, 108, 113, 119, 135, 141, 229, 309, 327, 335, 340, 359, 360, 361, 362, 363, 364, 365, 367, 369, 370, 371, 372, 373, 374, 379, 381, 385, 386, 400, 401
materials, viii, xi, 72, 126, 133, 134, 136, 141, 165, 183, 184, 189, 192, 193, 194, 195, 199, 200, 201, 202, 203, 205, 206, 207, 210, 211, 212, 238, 278, 279, 293, 296, 347, 351, 374, 377, 378, 387, 388, 394, 395, 399
matrix, x, xi, 12, 26, 45, 138, 177, 187, 206, 215, 223, 224, 229, 231, 239, 243, 244, 293, 297, 305, 307, 308, 310, 311, 312, 313, 326, 335, 336, 340, 342, 346, 349, 373
matter, vii, ix, 136, 147, 202, 213, 252, 348, 349, 351, 399
measurements, viii, 12, 16, 34, 42, 53, 58, 59, 74, 105, 133, 136, 138, 142, 144, 145, 146, 147, 149, 150, 159, 160, 165, 170, 171, 173, 175, 249, 259, 281, 283, 287, 348
media, ix, 51, 237, 238, 239
melt, 199
memory, 231
mercury, 72, 186, 189
metal ion, viii, 183, 185, 190, 278
metal nanoparticles, 206
metallic particles, viii, 183, 185, 190
metals, 184, 346, 347
methanol, 186
methodology, 215
methylene blue, 186
microscopy, 193
migration, ix, 23, 24, 237, 248, 249, 250, 251
misfit dislocations, 136
mission, 113, 164, 199
mixing, vii, x, xii, 1, 3, 65, 135, 140, 305, 306, 307, 308, 314, 315, 316, 317, 318, 337, 377, 378, 379, 384
models, 135, 211, 252, 293, 296, 379
modifications, 362
modulus, 241
molar ratios, 200
Moldova, 1
molecular structure, 209, 218, 220, 222, 287
molecules, 218, 278, 287, 306, 383, 389, 390, 393, 396
momentum, 12, 16, 19, 25, 26, 57, 70, 72, 137, 214, 215, 223, 226, 227, 239, 250, 251, 258, 325, 333, 336, 340, 372
monolayer, 136, 165
monomers, 281, 287
monotonic changes, 322
morphology, 189, 205, 210
Moscow, 128, 235, 253, 254, 255

multiwalled carbon nanotubes, 402

N

nanocomposites, 207
nanocrystals, viii, 133, 185, 191, 196, 205, 207, 208, 377
nanodevices, 388
nanomaterials, ix, 183
nanometers, 134
nanoparticles, viii, 135, 137, 138, 142, 165, 175, 183, 186, 192, 194, 205, 206, 207, 209
nanopowder materials, viii, 183
nanorods, vii, viii, 133, 135, 136, 137, 138, 140, 143, 164, 169, 170, 175, 177, 187, 208, 209
nanostructured materials, 136
nanostructures, ix, 134, 135, 137, 140, 167, 184, 186, 187, 206, 207, 208, 209, 257, 258, 399
Nanostructures, 135, 397
nanosystems, 399
nanowires, 186, 208
Nd, 199, 210, 211, 212
neodymium, 199
net electric charge, vii
neutral, vii, x, 2, 82, 86, 89, 90, 91, 277, 279, 281, 288
nickel, 184
nitrates, 189, 199, 200
nitrous oxide, 209
noble metals, 184
nodes, 218, 219, 226, 367
nonequilibrium, ix, 104, 213, 230, 235
non-linear equations, 29
nonlinear optical response, 158
non-polar, xi
normalization constant, 295, 312
novel materials, 134
nuclear normal modes, ix, 213
nucleation, 246
nuclei, 218, 219, 223, 227, 306
nucleus, 223, 224, 231

O

obstacles, 218
oil, 207
one dimension, 294, 297
operations, 47
opportunities, 212, 229
optical activity, 296
optical density, 138, 143, 144

optical gain, 137, 138, 146, 153, 154, 155, 164, 170, 174, 177, 338
optical properties, vii, viii, ix, 1, 28, 76, 133, 134, 135, 136, 138, 186, 207, 209, 210, 257, 258, 259, 281, 293, 295, 296, 378, 379, 382, 388, 396, 402
optimization, 206
optoelectronics, 208
orbit, 12, 14, 40, 44, 46, 47
organ, 278
organic compounds, 205
organic dendrites, xi
oscillation, xii, 57, 58, 87, 88, 89, 175, 176, 177, 210, 318, 321, 347, 399, 400, 401
oscillator strength, vii, 1, 4, 11, 33, 42, 49, 57, 58, 59, 62, 64, 65, 68, 113, 126, 169, 186, 225, 295, 378, 379, 380, 382, 385, 386, 396
overlap, 84, 158, 164, 169, 216, 217, 227, 229, 294, 378
oxidation, viii, 183, 185, 188, 190, 209
oxygen, 185, 187, 188, 189, 190, 195, 198, 205, 209, 210, 281, 288, 400, 401, 402

P

Pacific, 245, 246, 247, 248, 253, 254
pairing, 215, 363, 370, 374
parallel, 11, 16, 27, 31, 47, 48, 57, 59, 61, 66, 106, 113, 239, 240, 243, 245, 379, 383
parity, 12, 14, 314, 316, 317
passivation, 136, 138, 151, 164
path integrals, 361
pathways, 151
periodicity, 336, 342
permittivity, 29, 35
peroxide, 186, 189, 190, 192, 205, 206, 207, 208
pH, 190, 191, 200, 206, 209
phase transitions, 88
phonons, viii, xi, 14, 18, 19, 22, 24, 26, 34, 53, 66, 70, 71, 72, 74, 76, 78, 79, 80, 81, 99, 100, 101, 102, 103, 104, 105, 106, 107, 114, 117, 126, 133, 175, 176, 233, 347, 348, 359, 361, 362, 363, 367
phosphorus, 12
photoabsorption, viii, 133
photobleaching, viii, 133
photocatalysis, 209
photodegradation, 136
photodetectors, 278
Photoinduced Absorption (PA), viii, 133, 144
photoluminescence, viii, 10, 19, 39, 79, 80, 134, 136, 138, 145, 164, 185, 187, 197, 208, 212, 259, 278, 314, 318
photolysis, 184, 189, 190, 206, 207
photons, 3, 80, 102, 113, 189, 214, 279, 283, 333, 362
photovoltaic devices, 134
physical properties, 396
physics, xii, 207, 216, 258, 328, 395, 399
piezoelectricity, 208
PL quantum yield, viii, 134
PL spectrum, 16, 17, 20, 27, 188
Planck constant, 3, 400
planets, 252
platelets, 16, 61, 62
platinum, 184
polar, xi, 32, 74, 102, 103, 126, 226, 348, 359, 360, 361, 362, 363, 374
polariton effects, vii, 1, 11, 15, 26, 28
polariton waves, vii, 1, 58, 60, 62, 64
polarizability, 11, 48, 229
polarization, vii, viii, 1, 2, 4, 11, 13, 14, 15, 17, 19, 22, 27, 29, 30, 31, 32, 33, 34, 36, 37, 38, 40, 42, 43, 47, 50, 52, 57, 58, 59, 61, 62, 63, 64, 65, 66, 73, 74, 77, 78, 93, 102, 103, 104, 105, 106, 108, 112, 113, 114, 119, 252, 324, 326, 328, 329, 333, 334, 339, 348, 362, 381, 382, 383, 384
polarization operator, 384
polymer, 209
polymerization, x, 278, 281
polymers, 278
polypropylene, 189
polyvinyl alcohol, 189
population density, 81, 324, 326
potassium, 200, 205
precipitation, 185, 205, 212
principles, viii, 183, 238
probability, 12, 19, 21, 22, 26, 27, 72, 76, 79, 80, 157, 216, 229, 301, 363, 365, 371, 373
probe, viii, 133, 142, 143, 145, 146, 147, 153, 157, 159, 162, 164, 169, 170, 171, 174, 175, 176, 177, 325, 328, 333, 334, 347, 348, 351
project, 206
propagation, 2, 4, 16, 17, 19, 23, 49, 57, 58, 217, 308, 310, 311, 312, 313, 340, 342
propagators, 227, 230, 372, 373
proportionality, 229, 263, 280
purity, 185, 196
PVA, 189, 190, 196, 209

Q

quanta, 225
quantization, 12, 137, 215, 367
quantum chaos, 339
quantum confinement, 135, 138, 183

quantum dot, vii, ix, xi, xii, 135, 165, 257, 258, 260, 261, 262, 263, 265, 268, 272, 306, 377, 379, 380, 382, 383, 384, 388, 389, 390, 392, 393, 394, 395, 399, 400, 401, 402
quantum dots, vii, xi, 135, 165, 258, 260, 261, 268, 306, 377, 379, 380, 382, 383, 384, 388, 395, 399, 400, 401, 402
quantum dynamics, ix, 213, 232, 332
quantum field theory, 231, 363
quantum Hall effect, 370
quantum mechanics, 218
quantum state, 70, 229, 319, 333
quantum structure, 135
quantum well, x, xi, 185, 258, 265, 305, 359, 360, 367, 368, 378, 379, 394
quartz, 189, 206, 281
quasi-equilibrium, 71, 80
quasiparticles, iv, vii, 371
quasi-physical model, ix, 213
qubits, 339

R

Rabi frequency, 329
radiation, viii, 11, 16, 19, 66, 67, 74, 75, 77, 78, 79, 80, 82, 83, 84, 85, 88, 89, 90, 91, 102, 183, 184, 185, 187, 189, 190, 191, 193, 194, 196, 198, 200, 205, 206, 207, 208, 210, 226, 230, 235, 279, 325
Radiation, v, 66, 183, 184, 185, 186, 189, 200, 205, 207, 209, 212
radiation induced oxidation, viii, 183
radicals, 184, 189, 190, 205, 206
radioisotope, 187
radius, viii, xi, 28, 34, 38, 43, 59, 133, 134, 135, 138, 140, 165, 169, 183, 198, 199, 241, 244, 245, 310, 342, 377, 378, 379, 380, 381, 382, 383, 384, 385, 386, 387, 395, 396, 400
Raman spectra, 74
raw materials, 203, 205
reactions, 184, 185, 186, 200, 206, 226
reactivity, 209
real time, 227, 233, 348
reality, 401
recombination, viii, ix, x, 18, 39, 74, 102, 133, 134, 137, 149, 151, 152, 153, 155, 156, 160, 161, 162, 165, 167, 169, 174, 175, 211, 257, 258, 259, 262, 263, 264, 265, 267, 272, 278, 279, 280, 283, 286
recombination processes, viii, 133, 134, 137, 151, 211
reconstruction, 25
recrystallization, 192
recurrence, 242
recycling, 278
red shift, 165, 338
redistribution, 143, 258, 259, 261, 268, 269, 270, 272, 273
redshift, 145, 158, 258, 268
reflectance spectra, 281
reflectivity, vii, 1, 2, 4, 6, 7, 8, 9, 10, 11, 12, 13, 14, 15, 16, 17, 20, 23, 25, 27, 30, 31, 33, 34, 35, 36, 37, 38, 39, 40, 41, 42, 43, 44, 45, 46, 47, 50, 51, 52, 53, 54, 55, 56, 57, 58, 59, 60, 61, 62, 63, 64, 65, 66, 73, 76, 80, 81, 94, 95, 96, 97, 98, 99, 100, 101, 108, 113, 117, 118, 119, 121, 123, 347, 348, 350
refraction index, 61
refractive index, 26, 48, 51, 52, 58, 63, 64, 68, 138, 265
refractive indices, 47, 50
regional problem, 238
relative size, x, 278
relaxation, vii, viii, 1, 14, 16, 19, 21, 24, 26, 27, 59, 70, 103, 104, 105, 106, 126, 133, 134, 136, 137, 138, 139, 140, 142, 143, 147, 149, 150, 151, 152, 153, 154, 155, 156, 157, 159, 160, 162, 164, 165, 166, 167, 168, 169, 170, 171, 172, 173, 174, 176, 177, 202, 258, 259, 262, 263, 264, 283, 327, 347, 387
relaxation model, 24
relaxation process, 70, 134, 137, 139, 140, 143, 147, 149, 151, 152, 154, 157, 159, 160, 162, 164, 165, 168, 170, 171, 172, 173, 174, 176, 177, 283
relaxation processes, 70, 134, 137, 139, 140, 143, 147, 149, 151, 152, 157, 159, 160, 162, 164, 165, 168, 170, 171, 172, 174, 176, 177, 283
relaxation rate, 259
relaxation times, 19
relevance, 17
renormalization, 215, 234, 343
reparation, ix, 183
reproduction, 319
repulsion, 318
resolution, 17, 142, 160, 166, 167, 193, 211, 259, 307, 314, 316
Resonance Raman scattering spectra, viii, 2, 120, 121
resources, 235
response, 158, 199, 216, 268, 270, 313, 348, 350
restoration, 137
restructuring, 17
R-matrix theory, xi
rods, 134, 135, 140, 141, 161
Romania, 129
room temperature, 88, 125, 138, 139, 143, 164, 185, 187, 204, 205, 259, 283, 370
rotating medium, ix, 237, 239

rotation axis, 239, 240, 243
routes, 217
rules, 12, 22, 33, 103, 114, 135, 141
Russia, 237

S

sapphire, 142
saturation, 152, 159, 161, 378, 386
scaling, 135
scanning electron microscopy, 193
scattering, viii, xi, 2, 10, 18, 19, 21, 22, 24, 25, 26, 27, 33, 72, 74, 75, 76, 78, 79, 80, 84, 86, 87, 88, 89, 90, 99, 100, 102, 103, 104, 106, 107, 108, 109, 112, 113, 114, 115, 116, 117, 118, 119, 120, 121, 122, 123, 124, 125, 126, 133, 142, 149, 153, 160, 258, 261, 281, 305, 306, 308, 312, 325, 336, 342, 346, 348, 349, 350, 382, 388
Schrödinger equation, 141
Schröedinger picture, ix, 213
science, 351
scope, 251, 288, 325
seismotectonic problems, vii, 239
selected area electron diffraction, 193
selenium, 278
self-assembly, 258
self-organization, 258
semiconductor, ix, x, xi, 3, 11, 17, 28, 134, 135, 136, 137, 138, 142, 175, 183, 206, 257, 258, 305, 306, 307, 314, 318, 324, 327, 348, 349, 351, 362, 377, 378, 379, 380, 382, 384, 387, 388, 395, 396, 399, 402
semiconductors, xi, 10, 70, 92, 102, 126, 181, 183, 208, 278, 306, 332, 339, 340, 346, 347, 361, 377, 378, 379, 397, 402
semimetals, 346
sensing, 134, 208
sensitivity, 195, 347
sensors, 278
shape, vii, 1, 20, 27, 33, 38, 47, 48, 53, 54, 99, 100, 101, 118, 134, 135, 136, 137, 141, 175, 176, 187, 200, 205, 259, 273, 282, 286, 319, 329, 331, 333, 340, 382
shear, 241, 244, 248, 250, 251
showing, 45, 172, 177, 325, 346
signals, 140, 145, 146, 153, 155, 162, 164, 175, 210, 322, 328, 329, 330, 331
signs, 295
silica, 186, 187
silicon, 186, 306
silver, 184, 186, 206, 207
simulation, 265
Singapore, 234, 355, 396

single crystals, 11, 41, 42, 43, 72, 114, 115, 117, 118, 120, 121, 123, 124, 193, 199, 205, 206, 210, 212
sintering, 212
SiO_2, 206
slow tectonic waves, ix, 237, 251
sodium, 187
sodium hydroxide, 187
software, 189, 281
solar cells, 278, 289
solar system, 251
sol-gel, 186, 199, 205, 212
solid phase, 185, 187, 189, 190, 191, 192, 193, 196, 199, 200, 201, 202, 203, 204
solid solutions, 73, 96, 97, 98, 100, 101, 108, 113, 114, 115
solid state, 199, 211, 278
solitons, vii, ix, 237, 246, 247, 249
solution, 16, 21, 22, 23, 24, 47, 72, 114, 116, 119, 121, 125, 138, 143, 177, 184, 186, 187, 189, 191, 192, 194, 196, 199, 200, 202, 205, 206, 209, 239, 241, 250, 251, 298, 309, 342, 349, 360, 364, 365, 366, 368, 372, 373
sorption, 330, 338
space charge region (SCR), x, 277, 279
space-time, 252
Spain, 399
spatial-energetic distribution of LPB, vii, 1, 27
speciation, 189
species, 190
specific surface, 193
spectra of chalcopyrite crystals, vii, 1
spectroscopy, viii, x, 126, 131, 133, 134, 135, 136, 137, 138, 139, 142, 143, 164, 211, 305, 306, 307, 313, 318, 322, 347
speculation, 320
speed of light, 337, 340
spending, 215
spin, viii, 2, 12, 13, 14, 32, 33, 40, 44, 46, 47, 102, 113, 135, 186, 231, 239, 252, 262, 339, 370
SSA, 193, 194
St. Petersburg, 254
stability, 136, 184, 210, 238, 258, 278, 308, 314, 339
Stabilization of nanoparticles, viii, 183
Stark effect, 145, 322, 331, 395
Stimulated Emission (SE), viii, 133, 134, 144
stoichiometry, 195, 198, 200
storage, ix, 213, 258
stress, ix, 157, 237, 238, 239, 240, 241, 242, 246, 250, 251
stress fields, 250
stroke, 313
strong interaction, 99, 126

structure, vii, xi, 1, 11, 15, 17, 24, 28, 33, 41, 42, 43, 44, 45, 54, 55, 62, 65, 66, 88, 96, 115, 126, 135, 136, 138, 157, 159, 169, 186, 198, 205, 209, 211, 215, 218, 220, 222, 247, 252, 258, 278, 287, 306, 308, 313, 321, 332, 337, 338, 340, 368, 377, 379, 388, 394, 400
substitutes, 199
substitution, 364, 373
substrate, 28, 187, 259, 280, 283, 285, 286, 287, 288
substrates, 186, 281
succession, 321
sulfate, 185
sulfur, 46
superfluid, 360, 370
supervision, 177
surface area, 193, 194
surface layer, 51, 52, 53, 54, 55, 57
surface modification, 206
surface photovoltage (SPV) method, x, 277, 278
surface region, viii, 2, 28, 58, 67
surface structure, 136
surfactants, 199, 207
susceptibility, 337, 348, 350, 386, 387
symmetrical stress tensor, ix, 237
symmetry, vii, ix, 1, 3, 11, 12, 13, 14, 23, 33, 48, 65, 74, 90, 103, 104, 105, 106, 114, 137, 213, 225, 226, 309, 314, 364, 372, 392, 394
synthesis, 134, 184, 199, 200, 205, 206, 207, 209, 212, 388

T

Taiwan, 253, 257
target, x, 184, 229, 278, 281
techniques, x, 134, 138, 140, 142, 187, 235, 293, 294, 297, 347, 360, 361, 378
technologies, 134, 360
technology, 206, 252
temperature, viii, x, 2, 15, 20, 39, 70, 72, 75, 88, 101, 102, 123, 124, 125, 138, 139, 143, 164, 185, 187, 188, 189, 192, 193, 195, 196, 197, 199, 203, 204, 205, 208, 209, 218, 228, 230, 257, 258, 259, 261, 262, 263, 264, 265, 266, 267, 268, 269, 270, 271, 273, 281, 283, 360
temperature dependence, 20, 102, 258, 259
tetrapod, 134
theoretical approaches, 361
theoretical support, 178
theory of elasticity, ix, 237, 239
thermal decomposition, 186, 195
thermal energy, 184
thermal evaporation, 281, 287
thermal quenching, 264

thermal stability, 184
thermal treatment, 3, 186
thermalization, 150, 160, 259
thermoluminescence, 205
thin films, x, 186, 207, 208, 278, 281, 282, 289, 401, 402
time resolution, 142, 160, 167
time variables, 363
titania, 289
titanium, 278
titanyl phthalocyanine (TiOPc), x, 278
tonic, 307, 313, 322
total energy, 400
trajectory, 57, 218
transformation, 19, 92, 278, 308, 327, 340, 341, 350, 381, 389
transformations, 328
transition rate, 268
translation, vii, 1, 14, 36, 39, 40, 44, 56, 70, 108, 113, 137, 219, 364, 365, 372, 373
transmission, x, 11, 12, 13, 14, 16, 19, 20, 22, 23, 24, 25, 27, 78, 138, 142, 143, 144, 145, 146, 147, 149, 150, 153, 157, 158, 159, 160, 170, 171, 175, 193, 259, 260, 278
Transmission Electron Microscopy (TEM), 136, 138, 139, 187, 193, 202, 260
transmittance spectra, 187
transparency, 24, 88, 283
transport, vii, 16, 19, 258, 259, 262, 278, 332, 339, 348
transport processes, 262
treatment, 3, 74, 186, 189, 196, 198, 206, 221, 229
tunneling, xi, 135, 294, 305, 307, 308, 318, 322, 332, 339
Turkey, 293

U

USA, 213, 277, 291, 293, 359, 377
USSR, 127
UV, viii, 183, 184, 186, 187, 189, 190, 191, 192, 193, 194, 196, 198, 200, 201, 202, 203, 204, 205, 206, 207
UV irradiation, viii, 183, 186, 193, 194, 196, 198, 200, 201, 202, 203, 204, 205
UV light, 191
UV radiation, 184, 194, 200

V

vacancies, 187, 198, 203, 205
vacuum, 3, 22, 29, 51, 61, 63, 195, 215, 337

valence, viii, 2, 3, 4, 12, 20, 32, 33, 39, 40, 42, 44, 46, 47, 59, 65, 84, 105, 113, 325, 333, 349, 359
vapor, x, 16, 209, 257, 259, 288
variables, 21, 235, 327, 363, 370, 374
variations, 59, 135, 168
vector, 2, 3, 4, 11, 12, 13, 16, 19, 20, 22, 26, 47, 56, 57, 58, 59, 60, 61, 65, 70, 71, 79, 103, 104, 106, 214, 215, 218, 222, 223, 224, 225, 226, 227, 230, 231, 243, 294, 309, 335, 340, 361, 362, 370, 371, 381, 382, 383, 385
velocity, ix, 3, 19, 22, 24, 237, 238, 239, 242, 243, 244, 245, 246, 247, 248, 250, 251, 279, 286, 362, 370, 400, 402
vibration, 74, 76, 81, 106, 126, 217, 229
video games, 217

W

Wannier Mott exciton, xi, 377, 378, 380, 383
water, 184, 189, 190, 199, 206, 207, 212, 288
water vapor, 288
wave number, 246
wave vector, 2, 3, 4, 11, 12, 13, 19, 20, 25, 26, 47, 56, 57, 58, 59, 60, 61, 65, 70, 71, 79, 103, 104, 106, 114, 362, 374, 380, 381, 382, 385
wavelengths, 14, 33, 59, 61, 74, 113, 186, 189, 281, 289
wavelet, 308, 348, 350

wells, x, xi, 185, 232, 258, 368, 374
wetting, 258, 259, 261, 262, 263, 264, 265, 266, 267, 268, 272
wires, 134, 135, 185, 258, 379
Wisconsin, 293

X

xenon, 72, 75, 76, 77, 79

Y

Y-axis, 106, 113
yield, viii, 134, 145, 185, 190, 199, 200, 207, 209, 220, 298
yttrium, 199, 200, 201, 202, 203, 205, 210, 211, 212

Z

zinc, 11, 12, 186, 187, 188, 189, 190, 191, 192, 195, 196, 198, 205, 207, 208, 209, 210
zinc oxide (ZnO), 72, 183, 185, 186, 187, 188, 189, 191, 193, 194, 195, 196, 197, 198, 205, 207, 208, 209, 210, 399
ZnO nanorods, 209
ZnO nanostructures, 207